Resource Preview

■ *Fire Service Instructor: Principles and Practice, Second Edition*

With the release of the *Second Edition*, Jones & Bartlett Learning, the National Fire Protection Association (NFPA), the International Association of Fire Chiefs (IAFC), and the International Society of Fire Service Instructors (ISFSI) have joined forces to bring you a complete one-volume teaching and learning solution for Fire Service Instructors I, II, and III. With cutting-edge technological resources to engage candidates and assist instructors, the *Second Edition* takes training off the printed page. Covering the entire spectrum of the 2012 Edition of NFPA 1041, *Standard for Fire Service Instructor Professional Qualifications*, the one-volume *Second Edition* meets all of the Fire Service Instructor I, II, and III job performance requirements.

■ New and Exciting Changes

With the new edition of *Fire Service Instructor: Principles and Practice* our goal is to enhance the effectiveness of the Fire Service Instructor curriculum package. We understand that fire service instructors can benefit from, and use in their own courses, both traditional and the newest technological instructional methods. Additionally, instruction goes beyond what is found in a book or presented in class. With these concepts in mind, we have enhanced *Fire Service Instructor: Principles and Practice* with the following improvements:

- The Technology in Training chapter has been heavily revised to include detailed information on technology-based instruction and learning management systems.
- Material in each chapter has been clearly marked to indicate the levels of fire service instructor to which the material applies, facilitating lesson planning and course development and delivery.
- Training Bulletin and Incident Report features have been added to each chapter to foster discussion and to correlate the classroom experience to the students' own experiences and departments.
- Chapters on curriculum development, evaluation system management, and program management have been added to expand the scope of this book to include Fire Service Instructor III students.

■ Chapter Resources

Fire Service Instructor: Principles and Practice, Second Edition features reinforce and expand on fundamental information and focus on safe and efficient instructional methods. These features include:

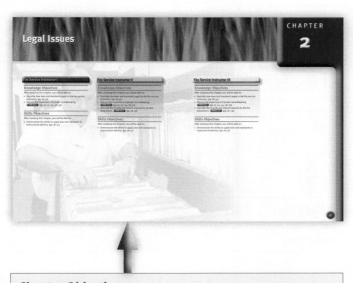

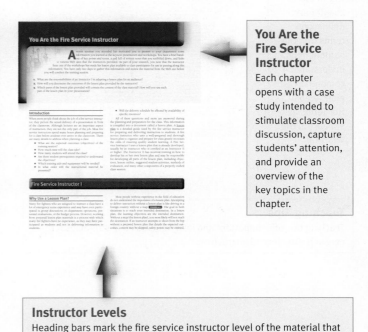

You Are the Fire Service Instructor
Each chapter opens with a case study intended to stimulate classroom discussion, capture students' attention, and provide an overview of the key topics in the chapter.

Chapter Objectives
NFPA 1041 is correlated to the level-specific Knowledge Objectives and Skills Objectives listed at the beginning of each chapter.
- Page references are included for quick reference to content.
- Knowledge Objectives outline the most important topics covered in the chapter.
- Skills Objectives map the skills achieved through implementation of the chapter content.

Instructor Levels
Heading bars mark the fire service instructor level of the material that follows.

Hot Terms

Hot Terms are easily identifiable within the chapter and define key terms that the student must understand. A comprehensive glossary of Hot Terms also appears in the Wrap-Up.

Ethics Tips

Ethics Tips are provided to help educate instructors on current ethical issues in the fire service.

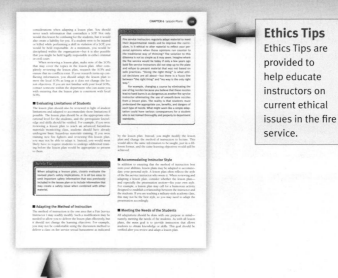

Teaching Tips

Teaching tips offer important and insightful teaching strategies for new and experienced instructors.

Safety Tips

Safety Tips reinforce safety-related concerns for fire service instructors.

Theory into Practice

Theory into Practice tips assist instructors in conceptualizing important theoretical topics as they relate to today's fire service. This feature answers many questions that students have on how chapter material relates to them or how it is intended to work in practice.

Voices of Experience

In the Voices of Experience essays, veteran fire instructors share their accounts of lessons learned on the job, while offering advice and encouragement. These essays highlight what it is truly like to be a fire service instructor.

Job Performance Requirements (JPRs) in Action

This unique feature appears in every chapter and provides:

- An understanding of the relationship among Fire Service Instructor I, II, and III
- Identification of the specific responsibilities of the Fire Service Instructor I, II, and III relating to how the job performance requirements (JPRs) work
- A discussion of how the instructor levels work together to achieve training goals

Training Bulletins

Training Bulletins appear in every chapter as opportunities for further discussion and exploration into how the chapter's material can be used in the students' own departments. These bulletins simulate a fire department training bulletin and provide the user with sample case applications for chapter content that can be used as a class activity or for continuing education for experienced instructors.

Incident Reports

Incident Reports of actual fire service training injuries and deaths are detailed in each chapter. Post-Incident Analysis summaries are offered as lessons learned. These can be used as introductory or practice student teaching presentations during the delivery of any level of Instructor course.

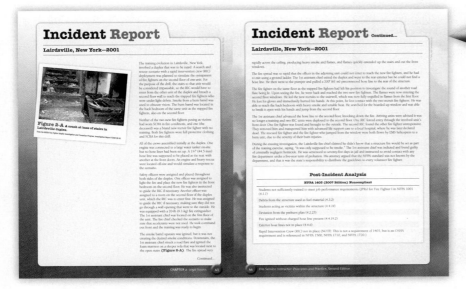

Wrap-Up
Chief Concepts
Chief Concepts highlight critical information from the chapter in a bulleted format to help students prepare for exams.

Hot Terms
Hot Terms are a collection of key terms and definitions from the chapter.

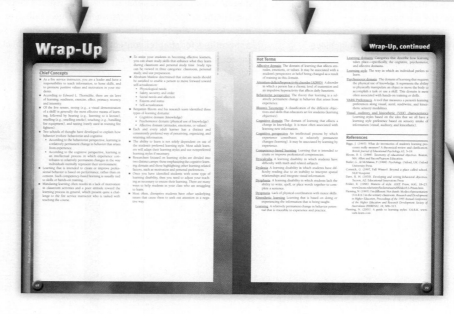

References
References list published material mentioned in the chapter, which may be helpful for further study.

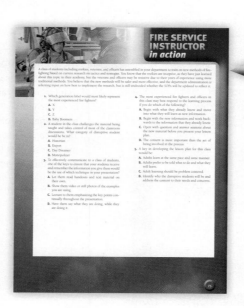

Fire Service Instructor in Action
This feature promotes critical thinking through the use of case studies and provides instructors with discussion points for the classroom.

Resources

■ Instructor's ToolKit

Preparing for class is easy with the resources on this CD. Instructors can choose resources for the Fire Service Instructor I, Fire Service Instructor II, Fire Service Instructor III, or a combination of the three. The CD includes the following resources:

- Adaptable PowerPoint® Presentations
- Detailed Lesson Plans
- Electronic Test Bank
- Skill Sheets
- Sample Fire Service Instructor I Presentations
- Fire Service Instructor II Student Application Package
- Fire Service Instructor III Student Application Package
- Student Workbook Answers
- Image and Table Bank

■ Student Workbook

This resource is designed to encourage critical thinking and aid comprehension of the course materials. The student workbook uses the following activities to enhance mastery of the material:

- Case studies and corresponding questions
- Matching, fill-in-the-blank, short answer, and multiple-choice questions

Acknowledgments

Jones & Bartlett Learning, the National Fire Protection Association, the International Association of Fire Chiefs, and the International Society of Fire Service Instructors would like to thank the current editors, contributors, and reviewers of *Fire Service Instructor: Principles and Practice, Second Edition*.

Authors

■ Forest F. Reeder, MS

Forest Reeder began his fire service career in 1978. He serves as Division Chief/Training & Safety with the Des Plaines (IL) Fire Department. Forest is the Fire Officer 1 and 2 Coordinator for the Illinois Fire Chiefs Association and also serves as the Eastern Regional Director of the International Society of Fire Service Instructors (ISFSI).

Additionally, Forest is the author of the weekly drill feature at www.firefighterclosecalls.com and at www.fireengineering.com. He has instructed at FDIC for over 10 years and trains both locally and nationally on fire service training, safety, and officer development related topics. Several articles on these programs have been published in major trade publications. Forest is a Director with the Illinois Society of Fire Service Instructors and has served as President. Forest served as the Fire Science Coordinator at Moraine Valley Community College for 10 years and continues to assist in course delivery and curriculum design projects at the College.

Forest holds many Illinois fire service certifications including Fire Officer 3, Training Program Manager, and Fire Department Safety Officer. He has an Associate's of Applied Science Degree in Fire Science Technology, a Bachelor's Degree in Fire Department Administration from Southern Illinois University–Carbondale, and a Master's Degree in Public Safety Administration from Lewis University. He was awarded the George D. Post Instructor of the Year by the International Society of Fire Service Instructors at FDIC in 2008.

■ Alan E. Joos, EFO, MS (ABD), FIFireE

Alan E. Joos serves as an Assistant Director of Certification and Training at Louisiana State University Fire and Emergency Training Institute (FETI). Prior to FETI, he worked at the Utah Fire and Rescue Academy for 12 years as the Assistant Director of the Certification System and Assistant Director of the Training division.

Mr. Joos received Associate's and Bachelor's Degrees from Utah Valley University in Fire Science and Business Technology management and a Master's degree from Grand Canyon University in Phoenix. He is currently a doctoral candidate in Human Resource Education/Workforce development at Louisiana State University. He is also a graduate of the Executive Fire Officer's Program at the National Fire Academy.

Alan's fire service background began in 1985 as a career fire fighter in a combination fire department and then as an on-call fire fighter/EMT-I in Utah. He has continued his fire service involvement with the City of Gonzales, Louisiana, as a contract fire fighter. During his fire service career he has served as a training officer, driver, shift officer, and EMT.

Alan has been involved with several national organizations, including the International Fire Service Congress (IFSAC), the ProBoard, and currently is on the NFPA Technical Committee for the Professional Qualification standards.

Alan is married to Carla Joos, and their family includes Nathan, his wife Shauntelle, Jordan (deceased) and Dallan, and granddaughter Chloe.

Editorial Board

Doug Cline
President, International Society of Fire Service Instructors
Assistant Chief, Horry County Fire Rescue
Conway, South Carolina

Robert Colameta, Jr.
Eastern Region Director, International Society of
  Fire Service Instructors
Owner, Public Safety Education Network
Everett, Massachusetts

Shane Cuttlers
Western Region Director
International Society of Fire Service Instructors
Raymond, Nebraska

Shawn Kelley
International Association of Fire Chiefs
Fairfax, Virginia

Thomas McGowan
Senior Fire Service Specialist
National Fire Protection Association
Quincy, Massachusetts

Kevin Milan, CFO, EFO
Director at Large
International Society of Fire Service Instructors
Parker, Colorado

Steve Pegram, CFO
First Vice President, International Society of Fire Service Instructors
Chief, Goshen Township Fire & EMS
Goshen, Ohio

Steven F. Sawyer
Senior Fire Service Specialist
National Fire Protection Association
Quincy, Massachusetts

Brian Ward
Board of Directors
International Society of Fire Service Instructors
Gray, Georgia

Devon Wells
Second Vice President, International Society of
Fire Service Instructors
Fire Chief, Hood River Fire & EMS
Hood River, Oregon

Contributors and Reviewers

■ Contributing Authors

Steve DiNolfo
Shareholder
Ottosen Britz Kelly Cooper Gilbert and
DiNolfo
Naperville, Illinois

Dr. Ben Hirst
Senior Vice President
Performance Training Systems, Inc.
Jupiter, Florida

Bill Kimball
Manager, Application Solutions
Ascend Learning
Burlington, Massachusetts

Bryant Krizik
Fire Chief (ret.)
Orland Fire Protection District
Orland Park, Illinois

Benjamin Andersen
Engineer
Loveland Fire Rescue Authority
Loveland, Colorado

Bill Guindon
Director
Maine Fire Service Institute
Brunswick, Maine

Kevin Hammons, MA, EFO, CFO
President
Revelations, Inc.
Franktown, Colorado

F. Tom Hand
Fire Training Coordinator (ret.)
Mesa Fire & Medical Department
Mesa, Arizona

Justin R. Heim, MPA, EFO
Eagle Fire Department
Eagle, Wisconsin

Matt Hinkle
Training Officer, Lafayette County Fire
Department
Adjunct Instructor, Mississippi State Fire
Academy
Oxford, Mississippi

Kathe Jones
Manager, Technical Rescue Program
Louisiana State University Fire and
Emergency Training Institute
Baton Rouge, Louisiana

Jason Loeb
Hoffman Estates Fire Department
Illinois Fire Service Institute
Champaign, Illinois

Diane May
Instructor Trainer
Maryland Fire and Rescue Institute
College Park, Maryland

Michael Ridley
Fire Chief (ret.)
Wilton Fire Protection District
Wilton, California

Robert D. Shaw Jr.
Chief, Du Quoin Fire Department
Director, Rend Lake College Fire Science
Department
Ina, Illinois

Lloyd H. Stanley
Guilford Technical Community College
Jamestown, North Carolina

Dan Sullivan
Massachusetts Firefighting Academy
Stow, Massachusetts

Brian Ward
Board of Directors
International Society of Fire Service
Instructors
Gray, Georgia

■ Reviewers

W. Chris Allan, MPA
Stanly Community College
Albemarle Fire and Rescue
Albemarle, North Carolina

Robert Jay Alley
Blue Ridge Community College
Flat Rock, North Carolina

Bryan Anthony Altman
Lieutenant/Assistant Training Officer
Worth County Fire/Rescue
Sylvester, Georgia

Benjamin Andersen
Engineer
Loveland Fire Rescue Authority
Loveland, Colorado

Mike Armstrong, EdD, EFO
Deputy Chief
Augusta County Fire-Rescue
Verona, Virginia

Mark Ayers, MPA, EFO
Fire Program Supervisor
Department of Public Safety Standards &
  Training
Salem, Oregon

Bernard W. Becker, III, MS, EFO, CFO, MIFireE
Cleveland State University
Cleveland, Ohio

Stephen Benson
Training Officer
Southaven Fire Department
Southaven, Mississippi

Janet A. Boberg, EdD, CPM, LPC, NCC
Glendale Fire Department
Glendale Arizona

Britt Brinson
Georgia Firefighter Standards and Training
  Council
Forsyth, Georgia

Christopher Connor
Bucksport Fire and Rescue
Bucksport, Maine

Ron Deadman, MSA, CFO, EFO
Deputy Fire Chief
Avondale Fire-Rescue
Avondale, Arizona

Jason Decremer
Program Manager
Connecticut Fire Academy
Windsor Locks, Connecticut

Jason D'Eliso
Scottsdale Fire Department
Scottsdale, Arizona
Assistant Chief, 944th Fire & Emergency
  Services
Luke AFB, USAFR

Ronald R. Dennis
Director of Training and Professional
  Development
Columbia Southern University
Orange Beach, Alabama

Neil R. Fulton
Town Manager/Deputy Fire Chief
Norwich Fire Department
Norwich, Vermont

Rob Gaylor
Deputy Chief of Operations
Westfield Fire Department
Westfield, Indiana

Joseph Guarnera, MEd
Fire Service Professional Development
  Coordinator
Massachusetts Firefighting Academy
Stow, Massachusetts

Bill Guindon
Director
Maine Fire Service Institute
Brunswick, Maine

Greg Hall, NRP
Lieutenant
Watertown Fire Rescue
Watertown, South Dakota

Kevin Hammons, MA, EFO, CFO
President
Revelations, Inc.
Franktown, Colorado

F. Tom Hand
Fire Training Coordinator (ret.)
Mesa Fire & Medical Department
Mesa, Arizona

Justin R. Heim, MPA, EFO
Eagle Fire Department
Eagle, Wisconsin

Matt Hinkle
Training Officer, Lafayette County Fire
  Department
Adjunct Instructor, Mississippi State Fire
  Academy
Oxford, Mississippi

Wesley Hutchins, EFO
Forsyth Technical Community College
Winston-Salem, North Carolina

Christopher Johns
CAL/FIRE Riverside County Fire Department
Moreno Valley, California

Joel Journeay, MA
Faculty, Department of Fire Science
Columbia Southern University College of
  Safety and Emergency Services
Orange Beach, Alabama

Brian P. Kazmierzak, EFO, CTO
Chief of Training
Penn Township Fire Department
Mishawaka, Indiana

Darin Keith
Captain
Rock Island Arsenal Fire & Emergency
  Services
Rock Island, Illinois

J. Nathan Kempfer, MBA, EMT-P
Lieutenant
Lawrence Fire Department
Lawrence, Indiana

Jeff King, AAS
Battalion Chief—Training
Flower Mound Fire Department
Flower Mound, Texas

Lew Lake
Illinois Fire Service Institute
Champaign, Illinois

Casey Lindsay
Garland Fire Department
Garland, Texas

Jeremy S. Linn
Clark State Community College
Springfield, Ohio

Joshua Livermore
Bullhead City Fire Department
Bullhead City, Arizona

Jason Loeb
Hoffman Estates Fire Department
Illinois Fire Service Institute
Champaign, Illinois

Charles Lott
Kentucky Fire Commission/State Fire
  Rescue Training Area 1
Paducah, Kentucky

Mark Martin, MPA
Fire Chief
Perry Township Fire Department
Massillon, Ohio

Diane May
Instructor Trainer
Maryland Fire and Rescue Institute
College Park, Maryland

Randy McCartney
President, Wisconsin Society of Fire Service
  Instructors
Moraine Park Technical College
West Bend, Wisconsin

Darryl Oliveira
Hawaii Fire Chiefs Association
Hawaii Community College
Hilo, Hawaii

Gary E. Patrick
Fire Chief
Silverhill Volunteer Fire Department
Silverhill, Alabama

Michael D. Penders, MPA
Massachusetts Firefighting Academy
Stow, Massachusetts

Rainier Perez
Battalion Commander
Albuquerque Fire Department
Albuquerque, New Mexico

Lawrence "Larry" Phillips, BS, MS, EMT-P
Adjunct Instructor—Fire Protection
  Technology
El Centro College (DCCCD)
Dallas, Texas

Lee Poteat, Jr.
Dekalb County Fire Rescue
Tucker, Georgia

Carl Raymond
Southern Crescent Technical College
Griffin, Georgia

Michael Ridley
Fire Chief (ret.)
Wilton Fire Protection District
Wilton, California

Dan Roeglin
Customized Fire Training Coordinator
Hennepin Technical College
Eden Prairie, Minnesota

Robert D. Shaw Jr.
Chief, Du Quoin Fire Department
Director, Rend Lake College Fire Science
  Department
Ina, Illinois

Kenneth B. Smarr
DeKalb County Fire Rescue
Tucker, Georgia

Joshua J. Smith
Statesville Fire Department
Statesville, North Carolina

Thomas Y. Smith, Sr.
West Georgia Technical College
LaGrange, Georgia

Lon Spencer MS, BS, AAS
Captain/Safety Officer
Daisy Mountain Fire Department
Phoenix, Arizona

Lloyd H. Stanley
Guilford Technical Community College
Jamestown, North Carolina

Lori P. Stoney
Lieutenant
Homewood Fire & Rescue Service
Homewood, Alabama

Dan Sullivan
Massachusetts Firefighting Academy
Stow, Massachusetts

Jimmy VanCleve
Kentucky Fire Rescue Training
Calhoun, Kentucky

Brad D. Weilbrenner, BS
NH EMS Instructor/Coordinator
New Hampshire Division of Fire Standards
  & Training and EMS
Concord, New Hampshire

Brent Willis
Training Chief
Columbia County Fire Rescue
Martinez, Georgia

William Wren
Instructor, New York State Fire Academy
ISO, New Hartford Fire Department
Clinton, New YorkResources

Introduction, Roles, and Responsibilities

PART
I

Today's Emergency Services Instructor

Fire Service Instructor I

Knowledge Objectives

After studying this chapter, you will be able to:

- Define the roles and responsibilities of the Fire Service Instructor I (NFPA 4.1.1 , NFPA 4.2.1 , NFPA 4.2.2 , NFPA 4.2.3 , NFPA 4.2.4 , NFPA 4.2.5 , NFPA 4.3.2 , NFPA 4.3.3 , NFPA 4.4.1 , NFPA 4.4.2 , NFPA 4.4.3 , NFPA 4.4.4 , NFPA 4.4.5 , NFPA 4.4.6 , NFPA 4.4.7 , NFPA 4.5.1 , NFPA 4.5.2 , NFPA 4.5.3 , NFPA 4.5.4 , NFPA 4.5.5). (pp 6–10)
- Identify physical elements of the classroom (NFPA 4.4.2). (pp 10–11)
- Discuss the importance of visioning and succession planning for the instructor. (pp 11–12)
- Identify instructor credentials and qualifications. (pp 5, 13)
- Identify four issues of ethics for the fire service instructor. (pp 15–16)
- Identify three ways to assist the instructor in managing multiple priorities. (pp 16–19)

Skills Objectives

After studying this chapter, you will be able to:

- Demonstrate the ability to manage the five roles of the fire service instructor. (pp 8–10)
- Demonstrate ethical behavior in the classroom. (pp 15–16)
- Demonstrate the ability to manage multiple priorities as a fire service instructor. (pp 16–19)

Fire Service Instructor II

Knowledge Objectives

After studying this chapter, you will be able to:

- Define the roles and responsibilities of the Fire Service Instructor II (NFPA 5.1 , NFPA 5.2.1 , NFPA 5.2.2 , NFPA 5.2.3 , NFPA 5.2.4 , NFPA 5.2.5 , NFPA 5.2.6 , NFPA 5.3.1 , NFPA 5.3.2 , NFPA 5.3.3 , NFPA 5.4.1 , NFPA 5.4.2 , NFPA 5.4.3 , NFPA 5.5.1 , NFPA 5.5.2 , NFPA 5.5.3). (pp 6–10)
- Discuss the importance of visioning and succession planning for the instructor. (pp 11–12)
- Identify instructor credentials and qualifications. (pp 5, 13)
- Identify four issues of ethics for the fire service instructor. (pp 15–16)
- Identify three ways to assist the instructor in managing multiple priorities. (pp 16–19)

Skills Objectives

After studying this chapter, you will be able to:

- Demonstrate the ability to manage the five roles of the fire service instructor. (pp 8–10)
- Demonstrate ethical behavior in the classroom. (pp 15–16)
- Demonstrate the ability to manage multiple priorities as a fire service instructor. (pp 16–19)

Fire Service Instructor III

Knowledge Objectives

After studying this chapter, you will be able to:

- Define the roles and responsibilities of the Fire Service Instructor III (NFPA 6.1 , NFPA 6.2.1 , NFPA 6.2.2 , NFPA 6.2.3 , NFPA 6.2.4 , NFPA 6.2.5 , NFPA 6.2.6 , NFPA 6.2.7 , NFPA 6.3.1 , NFPA 6.3.2 , NFPA 6.3.3 , NFPA 6.3.4 , NFPA 6.3.5 , NFPA 6.3.6 , NFPA 6.3.7 , NFPA 6.5.1 , NFPA 6.5.2 , NFPA 6.5.3 , NFPA 6.5.4 , NFPA 6.5.5). (pp 7–10)
- Discuss the importance of visioning and succession planning for the instructor. (pp 11–12)
- Identify instructor credentials and qualifications. (pp 5, 13)
- Identify four issues of ethics for the fire service instructor. (pp 15–16)
- Identify three ways to assist the instructor in managing multiple priorities. (pp 16–19)

Skills Objectives

After studying this chapter, you will be able to:

- Demonstrate the ability to manage the five roles of the fire service instructor. (pp 8–10)
- Demonstrate ethical behavior in the classroom. (pp 15–16)
- Demonstrate the ability to manage multiple priorities as a fire service instructor. (pp 16–19)

You Are the Fire Service Instructor

After a recent budget meeting, your chief asked you to develop a presentation outlining the requirements for promotions within the department and to specifically identify and outline the role of the three levels of instructor qualifications described in NFPA 1041, *Standard for Fire Service Instructor Professional Qualifications*. The city is looking to cut funding for training in every place possible, and without proper justification several department positions could be eliminated.

1. What are the differences among the Instructor I, Instructor II, and Instructor III roles?
2. How do the instructor levels fit into the certification process as identified in the NFPA standards?
3. Why are instructor qualifications so important to a department in providing emergency service to a community?

Introduction

Remember the first day you began your career in the fire service? Somewhere in your memory is probably a great fire service instructor. He or she just may be the person who introduced you to the greatest career. Fire service instructors are the guardians of knowledge, skills, and ability in the fire service; their knowledge is wrapped up in a long tradition that has served many generations of fire fighters well. As protectors of that tradition, instructors look to and rely on innovation and creativity while constantly striving to perfect training.

In years past, firefighting training was often left to on-the-job education. New recruits would be handed coats, boots, helmets, and gloves and told to jump on the back of the rig as it rushed to the scene of the emergency. In those days, the fire ground was the recruit's training ground—and the title of "fire service instructor" may have been handed out too generously. Formal instructors and, in some cases, formal instruction were left to the larger departments that had budgets large enough to hire specialized personnel and build training facilities.

Obviously, times have changed. The fire service is now being pushed to the limit by communities that expect more out of the types of services provided. Retired Phoenix Fire Chief Alan Brunacini often referred to today's recipient of that service as "Mrs. Smith." As a pioneer of customer service, Chief Brunacini knew the value of the instructor in preparing the troops to meet Mrs. Smith's demands. Today, fire departments rely on their instructors both to train new recruits and to maintain the skill levels of existing fire fighters.

What does it take to join this elite group of fire fighters on whose shoulders rest the success and safety of emergency operations? Although excellent fire service instructors possess many attributes, a few important qualities come to mind and rise to the top of the list. First, there is desire—a desire to be of assistance to those in need. In this case, we are not referring to those persons who require our assistance during times of trouble or emergency, but rather to those individuals who would require the knowledge needed to assist those in trouble.

These fire fighters must be trained so that they will gain the knowledge, skills, and abilities necessary to serve and respond to those in need of assistance. The fire service instructor gives freely without hesitation in the pursuit of excellence **FIGURE 1-1**. He or she should certainly have experience in the subject matter taught. Students have more confidence in fire service instructors who have demonstrated competence in the subject matter through experience; in turn, experience opens the door for the fire service instructor to reach the minds of the students. Even so, good fire service instructors also understand that they cannot rely on past experience alone: They must remain vigilant to their always dynamic environment and gather the newest information, technology, and skills to remain out front and stay current with an ever-changing work and instructional environment.

Good fire service instructors also demonstrate flexibility: They are able to work in a variety of environments while offering instruction on a variety of topics to a class full of students with a variety of talents, backgrounds, and experience levels. Instructing the adult learner can be challenging, and the fire

FIGURE 1-1 The fire service instructor gives freely without hesitation in the pursuit of excellence.

service instructor must be willing to alter the approaches to education used to reach all students. In some cases, the fire department's schedule provides the greatest challenge, as instruction time is often interrupted by calls for assistance. In other fire departments, it is the fire fighters' personal schedules that present the challenge, as training time competes with family time and full-time work schedules.

Motivation is one of the keys to bringing excitement to the training environment. Nothing breeds fire fighter motivation like a motivated instructor. The right kind of motivation leads to the attitude, "I can't wait to teach," and, just as important, "I can't wait to learn." Motivation is contagious; it can spread from the classroom to the station floor, and ultimately it can drive the quality of the service provided to the community. Motivated fire service instructors bring creativity and ingenuity to the classroom as a means of creating excitement about the learning process.

The skills to be an effective instructor can be learned by a person who has the desire to develop and perfect them. The ability to use these skills at the correct time and in a manner that helps others learn a topic is developed by teaching courses and being mentored, coached, and supervised by experienced trainers. The learning process never ends for those who desire to improve their knowledge and skills. All instructors start out at the Instructor I level; with additional training and professional development, they may reach the top of the qualification set owing to their desire to learn and develop their skills. Desire matched with knowledge creates great instructors.

Are you up for the challenge? Can you accept the responsibility? Can you be a steward of tradition while remaining open to teaching new ideas? Can you strive to maintain desire, experience, flexibility, and motivation as you pursue operational excellence? If your answers to these questions are all "yes," then you are on your way to joining the ranks of the greatest profession. Ideally, your curriculum will serve as the launching pad for excellent fire fighters and serve them well as a critical reference for years to come.

This book provides information to meet the job performance requirements (JPRs) outlined in National Fire Protection Association (NFPA) 1041, *Standard for Fire Service Instructor Professional Qualifications*, at the Instructor I, II, and III levels. These definitions resulted from a task analysis intended to validate these levels and to create specific requirements that would apply to each level. Although the three levels are identified as instructor qualifications, in many states they may become certification levels, with candidate prerequisites to be completed before a certification is granted.

Fire Service Instructor I, II, and III

Levels of Fire Service Instructors

According to NFPA 1041, the duties performed by specific instructors are broken down into three distinct levels. These three classifications build on one another and progressively give the fire service instructor additional skills, duties, and responsibility.

Instructor I is defined as follows (NFPA 1041, Section 3.3.2.1):

A fire service instructor who has demonstrated the knowledge and ability to deliver instruction effectively from a prepared lesson plan, including instructional aids and evaluation instruments; adapt lesson plans to the unique requirements of the students and authority having jurisdiction; organize the learning environment so that learning and safety are maximized; and meet the record-keeping requirements of the authority having jurisdiction.

Stated in the most basic terms, the Instructor I delivers instruction from prepared materials at the direction, and often under the supervision, of an Instructor II or higher. Emphasis of this level of instructor is on the ability to communicate effectively and to use various methods of instruction, including hands-on training and lecture.

The Instructor II is a fire service instructor who, in addition to meeting the Instructor I qualifications, satisfies the following criteria (NFPA 1041, Section 3.3.2.2):

Has demonstrated the knowledge and ability to develop individual lesson plans for a specific topic including learning objectives, instructional aids, and evaluation instruments; schedule training sessions based on overall training plan of authority having jurisdiction; and supervise and coordinate the activities of other instructors.

The Instructor II functions at a higher level of authority and responsibility than the Instructor I; he or she is responsible for all duty areas of the Instructor I, and is also able to create the training materials. In the purest sense, the Instructor II will create the training materials for distribution to the Instructor I to present to the students. In reality, both tasks may be completed by the same person.

An Instructor III is defined as follows (NFPA 1041, Section 3.3.2.3):

A fire service instructor who, in addition to meeting Instructor II qualifications, has demonstrated the knowledge and ability to develop comprehensive training curricula and programs for use by single or multiple organizations; conduct organization needs analysis; design record keeping and scheduling systems; and develop training goals and implementation strategies.

The Instructor III typically works as an overall training program manager and oversees the entire spectrum of a comprehensive training program. This includes the planning, development, and implementation of curricula and development of an evaluation plan and training program budget.

Theory Into Practice

The relationship between training and education is often confusing. Education is the process of imparting knowledge or skill through systematic instruction. Education programs are conducted through academic institutions and are primarily directed toward an individual's comprehension of the subject matter. Training is directed toward the practical application of education to produce an action, which can be an individual or a group activity. There is an important distinction between these two types of learning.

Within the fire service, training has been considered essential for many years. In contrast, the emphasis on fire fighter education is a much more recent development. The First Wingspread Conference on Fire Service Administration, Education and Research was sponsored by the Johnson Foundation and held in Racine, Wisconsin, in 1966. This conference brought together a group of leaders from the fire service to identify needs and priorities. They agreed that a broad knowledge base was needed and that an educational program was necessary to deliver that knowledge base. Their conclusions became the blueprint for the development of community college fire science and fire administration programs, as well as the degrees-at-a-distance program. The transition from training to education had begun.

In 1998, the U.S. Fire Administration hosted the first Fire and Emergency Services Higher Education (FESHE) conference. That conference produced a document, "The Fire Service and Education: A Blueprint for the 21st Century," that initiated a national effort to address and update the academic needs of the fire service. Participants at the 2000 FESHE conference began work to develop a model fire science curriculum that would span from the community college through graduate school levels. At the 2002 conference, U.S. Fire Administration Education Specialist Edward Kaplan compared the results of the FESHE effort with the Wingspread higher education curriculum. The FESHE work affirmed the soundness of the original Wingspread model, with information technology being the only new knowledge item added to the curriculum.

Roles and Responsibilities of an Instructor

Throughout our lives, we are asked to conform to someone else's idea of how we should act or behave. Child, adolescent, adult, parent, spouse, employee, supervisor, owner, citizen— all are examples of roles that we fill while negotiating through life. With each of these roles, expectations direct our actions and allow us to evaluate our success or failure at that role.

The fire service mirrors life in that it also contains various roles, each with its own expectations and responsibilities. At every step along the fire service path from recruit to fire chief, we strive to meet these expectations and responsibilities. The fire service instructor is one of those important roles in the fire service that requires dedicated and competent individuals who can positively influence the entire fire department.

The roles and responsibilities of a fire instructor vary greatly by fire department according to the size, make-up, and delivery system used. Understanding the roles and responsibilities for Instructors I, II, and III is essential for success in these key positions.

■ Roles and Responsibilities of the Fire Service Instructor I

The roles and responsibilities of the Fire Service Instructor I include the following:

- Manage the basic resources and the records and reports essential to the instructional process.
- Assemble course materials.
- Prepare training records and report forms.
- Prepare requests for training resources.
- Schedule instructional sessions.
- Review and adapt prepared instructional materials.
- Deliver instructional sessions using prepared course materials.
- Organize the classroom, laboratory, or outdoor learning environments.
- Use instructional media and materials to present prepared lesson plans.
- Adjust presentations to students' different learning styles, abilities, and behaviors.
- Operate and utilize audiovisual equipment and demonstration devices.
- Administer and grade student evaluation instruments.
- Deliver oral, written, or performance tests.
- Grade students' oral, written, or performance tests.
- Report test results.
- Provide examination feedback to students.

■ Roles and Responsibilities of the Fire Service Instructor II

The Instructor II must meet all of the requirements for, and perform all of the duties of, the Instructor I. In addition, the Fire Service Instructor II is responsible for performing the following tasks:

- Manage instructional resources, staff, facilities, and records and reports.
- Schedule instructional sessions.

- Formulate budget needs.
- Acquire training resources.
- Coordinate training recordkeeping.
- Evaluate instructors.
- Develop instruction materials for specific topics.
- Create lesson plans.
- Modify existing lesson plans.
- Conduct classes using a lesson plan.
- Use multiple teaching methods and techniques to present a lesson plan that the instructor has prepared.
- Supervise other instructors and students during training.
- Develop student evaluation instruments to support instruction and evaluation of test results.
- Develop a class evaluation instrument.
- Analyze student evaluation instruments.

The Instructor I delivers instruction from prepared material, while the Instructor II develops course materials.

■ Roles and Responsibilities of the Fire Service Instructor III

The Instructor III must meet all of the requirements for, and perform all of the duties of, the Instructor II. In addition, the Fire Service Instructor III is responsible for performing the following tasks:

- Administer agency policies and procedures for training.
- Administer a training record system for an agency.
- Develop recommendations for policies that support a training program.
- Select instructional staff based on qualifications and agency requirements.
- Construct a performance-based instructor evaluation plan.
- Write equipment purchasing specifications based on curriculum information and needs.
- Present agency evaluation findings to agency administration.
- Plan, develop, and implement curricula.
- Conduct an agency needs analysis.
- Design curricula given needs analysis findings and agency goals.
- Write program and course goals given analysis information and JPRs.
- Modify an existing curriculum to meet the needs of the agency.
- Construct a course content outline given course objectives.
- Write course objectives that are clear, concise, and measurable.
- Develop a system for the acquisition, storage, and dissemination of evaluation results.
- Develop a course evaluation plan.
- Create a program evaluation plan.
- Analyze student evaluation instruments.

■ Where Do I Fit in?

Much has been written about the roles and responsibilities of positions within various types of organizations, and the fire service is no different. Fire chiefs have their role, rooted in visions of leadership, as the commander-in-chief of the organization. They have a defined responsibility that not only allows them to assume this role, but also gives them the authority to shape their own destiny.

Within the fire service's remaining command structure lie the positions of supervision, including front-line lieutenants, captains, and various chief officers. These ranks have had their roles defined through a history of developed policies and job descriptions created by necessity, the chief's vision, and, yes, tradition. These positions are often referred to as middle management, as they are created and reside in the middle of the chain of command. They are pushed and tested by fire fighters at the bottom and by fire chiefs at the top.

As tough as these positions are, there is one job that still finds itself in a somewhat more precarious spot: the fire service instructor. All too often, a fire service instructor may ask, "Where do I fit in?" Is the fire service instructor a line officer, a training officer, a fire fighter, or someone with a specific specialty? In some fire departments, line officers are assigned to the training division as part of their duties. In volunteer organizations, the training officer often volunteers for the job; alternatively, he or she may be assigned to it. Regardless of how you come to be assigned to training, you may feel challenged if you are not given the proper authority to carry out that job.

In examining fire departments' organizational charts, you will find positions identified for operational activities, clearly outlining the chain of command and the corresponding responsibilities for firefighting personnel and their fire officer counterparts. Fire service instructors, if shown at all, are usually found off to the side on the organizational chart, preferably within a training division. This disconnect may create a sense of isolation from the fire fighters whom fire service instructors are expected to train. It may leave the fire service instructor with a void, feeling unfulfilled in terms of a management role that demonstrates the fire department's lack of respect for the role FIGURE 1-2 . Fire fighters and even some fire officers are

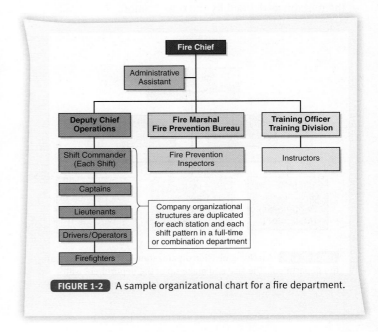

FIGURE 1-2 A sample organizational chart for a fire department.

often seen as resisting or even obstructing the training process, because all too often those within the operations staff do not view training as a high priority.

As a fire service instructor, you must remain focused, upbeat, positive, proactive, and true to your role **FIGURE 1-3**. You must remind yourself that you hold the power to shape the future fire service. Every member of the fire service—from fire fighter through fire officer—has had a fire service instructor affect their careers. That fact demonstrates the importance of your position within the fire department.

Managing the Fire Service Instructor's Role

As a fire instructor, you must look beyond charts and titles and instead focus on those important, yet sometimes invisible, roles that produce lasting contributions to overall fire department organizational health and success. Today's fire service instructor is asked to fulfill these roles. These roles are not unique to the fire service; indeed, examples of each can be found in many different professions. Each role is important, however, and will help you develop the talents found within the organization. Understanding each of these roles can assist you in creating and building your own tradition within the fire department.

Fire instructors at all levels must learn to manage the following roles to be effective in carrying out their duties in a department **FIGURE 1-4**:

- Leader
- Mentor
- Coach
- Evaluator
- Teacher

The Fire Service Instructor as a Leader

Fire service instructors are often asked to lead fire departments into the future by preparing fire fighters for new missions. Fire service instructors set the example for all fire fighters to follow in terms of performance excellence. As such, they must remain true to the direction set by upper management and supportive of the mission as defined by the organization.

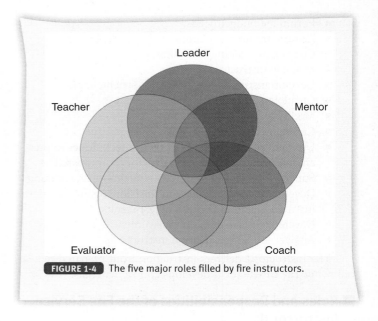

FIGURE 1-4 The five major roles filled by fire instructors.

Your place in front of the classroom puts you in a visible—and sometimes vulnerable—position where you may have to endure the anxieties of an impatient rank and file. Do not waver from the training mission by letting the instructional environment be turned into a debate on issues of management within the fire department. The most powerful leadership tool is the one over which you have the most control: Lead by example.

Teaching Tip

Leading others is a major responsibility. Part of that responsibility is leading in a positive manner. Belittling a new recruit over a mistake may cause the recruit to hide future mistakes, which places both the recruit and others at risk. Instructors should foster an environment that allows for error and turns mistakes during training into a teaching opportunity.

The Fire Service Instructor as a Mentor

Good leaders are also good at identifying future talent. Fire service instructors are in the best position to observe first-hand both the raw talent of recruits and the ongoing growth of department personnel. They can support and enhance the careers of fire fighters by identifying their talents and mentoring them appropriately. Fire service instructors with mentorship abilities enhance the fire department's succession planning by evaluating and developing the talent pool of future instructors and officers. For example, established fire service instructors assist future fire service instructors by showing confidence in their abilities and recommending additional training and opportunities. In short, good fire service instructors mentor future good fire service instructors and leaders.

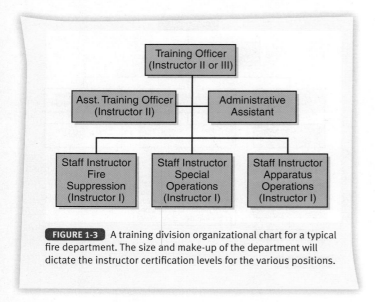

FIGURE 1-3 A training division organizational chart for a typical fire department. The size and make-up of the department will dictate the instructor certification levels for the various positions.

Theory Into Practice

As a fire service instructor, you must have the respect of the trainees. Respect may come from a title or position, but it can also come from the individual, based on his or her expertise and knowledge. If you find yourself charged with the task of teaching those of higher rank, be prepared to work for their respect.

Earning someone's respect can be done in a number of ways. Do not wait to begin building credibility until you are asked to teach—be a professional at your job from the first day you join the fire department. Read books, trade magazines, and other material so that you will become well versed in the trade. Take additional courses and training, and learn how to use the Internet to research the latest information on topics that you are assigned to teach or that are of interest to you. Get out and practice what you know: For example, organize the medical kit. Demonstrate that you are an avid learner and practitioner of your craft. Other fire fighters will notice and respect your thirst for knowledge, which builds your credibility when you are teaching any topic.

Credibility is a direct result of your knowledge of the subject. Rank does not always equate with knowledge— but hard work does. Go over the material so you can determine the flow of the course. Read the activities carefully so you have the right materials on hand and you understand the intended result. Write notes in the margins of the instructor's guide about comments you want to bring up and examples of the practical application of the material. An ill-prepared instructor is doomed to fall short when it comes to gaining credibility.

In addition to your before-class preparation, be ready to build credibility in the classroom. Respect the knowledge that the students possess. Take advantage of those individuals with specialized knowledge to add value to the course. Do not allow yourself to be chastised. Be confident in your skills—obviously whoever asked you to teach believes in you. Take control of the class and be a professional. In the long run, you will be respected for your teaching skills if you have the credibility to support them.

Building respect and credibility with students is essential for those who do not have a formal position in the fire department or when teaching those with greater rank. Although it can be more challenging, success can be achieved with proper preparation and implementation.

The Fire Service Instructor as a Coach

In their book *A Passion for Excellence*, Tom Peters and Nancy Austin define coaching as follows (Austin 1986, pp 325–326):

> Face-to-face leadership that pulls together people with diverse backgrounds, talents, experiences and interest; encourages them to step up to responsibility and continued achievement; and treats them as full-scale partners and contributors.

In the sports world, the coach receives the majority of attention and pressure because he or she is cast in the role of determining the success or failure of the team. In the fire service, we often stress the need for teamwork in all aspects of the job. Teamwork is an important ingredient on the fire ground as the incident commander orchestrates the company's actions through various tactical assignments. Even in the fire station, much of the nonemergency activity is accomplished not by an individual, but rather by a dedicated team of professionals. Whether on the fire ground or in the station, the incident commander acts much like the quarterback of a football team, calling out the formations and plays. In the training environment, the instructor plays a role analogous to that of the football coaching staff: Whereas the coaching staff works on drills to prepare the players for the next game, the instructor works to prepare the fire fighters for the next incident response.

The fire service instructor's job within the fire service corresponds to that of a coach in the sports world in many ways. It is the fire service instructor who must prepare the team for battle. An exterior attack can be as challenging as an interior one; students have to be prepared for both equally. The fire service instructor studies the team's opposition (the fire) and prepares the team for what they may face on game day. He or she builds team skills through practice and repetition on the drill ground **FIGURE 1-5**. The fire service instructor, much like

FIGURE 1-5 The fire service instructor builds team skills through practice and repetition on the drill ground.

a coach, must enthusiastically encourage fire fighters as they struggle with new ideas and techniques. Finally, the instructor must remember to remain positive, even in tough times, and to support the direction in which the team is headed as determined by the general manager (fire chief) and the owners (residents). Coaches and instructors lead, train, and drive their teams' performance.

Ethics Tip

Have you ever gone through a class, only to find out that there is a written exam at the end? The mere thought of this kind of test may make you nervous. Too often, some instructors go over the exam questions just prior to the exam to reduce this apprehension. But is this practice ethical?

Many new fire service instructors believe that they need to be liked to be effective instructors. This mindset leads them to preview answers to written exams or intentionally overlook mistakes during practical evaluations. A certificate is designed to show that the student has demonstrated a level of knowledge, skills, and abilities. Handing out a meaningless certificate is detrimental to the student, undermines the long-term credibility of the instructor, and places the instructor and the fire department at risk of liability should an accident occur. It can also raise concerns about the credibility of the certification.

Be upfront at the beginning of the course about your expectations for your students, and remain firm in that stance. Examinations encourage students to be active participants rather than just passive observers. Students take great pride in the certificates that they actually earn.

The Fire Service Instructor as an Evaluator

Who is in the best position to evaluate the capabilities of a fire department? The answer is found standing in the front of the classroom: the fire service instructor. As obvious as that answer is, how many fire chiefs have actually taken advantage of the knowledge fire instructors have assembled through their daily interactions with the troops? Fire service instructors must evaluate students' learning and competencies at many points during the training process. To do so, they must sharpen their skills in evaluating the proficiency and knowledge retained by their students. The ability of the fire service instructor to assess the comprehension of the class or student accurately can determine the pace of the training. Determining the right time to introduce new techniques or to review previously taught topics cannot be done without the ability to evaluate students honestly. Proper evaluations done in the course of training save the fire department from the heartache of critical evaluations performed by the public at the scene of a mistake.

Fire service instructors are in a unique position to evaluate the capabilities and limitations of fire fighters. In many fire departments, input from the training division and fire service instructors is regularly sought by fire department leadership to help them develop policies and procedures. The fire service

instructor can evaluate fire fighters, suggest operational directives, and field-test new standard operating procedures (SOPs).

Fire service instructors should also be evaluators of the classroom. That is, they must always be able to evaluate their effectiveness, recognizing what the students need and what the students are actually learning. They must evaluate themselves as well, looking for ways to improve their own teaching techniques. Classroom and self-evaluations are essential in keeping the training program proactive and on the road to continuous improvement.

The Fire Service Instructor as a Teacher

As simple as it sounds, the fire service instructor must be a teacher. What comes to mind as you hear the word "teacher"? Perhaps it is someone you remember from your elementary school days. Why does that name come to mind? Teachers affect students in many ways—many times positively, but unfortunately sometimes negatively. Fire service instructors must teach students new skills and abilities, thereby shaping the abilities of the team. They must remember the potential influence they have with their students. It is not just what the students learn that is important; rather, the shaping of a fire fighter's demeanor, attitudes, and desires ultimately determine how the instructor will be remembered.

Being an effective teacher should be the goal of each fire service instructor. Mastering the interpersonal side of the classroom goes a long way toward reaching that goal. The best and most remembered teachers shape and mold the total person to become part of the team and contribute the best they have to offer. A true teacher helps others see their potential and develop a path to accomplish those goals.

■ Setting Up the Learning Environment

Every fire service instructor should understand the effect that the learning environment has on the ability of students to grasp the material being presented. You need to take command of that environment and use it to your advantage. The learning environment includes those environmental factors that influence the learning process and can include multiple elements, both physical (e.g., lighting, temperature, furniture configuration) and emotional (i.e., attitudes, comments, learning).

Physically, the classroom and drill ground must provide a safe, comfortable, and distraction-free environment. Emotionally, students' minds must be kept focused on the instructional material in front of them and away from the sometimes-contentious issues found in many fire departments. To make matters even more challenging, some of the factors affecting the learning environment are outside the fire service instructor's control. If you are not careful, you may bring some frustrations into the classroom, thereby impeding the learning process. To avoid these mistakes, familiarize yourself with the elements that affect the learning environment, including the audience.

Physical Elements Affecting the Learning Environment

Imagine yourself as you walk into a classroom. What do you notice first? Lighting? Cleanliness? Temperature? Or perhaps how far from the front of the room you can sit? Every student who has ever taken a class has evaluated the learning environment. The simple arrangement of tables and chairs can affect how we interact and ultimately learn. The best arrangements allow for the free exchange of information, both from student to student and from student to instructor.

Traditional school classroom setups can lead to the formation of groups in distinct areas of the room, as students tend to sit next to those with whom they are most comfortable. Both the quiet and the uninterested students vie for the seats farthest from the front, not wanting to become involved in the learning discussion. Arrangements such as squares or U-shaped tables, which allow face-to-face interactions, can sometimes neutralize the hierarchy of traditional settings **FIGURE 1-6**. These arrangements also allow you to move among the students freely, improving the exchange of information and increasing the attentiveness of students.

In setting up the classroom, you must also take into account both the natural light and the installed lighting. Lighting affects many aspects of the learning environment. If projectors are used in the presentation, lighting becomes critical to the ability of the students to view the projected information. If natural light is a problem, a simple rearrangement of the room may allow the movement of a screen to a better location. If that is not possible, perhaps the addition of window blinds can correct the problem.

You must also evaluate the installed lighting in the room. Improperly installed lights may make it difficult to dim the lights enough to ensure quality projection. Conversely, dimming the lights too much may create reading or note-taking difficulties for students. If installed lighting is causing problems, your supervisor might be able to recommend improvements to the system through the annual department budget process.

Room temperature can also create problems for the learning environment. Take the entire class into account when setting the classroom temperature. Finding a comfortable compromise for all may turn you into a negotiator. Of course, some classrooms have environmental controls that cannot be changed by those using the facility. In those cases, you may have to alter

lecture and break times if temperatures are uncomfortable. You may also have to become a student advocate and seek improvements to the environmental controls. More information on the physical learning environment can be found in the chapter *The Learning Environment*.

Teaching Tip

Conduct a review of your classroom, drill-ground facilities, and equipment. Submit your results to department administration in the form of a recommended capital improvement plan so that they can be considered in the budget planning process.

Emotional Elements Affecting the Learning Environment

You must also protect the individual student from the emotional letdown that comes with the inability or difficulty to learn a new task or subject. The fire service training environment includes both rookies and senior personnel, adult learners and students fresh out of school. In addition, it is populated by both students who know what to do and those who think they know what to do. Fast learners and those requiring more individual efforts will challenge even the most seasoned instructors.

The best fire service instructors can present material effectively for all types of learners. Instructors who protect those who struggle by placing them in positions where they can make progress without the embarrassment of failure in front of the greater group will be successful. Good fire service instructors also learn how to use the more talented department members to coach and teach those with lesser skills.

Learning to work within the environment challenges you to be flexible, loyal, confident, and fair. Building a bond of trust with your students will enable you to present new ideas and make needed revisions to old traditions. As the fire service instructor, you are the visionary of the fire service.

■ Staying Ahead of the Curve: Vision

Have you ever thought about what drives an organization? Why do some survive and others struggle as time forces change?

In examining the structure of a typical fire department, it is easy to point out the formal leader: It is the individual at the top of the pyramid, the chief who provides the formal direction for the troops to follow. The fire chief is responsible for the ultimate success or failure of the fire department. Of course, organizational charts are filled with many other positions as well, each of which has its own responsibility for providing direction in support of organizational objectives. It is within these ranks that you find assistant, deputy, battalion, and division chiefs; line officer positions including captains and lieutenants; fire marshals; and inspectors.

Where does the fire service instructor fit within this scheme? In many fire departments, the fire service instructor

FIGURE 1-6 Arrangements such as squares or U-shaped tables allow face-to-face interactions.

is treated as an operational support assignment more than an official rank or position. The fire service instructor position is often viewed from one of two extremes: as vitally important to the overall operation or as unnecessary by the administration. They may also have a rank, be scene safety officers, or hold other responsibilities in the department. Wise leaders understand that fire service instructors are important members of the team. They use their instructors to maintain their department's state of readiness and prepare their fire fighters for future missions.

It is this role—preparation for future activities—that requires you to keep ahead of the curve. To fulfill it, you must become a visionary force within the fire department. It is vision that drives fire service instructors to keep abreast of the ever-changing world of firefighting. Changes and advancements in firefighting tactics must be reviewed and implemented through revisions in training curriculum. If the fire department desires a change in operations, the fire service instructor and training division will be charged with educating and training fire fighters to make that change. New techniques and advancements must be tested and made applicable to each organization, because many changes are not "one size fits all" measures. You will be challenged as you introduce new ideas that seem to conflict with established traditions.

As the mission of the fire service continues to evolve, you must be able to prepare the troops to carry out the new mission. The history of the fire service is full of pertinent examples, as the fire service instructor has had to evolve to provide training for medical, hazardous materials, and technical rescue services. Today the threat of terrorism and the use of weapons of mass destruction (WMDs) present their own unique challenges for the fire service, with fire service instructors once again being called upon to lead the troops into these new areas of service.

Fire service instructors must monitor the ever-changing learning environment as well. Struggles with budgets and staffing present ongoing problems for fire service instructors, who must continually seek to keep training at the forefront of nonemergency operations. In the instructor role, you must find new and creative ways to reach students and present training. Today, visionary fire service instructors are turning to the cyberworld as online training programs gain a foothold in the instructional world. Technology will continue to advance and, consequently, affect the delivery of training. Fire service instructors with the vision to see how this new technology can be used for training purposes are establishing virtual classrooms and using satellite and video classrooms as effective learning media.

Given the unique challenges apparent in the modern-day fire service, the fire chief would be wise to select the very best personnel for the position of fire service instructor. A fire service instructor with vision can greatly assist the fire chief in meeting the fire department's future challenges.

■ The Fire Service Instructor's Role in Succession Planning

Preparing fire fighters for battle is not the only job of the fire service instructor. For any organization to survive and

grow, a continuity plan must be in place. Continuity of the organization provides security for the community that the organization serves. The fire service instructor can assist in that regard by becoming involved in succession planning.

As trainers of fire fighters, fire service instructors are often in the best position to recognize potential talent and leadership qualities in the fire fighters and officers they train. The fire service instructor's role is to nurture and challenge these fire fighters through the training program. During the course of training, the instructor may use some fire fighters to assist in training other members. By placing fire fighters in leadership positions within the training environment, you allow them to refine, enhance, and demonstrate their leadership qualities.

The fire service instructor may also be in the position of providing input to ranking officers on the performance of fire fighters in training. By recommending high achievers to fire department administrators, you assist in the identification of future fire officers and fire service instructors. You may also be in the best position to identify those fire fighters who do not fit the model fire officer that the fire department has established.

Succession planning also means looking for future instructors who someday will take your place as the lead instructor in your department. As difficult or as uncomfortable as it might seem, identifying, nurturing, and training their replacements is something successful instructors need to learn to do. Another way to approach this goal is to identify someone who is capable of building upon the foundation that you, as an instructor, have established, and can move it to the next level. Because you have worked hard to develop a successful training program, the last thing you would like to see happen is to have the program fail. Herein lies the opportunity to find another fire fighter who is passionate, skilled, and has the desire to train the other members of the department.

Theory Into Practice

You can increase your visibility within the fire department by volunteering to assist in the hiring and promotional processes. Sell your fire chief on the knowledge you have of each individual fire fighter. You can share information about the training performance of fire fighters vying for promotion with the fire chief.

The fire service instructor walks a fine line between operations and training. In organizations with a weak operations leadership, the instructor may become absorbed with setting the operational direction and standards simply because he or she has expertise in specific areas. In other cases, the fire service instructor has a formal role in establishing policy. Of course, you should be careful to ensure that those policies exist to support the direction of the operations division—not the other way around. Although you may not always agree with the standards chosen by the operations chief, you must accept that direction and train fire fighters accordingly.

Theory Into Practice

A degree is awarded by an institution of higher learning after a person has completed acquisition of the required knowledge in a particular field. A certificate is given after attendance at a course. A certification is awarded after a person has passed an examination process that is based on a set standard such as the NFPA.

Teaching Tip

If you are interested in becoming a fire service instructor for your fire department, talk to current instructors and offer to assist them in their classrooms. Better yet, offer to deliver a lesson plan, with permission of the training officer.

Fire Service Instructor Credentials and Qualifications

You've heard it before: "Walk the talk." It's a phrase that can certainly be applied to the fire service instructor. Your best friend is the confidence that your students have in your abilities and knowledge. This is a quality that cannot be learned from a book. Often, fire service instructors are born in the classroom from energetic students. Others vow to develop their instructional skills after attending a highly charged training session or a lecture delivered by an impassioned instructor. As a fire service instructor, you must be aware that your success or failure might rest with the degree of effort and preparation you put forth at the beginning of your career in the fire service. It is here that dividends are paid.

■ Laying the Groundwork

Good fire service instructors are born from good students—students who thrive on the knowledge gained through training, and those who actively participate in training activities and are not afraid to learn from their initial failures. These individuals are the fire fighters who understand the value of education and refinement of skills. They continually place themselves in learning situations, looking to upgrade their skills and knowledge.

Think for a moment about the fire service instructors who have influenced you. Students attending any training program immediately focus their attention on the instructor as they begin looking for clues about the quality of the program. They might ask what experience the instructor has in the subject area. What is the instructor's firefighting experience level? If the training is being held in-house, your students might remember the days when you were a student. Your credibility in the classroom depends on your past behavior. The past always has a way of finding the future, so it is always wise to protect your future by engaging in proper behavior in the present. Laying the groundwork for becoming a good fire service instructor begins the day you join the profession. While some mistakes will be made, you must always guard your credibility and integrity.

■ Meeting Standards

Standards dictate many things within the fire service. Training programs are not immune from national standards. Standards set the bar for fire service instructors' proficiency and knowledge. They seek to establish uniformity for fire service instructors across the entire profession. NFPA 1041 outlines the instructional levels in the fire service and guides fire service instructor trainer programs; this book is written to be in compliance with this standard. Fire fighters seeking local, state, provincial, or national instructor certifications may be asked to demonstrate compliance with NFPA 1041 as well, though it may not be the only standard that the fire service instructor must meet.

Fire chiefs and fire departments are free to establish their own set of qualifications for fire service instructors. If the fire service instructor is asked to teach only one class in one department, he or she may not be asked to obtain a formal certification from a certifying agency. Instead, the instructor may simply receive the training that the fire department deems necessary from other in-house instructors. In some cases, fire service instructors may be asked to have a certain number of years of experience prior to taking on an instructional assignment. In other cases, instructor positions might be reserved for those with command authority, holding a line officer's or chief officer's position. Some fire departments attach little or no additional requirements in an effort to find a fire fighter willing to take on the extra responsibilities of the fire service instructor.

Whatever the case, it is wise to meet these challenges head on. Becoming a fire service instructor may open doors for future promotions. Meeting the qualifications of this position will also prepare you for future leadership assignments; for example, Fire Instructor I is a prerequisite for Fire Officer I in the NFPA standards system.

■ Continuing Education

Meeting requirements, whether set by a national standard or through fire department policy, is just the beginning for the fire service instructor. Working within the dynamic world of the fire service means you need to continue your own professional growth and development. While some individuals dread the idea of continuing education, you need to understand the need to improve your knowledge, as it allows you to provide the very best and up-to-date information in the classroom. The idea of requiring continuing education is not a new one. Many professions, including health care, education, and inspections, require their members to participate in continuing education to remain licensed or certified.

For the fire service instructor, the organization that issues the initial certification decides what, if any, continuing education is necessary. Some states may require only proof of continued

VOICES
OF EXPERIENCE

When I first became an instructor at Louisiana State University Fire and Emergency Training Institute, I was petrified. I quickly learned that preparation, confidence, and honesty were the most important qualities of today's fire service instructor. I believe that good preparation will always win the day. It is important for your students to know that you have invested your time and trust into the material you are presenting. Many times I was asked to teach a topic that I wasn't an expert in. In these situations, I gathered and studied all of the information available, including texts, case studies, lesson plans, and instructor guides. This preparation allowed me to impart information to my students that would assist them in their daily duties.

"Honesty with your students is paramount"

If you are familiar with the information you are presenting, your confidence will come naturally. If you don't have confidence in yourself or your abilities, your students will not have confidence in you! It is crucial to your future as a fire service instructor that you walk into a room of students confident in yourself and the information you will be presenting. While everyone can get nervous in front of groups, it is imperative that your students know how much time and energy you have put into being an effective trainer.

When I first began my journey as an instructor, I rode with local emergency responders to gain practical knowledge and experience. This enhanced my teaching abilities because I could relate to both the textbook information and practical response information. Although I invested copious amounts of time into my education and teaching style, I quickly learned that I did not know everything about every subject I taught. As an instructor, you are often expected to know the answers to every question asked. If you don't know the answer to a question, be honest with the student, and then do the research necessary to find the correct answer. Honesty with your students is paramount.

In my opinion, it is a great honor to train emergency response professionals. These three qualities—preparation, confidence, and honesty—are important to any fire service instructor. The information you give your students will allow them to perform their duties effectively and safely.

Kathe Jones
Manager, Louisiana State University Fire and Emergency Training Institute
Technical Rescue Program
Baton Rouge, Louisiana

instructional activity. Do not rely on the requirements of outside entities as a motivational force to professional growth. You should always strive to be on the cutting edge of the fire service. Attend outside trainings, seminars, and instructional conferences to stay up-to-date with the most current fire-ground tactics, management practices, and instructional techniques available.

> **Teaching Tip**
>
> Look for state and national instructor organizations to join that will enable you to gain access to the most up-to-date training information (e.g., the International Society of Fire Service Instructors [ISFSI]). These organizations are also helpful in building networks from which to receive and share knowledge and experiences.

■ Building Confidence

By keeping abreast of the latest information, you demonstrate the very value of education. For learning to occur, students must believe in your knowledge of the subject—which is not to say that you will always know everything about a particular topic.

There is a saying that in the classroom, "The instructor is always right." Believing in this old adage could be a fatal mistake and result in the eventual loss of your credibility. It does not take many times of being proven wrong in the classroom to lose the confidence of your students.

Building and maintaining student confidence in both the instructor and the training program go hand-in-hand. Instructors can build high levels of confidence in several ways. For example, you can demonstrate your own commitment to lifelong learning through continuing education. Be willing to admit freely if you are in doubt about a particular question and look to find real answers for the inquisitive student instead of trying to make something up on the fly in an attempt to impress your pupils. Be open to the suggestions and ideas of the students. Because fire service instructors often serve in operational roles as well, be very careful about following all teachings when working the emergency scene or when in the supervision role. One sure-fire way to lose credibility is to project a "Do as I say, not do as I do" attitude.

> **Teaching Tip**
>
> Keep current by subscribing to magazines and credible Web sites, participating in e-mail lists, and tracking blogs that relate to the topics you teach.

Issues of Ethics in the Training Environment

Firefighting is perhaps one of the most respected professions in today's society. The task of maintaining the public's trust in its fire service is held in high regard by all those who wear the uniform.

Ethics reaches beyond laws and standards to define behavior. Codes of ethics spell out what is acceptable and what is unacceptable within a fire department. For many individuals, ethics is rooted in their own personal assessment of right versus wrong (or values), which begins during their early upbringing. Ethics can reflect the attributes of the people we respect and interact with in our own profession. In the end, it is perhaps that feeling in the pit of your stomach—your "gut" instinct—that gives you the best clue about whether a certain behavior is ethical.

The issue of ethical behavior also affects the training program. Fire service instructors are often the ones who are asked to judge which candidates are ready to provide public service. Failing to hold students accountable for their learning objectives or simply passing fire fighters through the training program in the interest of moving on is unacceptable by any standard. This failure to hold to agency standards will affect not only the fire fighter who "slips" through, but also those with whom this fire fighter works, and ultimately the instructor who did not maintain the standards of the agency.

Another aspect of ethical conduct from an instructor's point of view is to ensure that your teaching methods and classroom behavior are acceptable to all students **FIGURE 1-7**. The fire service has evolved over the years to include members from across all demographics and socioeconomic levels of our society; as an instructor, you have a responsibility to be aware of and sensitive to the diverse backgrounds of your students. This might sound challenging or difficult, but the solution is actually quite simple—treat everyone equally and fairly. It is important to acknowledge the diversity that is found in our classrooms and fire departments. Effective instructors are aware of who their students are and strive to teach all of

FIGURE 1-7 Ensure that your teaching methods and behavior are ground in solid classroom and teaching ethics.

them with compassion, fairness, and a desire for all of them to be successful in whatever subject is being taught.

Leading by Example: Do as I Say and as I Do

Ethics in training must begin with the fire service instructor. Instruction must extend beyond the classroom and into everyday operations. Fire service instructors demonstrate ethical behavior through their "Do as I say and as I do" lifestyles. Leading by good example lays the cornerstone for training ethics. If you are responsible for teaching the confidentiality of disciplinary measures in a leadership class but are later seen discussing an employee's behavior around the coffee table, you will lose credibility.

Accountability in Training

Another ethical issue related to the training program is the need to maintain student accountability for training. As an example, if 10 fire fighters attend training on pump operations but only four fire fighters actually operate the pump while the remaining six chat away, who should receive credit for pump training?

Never put yourself in a position of simply passing a student by rote. It is important to document students who fail to meet training objectives. Because the training program is tasked with preparing fire fighters for critical life-safety operations, you must be dedicated to protecting the program's integrity. The classroom or drill ground is no place for favors and friendship to affect a student's achievement record. At the end of the day, the fire fighter's performance ultimately judges your ability as an instructor.

Recordkeeping

Just as you must maintain accountability in individual training accomplishments, you must also maintain accurate records of program achievements. Training that is not fully completed owing to interruptions should never be recorded as accomplished. Good records are important to any training program, but only if those records are accurate.

Less-than-honest fire service instructors have been known to falsify records by simply recording trainings or skill attainments that did not occur for one reason or another. This act of falsification, also known as "pencil whipping an exercise," might look good on paper, but can lead to disastrous consequences once fire fighters actually have to perform the task under the stress of a true emergency or during a legal issue.

All training sessions must be accurately recorded, along with a factual listing of objectives accomplished. Whenever training records are found to be inaccurate, whether due to negligence or just because of an error, your credibility will be called into question. Accurate recordkeeping builds confidence in the training program and sets ethical expectations for all to follow.

The accuracy of training records is not only an ethical issue, but also a legal issue. In recent court cases dealing with fire fighter fatalities, training records have been used as evidence both for and against the authority having jurisdiction (AHJ). Clearly, accurate and correct training records will have a tremendous influence in the outcome of such lawsuits. Recordkeeping as it pertains to the law is discussed in the *Legal Issues* chapter.

Sharing the Knowledge Power Base

Much has been written about the use and misuse of power within organizations. The unethical use of power wielded by a supervisor over an employee has been the catalyst for many lawsuits. Expertise in a subject is considered a power, and you should be aware of its proper use. Individuals increase their expert power by increasing their own knowledge.

Trust and Confidentiality

Trust and confidentiality go hand-in-hand: Lose one and you risk losing both. Both of these attributes have tremendous influences in terms of how well you are able to lead in the classroom and maintain an effective teaching ability.

Fire service instructors are often one of the first points of contact fire fighters make when joining the fire department. As such, you are placed in a position where your trust and confidentiality are critical to students. Struggling students may relay personal information referencing problems in their personal lives to explain why they are having difficulty in completing a task. A student might also share information about learning disabilities that require special considerations in the classroom. Conversations such as these require that you make every effort to maintain confidentiality. If circumstances require that fire department management become involved in the issue, then you should make that fact known first to the student. You may also provide guidance to students about where to go for assistance within the fire department's command structure.

Another area of concern is the need to maintain the confidentiality of student performance in the program. Information dealing with test scores, student evaluations, attendance, and behavioral issues should always be protected. Guidelines for the recording and sharing of this type of information should be written into policy to protect students' rights. Fire service instructors are often required to handle sensitive student information and must consistently demonstrate the ability to do so with professionalism.

Managing Multiple Priorities as a Fire Service Instructor

In today's fast-paced world, we are bombarded with many priorities, each demanding our attention and time. As a consequence, success or failure is often determined by the ability to understand and prioritize tasks. Recognition, planning, and delegation are skills that can assist those facing multiple tasks while maneuvering through sometimes hectic and complicated assignments.

Training Priorities

Managing today's modern fire service creates multiple priorities for the training program. Deciding what and when to teach is just one of the challenges. The direction that the program takes may be decided at different levels depending on the fire department's organization. In some cases, upper management may choose to lay out the training schedule for the period and then leave it to the fire service instructor to decide how to accomplish it. If the fire chief takes a hands-off approach to training, then the fire service instructor may be asked to develop both the schedule and the training program's objectives.

Training priorities may also reflect specific community characteristics. For example, a community with a heavy chemical industrial presence may place additional emphasis on hazardous materials training. Large cities with high-rise construction may require extra training on high-angle rescue and high-rise fire operations. Rural communities may require training on water supply shuttle operations and farm rescue. Training priorities are driven by mandatory training requirements to maintain certifications or Occupational Safety and Health Administration (OSHA) regulations. Clearly, the task of setting training priorities is not a simple one.

Decisions about how precious training time should be spent must include input from all levels of the fire department. Fire company officers may be able to provide valuable insight into fire-ground performance issues affecting operations; perhaps some problematic issues could be corrected through additional training. Fire chiefs may have knowledge of changes in future missions that will require fire fighters to learn new skills. Training committees may need to be established to provide a broad-based perspective regarding the fire department's training needs, tackling the question, "Should we train more on our most serious hazards or on those hazards we respond to most often?" Both elements deserve consideration in the planning process **FIGURE 1-8**.

A. An example of a low-frequency, high-risk incident might be a large industrial fire.

B. An example of a high-frequency, low-risk incident might be a motor vehicle collision.

FIGURE 1-8 Training committees need to determine how to balance the training needs for low-frequency high-risk incidents and high-frequency low-risk incidents.

■ Planning a Program

The level of success of any training program is directly proportional to the planning efforts. In career settings, the schedules of on-duty personnel must be considered when setting up training programs. By contrast, fire service instructors who are dealing with volunteer and part-time fire fighters must take into account the availability of personnel to leave their full-time jobs and to balance family time with the many department requirements. Holidays, vacations, multiple shifts, injuries, and illnesses all affect the need to reschedule or make up lost training opportunities.

The program must also plan for the use of training facilities and equipment. In some cases, outside instructors with special expertise may need to be scheduled.

No employee likes surprises, so consideration of the fire fighter's efforts to be put forth during training and other personal commitments require that training program schedules and completion requirements be established well in advance and communicated clearly to all. Once established, training schedules need to be followed as closely as possible. Although changes will inevitably need to be made, any changes should be implemented with as much advance notice as possible and with sufficient time planned for rescheduling of the training so as to allow fire fighters to readjust their own schedules.

In some locations, a classroom may be shared by the city government, community access groups, and fire department training programs. Nothing is more frustrating to both fire fighters and fire service instructors than to have an overlap of scheduling or other unforeseen event force a last-minute cancellation or location change of a training session. Given the very nature of emergency services, it is always difficult to schedule training and actually execute it according to schedule without interruption. In a combination or volunteer organization, conflicting priorities of personal and professional life outside the fire station make the program management portion of fire service training all the more important.

Program management aspects of running a training division and training program include many basic management skills learned in fire officer training, albeit focused in this case on the goals and objectives of the training division. Administration of training policy and procedures, training recordkeeping systems, selection of instructional staff, review of curricula, and design of programs and courses that meet the current and future needs of the organization require a training team to accomplish many tasks. The *Scheduling and Resource Management* chapter of this text provides more information and resources to help plan and organize the training division's responsibilities.

JOB PERFORMANCE REQUIREMENTS (JPRS)
in action

As a fire service instructor, you are charged with a tremendous responsibility: training and educating personnel on the diverse aspects of the job that ultimately can affect their safety and survival. That is a lot to ask of one person. In many cases, an entire training team is charged with carrying out this mission. Instructors function in many capacities and at different levels of professional qualifications. Understanding the relationships between the levels of instructor qualifications is an essential task in the application of job skills at all levels. It requires understanding both the process of instruction and the process of the delivery of instruction.

Instructor I

The relationship between an Instructor I and an Instructor II is based on an understanding of the role each person plays in the overall delivery process. The Instructor I must demonstrate an ongoing desire to improve his or her instructional skills to enhance the instructor's ability to get the training message across.

Instructor II

The Instructor II will often develop the instructional materials by developing lesson plans, class content, and evaluation materials. Staying on top of the latest changes in instructional delivery and current subject matter allow such instructors to maximize their ability to train fire fighters.

Instructor III

The Instructor III is the person who oversees the instructional development process. This process starts with a needs assessment of the agency, matching resources with budget needs. It also includes instructor selection for curriculum, developing curriculum, and the development of an instructor and program evaluation plan and process.

JPRs at Work

It is important to note the importance of staying up-to-date on the current events and issues facing instructors today.

 ### Bridging the Gap Among Instructor I, Instructor II, and Instructor III

Both the Instructor I and the Instructor II must collaborate on challenges they face in the classroom and on the drill ground, making sure that they account for any successes, failures, obstacles, and new techniques that allow for better delivery of training. Consider scheduling frequent planning meetings between training members or creating an e-mail/bulletin board system to facilitate good communication flow between positions.

Fire service instructors also need to plan the content of individual training sessions. Without planning, conflicts between competing groups can arise in the use of facilities and lead to a delay or cancellation of a scheduled training session, further eroding confidence in the fire service instructor's abilities. Contingencies for the use of specialized equipment needed in training evolutions must be built in if dedicated training equipment is not available. The fire service

instructor should also plan how a missed training session will be made up and ensure that make-up sessions accomplish the same learning objectives as the original training. Training time should be planned so as to eliminate distractions and interruptions. This may not always be possible when students attend training while on duty and as part of in-service training, as the need to respond to an emergency can interrupt even the best-laid plans.

■ Training Through Delegation

Fire service instructors are faced with many tasks in managing the fire department's training program. Planning and scheduling for a multitude of priorities, developing and reviewing curricula, evaluating program effectiveness, coaching and mentoring students, and instructing all place enormous demands on the fire service instructor's time. Without assistance, it is easy to become burned out, even for the most dedicated fire service instructor.

One way to minimize the potential for burnout is through delegation. Be alert to those fire fighters who have shown through recognized efforts the ability to handle additional training assignments. Be observant in the classroom for students who show an interest in instructional activities. It is often said that there is no better way to learn a subject than to have to teach it. Learning can be achieved through delegation, and you can obtain valuable assistance through the assignment of certain instructional tasks to competent fire fighters.

Ethics Tip

Have you ever been to a class where a student has been late to the course? Maybe it was only 5 minutes. Maybe it was 15 minutes. But what if it is 30 minutes or even an hour and 30 minutes? Would you let it go, require the student to make up the time, or fail the student?

You will inevitably face this issue as a fire service instructor. Unfortunately, there is no clear-cut rule for handling the issue, although some principles exist to help guide your decision about how to deal with this problem. The best method is to strive to avoid the situation by setting clear expectations at the beginning of class. Even with that step, however, at some point student tardiness will occur.

Use the following steps to aid your decision-making process. First, determine whether a departmental policy on the issue exists. Next, determine whether the student missed any essential information or whether the missed content would place him or her at risk later. Consider the impact on the other students and their perceptions of the student's nonattendance, and make sure you are consistent. If you would be uncomfortable explaining your decision to the fire chief and other students, it is probably not the right choice.

Safety Tip

Delegate only those tasks where a student has shown competency and where such a hand-off of responsibilities will not jeopardize the safety of other students.

Fire Service Instructor I

Summary of Instructor I Duties

At the Instructor I level, the fire service instructor is beginning to develop the basic skills to present a lesson plan to a classroom of students or a crew of fire fighters sitting around a break room table. Even at this basic level, the instructor must have the basic skills to be effective in delivering the assigned material. The instructor at this level exercises leadership in presenting the material in a way that instills confidence in students and helps build the team. In the process of presenting lesson material, the instructor uses mentoring and coaching skills that will help students obtain

new knowledge and build the necessary skills to perform their jobs as part of the organization. As part of the teaching role, the Instructor I evaluates student performance to ensure competency in learning and demonstrating the new knowledge and skill sets.

Another important task at this level is accurate record-keeping, along with reporting and recording test scores and other evaluative information from the course. This type of documentation might not seem important at the time, but future events might require documentation of course material presented, students' scores and evaluation sheets, and the instructor who presented the material.

Fire Service Instructor II

Summary of Fire Service Instructor II Duties

As an Instructor II, your skills as a leader, mentor, coach, and evaluator take on a different focus. For an Instructor I, the focus is on the student. At the Instructor II level, your focus now takes on a supervisory role; thus, in addition to using your skills with your students when teaching a class, you are now using your skills as a leader to assist those instructors assigned to you as their supervisor. Now you need to mentor and coach those you supervise to be better as instructors, and your role as an evaluator moves from the student to your staff. When necessary, you take on the role of "teaching" your instructors how better to perform their roles as instructors; this is determined while "evaluating" your staff.

Fire Service Instructor III

Summary of Fire Service Instructor III Duties

At the Instructor III level, the fire service instructor is at the top of this supervisory role within the organization and is expected to embrace all the skill sets and knowledge of an Instructor I and an Instructor II. The main focus at this level is that of truly demonstrating leadership in directing the affairs of a training division, providing direction for curriculum development, developing budgets, and supervising all staff members.

To function effectively in this role, the Instructor III must exercise strong leadership skills that provide clear direction and support for the persons he or she is supervising. He or she must exercise vision in looking for those fire fighters and instructors who share the vision of the organization and support their development; this, in turn, supports the constant need for succession planning, which is important for an organization to continue to move forward into the future.

The Instructor III must provide direction for the organization in the development of entirely new programs and course material. Whereas the Instructor II develops lesson plans for a curriculum, the main focus at the Instructor III level is providing the entire course curriculum package.

At this level, the skills of mentor, coach, and teacher are focused within the training division to maintain a high quality of performance for those individuals assigned to teach the entire organization. Personnel at the Instructor III level must set a good example by conducting their activities in support of the department's mission statement and the direction set by the chief, thereby showing leadership in action.

Jones & Bartlett Fire District

Training Division
5 Wall Street, Burlington, MA, 01803
Phone 978-443-5000 Fax 978-443-8000
www.fire.jbpub.com

Instant Applications: Today's Emergency Services Instructor

Drill Assignment

Apply the chapter content to your department's operation, training division, and your personal experiences to complete the following questions and activities.

Objective

Upon completion of the instant applications, fire service instructor students will exhibit decision making and application of job performance requirements of the fire service instructor using the text, class discussion, and their own personal experiences.

Suggested Drill Applications

1. Identify the changes in training you have seen or experienced since you began your career in the fire service.

2. Check industry Web sites and reference materials to identify at least three hot topics facing fire service instructors today.

3. Identify a new piece of equipment purchased by your department. Discuss how the initial training was accomplished on that equipment before it was placed in service. How will your department ensure that the members remain proficient in the equipment's operation after the initial training?

4. Ask the newest or youngest member of your organization about his or her training. Identify similarities and differences between current fire service training and the traditional education of a fire fighter.

5. Review the Incident Report in this chapter and be prepared to discuss your analysis of the incident from a training perspective and as an instructor who wishes to use the report as a training tool.

Incident Report

© Greg Henry/ShutterStock, Inc.

Greenwood, Delaware—2000

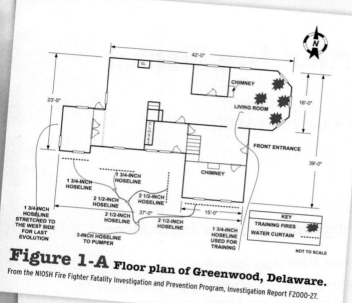

Figure 1-A Floor plan of Greenwood, Delaware.
From the NIOSH Fire Fighter Fatality Investigation and Prevention Program, Investigation Report F2000-27.

A volunteer fire department was conducting a live fire training exercise in a 100+-year-old house **(Figure 1-A)**. Upon completing interior operations, personnel were preparing for the "final burn down" of the 2½ story house.

Three officers were in the attic. One officer was in full protective clothing and self-contained breathing apparatus (SCBA). He was using a sprayer can of diesel fuel to spray fuel. Two other officers, without SCBA, lit multiple fires. The fire spread rapidly and as conditions worsened, the two ignition officers exited to the top of the staircase and told the remaining officer to follow. He agreed to follow, but heat and fire conditions in the attic became untenable very rapidly. The others escaped down the stairs, but fire blocked their colleague's escape.

Multiple rescue attempts were unsuccessful, and the roof quickly collapsed into the attic.

Post-Incident Analysis: Greenwood, Delaware

NFPA 1403 (2007 Edition) Noncompliance

Interior personnel without hose line (*4.4.6*)

Flashover and fire spread unexpected (*4.3.9*)

Flammable/combustible liquids used in large quantities (*4.3.6*)

Multiple fire sets (*4.4.15*)

Interior stairs only exit other than windows without ladders in place (*4.2.12.1*)

Participants not wearing PPE or SCBA (*4.4.18*)

Chief Concepts

- NFPA 1041 categorizes instructor duties into three classifications.
  - The Fire Service Instructor I is able to deliver presentations from a prepared lesson plan, adapt lesson plans to meet the needs of a jurisdiction, organize the learning environment, and provide appropriate recordkeeping.
  - The Fire Service Instructor II meets the requirements of Instructor I and is able to develop lesson plans, schedule training sessions, and supervise other instructors.
  - The Fire Service Instructor III meets the requirements of Instructor II and is able to develop training curricula for single or multiple organizations, conduct needs analysis, develop training goals and implement strategies, and develop a comprehensive program and instructor evaluation plan.
- Fire service instructors can have a wide range of ranks, ranging from fire fighter to chief of department.
- Fire service instructors will act in a variety of roles, including leader, mentor, coach, evaluator, and teacher.
- Both physical considerations and emotional issues influence the learning environment.
- Fire service instructors have a responsibility to stay current on topics and aid in succession planning.
- Fire service instructors regularly face ethical dilemmas that require them to make value-based judgments.

Hot Terms

Certificate Document given for the completion of a training course or event.

Certification Document awarded for the successful completion of a testing process based on a standard.

Continuing education Education or training obtained to maintain skills, proficiency, or certification in a specific position.

Degree Document awarded by an institution for the completion of required coursework.

Delegation Transfer of authority and responsibility to another person for the purpose of teaching new job skills or as a means of time management. You can delegate authority but never responsibility.

<u>Ethics</u> Principles used to define behavior that is not specifically governed by rules of law but rather in many cases by public perceptions of right and wrong. Ethics is often defined on a regional or local level within the community.

<u>Learning environment</u> A combination of a physical location (classroom or training ground) and the proper emotional elements of both an instructor and a student.

<u>Organizational chart</u> A graphic display of the fire department's chain of command and operational functions.

<u>Power</u> The ability to influence the actions of others through organizational position, expertise, the ability to reward or punish, or a role modeling of oneself to a subordinate.

<u>Standards</u> A set of guidelines outlining behaviors or qualifications of positions or specifications for equipment or processes. Often developed by individuals within the regulated profession, they may be applied voluntarily or referenced within a rule or law.

<u>Succession planning</u> The act of ensuring the continuity of the organization by preparing its future leaders.

<u>Vision</u> Having an alertness to the future, recognition of potential, and expectations of improvement.

Reference

Austin, Nancy, and Tom Peters. (1986). *A Passion for Excellence*. New York: Grand Central Publishing. pp 325–326.

National Fire Protection Association. (2012). *NFPA 1041: Standard for Fire Service Instructor Professional Qualifications*. Quincy, MA: National Fire Protection Association.

National Fire Protection Association. (2007). *NFPA 1403: Standard on Live Fire Training Evolutions*. Quincy, MA: National Fire Protection Association.

FIRE SERVICE INSTRUCTOR *in action*

You have just completed your fire service instructor training and are collecting your thoughts about all the skills and knowledge you gained throughout the course. The department training officer has asked to discuss and review the course material with you and to schedule you for your first department training session. You have completed your Instructor I training and know that you have several new skills within your toolbox, but other skills will need more attention and mentoring from the department's instructor staff. As an Instructor I, you known that you have the ability to work with established curricula and to adjust presentations to the learning environment and students. The course and your previous experiences in fire training classes have demonstrated the responsibilities of the instructor and the attributes of today's fire service instructor. You also now believe that as you prepare for a training session, your skills and knowledge will improve. You hope you can bring fresh ideas and enthusiasm to every presentation you are assigned.

1. What are common attributes of today's fire service instructor?
   A. Desire to teach
   B. Experience in subject matter
   C. Motivational skills
   D. All of the above

2. Which of the following is *not* a responsibility or duty of the Instructor I?
   A. Develop lesson plans
   B. Administer evaluation tools
   C. Adjust classroom seating so all students can view visual aids
   D. Document training on a training record report

3. Which NFPA standard identifies the levels of instructor responsibility and job duties?
   A. 1001
   B. 1021
   C. 1041
   D. 1500

4. Of the following roles that a fire service instructor plays, which one puts the instructor in a position to identify future talent in fire service personnel?
   A. Mentor
   B. Evaluator
   C. Teacher
   D. Coach

5. As you review the responsibilities of Instructor I and Instructor II, you notice one key difference. Which of the following identifies the key difference?
   A. An Instructor II delivers prepared materials.
   B. An Instructor I develops and modifies existing materials.
   C. An Instructor I and Instructor II collaborate to develop program budgets and program evaluation reports.
   D. An Instructor II develops all materials used in the instructional area.

Legal Issues

Fire Service Instructor I

Knowledge Objectives

After studying this chapter, you will be able to:

- Describe how laws and standards apply to the fire service instructor. (pp 28–41)
- Discuss the importance of proper recordkeeping. (**NFPA 4.2.5**) (pp 30–31, 34, 38–39)

Skills Objectives

After studying this chapter, you will be able to:

- Demonstrate the ability to apply laws and standards to instructional delivery. (pp 28–41)

Fire Service Instructor II

Knowledge Objectives

After studying this chapter, you will be able to:

- Describe how laws and standards apply to the fire service instructor. (pp 28–41)
- Discuss the importance of proper recordkeeping. (**NFPA 5.2.5**) (pp 30–31, 34, 38–39)
- Describe the records and reports required by the fire department. (**NFPA 5.2.5**) (pp 38–39)

Skills Objectives

After studying this chapter, you will be able to:

- Demonstrate the ability to apply laws and standards to instructional delivery. (pp 28–41)

Fire Service Instructor III

Knowledge Objectives

After studying this chapter, you will be able to:

- Describe how laws and standards apply to the fire service instructor. (pp 28–41)
- Discuss the importance of proper recordkeeping. (NFPA 6.2.2) (pp 30–31, 34, 38–39)
- Describe the records and reports required by the fire department. (NFPA 6.2.2) (pp 38–39)

Skills Objectives

After studying this chapter, you will be able to:

- Demonstrate the ability to apply laws and standards to instructional delivery. (pp 28–41)

You Are the Fire Service Instructor

During a recent emergency in a neighboring fire district, several on-scene fire fighters took pictures and videos of the incident and then posted them on a popular social media site on the Internet. This situation generated a great deal of discussion in the local media about patient privacy and the conduct of the agency involved.

Today at the weekly staff meeting, the chief asks those present to discuss this issue and to develop a policy that can be adopted by the department as a whole. After an intense discussion among members of the staff, all eyes turn to you, as the training officer, to develop a draft policy to address this issue and bring it to the next staff meeting for further discussion.

1. Which resources would you seek out to help you in developing this initial draft?
2. Which legal issues do you need to consider in framing this new policy?
3. How will this new policy likely affect your agency?

Introduction

In the fire service, training is designed both to keep existing skills sharp and to introduce new skills. Every training session exposes the fire service instructor and the fire department to potential liability. For example, inappropriate comments voiced in a training session based on religion, nationality, gender, or sex—even if made in jest—can subject the fire service instructor and the fire department to liability. A good working knowledge of the law, standard operating guidelines, and the department's own rules and regulations will reduce the potential exposure to liability.

An understanding of the legal issues pertaining to training is critical for fire service instructors. This chapter explores legal issues inherent in instruction and provides you with a working knowledge of the laws you will most likely encounter in your role as fire service instructor. The chapter also describes recordkeeping strategies and offers practical advice on how to deal with commonly encountered difficult situations.

Fire Service Instructor I, II, and III

Types of Laws

Laws are generally established by one of two possible methods. The first method is through the legislative process (federal or state), which usually results in the establishment of statutes **TABLE 2-1**. The second method usually is performed at the administrative or local level and results in regulations or codes.

The Law as It Applies to Fire Service Instructors

As a fire service instructor, your conduct is governed by three sources of law:

- Federal law, which is made up of statutory laws
- State law, which is established by each state's legislature
- Policies and procedures crafted by each individual fire department

All three sources set expectations for and restrictions on the conduct of fire service instructors **FIGURE 2-1**. Nevertheless, federal and state laws differ from a fire department's policies and procedures in an important way. Federal and state laws apply evenly to all citizens and are the standard by which all fire service instructors are measured. The Americans with Disabilities Act (ADA) and the Age Discrimination in Employment Act (ADEA) are two examples of federal laws that govern all citizens across the nation. By contrast, policies and procedures are specific to a particular fire department and vary from region to region within a state, because each fire department must take into consideration numerous variables, such as the size of the community it serves, its locale, and its resources.

A fire department's policies and procedures must not violate any state or federal laws. Indeed, they do not have the same status in the law as state and federal statutes. A fire department's policies and procedures represent the minimum expectations of the fire department and do not necessarily impose a legal duty, as does a state or federal law. The only true benchmark against which the departmental policies and procedures are measured is the national standards or guidelines set within the

Table 2-1	Types of Laws	
Type of Law	**Definition**	**Example**
Statutes	Created by legislative action and embody the law of the land at both the federal and state levels.	Americans with Disabilities Act Title VII Motor vehicle codes
Codes/regulations	Can be established by legislative action, but are most commonly created by an administrative agency or a local entity with the authority to do so. Codes and regulations usually deal with narrower issues and enhance existing laws.	National Fire Protection Association (NFPA) codes Building codes
Standards	Any rule, guideline, or practice recognized as being established by authority or common usage. Standards take on the effect of law if they are adopted by a local entity.	NFPA standards Occupational Safety and Health Administration (OSHA) standards

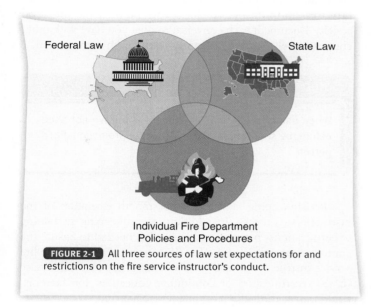

FIGURE 2-1 All three sources of law set expectations for and restrictions on the fire service instructor's conduct.

fire service. Departmental policies and procedures can place obligations upon a fire service instructor that are not otherwise required by federal and state laws. For instance, a fire department's policies and procedures may require the department's fire service instructor to keep all department licenses and permits up-to-date, even though there is no legal requirement for the fire service instructor to do so. The policies and procedures may also specify additional recordkeeping requirements not set forth by state or federal laws.

This chapter examines the laws that apply to fire service instructors—from those with the broadest coverage to those with the narrowest scope. Federal law is the broadest, in that it applies to all persons and requires the same expectations across the board regardless of your chosen profession or the state in which you live or work. For example, the ADA, the ADEA, and Title VII govern not only fire departments, but private employers as well. State laws are somewhat narrower, in that they apply only to the state in which they are enacted. State laws set some expectations for that state, but are not necessarily specific to a profession. The policies and procedures of a fire department have the narrowest scope, being designed specifically for an individual department and the professionals who work in it.

Although the law is designed to set certain standards and expectations, it also can be used to protect fire service instructors from lawsuits. Fire service instructors must have a working knowledge of federal and state laws. In addition, they must be well versed in the policies and procedures of their particular fire department, which requires that they possess a strong working knowledge of national guidelines. Although national guidelines are not legally binding on a fire department, they clearly set out the consensus of the fire service across the United States. The NFPA standards, for example, are the national standards upon which most fire departments model some of their own standards.

Should a fire service instructor's conduct be called into question, the national guidelines serve as the backdrop against which that conduct will be judged. For example, if a live fire training session gets out of control, the fire service instructor's methods will be compared to NFPA 1403, *Standard on Live Fire Training Evolutions*, to determine whether the instructor's conduct was reasonable. NFPA 1403 outlines a checklist of safety activities when performing a live fire training session, which includes activities such as ensuring that all utilities are disconnected and that proper permits have been obtained. If a fire service instructor does not follow this checklist, his or her conduct may be considered unreasonable and result in liability.

Safety Tip

Every fire service instructor who conducts live fire training must be thoroughly familiar with NFPA 1403 to ensure the safety of the participants.

Fire service instructors have an immense responsibility to conduct, manage, and administer training programs. With that responsibility comes the potential for liability in the event that an injury, accident, fatality, or lawsuit occurs. In our increasingly litigious society, the instructor must be aware of the laws that pertain to each level of instructor responsibility. All certification levels of instructor will have different applications of laws and statutes in the performance of their duty areas, and professional development as an instructor will be necessary throughout your career to review trends and keep up-to-date on cases that have occurred and judgments for and against a fire department and its training program.

You will likely encounter situations during training evolutions, classroom participation, and your daily duties when issues pertaining to federal law will arise. You may face questions about compliance with the ADA, hostile work environment claims, and complaints of sexual harassment. A strong familiarity with these federal laws will prepare you to fulfill your obligations properly. State law could come into play on issues of privacy, indemnification, injuries to training participants, and equipment failures.

It is only through familiarity with the appropriate laws that you can insulate yourself and the fire department from potential liability. Directly associated with familiarity is proper record-keeping. Maintaining an appropriate documentation system affords you some level of protection; conversely, inadequate or incomplete records may expose you to increased liability.

Teaching Tip

Have examples of local laws and ordinances ready to discuss in class, along with any judgements or lawsuits brought against your organization. Use them as valuable learning tools.

Legal Protection

Once you understand the laws that apply to you as a fire service instructor, the next logical question is this: How do you perform your job so that you abide by the law and have protection should a lawsuit be filed? The first line of defense is always proper and thorough recordkeeping. Creating and maintaining records, reports, memoranda, e-mails, and other records in the course of your duties places you in a good position. These records should be constructed contemporaneously and maintained for a reasonable period of time and in accordance with any local law.

The recordkeeping process starts at the outset of each activity. For example, when you are organizing a training evolution, the materials you prepare should set forth the purpose of the training, the activities to be performed, the equipment to be used, and the goals of the training. You should have available any sources that were relied upon in creating the evolution and should confirm that the training complies with all policies and procedures of your fire department.

If the training will involve participants from other fire departments, additional documentation should be obtained for those fire fighters. Moreover, your fire department should enter into a hold harmless and indemnification agreement with any other department or agency involved in the program. Such an agreement provides that your fire department will not be held monetarily or legally liable for injuries or conduct of the other department or agency's employees. This document is necessary to protect your fire department from individuals over whom you have no control. As a representative of the

hosting fire department, you should also verify that all outside participants are medically cleared to participate in the training, have met the necessary prerequisites, and possess the proper credentials. While this approach might seem somewhat burdensome, the protection it affords is priceless.

Off-site training is filled with additional risks not encountered with on-site training. You should perform a site inspection well in advance of scheduling the training and prior to development of any handouts to identify any unusual or potentially unsafe conditions. In addition, contact should be made with all other entities (town, city, utility companies) to verify that the proposed training does not violate any law or create any unreasonable risks. For example, you should confirm that no ordinances or local laws prohibit the proposed activity. Document each conversation that you have with other entities as well as the steps you take after each conversation.

Safety Tip

When using facilities or equipment you have not used before, be sure to get proper instruction from a qualified person.

Recordkeeping is vital to the smooth operation of the instructional process. In addition, recordkeeping pertaining to certifications, permits, and licenses is needed to ensure the continued operation of the fire department. Failure of the fire service instructor to keep track of emergency medical service (EMS) recertification or continuing education, for example, could impair the staffing of the fire department.

Theory Into Practice

Generalized information regarding the main legal topics can be found through a basic online search. In contrast, for specific issues involving a specific fact pattern, fire service instructors should contact their respective fire protection district or city attorney.

Federal Employment Laws

The U.S. federal government has created an expansive body of laws designed to provide equality in employment. The scope of this chapter is too limited for a complete discussion of all federal employment laws, but you should always ensure equality of treatment for all participants.

As a fire service instructor, you are not expected to know all the laws of the land, but to perform your job correctly you do need to have an understanding of three laws: the Americans with Disabilities Act, Title VII, and Section 1983 of the Civil Rights Act of 1964. In all likelihood, you will be called upon to teach or arrange instruction on each of these laws. You will also encounter situations that will require you to act or react to situations involving the practical application of these three laws. Here is a brief summary of their content:

- **Americans with Disabilities Act (ADA):** Prohibits discrimination against a qualified person because of a disability, where discrimination may include hiring, firing, promotions, and compensation.
- **Civil Rights Act of 1964:** Prohibits discrimination based on race, sex, or national origin.
- **Title VII of the Civil Rights Act:** Prohibits discrimination that creates a hostile work environment.

In addition to those laws, you should be familiar with the following legislation:

- **Age Discrimination in Employment Act (ADEA):** Prohibits discrimination against persons older than age 40 unless there exists a recognized qualification requirement.
- **Equal Pay Act:** Requires that all persons, regardless of gender, be paid at the same rate when performing the same job.

■ Americans with Disabilities Act

The Americans with Disabilities Act of 1990 (ADA) is designed to protect individuals with disabilities from discrimination in the workplace. A disability is a physical or mental impairment that substantially limits one or more major life activities. In 2008, the ADA was amended with the primary purpose of expanding what constitutes a disability. Determining whether an impairment substantially limits a major life activity should not take into account possible mitigating measures. Mitigating measures include medication, medical supplies, prosthetics, hearing aids, mobility devices, and the like. The mitigating measures of ordinary eyeglasses and contact lenses, however, are considered in determining whether an impairment substantially limits a major life activity. While the definition itself did not change with the 2008 revision of the ADA, the amendment added instructions to follow when determining what constitutes a disability. Major life activities include functions such as caring for oneself, performing manual tasks, walking, seeing, hearing, speaking, breathing, learning, and working. Conditions that previously would not have qualified as a disability—such as compulsive gambling, kleptomania, pyromania, and current illegal use of drugs—may now be considered a disability. It should also be noted that an impairment that is episodic or on remission is considered a disability if it would limit a major life activity when active.

In the fire service, it is unlikely that you will encounter physical disabilities such as blindness or the inability to walk. Nevertheless, you may encounter many other types of disabilities, including psychological disorders, neurological disorders, cardiac issues, respiratory issues, diabetes, epilepsy, dyslexia, drug and alcohol addiction, and prosthetic devices.

If you are approached by an individual alleging a disability, you should undertake a two-step analysis. First, you must determine whether an actual disability exists. If the answer to that question is yes, then you must determine whether the disability can be reasonably accommodated. The ADA defines reasonable accommodation as a modification or adjustment to a job or work environment that will enable a qualified applicant or employee with a disability to perform the essential functions of the job or enjoy the benefits and privileges of employment, and that does not create an undue hardship for the employer. For example, if a fire fighter has diabetes, a reasonable accommodation may be to allow for periodic blood glucose testing during the day or during training FIGURE 2-2 .

Examples of reasonable accommodations found to be in compliance with the ADA include modifying facilities to make them accessible to employees with disabilities, restructuring a job, modifying the work schedule, acquiring or modifying equipment, and reassigning a disabled employee to a vacant position for which the individual is qualified. Accommodation is not automatic—that is, you are not required to make an accommodation if it would impose an undue hardship on the operation. An undue hardship exists if the accommodation would require significant difficulty or expense relative to the size, resources, and nature of the employer.

Further, a reasonable accommodation does *not* have to be provided if doing so would present a direct threat to the health or security of others. Four factors are considered when assessing whether a direct threat exists: the duration of the risk, the nature and severity of the potential harm, the likelihood

FIGURE 2-2 A reasonable accommodation may include allowing for periodic blood glucose testing during the day or during training.

that potential harm will occur, and the imminence of the potential harm.

Like many legal situations, ADA issues are fact driven. There is no "bright line" test; each case must be analyzed based on the particular facts at hand. Courts have ruled on a number of cases involving fire fighters and the ADA, and those rulings can provide guidance for you in similar situations. For example, the U.S. Court of Appeals for the Fifth Circuit has held that a fire department is not required to accommodate a fire fighter who is unable to perform the essential duties of firefighting owing to a back injury when no light-duty position exists in a small fire department (*Burch v. City of Nacogdoches*, 5th Cir.). In Pennsylvania, a fire department was not required to provide a fire fighter with a second chance at the academy course after he failed due to depression. In Maryland, the court held that it was proper for a fire department to prohibit a fire fighter from using an inhaler when the brand used was highly flammable (*Huber v. Howard County Maryland*, 4th Cir.).

It is important for the fire service instructor and managers up the chain of command to be aware of both the requirements of the ADA and those personnel under their command who qualify for reasonable accommodation under the ADA. In situations when an accommodation is requested during a training evolution, you should take great care to evaluate the request and, if needed, contact a higher authority for advice. Once a decision is made, it should be documented immediately to protect both you and the fire department.

Theory Into Practice

Not all medical and psychological issues qualify as disabilities under the ADA. Each issue must be examined separately. Also, any accommodation request must be reasonable.

◼ Title VII

Title VII prohibits employment discrimination based on race, color, religion, sex, or national origin. Discrimination does not have to be intentional to violate Title VII. Any practice that has the effect of discriminating against a person because of his or her race, color, religion, sex, or national origin also violates Title VII. Title VII applies to state and local units of government with 15 or more employees. Although issues such as hiring, firing, promotion, and compensation do not generally arise in the context of training, harassment on the basis of race, color, religion, sex, or national origin may occur in training. You must be on the lookout for harassing behavior and know how to address it if it occurs.

In addition to your primary role of educating employees, you have a responsibility to address harassment and discrimination that you observe or that is reported to you. If you witness such conduct, you must take corrective steps immediately. Be sure to follow any grievance procedure set by policy or agreement in such a case. You should document the event, including statements from any witnesses of the event.

All documentation should be completed as soon after the incident as possible.

If a report is made to you about harassing conduct but you did not witness that conduct firsthand, you must follow the steps set out in the fire department's harassment policy. If you are the person responsible for investigating claims, begin the investigation and document all relevant information regarding the allegation. If someone other than you is responsible for investigating the incident, prepare a report to that person documenting the information provided by the complainant and forward it to the appropriate parties **FIGURE 2-3**. A prompt and thorough response to all claims should be made. Finally, follow up with the complainant when and if appropriate.

Race and Color Discrimination

Race and color harassment includes ethnic slurs, jokes, and derogatory or offensive comments based on a person's race or color if the behavior creates an intimidating, hostile, or offensive work environment or interferes with a person's work performance.

Religious Discrimination

Religious discrimination can also take the form of harassing slurs, jokes, and derogatory or offensive comments. In addition to harassment, a religious discrimination issue may arise based on the scheduling of training classes. If training classes are scheduled only on days that conflict with an employee's religious needs, classes may need to be rescheduled or repeated on a different day unless doing so would create an undue hardship for the employer.

Sexual Discrimination

The most common form of sexual discrimination experienced in the workplace is sexual harassment. Sexual harassment consists of sexual advances, requests for sexual favors, and other verbal and physical conduct of a sexual nature that affects an individual's employment, either explicitly or implicitly; unreasonably interferes with a person's job performance; or creates an intimidating, hostile, or offensive work environment. Sexual discrimination also includes discrimination based on pregnancy.

FIGURE 2-3 Properly document all information provided by the complainant when an allegation of discrimination is made.

Every fire department must have in place a policy prohibiting sexual harassment in the workplace. Generally, two categories of sexual harassment are distinguished: *quid pro quo* harassment and creation of a hostile work environment.

Quid pro quo sexual harassment usually occurs in a supervisor–subordinate relationship, where one party controls the other party's future in the organization. It ties employment decisions to the submission to or rejection of some form of sexual conduct. An employment decision might include a promotion, demotion, raise, transfer, termination, or other benefit of employment. For example, a supervisor would be guilty of *quid pro quo* harassment if he or she said to a subordinate, "If you have sex with me, I will give you a passing grade on the training evolution." Whether the employment decision is positive for an individual or has adverse consequences for the person is irrelevant if the decision is based on submission to or rejection of some form of sexual conduct.

The second form of sexual harassment is sexual conduct that creates a hostile work environment for an individual. Often—but not always—the inappropriate conduct occurs between co-workers who do not necessarily have the power to affect a person's job. The conduct in this scenario consists of a pattern of behavior that is sexual in nature and that interferes with an employee's work performance. Conduct that can rise to the level of sexual harassment includes inappropriate touching, sexually explicit language, sexually oriented jokes, comments about a person's appearance, and repeated requests for a date.

Some misconceptions exist concerning sexual harassment. For example, one misconception is that a victim must be of the opposite sex from the alleged harasser. In reality, the sex of the victim or the harasser has nothing to do with the application of Title VII.

Another misconception is that the harasser must be an employee of the fire department and the victim's supervisor. In fact, individuals who are not employed by the fire department can create a hostile environment, which can subsequently create a cause of action against your fire department. For example, an outside instructor who is brought in to provide training can create a hostile environment for department employees. Likewise, fire fighters or instructors from other fire departments involved in joint training can create a hostile work environment.

A third misconception is that only the person who is the object of the behavior can be harassed. In actuality, anyone affected by the harassment—including bystanders who observe the conduct—can make a claim under Title VII.

To constitute harassment under Title VII, the conduct must be unwelcome. Conduct that might seem appropriate to you might constitute harassment to others. Off-color jokes can be harassment if they are unwelcome and pervasive. Horseplay and comments about a person's appearance can constitute harassment. Inappropriate pictures, calendars, and posture can create a hostile work environment.

During training evaluations, especially those involving physical agility, you must make sure that comments and ribbing about each participant's performance are not sexual in nature. While good-natured ribbing may be acceptable, comments based on sex, gender, or stereotypes can create a hostile work environment in violation of Title VII.

If you, as a fire service instructor, witness conduct that may be a violation of Title VII, or if you receive a complaint from someone that a hostile environment is occurring, you have an obligation to take steps to rectify the situation. Those steps could include taking immediate corrective action if you witness the event or conducting an investigation if a complaint is brought to you. Your obligations should be set out in departmental policies and procedures. Taking no action or ignoring a complaint could expose you and the fire department to liability. As always, document all events associated with the matter, including summaries of statements made and steps taken in response to any potential violation.

National Origin Discrimination

National origin harassment can take the form of ethnic slurs or other verbal or physical conduct based on a person's nationality. "English only" rules can also constitute national origin discrimination unless they are necessary for business purposes and employees are notified in advance of the rule's enforcement.

Antidiscrimination Policies

As a fire service instructor, you must take the necessary steps to provide an environment that is free of discrimination. The best way to achieve this goal is to ensure that your department has in place a proper policy prohibiting discrimination in the workplace. Once such a policy is in place, all members of the department should be given a copy of the policy and sign an acknowledgment of its receipt and a commitment to abide by the policy. In addition, annual training on the policy should be provided for all employees, and documentation should be maintained indicating which employees attend the training.

You should also take steps to keep the antidiscrimination policy current. As the law changes, the policy should likewise

Ethics Tip

At some point, you will likely find yourself in a situation where students are harassing other students, yet it is not an illegal form of harassment. For example, a group of fire fighters from a career department may tease fire fighters from a volunteer department when they are slower at the evolutions. Would you prohibit such actions? Would you take the same steps in the reverse situation—that is, if the volunteers were teasing the career fire fighters about their behavior?

These situations arise on a regular basis. As a fire service instructor, it is your responsibility to provide an environment that is conducive to learning for *all* students. Doing so requires a subjective evaluation of the effects of the teasing. It does not matter whether the targets "deserve it" or whether the teasing is meant in good nature. If any doubt arises as to whether the behavior interferes with learning, it must be prohibited. Obviously, such behavior must always be prohibited if it is illegal, whether it interferes with learning or not.

evolve to reflect these changes. Whenever the policy is modified, all employees should be given a copy of the new policy and sign an acknowledgment of its receipt as discussed previously.

Retaliation

Title VII also prohibits retaliation against a person for filing a charge of discrimination, participating in a discrimination investigation, or otherwise enabling discrimination. Retaliation occurs when an employer takes an adverse action against a covered individual because that person engaged in a protected activity.

- An *adverse action* is any substantive action intended to discourage a person from exercising his or her rights under Title VII. Minor slights and comments or actions that are justified by an employee's poor work performance do not necessarily rise to the level of a retaliatory adverse action.
- A *covered individual* is anyone who has opposed unlawful discriminatory practices, and those who have a close association with such people. For example, an employer cannot take adverse action against a person because his or her spouse filed a discrimination complaint.
- *Protected activities* include opposing a practice that a person believes to be unlawful discrimination and participating in a discrimination proceeding.

Simply put, the complainant and any witnesses must not be subjected to activities that are designed to punish them for reporting the misconduct. Activities such as not including them at dinner, threatening not to back them up during a call, or refusing to speak to the complainant or witness during work hours can constitute retaliation.

Section 1983 of the Civil Rights Act of 1964

Title VII applies to all employers, whether public or private. By comparison, Section 1983 of the Civil Rights Act of 1964 applies only to public officials, and it has a much broader reach than Title VII. Whereas a Title VII suit is directed only at the employer, a claim under Section 1983 allows a complainant to sue not only the department, but also the alleged aggressor personally. Personal liability is claimed where the alleged conduct falls outside the scope of employment. Inappropriate touching of a co-worker, for example, is not part of one's job description and, therefore, is a personal act. To be sued personally means that a complainant can seek damages directly against the alleged harasser and the assets of the alleged harasser. The complainant

Safety Tip

As a fire service instructor, you have a responsibility to provide for the safety of your students. Generally, the hazards in such training arise from the activities being performed. Sometimes, however, the hazards may come from other students. As a fire service instructor, if you observe or are notified of threats by a student, you have an obligation to take action to provide for the protection of the class.

can also seek punitive damages against an individual that are barred against units of local government.

Freedom of Information Act

Each state has freedom of information laws that govern public access to government records, documents, and other information kept by government bodies. The purpose of the Freedom of Information Act (FOIA) is to promote government accountability and the public's right to know. Although each state has its own variation on FOIA, generally the laws require the following records and information to be made available to the general public upon request:

- Administrative manuals and procedures
- Instructions to staff
- Substantive rules
- Statements and interpretations of policy
- Final planning policies
- Factual reports (such as inspection reports)
- Information about accounts, vouchers, and contracts involving the expenditure of public funds
- Names, salaries, titles, and dates of employment for all employees and public officers
- Meeting notes or recordings
- All records, reports, forms, writings, letters, memoranda, books, and papers (even if in electronic form)

Generally, members of the public can make a FOIA request through designated public officials. Depending on the applicable state law, there may be exceptions to the disclosure of certain documents or the government body releasing the documents may be required to redact personal information from certain documents. In addition, government entities may be required to retain public records for a specific period of time. In most cases, state record retention laws will dictate the minimum length of time public records must be retained by government entities. Please review your state law and consult with your attorney for more specific information about the requirements of FOIA, the exceptions to FOIA disclosure, and the record retention requirements in your state.

Equal Employment Opportunity Commission

The Equal Employment Opportunity Commission (EEOC) is the federal agency charged with investigating and enforcing federal laws pertaining to discrimination in the workplace. It requires that employers take all steps necessary to prevent workplace discrimination from occurring. Actions such as raising the subject of preventing discrimination, expressing strong disapproval for the discriminatory conduct, and creating a policy that informs employees of their rights and states that investigations will be performed for all complaints should be discussed and reviewed. Most of the EEOC goals can be accomplished through education of fire department employees by fire service instructors from within the department or from outside the department who are familiar with that area of the law.

In addition to the federal laws enforced by EEOC, many state and local governments have statutes prohibiting discrimination and agencies that enforce these statutes. These agencies generally cooperate with the federal EEOC in work-sharing arrangements to investigate both federal and state claims of discrimination. Discrimination complaints are typically investigated by the agency with which the complaint is filed.

Theory Into Practice

Many new fire service instructors have never considered the possibility that they might face a complaint dealing with an area governed by federal law. In reality, every fire service instructor who teaches long enough must deal with this prospect. For that reason, it is essential for all fire service instructors to understand their responsibility in such cases. Failure to respond appropriately can place you at risk.

To prepare for this possibility, it is important to keep current on federal and state laws. Understand to whom the laws apply and how they fit into the educational environment. Find out your role and responsibility within the organization and then follow the mandated procedures precisely if a complaint arises. As a fire service instructor, you have both moral and legal obligations to the students. You will be held accountable for your actions.

Negligence, Misfeasance, and Malfeasance

Negligence, *misfeasance*, and *malfeasance* are terms you are likely to encounter in cases where a civil wrong is alleged. Generally, two types of wrongs can be alleged in civil cases when an injury occurs:

- Unintentional conduct, which is best equated to the term *accident* (negligence, misfeasance)
- Conduct that includes an element of intent or knowledge that a wrong is being committed (malfeasance)

Which conduct constitutes negligence/misfeasance and which conduct constitutes malfeasance depend on the facts of each situation. This determination is based on the conduct involved, the fire department's rules and regulations, prior training, and national guidelines.

■ Negligence/Misfeasance

Negligence and misfeasance are kindred spirits: They are void of the element of intent and are based on a duty to prevent harm or injury to others by exercising a degree of care commensurate with the task being performed. Negligence exists when there is a breach of the duty owed to another, there is an injury, and the breach is the proximate cause of the injury.

Once these three elements are established, you can become legally liable for your conduct absent some type of immunity. Liability in its simplest form means responsibility. If a person or entity is found liable, that person or entity is responsible for paying for the damages caused by its actions or inactions.

A fire service instructor's conduct is judged according to the reasonable person standard, which is an objective (not subjective) standard. In other words, the question is asked: What would a reasonable person have done in the same or similar situation? In the case of fire service instructors, what constitutes "reasonableness" is gleaned from the department's policies and procedures, national standards, and federal and state law. For this reason, it is important that you remain apprised of changes in the laws that affect the fire service. It is also important that any changes in the law, technology, and the national guidelines be considered and incorporated into the policies and procedures of the fire department when appropriate.

Misfeasance is the most commonly encountered type of wrong in the performance of otherwise lawful acts. Simply defined, misfeasance consists of improper or wrongful performance of a lawful act without intent through mistake or carelessness. An example would be the hiring of a family member—a practice that, unbeknownst to you, is a violation of the department's nepotism policy.

Most states offer some level of protection to public employees against lawsuits premised on claims of negligence or misfeasance. These protections often take the form of immunity laws that bar certain causes of action outright or require a complainant to allege and substantiate a higher level of proof. The higher standard of proof usually requires that an element similar to intent be proved to prevail on a claim. Quite often, the highest standard requires proof of willful and wanton conduct or establishment of gross negligence. Gross negligence could be found, for example, if a fire service instructor fails to follow a written protocol or a standing order. For example, a standing order might require personnel to wear protective equipment when dealing with a live fire; failure to do so could constitute gross negligence. The purpose of the higher standard of proof is to allow public employees to perform their jobs without fear that every little mistake will expose them to potential liability.

Willful and wanton conduct is defined by some courts as an act that, if not intentional, shows an utter indifference or conscious disregard for the safety of others. For example, it might include the failure to follow a written protocol or a standing order such that this omission creates a substantial risk to another person. That standard can be established if it is shown that, after having knowledge of an impending danger, a person failed to exercise ordinary care to prevent the injury. In addition, if, through recklessness or carelessness, a danger is not discovered by the use of ordinary care, the willful and wanton conduct standard may be satisfied.

Similarly, gross negligence is defined as an act (or a failure to act) that is so reckless it shows a conscious, voluntary disregard for the safety of others. Gross negligence involves extreme conduct but not intentional conduct. It falls somewhere between a mere inadvertent act that causes harm and acting with an intent to harm.

Whether a particular act constitutes willful and wanton conduct or gross negligence depends on the facts of the situation. The most common evidence considered when analyzing whether an act is willful and wanton are your departmental policies and procedures, surrounding departments' policies and procedures, training records, and national guidelines or standards. It is

JOB PERFORMANCE REQUIREMENTS (JPRS)
in action

Legal exposures exist in many training sessions that you will teach or in which you will participate. In the event of almost any accident or injury, the training records and training history of the participants may be examined by attorneys and investigators in an attempt to identify any potential errors or omissions in the training and educational process. Many fire service instructors fail to realize the importance of the documentation part of the training process. When shoddy records are reviewed during investigations, the professional reputation of the fire service instructor may suffer—not to mention that such lackluster recordkeeping presents the potential for litigation under the heading of "failure to train" or "inadequate training." Training development from established reference sources based on up-to-date standard operating procedures (SOPs) is key to ensuring legally defensible training development.

Instructor I

The Instructor I must have an understanding of the legal system and process used to validate training by referencing established standards and procedures. Understanding potential legal exposures in the delivery, recordkeeping, and evaluation of training are tasks involved in every training session.

JPRs at Work

Know the components of the lesson plans you use as part of your instruction so you can identify references and evaluation criteria. Be able to complete training record report forms correctly, documenting each training event.

Instructor II

The Instructor II serves in many cases as the "author" of training materials that are often used by others within their department. Knowledge of areas of exposures, identification of reference sources, and sound objective and lesson plan development reduce potential exposure to the instructor.

JPRs at Work

Understand the laws and standards that apply to the development and delivery of training. Prepare lesson plans, evaluate materials, and collect records in accordance with local training policies and industry standards.

Instructor III

The Instructor III is responsible for the creation or modification of curricula and must be knowledgeable of laws and standards that apply to its design and delivery.

JPRs at Work

Apply the laws and standards in your curriculum development, making sure you are aware of the federal, state, and local impact of them. Curricula should be based on a course goal, with accompanying objectives and properly formatted lesson plans that have a fair and objective evaluation plan.

Bridging the Gap Among Instructor I, Instructor II, and Instructor III

The Instructor III will design and develop curricula based on an agency needs analysis. Many laws influence this design process, especially in the areas of recordkeeping, evaluation, and performance evaluations and review. When an Instructor II develops a lesson plan, he or she should always keep in mind these laws and standards as they apply to the audience to whom the lesson plan will be taught. The Instructor I must understand the requirements of proper recordkeeping, including how a completed training record report should look and how to evaluate students properly after training. The Instructor I should also make sure that any inconsistencies in the delivery of training and any recordkeeping irregularities are brought forward to the instructor in charge immediately. Any potential areas of legal exposure need to be immediately documented and passed through the appropriate chain of command.

vital that your policies and procedures are reviewed annually to ensure that they are current and up-to-date with national standards set for the fire service. The same is true for the training evaluations used. Prior to revising an evaluation, it is important to verify that it still satisfies the standards of the fire service.

■ Malfeasance

Malfeasance, unlike negligence and misfeasance, is more than a simple mistake or accident. Most often malfeasance is associated with public officials who partake in unlawful conduct even though they know that the conduct is illegal and contrary to their duties as public employees. Such conduct usually results in criminal charges, but could be made part of a civil suit. An example of malfeasance would be the acceptance of a bribe by a fire service instructor in return for a passing grade.

Theory Into Practice

Make sure that you have all standards, policies, or rules relied upon in creating a training session readily available. The resources used to develop a training program should always be current and consistent. Provide copies of any applicable policies to the participants during training.

Social Media

As more of the world goes online, the line between public and private, and professional and social, has become increasingly blurred. Social media usage has skyrocketed in recent years, which has in turn opened up a new area of potential liability. Examples of social media include Facebook, Twitter, LinkedIn, MySpace, Flickr, Foursquare, Pinterest, Instagram, blogs of all sorts, and a variety of other Web sites. To avoid this increased liability, fire departments are encouraged to adopt and implement smart and proactive social media policies.

Technology can be a very valuable tool for both the fire department and the fire service instructor. However, the department and personnel must thoroughly understand the concerns and issues raised when information placed on social media sites violates privacy concerns or portrays the department to the greater public in an illegal or otherwise negative manner. In crafting an effective and appropriate social media policy, it is important to consider the host of legal issues that may arise:

- First Amendment concerns (free speech)
- Health Insurance Portability and Accountability Act (HIPAA) and medical confidentiality
- Privacy rights of employees and citizens
- Collective bargaining rights and the right to engage in concerted activities
- Copyright law
- Laws intended to restrict photo taking and dissemination by emergency personnel at incident scenes

Many social media policies prohibit the posting or sharing photos or videos in any form when those photos or videos are gathered while completing fire department business.

Fire department business includes emergency calls, meetings, drills, details, training, various organization functions, and almost any activities taking place on the organization's property. However, a social media policy must be carefully tailored so as not to offend federal and state laws. The current federal law encourages discussion of the terms and conditions of employment and forbids any action taken by the employer to quiet its employees engaged in such discussions. An overly broad social media policy may have just that forbidden effect. For example, a fire department's social media policy cannot prohibit fire fighters from identifying themselves as employees of the department or from making comments about the conditions of their employment, even if those comments are disparaging.

A carefully drafted social media policy and training exercises will remind fire fighters of their obligations not to take or post photos while completing fire department business, including via cell phone cameras. Additionally, the policy should strongly caution against, but not explicitly prohibit, posting disparaging remarks about the fire department, by reminding fire fighters that the department and its employees are carefully scrutinized by the public. Although fire departments have a legitimate interest in protecting against disclosure of private information, they are not permitted to restrict fire fighters' use of social media to discuss their employment.

With regard to social media issues, instructors should emphasize to students the importance of adhering to the established social media policies. In some departments, failure to adhere to the social media policy carries severe disciplinary action up to and including termination of employment.

Copyright and Public Domain

Many materials you may want to use in training are protected by copyright. More than three decades ago, Congress passed the Copyright Act of 1976, which governs copyright issues. The act divides work into two general categories: those protected by the law and those considered to be part of the public domain. Works include written words, photographs, and some artwork. The Copyright Act sets out some specific time frames pertaining to copyrighted material. Any work published before 1923 is considered to be in the public domain. Any works published between 1923 and 1978 have a 95-year copyright protection from the date of actual publication. For any works published after 1977, the author, artist, or photographer has a copyright for life plus 70 years. Violation of the copyright law can result in liability, with the creator of the work being eligible to recover any damages suffered or lost profits.

When providing training materials for fire service employees, many of the handouts used are photocopied from copyrighted materials. The Copyright Act does not prohibit their use outright or require that permission be sought before they are used. Indeed, the fact that some material is protected by a copyright does not necessarily prohibit you from using it in your training, because a "fair use" exception exists relative to the protection afforded by copyrights. Examples of fair use include copying materials for the purpose of criticizing, commenting, news reporting, teaching, and research. Copies for the class in a training session generally fall into this category. Ultimately, fair use depends on the nature of the copyrighted work and how much of the work is copied. For example, it would not be fair use to purchase one copy of a textbook and make copies of the entire book for the class. By contrast, if you want to use an article from a magazine or newspaper for classroom purposes, that would likely fall within the fair use exception.

If a particular use does not constitute fair use, you may still be able to use the material in a classroom setting by contacting the copyright holder and seeking permission in advance to use that material. You should obtain permission in writing prior to using the material.

Whether a work is copied under the fair use exception or with permission, it should include a notice indicating who holds the copyright to the work.

Ethical Considerations

Ethical conduct should be the goal for all those involved in the fire service. Society has established certain socially acceptable conduct that matches the morals and values held by its citizens. Simply put, ethics is the difference between right and wrong. Ethical issues must be considered in all aspects of the fire service instructor's job. Ethical considerations include those held by the fire department as well as your own personal ethics.

Unethical conduct by any member of the fire service has a chilling effect on how society views fire fighters. During training, it is important that you reinforce the ethical standards by which members of the fire service must abide. To do so, first ensure that your department has a clear set of standards that sets forth expectations for conduct by fire fighters not only in terms of their job performance, but also in terms of their learning environments. Establish the tone for such ethical behavior by acting as a role model for your students. By showing respect to individuals who

Ethics Tip

The electronic age has greatly changed the world we live in. Cameras are everywhere, recording everything we do—from walking through a checkout line to driving through an intersection. Many of these videos are readily available on the Internet—only a mouse click away. When you are creating a presentation, these images provide powerful lessons for students. Images such as a fire engine demolishing a car as it runs through a red light will have a lasting impact on students who will soon be making the decisions behind the apparatus wheel, for example. It is often impossible to know whether these videos and images are protected by copyright, although sometimes their copyright status is readily apparent. If you wanted to use video from a TV news program for your class, would you contact the TV station to get permission for its use? What about videos found on YouTube or various fire department Web sites?

are different in gender, race, or religious belief, for example, you will foster an environment that is both ethical and positive.

As part of your ethical conduct, place the education of your students above all else, except safety. This is accomplished by creating an environment that promotes growth through challenging evolutions. It is also done by following the law as it pertains to discrimination, copyright issues, and criminal conduct such as taking bribes.

Records, Reports, and Confidentiality

Lawsuits evolve over long periods of time, which means that the thoroughness and accuracy of your records is vital in mounting a defense against such litigation. Proper recordkeeping is as important as the training you provide to your department. In cases where an injury occurs to a participant in or an observer of a training session, it is crucial to document what occurred, who was present, and what each person observed. The same holds true for claims of sexual harassment: It is absolutely necessary for the person who receives the report to document what was said and which actions were taken once the complaint was received.

Proper documentation is essential to the proper performance of your job, as is storage of the records in an organized and secure setting. Both state and federal laws (Privacy Act of 1979) prohibit the disclosure of certain information. Most likely, your department has a policy or procedure that requires the protection of personal information that is kept in the control of the fire service instructor. Failure to abide by the confidentiality provisions required by law or your department could lead to liability or adverse action against you.

Records that should be kept secure and confidential include the following:

- Personnel files: Usually include information such as date of birth, Social Security number, dependent information, and medical information.
- Hiring files: Include test scores, pre-employment physical reports, psychological reports, and personal opinions about the candidate.
- Disciplinary files: Any report or document about an individual's disciplinary history and related reports.

Regardless of the type of training being provided, recordkeeping should be a top priority as part of risk management. Records with sign-in sheets and the topic presented should be maintained for all training sessions for a period of at least five years, or longer as required by law. You should also keep copies of the handouts, training outline, and other training materials presented with the sign-in sheet. Records, reports, or memoranda related to any unusual occurrences that happen during training or that are reported to you should be kept for at least five years. Because many lawsuits are filed long after the event occurred, the only real protection from this risk for the fire department may be the records kept. Incomplete or missing records often make defending suits more difficult.

Risk Management

A vital function of the fire service instructor is to minimize risk. Instructors should incorporate risk management into all aspects of their job. In particular, they need to focus on minimizing risk as it pertains to fire fighters performing on fire or EMS scenes **FIGURE 2-4** . This is accomplished by ensuring training is completed on all pieces of apparatus and equipment used by the fire department, as well as by providing adequate training related to fire suppression tactics and EMS standards. Risk management techniques in fire suppression tactics and EMS standards include repetitive training in all areas to lower the risk of liability. The goal is to make the task so routine that when an actual event occurs, responders to the incident are thoroughly familiar with their responsibilities and the equipment being used. The goals of all training should be to educate, eliminate mistakes, and refine existing skills.

As important as practical training is in the fire service, equally—if not more—important is the non-operational training and duties of the fire service instructor. First and foremost, you must take steps to educate all members of the fire department about certain aspects of the law—specifically, hostile work environment (Title VII) claims and sexual harassment claims. Annual training should be provided on these topics, and you should ensure that up-to-date policies on both issues are maintained. In addition, you should provide training on all policies and procedures in place at the fire department—if not annually, at least every two years. Lastly, you should review all

areas of the department to eliminate any potential risks to the fire department and its members, including lifting procedures, apparatus storage, and daily duty requirements.

On an ongoing basis, you should take the steps needed to keep the fire department and its training up-to-date. Changes are constantly occurring in the fire service pertaining to operational activities, as are changes in laws and national guidelines such as those formulated by the NFPA. You may need to modify the training presentation or policies in place to ensure that they continue to reflect the latest laws and standards.

■ Instructor Liability

Firefighting is an inherently dangerous job. In an effort to prepare fire fighters for the dangers of the job, fire service instructors often simulate dangerous situations. Being in those dangerous situations creates the risk of injury, or even death. No fire service instructor intentionally wants any of the training participants to become injured, but a fire service instructor may still be found civilly liable for negligence or possibly charged criminally if he or she does not fulfill the necessary duty of care to the trainees. In addition, the instructor's employer or agency can be held vicariously liable for trainee injuries.

Civil liability for negligence usually occurs when it is alleged that the instructor owed a duty of care to the injured party, the instructor breached that duty of care, and the breach was the cause of the injury. Generally, a claim does not rise to the criminal level unless some gross deviation from the standard of care occurs, such as actions that are intentional or reckless, and a criminal statute exists under which the local, state, or federal prosecuting attorney can file charges. The remedy for a criminal charge is usually jail time; however, a criminal prosecution will be successful only if guilt is proved beyond a reasonable doubt.

The instructor's employer or agency can also be held vicariously liable, meaning that the employer is held accountable on a secondary level for the actions of the instructor. For example, civil liability for negligence can arise for both the employer or agency and the instructor if an instructor fails to train students properly during a rescue training exercise that results in the injury of a participating student. Some employers and agencies maintain liability insurance to protect instructors and themselves from any potential liabilities; however, if the employing agency of the instructor does not have this insurance, instructors can purchase personal liability or errors and omissions insurance. Although an instructor is not required to purchase personal liability or errors and omissions insurance, it is an option and should be considered by instructors.

The following stories from the Point Edward Fire Department from Ontario, Canada, and the City of Kilgore Fire Department from Kilgore, Texas, provide cautionary tales about the dangers of training exercises.

Point Edward Fire Department

A fire service instructor was charged under the Canadian Occupational Health and Safety Act for failing to take necessary precautions during a training exercise that resulted in the death of one of the trainees.

> ### Teaching Tip
>
> Good recordkeeping is an essential part of risk management. Keep all reports and records in a safe and secure area. All records should be kept for a minimum of five years, or longer as required by law.

Photo by David Traiforos, luvfireimages.com

FIGURE 2-4 Minimize risk at training evolutions by using effective instruction and proper safety equipment.

VOICES
OF EXPERIENCE

Legal issues are everywhere around us and, although you may think it can't happen to you, sooner or later all leadership personnel become involved in some form of legal concerns. Due to the elevated publicity involved with all government agencies, everything we do can be under scrutiny. The scrutiny may come from citizens after an incident or regulatory agencies such as the state fire marshal, Occupational Safety and Health Agency (OSHA), National Institute of Occupational Safety and Health (NIOSH) or an outside agency such as the Insurance Services Office (ISO).

"On one particular day, the state's governing agency made a surprise visit and asked to review our training records."

The one issue as an instructor that becomes vital for documentation in relation to investigations of fire fighter injuries and/or deaths is accurate training records. Instructors typically are entrusted to ensure that all employees are properly trained and that records are kept up-to-date for the employee and the organization for which the instructor is employed. It is imperative that the instructor research and understand all requirements of the governing agency(s) and retain backup records in case something happens to the originals.

As part of the leadership it was one of my responsibilities to ensure that all training records were obtained and kept on file for review by internal and external stakeholders. On one particular day, the state's governing agency made a surprise visit and asked to review our training records. They typically performed random visits to ensure that fire departments were properly training their fire fighters in order to maintain state certification. The state agency selected random topics and random employees for training record review. They generally do not give a particular time frame other than stating you have a "reasonable amount of time" to deliver the training records. In essence, having records stored in an orderly fashion for a quick recovery and having a backup system—whether it's hard copies or electronic files, such as in a learning management system—is very beneficial. Due to having a strict regimen of accurately filing course and individual records, this was not an issue; however, one misplaced record could be cause for a more in-depth review, regardless of whether the training was completed.

Items that the agency was looking for were particular life safety topics, contact hours, skill completion, some form of assessment, and instructor credentials. From time to time they may ask for lesson plans, PowerPoints®, and roster sheets. It is not only imperative to keep the individual's records: the instructor should be sure to keep the roster, written assessments, skill checks, and any other forms of documentation that could be beneficial during a review.

Record retention is a large undertaking in most departments. Regardless of who may ask to review the records, keeping records will provide protection to the individuals, fire fighters, the organization, and most importantly the instructor.

Brian Ward
Board of Directors
International Society of Fire Service Instructors
Centreville, Virginia

The fire department was doing a training exercise in the St. Clair River that involved swimming to an ice floe, climbing on top of it, and riding it downriver. In the course of that training exercise, one of the fire fighters was unable to climb onto the ice floe and eventually became trapped beneath the ice. One of the fellow trainees saw the fire fighter go under and yelled for help, but was unsure if anyone heard him. At the time of the training, only the fire chief and one other fire fighter were on the shore. Neither of them had any rescue equipment because the equipment was in a Point Edward fire truck waiting in the parking lot.

The charges against the fire service instructor include failing to have adequate rescuers and rescue equipment available. They also included failing to have an adequate training plan in general. There did not seem to be a plan in place in the event that something went wrong with the training.

The takeaway message from this case is clear: If a training exercise is particularly dangerous, such as one in moving water with ice floes, the fire service instructor must be prepared to respond to life-threatening situations. Different training exercises may require differing levels of precautions. The fire service instructor must take those type of considerations into account when planning for the training. Failing to do so not only endangers the safety and well-being of the fire fighters participating in the training, but also opens up the fire service instructor to legal liability.

City of Kilgore Fire Department

After the Kilgore Fire Department replaced one of its aerial trucks with a newer model, it tried to train its fire fighters on the differences between the old and new models. All of the fire fighters did the classroom and practical exercises as part of the factory-authorized training. The fire fighters were then encouraged to continue doing practical exercises to familiarize themselves further with the new truck. It was during one of these training exercises that two fire fighters fell more than 80 feet to their deaths. The subsequent lawsuit not only alleged negligence on the manufacturer's part for certain defects in the aerial platform, but also negligence on the part of the City of Kilgore Fire Department for allowing fire fighters to operate the aerial lift in a high-risk training scenario before they were adequately trained on the equipment.

On the day of the accident, one of the shifts at the Kilgore Fire Department took its new truck to an eight-story building—the tallest building in its jurisdiction—to do training exercises with the aerial platform. In the course of the training, the platform became stuck on the inside edge of the parapet on the roof of the building. The fire fighter operating the aerial platform attempted to move it, but, because it was stuck, tension began building up on the platform instead of it moving. Once the platform broke free of the parapet, it swung back and forth violently. Of the four fire fighters in the aerial platform at the time, two were thrown through the platform's doors and fell to their deaths.

None of the fire fighters who participated in the training wore the safety harnesses designed for that aerial platform. In fact, the manufacturer never provided the safety harnesses, and throughout the factory-authorized training, use of safety harnesses was neither encouraged nor applied. Not only were the fire fighters going through training exercises without the proper safety equipment, but they did so at the tallest building in their jurisdiction.

There are obvious things that the fire department could have done to minimize the risk to the fire fighters in the training exercises. Requiring safety harnesses to be worn while in the aerial platform is certainly one such step. A fire service instructor should always use and encourage safety equipment during training exercises as a precautionary measure. Ignoring the need for safety equipment could show a disregard for the safety and well-being of the trainees. Providing training exercises at smaller buildings before progressing to taller ones is another step that could have been taken in the Kilgore case. It would have allowed the fire fighters to become more familiar and comfortable with the equipment before putting them in a high-risk situation. Ultimately, it is the fire service instructor's responsibility to minimize any unnecessary risks because firefighting has so many risks that cannot be avoided. Failure to do so puts the fire fighters in danger and makes the fire service instructor susceptible to legal liability.

Despite the existence of potential liabilities and the need for instructors to use caution, most instructors who exercise due care and act professionally at all times avoid these legal issues. Instructors who follow all regulations and policies should not let the possibility of legal action keep them from becoming an instructor. Being an instructor is a rewarding experience, and with a little extra vigilance most legal liabilities can be avoided.

Jones & Bartlett Fire District

Training Division
5 Wall Street, Burlington, MA, 01803
Phone 978-443-5000 Fax 978-443-8000
www.fire.jbpub.com

Instant Applications: Legal Issues

Drill Assignment

Apply the chapter content to your department's operation, training division, and your personal experiences to complete the following questions and activities.

Objective

Upon completion of the instant applications, fire service instructor students will exhibit decision making and application of job performance requirements of the fire service instructor using this text, class discussion, and their own personal experiences.

Suggested Drill Applications

1. Review your training policy for requirements in documentation of training sessions and recordkeeping.

2. Identify local ordinances, laws, and standards that apply to the development, delivery, and recordkeeping of training.

3. Review the Incident Report in this chapter and be prepared to discuss your analysis of the incident from a training perspective and as an instructor who wishes to use the report as a training tool.

Incident Report

Lairdsville, New York—2001

Figure 2-A A couch at base of stairs in Lairdsville duplex.

From the NIOSH Fire Fighter Fatality Investigation and Prevention Program, Investigation Report F2001-38 NY.

The training evolution in Lairdsville, New York, involved a duplex that was to be razed. A search and rescue scenario with a rapid intervention crew (RIC) deployment was planned to simulate the entrapment of fire fighters on the second floor of one unit. For the purpose of the drill, the stairs to that unit would be considered impassable, so the RIC would have to enter from the other unit of the duplex and breach a second floor wall to reach the trapped fire fighters who were under light debris. Smoke from a burn barrel was used to obscure vision. The burn barrel was located in the back bedroom of the same unit as the trapped fire fighters, also on the second floor.

Neither of the two new fire fighters posing as victims had worn SCBA in fire conditions, and one (the deceased) was a brand new recruit fire fighter with no training. Both fire fighters wore full protective clothing and SCBA for this drill.

All of the crews assembled initially at the duplex. One engine was connected to a large water tanker on-site but no hose lines had been set up. A 1¾" (44.5 mm) hose line was supposed to be placed at the rear, with another at the front doors. An engine and heavy rescue were located off-site and would simulate a response to the scenario.

Safety officers were assigned and placed throughout both sides of the duplex. One officer was assigned to light the fire and place the new fire fighters in the front bedroom on the second floor. He was also instructed to guide the RIC if necessary. Another officer was assigned to a room on the second floor of the duplex unit, which the RIC was to enter first. He was assigned to guide the RIC if necessary, making sure they did not go through a wall opening that went to the outside. He was equipped with a 20-lb (9.1-kg) fire extinguisher. The 1st assistant chief was located on the first floor of the unit. The fire chief checked the scenario to make sure that accelerants were not used. He took command out front and the training was ready to begin.

The smoke barrel upstairs was ignited, but it was not creating the desired smoke conditions. Downstairs, the 1st assistant chief struck a road flare and ignited the foam mattress on a sleeper sofa that was located next to the open stairs **(Figure 2-A)**. The fire spread very

Continued...

Incident Report Continued...

Lairdsville, New York—2001

rapidly across the ceiling, producing heavy smoke and flames, and flames quickly extended up the stairs and out the front windows.

The fire spread was so rapid that the officer in the adjoining unit could not enter to reach the new fire fighters, and he had to exit using a ground ladder. The 1st assistant chief exited the duplex and went to the rear exterior but he could not find a hose line. He then went to the pumper and pulled a 200' (61 m) preconnected hose line to the rear of the structure.

The fire fighter on the same floor as the trapped fire fighters had left his position to investigate the sound of another road flare being lit. Upon seeing the fire, he went back and reached the two new fire fighters. The flames were now entering the second floor windows. He led the new recruits to the stairwell, which was now fully engulfed in flames from the first floor. He lost his gloves and immediately burned his hands. At this point, he lost contact with the two recruit fire fighters. He was able to reach the back bedroom with heavy smoke and tenable heat. He searched for the boarded up window and was able to break it open with his hands and jump from the second floor.

The 1st assistant chief advanced the hose line to the second floor, knocking down the fire. Arriving units were advised it was no longer a training and two RIC units were deployed to the second floor. One RIC forced entry through the involved unit's front door. One fire fighter was found and brought to the outside. The second RIC found the other fire fighter unresponsive. They removed him and transported him with advanced life support care to a local hospital, where he was later declared dead. The rescued fire fighter and the fire fighter who jumped from the window were both flown by EMS helicopters to a burn unit, due to the severity of their burn injuries.

During the ensuing investigation, the Lairdsville fire chief claimed he didn't know that a structure fire would be set as part of the training exercise, saying, "It was only supposed to be smoke." The 1st assistant chief was indicted and found guilty of criminally negligent homicide. He was sentenced to seventy-five days in jail and instructed to avoid contact with any fire department under a five-year term of probation. His attorney argued that the NFPA standard was not known by the department, and that it was the state's responsibility to distribute the guidelines to every volunteer fire fighter.

Post-Incident Analysis

NFPA 1403 (2007 Edition) Noncompliant

Students not sufficiently trained to meet job performance requirements (JPRs) for Fire Fighter I in NFPA 1001 (4.1.1)

Debris from the structure used as fuel material (4.3.2)

Students acting as victims within the structure (4.4.14)

Deviation from the preburn plan (4.2.25)

Fire ignited without charged hose line present (4.4.19.2)

Exterior hose lines not in place (4.4.6)

Rapid Intervention Crew (RIC) not in place (NOTE: This is not a requirement of 1403, but is an OSHA requirement and is referenced in NFPA 1500, NFPA 1710, and NFPA 1720.)

Chief Concepts

- Every training session exposes the fire service instructor and the fire department to potential liability.
- Laws are established either through the legislative process (statutes) or at the administrative or local level (regulations or codes).
- As a fire service instructor, your conduct is governed by three sources of law: federal law, state law, and your fire department's procedures.
- Documentation and recordkeeping are essential to protect yourself and your fire department in case of a lawsuit.
- As a fire service instructor, you must be familiar with at least three federal employment laws: the Americans with Disabilities Act, the Civil Rights Act of 1964, and Title VII of the Civil Rights Act.
- The Equal Employment Opportunity Commission is a federal agency charged with investigating and enforcing federal laws pertaining to discrimination in the workplace.
- Negligence/misfeasance is unintentional conduct. Malfeasance is conduct that includes an element of intent or knowledge that a wrong is being committed.
- Social media usage has skyrocketed and has opened up a new area of potential liability.
- Many fire departments' social media policies prohibit the posting or sharing of photos or videos in any form when these photos or videos are gathered while completing fire department business.
- Some materials that you may want to use in training are protected by copyright laws.
- Ethical conduct should be the goal for everyone involved in the fire service.
- Lawsuits evolve over long periods of time, which means that the thoroughness and accuracy of your records is vital.
- A vital function of the fire service instructor is to minimize risk.

Hot Terms

Americans with Disabilities Act of 1990 (ADA) A federal civil rights law that prohibits discrimination on the basis of disability.

Code *See Regulation*

Confidentiality The requirement that, with very limited exceptions, employers must keep medical and other personal information about employees and applicants private.

Direct threat A situation in which an individual's disability presents a serious risk to his or her own safety or the safety of his or her co-workers.

Disability A physical or mental condition that interferes with a major life activity.

Gross negligence An act, or a failure to act, that is so reckless that it shows a conscious, voluntary disregard for the safety of others.

Hold harmless An agreement or contract wherein one party holds the other party free from responsibility for liability or damage that could arise from the transaction between the two parties.

Wrap-Up, continued

Hostile work environment A general work environment characterized by unwelcome physical or verbal sexual conduct that interferes with an employee's performance.

Indemnification Agreement An agreement or contract wherein one party assumes liability from another party in the event of a claim or loss.

Liability Responsibility; the assignment of blame. It often occurs after a breach of duty.

Major life activity Basic functions of an individual's daily life, including, but not limited to, caring for oneself, performing manual tasks, breathing, walking, learning, seeing, working, and hearing.

Malfeasance Dishonest, intentionally illegal, or immoral actions.

Misfeasance Mistaken, careless, or inadvertent actions that result in a violation of law.

Negligence An unintentional breach of duty that is the proximate cause of harm.

Quid pro quo sexual harassment A situation in which an employee is forced to tolerate sexual harassment so as to keep or obtain a job, benefit, raise, or promotion.

Reasonable accommodation An employer's attempt to make its facilities, programs, policies, and other aspects of the work environment more accessible and usable for a person with a disability.

Regulation A law that can be established by legislative action, but is most commonly created by an administrative agency or a local entity.

Sexual harassment Unwelcome physical or verbal sexual conduct in the workplace that violates federal law.

Statute A law created by legislative action that embodies the law of the land at both the federal and state levels.

Title VII The section of the Civil Rights Act of 1964 that prohibits employment discrimination based on personal characteristics such as race, color, religion, sex, and national origin.

Undue hardship A situation in which accommodating an individual's disability would be too expensive or too difficult for the employer, given its size, resources, and the nature of its business.

Willful and wanton conduct An act that shows utter indifference or conscious disregard for the safety of others.

References

Burch v. City of Nacogdoches. 174 F. 3d 615, Court of Appeals, 5th Circuit. 1999.

Civil Rights Act. Title VII. 1964. (Pub. L. 88-352.)

Huber v. Howard County Maryland, 4th Cir. 849 F.Supp. 407 (1994).

National Fire Protection Association. (2012). *NFPA 1041: Standard for Fire Service Instructor Professional Qualifications.* Quincy, MA: National Fire Protection Association.

National Fire Protection Association. (2012). *NFPA 1403: Standard on Live Fire Training Evolutions.* Quincy, MA: National Fire Protection Association.

National Fire Protection Association. (2007). *NFPA 1403: Standard on Live Fire Training Evolutions.* Quincy, MA: National Fire Protection Association.

U.S. Copyright Office. Copyright Act of 1976.

U.S. Department of Justice. Americans with Disabilities Act. www.ada.gov.

U.S. Department of Justice. Privacy Act. 5 U.S.C. § 552a. 1974.

U.S. Equal Employment Opportunity Commission. Age Discrimination in Employment Act (ADEA). 1967. (Pub. L. 90-202.)

U.S. Equal Employment Opportunity Commission. Equal Pay Act. 1963.

FIRE SERVICE INSTRUCTOR *in action*

You are in charge of a department training session involving hose training. The participants, both male and female, must attach lines to a hydrant and pull charged lines as part of this training evaluation. During the training, one of the participants asks to be excused to check his blood sugar just as his turn to perform the evaluation approaches. Some of the other fire fighters hear his request and begin to call him derogatory names and question his sexuality. One of the female fire fighters complains to you about the language of the jeering fire fighters.

1. Prior to beginning the training evaluation, which preliminary steps must you take to properly prepare for the training?
   - **A.** Prepare an outline with goals listed
   - **B.** Ensure that all participants have undergone physical examinations
   - **C.** Have each participant sign a waiver of liability
   - **D.** Provide no input prior to evaluation

2. To deal with the participant with the blood sugar issue, what should you do?
   - **A.** Excuse the participant only if presented with a doctor's note
   - **B.** Allow the participant reasonable time to check his blood sugar prior to performing the evaluation
   - **C.** Allow the participant reasonable time to check his blood sugar after the evaluation
   - **D.** Ignore the request

3. How should you deal with the inappropriate language being used by some of the participants?
   - **A.** Tell the participants to stop the conduct
   - **B.** Make a report of the conduct
   - **C.** Ignore the conduct
   - **D.** Join in the conduct

4. Once you have been approached by a fire fighter indicating that she found the language to be offensive and hostile, what should you do?
   - **A.** Follow the steps set forth in the department's policy
   - **B.** Meet with the complainant to document her complaint
   - **C.** Notify higher members of the chain of command about the complaint
   - **D.** All of the above

5. After completion of the evaluation, a student asks for his friend's results because the friend had to leave early. What is the most compelling reason why you should not provide the information?
   - **A.** It is not any of his business.
   - **B.** It may be illegal.
   - **C.** You do not have time to do that for everyone.
   - **D.** The two may not be friends.

6. During the course, you distributed a copy of an article from *Fire Engineering* magazine. Is this legal?
   - **A.** Probably not, because it is copyrighted material
   - **B.** Probably not, because it is patented
   - **C.** Probably yes, because it is a limited amount of material and is being used for educational purposes
   - **D.** Probably yes, because it is in the public domain

7. When documenting the training, you include the title of the course and indicate who attended but nothing else. Which statement is most correct?
   - **A.** This is adequate information should a legal question later arise.
   - **B.** This is inadequate information should a legal question later arise.
   - **C.** It is a departmental decision on which information is important to retain.
   - **D.** There is no need to document training.

8. If you accepted a student's offer of free concert tickets to let him pass the exam, you might be guilty of which offense?
   - **A.** Malfeasance
   - **B.** Misfeasance
   - **C.** Negligence
   - **D.** Creating a hostile work environment

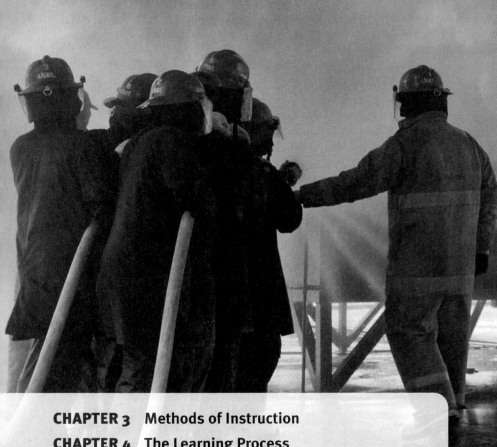

Instructional Delivery

© Surachai/ShutterStock, Inc.

© Photos.com

Methods of Instruction

© 2003, Berta A. Daniels

Fire Service Instructor I

Knowledge Objectives

After studying this chapter you will be able to:

- Describe motivational techniques. (NFPA 4.4.5) (pp 52–54)
- Explain how to adjust the classroom presentation and still meet the objectives of the lesson plan. (NFPA 4.4.4) (pp 61, 65)
- Describe the laws and principles of adult learning. (NFPA 4.4.5) (pp 54–55, 57)
- Describe methods of instruction typically used in adult and fire service education. (NFPA 4.4.1) (pp 61–68)
- Identify characteristics of Generation X, Generation Y, Generation Z, and the baby boomers. (NFPA 4.4.5) (pp 57–59, 61)
- Describe communication techniques that will improve your presentation. (NFPA 4.4.3) (pp 61–68)
- Explain how to deal with disruptive and unsafe behaviors in the classroom. (NFPA 4.4.5) (pp 66–68)

Skills Objectives

After studying this chapter you will be able to:

- Demonstrate basic coaching and motivational techniques. (NFPA 4.4.5) (pp 52–54)
- Deliver an instructional session utilizing prepared course material. (NFPA 4.4.1), (NFPA 4.4.3) (pp 61–68)
- Demonstrate professionalism during the learning process. (pp 63–68)
- Manage disruptive and unsafe behaviors in the classroom. (pp 66–68)

Fire Service Instructor II

Knowledge Objectives

After studying this chapter you will be able to:

- Explain how to adjust the classroom presentation and still meet the objectives of the lesson plan. (NFPA 5.4.2) (pp 61, 65, 68)
- Describe methods of instruction typically used in adult and fire service education. (NFPA 5.4.2) (pp 61–68)
- Describe communication techniques that will improve your presentation. (NFPA 5.4.2) (pp 61–68)

Skills Objectives

After studying this chapter you will be able to:

- Demonstrate basic coaching and motivational techniques. (pp 52–54)
- Demonstrate professionalism during the learning process. (pp 63–68)
- Demonstrate transitioning between methods of instruction to facilitate effective learning. (NFPA 5.4.2) (pp 64–65, 68)
- Manage disruptive and unsafe behaviors in the classroom. (pp 66–68)

Fire Service Instructor III

Knowledge Objectives

There are no knowledge objectives for Fire Service Instructor III students.

Skills Objectives

There are no skills objectives for Fire Service Instructor III students.

You Are the Fire Service Instructor

You are assigned to teach a basic incident command class at your training center that will include fire fighters from different agencies from surrounding departments. As you begin your presentation and discussion, it becomes apparent that some students are not following the handout material or participating in the classroom discussion. During the first break, an officer from another department approaches you and informs you that several members of his department struggle with reading and understanding what is written on the slides or handouts. The officer makes it clear that these are good fire fighters who perform well on the fire ground—they just struggle with reading and writing. He also informs you that these fire fighters are sensitive to their reading challenge and tend to shy away from class because of their situation, but attended today's session because they need to learn the material. Later in the lesson plan, you have included group activities that require students to work in small groups while reviewing fire-ground scenarios and applying what has been learned in class.

1. Taking into consideration what the officer has told you about some class members and your desire for them to be successful, how would you address this situation?
2. How would you adjust your lesson plan to accommodate those students in your class who struggle to read?

Introduction

As a fire service instructor, you must constantly ask, "How can I best help my students to learn?" It is your responsibility to present information and have a clear understanding of the course objectives. Instruction is the single most important factor in students' success: Teaching has a greater impact on achievement than all other factors combined.

Whether the class is a large introductory course for recruits or an advanced course for fire officers, it makes good sense to start off well. Students decide very early—maybe even within the first class—whether they will like the course, its contents, the instructor, and their fellow students.

Fire fighters are adults. This point may seem obvious, but sometimes fire service instructors can lose sight of that fact. Some of the strategies that work with second graders also work well with adults, but there are important differences in the way adults and children learn. Never treat your fire service students like children—you are teaching adults. Remember, too, that although you are teaching to a class, the individual may require a particular learning strategy.

Fire Service Instructor I and II

Motivation and Learning

Motivational factors make students feel enthusiastic and eager to learn. Motivation affects how students practice, what they observe, and what they do—all of which are critical to learning. As a fire service instructor, you need to understand how learning and motivation work together to make a great fire fighter. Instructors will face various motivations themselves that will require them to become more vocal, animated, or intense to emphasize why information is important to the student.

Learning is the potential behavior, whereas motivation is the behavior activator. Learning and motivation combine to determine performance in training. Without a motivated student and an expert instructor, learning will not take place. If the student's motivation is poor, then his or her performance in training will be sloppy and inaccurate. If the student's performance in training is sloppy and inaccurate, then the fire fighter's response at an incident will be sloppy and incorrect. This sort of shortcoming is not acceptable on the fire ground.

■ Motivation of Adult Learners

Cyril Houle (1961) identified three types of adult learners: learner oriented, goal oriented, and activity oriented. Each type is associated with different motivational issues, which in turn affects how learning takes place (Draper, 1993).

Goal-oriented learners are the most common type of adult learner. Their learning decisions are guided by clear-cut goals, as the learner takes a "utilitarian" or "pure" stance toward education. An example of a goal-oriented learner is a fire instructor attending a workshop on using distance learning to enhance a department's training program.

The second type of adult learner is the activity-oriented learner. For these individuals, participation is based on reasons other than the stated purpose of the learning activity. Examples of activity-oriented learners include a full-time worker who attends evening noncredit courses with a friend and a fire fighter attending a conference with other department members just to attend social events after the conference.

The third type of adult learner, the learning-oriented learner, seeks knowledge for the purpose of learning something new. For this learner, education is a way of life and, as such, is an intrinsically valued activity. Because this type of learner focuses on the content of a learning activity, he or she can participate either within a group or individually (Houle, 1961).

Consider the example of two fire fighters who are attending a cardiopulmonary resuscitation (CPR) training class. A fire fighter who is a goal-oriented learner might attend the class in an effort to improve her job skills, whereas the fire fighter who is an activity-oriented learner might take the class to spend time with his friends and possibly learn CPR at the same time. Although the learning agendas and goals of these fire fighters are quite different, both are in the same class. If the goal-oriented learner has an auditory learning style (i.e., prefers listening to a lecture), he or she may have to alter his or her learning style during that class to a kinesthetic style (i.e., physically performing the skills) to meet the goal of learning CPR. By contrast, the activity-oriented learner does not have to adjust his style to achieve his goal and can readily learn CPR by actually performing the skill. Learner-oriented adults focus on the act of being involved in learning. While often adult learners are able to adjust their learning styles for a class, this type of transformation is not always successful or even possible for some learners (McClincy, 1995).

Another factor that greatly influences success in the classroom is whether the adult learners grasp how the material relates to the real world and how this learning can be beneficial to job performance. This factor is well documented as affecting the learning outcome (Wlodkowski, 1999). Experienced instructors use real-world applications and can point directly to how information translates to use in the workplace. *Theory into practice* is a concept that can work to an instructor's advantage in this sense. For example, as part of their training, recruits learn that a gallon of water weighs a little more than 8 pounds. That statistical information is fairly simple to recall, but if the instructor shows recruits that a master stream flows 500 gallons per minute × 8 pounds per gallon, they can quickly see that when they use a master stream in the real world they are putting 4000 pounds of live load per minute into a structure, which may cause collapse.

Yet another key factor is the adult learner's psychological state. Mild anxiety can help the adult learner to focus on a particular task. Conversely, intense negative emotions (e.g., anxiety, panic, rage, insecurity) may detract from motivation, ultimately either interfering with learning or contributing to a low performance. To counteract this possibility, make your students feel welcome in the class session or at the drill ground by explaining the context and content of the training and allowing for a positive outcome-based environment.

The current position of adult learners in the life cycle also influences the quality and quantity of learning. Research suggests that each phase of life is characterized by the need for learning certain behaviors and skills. Because they are adults, adult learners invariably want to be treated as autonomous, independent learners.

Geographical changes (moving from one location to another), physical changes (reaction times), visual and audi-tory acuity, and intellectual functioning influence the adult learner's performance as well. The adult learner's internal world of ideas, beliefs, goals, and expectations for success or failure can either enhance or interfere with the quality and quantity of thinking and information processing. In addition, the adult learner's attitudes about his or her ability to learn and judgment as to whether learning is an important life goal affect motivation (Merriam & Caffarela, 1999). Moreover, motivational and emotional factors influence the quality of thinking and the processing of information.

No matter where the adult learner is and what he or she believes about his or her abilities, the adult learner continues to acquire knowledge both intentionally and unintentionally (Butler & McManus, 1998). The effectiveness of the learning process relates directly to the learning environment and the adult learner's interactions with that environment. In short, the adult learner's mental and emotional processes for learning can either help or hinder the fire fighter during training. Understanding each adult learner's preferred learning style will help you tap into the strengths of each adult learner in your classroom.

■ Motivation as a Factor in Class Design

Motivation is the primary learning component for adult learning. It generates the question "Why?" Why is the adult learner in the class? Why does the adult learner want to take (or not want to take) this class? When you can identify the *why*, it will allow you to tap into the *how*—the precise motivation and techniques that will help the adult learner achieve the goal of learning.

For example, knowing that attendance at a training class is mandatory allows you to design the lesson plan for the captive audience to maximize motivation. Unfortunately, most adult learners who are mandated to attend a class and forced to learn the information do poorly during such training, simply because they are unmotivated. Some adult learners may not know why they must learn the information or what is expected from the training or educational program, whereas other adult learners may have high expectations of the class. The unmotivated adult learners, the questioning adult learners, and the adult learners with high expectations all share the same classroom.

Find out from your students why they are in your class and what they expect. This information will enable you to design the lesson plans to meet every student's needs and sow the seeds of self-motivation in every student. You will inevitably need to employ a variety of motivational techniques to stimulate students. To learn, a student must be stimulated by the subject (McClincy, 2002).

If you are motivated, your enthusiasm may prove contagious in the classroom. When you are dynamic, outgoing, charismatic, and generally interested in the topic, your lessons will be taught with energy. Your personality will shine through in the classroom and will serve as a motivational factor for your students.

The use of audiovisual aids, teaching strategies, teaching methods, and even the classroom itself may serve as motivational factors. How you incorporate any or all of these motivational factors into the classroom will be reflected in the success of the learning process. For example, live-action video used to highlight a key point or term will motivate and engage the learner to observe the action much more than a still photo will, and even more than text on a slide will. Nevertheless, you must ensure that your audio or video enhances your presentation but does not *become* the presentation. The chapter titled *The Learning Environment* will expand on this topic.

Motivational factors are part of each level of instruction, from design to delivery to evaluation. The best courses you have attended motivated you to learn and provided you with the topic's relevance and application. Take time to consider ways to engage your learners throughout the presentation. Students are motivated to learn when they can understand how the material applies to them and how they will use it.

Teaching Tip

An adult learner's motivation may change throughout a session. The number of factors that can affect motivation is limitless. You need to be able to recognize the changing motivational needs of adult learners. Frequently cited course improvement suggestions include more periodic input on student grades, ongoing monitoring of progress in a course, and the ability to provide more feedback between the instructor and the student.

Safety Tip

Although a lack of sleep may be problematic when it comes to instilling motivation in students, it creates an even bigger problem in terms of safety. Do not allow students to participate in training without proper sleep. Even fatigue or sleeping through part of a presentation could prove deadly later on the training ground.

The Learning Process

Learning is an interactive process in which the adult learner encounters and reacts to a specific learning environment. That learning environment encompasses all aspects of the adult learner's surroundings—sights, sounds, temperature, other students, and so on. In addition, each adult learner has a distinct and preferred way of perceiving, organizing, and retaining these experiences—that is, a learning style (Merriam & Caffarela, 1999). Some adults learn by listening to and sharing ideas with others, others by thinking through ideas themselves, and still others by testing theories or synthesizing content and context.

Learning takes place in three domains: cognitive (knowledge), psychomotor (physical use of knowledge), and affective (attitudes, emotions, or values). These learning domains are discussed in the chapter titled *The Learning Process*.

Effective learning follows a natural process:

1. Adult learners begin with what they already know, feel, or need, based on the groundwork that was laid prior to what is currently taking place. In other words, learning does not take place in a vacuum.
2. Seeing that "newly" learned information has real-life connections prepares the adult learner for the next step—learning something new.
3. Adult learners use the new content, practicing how it can be applied to real life.
4. Adult learners take the material learned in the classroom and apply it to the real world, including in situations that were not always covered in the original lesson.

What Is Adult Learning?

Fire service training and educational programs are designed with the adult learner in mind. An adult learner is defined as someone who has passed adolescence and is out of secondary school. Self-motivation is rarely a problem for these individuals, who typically see the courses as a way both to advance their careers and to better themselves as people **FIGURE 3-1**.

According to Bender (1999), adult learning is the integration of new information into existing values, beliefs, and behaviors. New information is learned and synthesized so that students will be able to consider alternative answers to never-before-asked questions, new applications of knowledge, and creative ways of thinking. Adult students need realistic, practical, and knowledgeable lessons that deliver information in a variety of ways to accommodate their differing learning styles. In particular, adult learners may have difficulty imagining a piece of equipment, responding to a written scenario, or performing a skill if they do not have a realistic application for the learning. The more realistic the lesson, the more likely the adult learner will generalize the information and be able to apply it in the field.

FIGURE 3-1 Fire service training and educational programs are designed with the adult learner in mind.

■ Similarities and Differences Among Adult Learners

There are two general lines of thought in the realm of the learning process: pedagogy and andragogy. Pedagogy has been defined as the art and science of teaching children. In the pedagogical model, the teacher has full responsibility for making decisions about what will be learned, how it will be learned, when it will be learned, and whether the material has been learned. Pedagogy—or teacher-directed instruction, as it is commonly known—places the student in a submissive role requiring obedience to the teacher's instructions (Knowles, 1984). This process may occur in young adults who are members of an entry-level fire academy or young volunteer fire fighters who have not entered the workforce or higher learning environments. Andragogy (adult learning), in contrast, attempts to identify the way in which adults learn and help them in the process of learning. According to this theory's originator—professor Malcolm Knowles—although adult learners differ, some commonalities can be found (Knowles, 1990). For example, adults typically need to know why they need to know something and how they can use it. Adult learners also tend to focus on life-centered (practical) tasks and activities. They like to learn at their own pace and use their preferred learning style. Successful learning also takes advantage of adult learners' rich history of life experiences.

Not all adult learners are alike. Two early theorists in behavioral science, Erikson and Levinson, focused on age to help understand how adult learners differ. According to these researchers, young adults seem to be primarily focused on issues surrounding the development of a sense of identity, establishing a career, pursuing a life dream, having an intimate relationship, getting married, and achieving parenthood. By comparison, middle-aged adults focus more on caregiving for children and perhaps older family members, career changes, physical and mental changes, role changes, and planning for retirement. Older adults, by contrast, often face a series of upheavals—loss of their jobs (via retirement); the transition from parent to grandparent; age-related physical and mental changes; and the deaths of friends, spouses, and eventually themselves. Stage of life, in turn, directly affects the adult learner in the classroom. Many training sessions in the fire station will involve experienced (older) and less experienced (younger) fire fighters sitting side-by-side or working together on the same hose line. Youthful enthusiasm can be tempered by wisdom and naturalization of skills in this setting; when the two are combined, an effective firefighting team will emerge. A young fire fighter in an experienced company may be a motivation for the older fire fighters to engage at a faster and more educated pace than before the young member was introduced.

As discussed previously, Houle (1961) identified three types of adult learners:

- For learner-oriented adults, the content is less important than the act of being involved in learning.
- Goal-oriented learners acquire knowledge with the intention of improving their job prospects or learning a new skill. Members of this group view the instructor as being responsible for disseminating knowledge that they need.
- Activity-oriented learners learn by doing. To succeed in learning, they require personal productive time, with control over content and learning style, with the goal of improving social contact rather than the acquisition of knowledge.

With all of these orientations, the adult learners set goals, identify objectives, select relevant resources, and use the instructor as a facilitator for learning.

■ Influences on Adult Learners

Adult learning theories are based on the premise that adults have had different experiences than children and adolescents, and that these differences are relevant to learning. Although age is certainly a factor that can affect learning, other factors—such as motivation, prior knowledge, the learning context, and influences exerted by situational and social conditions—are influential as well. For example, in the classroom an adult learner might not pay close attention to the topic because a previous experience left the impression that the subject is not important. This view is merely reinforced for the adult learner when others in the class share the same feelings about the subject.

Adult learners are more apt to learn and retain information when it is meaningful to their lives. For instance, information directly related to a new piece of equipment will capture the interest of fire fighters working on that apparatus. In contrast, material that is outside the realm of the adult learner's experience and knowledge is viewed as less meaningful. The prior knowledge and experiences of the adult learner are often more diverse than those of a child or adolescent, yet you must always be aware that some adult learners may lack the expected prerequisite knowledge needed to learn new information.

What influences adult learning? Here is one answer to that question (Merriam & Caffarela, 1999, p. 107):

Adults often engage in learning as one way to cope with the life events they encounter, whether that learning is related to or just precipitated by a life event. Learning within these times of transition is most often linked to work and family.

Fire fighters work in a very intense environment. What is learned during times of intense stress inevitably influences learning in the classroom. If a fire fighter was just on an intense call, he or she may have trouble concentrating or sitting still in the classroom. Coping mechanisms (e.g., deflection of emotions, exaggeration of dark humor, being talkative or withdrawn) may affect behavior in the classroom. Always be on alert for signs of critical incident stress. Fire fighters need to discuss incident events in a proper environment.

Other influences on adult learners include social and cultural perspectives, such as social roles, race, and gender. Some adult learners, for example, may feel uncomfortable with classmates who are more educated or who hold a higher rank. You must ensure that in the classroom everyone is equal.

JOB PERFORMANCE REQUIREMENTS (JPRS)
in action

Effective training has many components. The audience, the learning environment, the instructor, and the expectations of the training process are all pieces of the puzzle that combine to make training effective. The methods of instruction chosen by the instructor are among the most vital parts of the puzzle, as they must match the audience, the environment, and the expectations or outcomes of the training session. The wrong method may have disastrous effects on the outcome and lead to poor performance or even safety and survival issues on the fire ground. Students' motivation can be enhanced by varying approaches to instructional delivery and choosing the appropriate setting for training delivery.

Instructor I

The Instructor I is responsible for delivering training from prepared lesson plans. Knowing how to identify the factors relating to the appropriate methods of instruction for the lesson plan will assist in the delivery of effective training.

Instructor II

Conducting an audience analysis and evaluating the learning environment help the Instructor II develop lesson plan materials that are effective for teaching students. The best learning process is interactive and student centered. Varied approaches to your training methods make this work.

Instructor III

There are no specific Instructor III requirements for this chapter. The Instructor III is responsible for the overall success of a program/curriculum. Depending on the material to be developed and delivered, the Instructor III should consider the proper method of instruction to be used in delivering a course.

JPRs at Work

Deliver the prepared lesson plan with the ability to adjust and adapt the lesson plan to the learning conditions presented by the students and the training location. Correct unproductive student behavior and control the delivery of training relating to the students' performance.

JPRs at Work

Be able to adjust the delivery of a lesson plan by varying the methods of instruction to keep students motivated and interested while ensuring that all objectives prepared for the lesson plan are covered and student performance is improved.

JPRs at Work

Consider the best method and format for delivering a class or curriculum. As a supervisor, ensure that those teaching the course follow the material as outlined.

 Bridging the Gap Among Instructor I, Instructor II, and Instructor III

The lesson plan content and the instructor's ability both play large roles in the method of instruction that is used to deliver a training session. Certain methods are required for specific types of objectives and intended outcomes. The Instructor II will prepare the lesson plan with a specific setting in mind for the training delivery. The Instructor I, based on his or her experience, comfort, and background, may vary the training delivery according to these factors. The Instructor III would need to monitor overall course delivery and ensure that the material presented fits the learning environment and method of delivery.

A unique feature of adult learning is that adults often desire to be self-directed learners—that is, they want to take their own direction within the learning process. Self-directed learning is related to humanism, the concept that the adult learner can make "significant personal choices within the constraints imposed by heredity, personal history, and environment" (Elias & Merriam, 1980). Although adults may have the desire to be self-directed in their learning, their prior experience with didactic classroom-based instruction may not have prepared them sufficiently for self-management in the learning environment. As the fire service instructor, it is your responsibility to provide the necessary guidance that keeps your students on track.

Critical reflection—that is, the ability to reflect back on prior learning to determine whether what has been learned is justified under present circumstances—is often a unique advantage attributed to adult learners, and a key asset that you should utilize when teaching. This kind of reflection involves "meta-thinking" about the strategies required to achieve a goal. Such reflection often occurs when adult learners compare their ideas to those of their mentors or those of their classmates. Over a period of time, an adult acquires basic cultural beliefs and ways of interpreting the world. Often such learning and beliefs are assimilated uncritically. In the classroom, however, adult learners may reflect critically on the assumptions and attitudes that underlie their knowledge. Such reflection helps them become deeper thinkers and can enhance their processes of thinking.

The benefits of this evolution in thinking are readily evident in the fire service. How many times have fire fighters gone to training with an attitude of "I really don't want to be here"? These students would rather be doing something else, but they go because the education is mandated. This disdain for training may represent an assimilated attitude—perhaps others in the station feel the same way. It may not indicate the way that a particular fire fighter truly feels about training or education; fire fighters naturally want to fit in socially, so they tend to adopt the attitudes that others around them have. As the fire service instructor, part of your job is to help all students become motivated to learn the information presented.

Factors such as participation in formal educational programs or training programs since high school will also affect the adult learner's ability to learn. For many adult learners, traditional learning skills (e.g., reading and understanding textbook material, taking notes during lectures, and studying for tests) may have been forgotten or, in some cases, were never developed properly. Many adult learners have developed a learning style that is limited to "doing" (McClincy, 1995). Adult learners with this learning style learn best when they can perform the skill. Difficulty with developing study skills, and problems with unlearning and relearning, present challenges to both the adult learner and you, as the fire service instructor.

Today's Adult Learners: Generation X, Generation Y, Generation Z, and Baby Boomers

Many types of learners exist, and many groups of students emerge with similar traits and skill sets that the instructor must take into account when developing or presenting a lesson plan.

This type of assessment, known as an *audience analysis*, is part of all levels of instructor responsibility when planning instruction. Audience analysis is the determination of characteristics common to a group of people—in this case, students. An audience analysis can be used to choose the best instructional approach. Factors considered in an audience analysis include the following:

- Age and experience in the subject matter of the students
- Who the lesson is intended for and how students will use the material
- Number of students in the class
- Level of previous knowledge of the class material

Public speaking expert Lenny Laskowski devised the following use of the term *audience* as an acronym to help identify this essential element in presentation preparation and delivery:

Analysis: Who is the audience?

Understanding: What is the audience's knowledge of the subject?

Demographics: What are the ages, genders, educational backgrounds, and other distinguishing characteristics?

Interest: Why are the students attending your class?

Environment: Where will this presentation be delivered, and which possible barriers to effective learning will exist?

Needs: What are the audience's needs associated with your presentation, and how will they use the material in the future or as part of their job?

Customization: Which specific needs and interests should you, as the instructor, address relating to the specific audience? Are appropriate department SOPs and terminology being used?

Expectations: What does the audience expect to learn from your presentation? The audience members should walk away having their initial questions answered and explained, or they should be able to perform a task.

The following are some generalized categories of learners divided by the generation in which they were born. Most share common characteristics of behavior, knowledge, and skills. Again, understanding your students is a key behavior for any instructor.

■ Baby Boomers

The baby boomers were born between 1946 and 1964. At the end of World War II, soldiers returned home, married, and began families in the midst of an economic boom. The baby boom ended when the increased use of birth control caused a sharp decline in the birth rate in 1964. Generalized characteristics of baby boomers include the following:

- The baby-boomer generation was raised on television, which had a tremendous impact on them and caused them to see world events from a single source **FIGURE 3-2**.
- This generation was raised on rock and roll.
- This generation has high standards concerning work, value of education, and the expectation that others will do the same.

© Courtesy of Fire Chief Gordon J. Nord Jr.

FIGURE 3-2 The baby boomer generation of fire fighters are the backbone of many departments and their trust must be earned by new instructors.

FIGURE 3-3 Instructors must learn techniques to bridge the generation gaps between their students. Generation X firefighters were the first generation of technology literate students to enter the classroom.

- This generation is independent, confident, and self-reliant.
- This generation spoke out against social injustice.

How do you teach to baby boomers? This generation of individuals makes up the backbone of most organizations today. Look at the senior members of your fire department and observe how you interact with them. As with all students, you must first gain baby boomers' trust and show them respect. As a fire service instructor, use them to assist you in teaching your course. Baby boomers will want to share their experiences and knowledge. This generation is known for both accepting change and resisting it because it pushes them out of their comfort zones. An effective method for teaching baby boomers is using group activities that will allow them to take a lead role and share their knowledge.

■ Generation X

As B. L. Brown points out in *New Learning Strategies for Generation X*, the "gap between Generation X and earlier generations represents much more than age and technological differences. It reflects the effect of a changing society on a generation" (Brown, 1997, p. 4). The term Generation X has been used to describe the generation consisting of those people whose teenaged years were touched by the 1980s, although this definition excludes the oldest and youngest Gen Xers. In broad terms, members of Generation X are those persons born in the late 1960s to late 1970s. This was the first generation for whom it was common for both parents to work outside the home. Their parents' absence required children to learn to take care of themselves and their siblings until Mom and Dad returned home. While their physical development was the same as earlier generations, the psychological and learning development of Generation X was not. This generation was the first to abandon the traditional family models, to consider divorce to be the norm, and to promote single-parent families. Many Gen Xers prefer to stay childless **FIGURE 3-3**.

Many members of this generation were raised in single-parent homes, growing up with "fast food, remote controls, entertainment and quick response devices such as automatic teller machines and microwave ovens, all of which provide instant gratification" (Brown, 1997, p. 3). These life experiences shaped the way they learned and resulted in certain learning characteristics that require some modifications of traditional classroom tactics and new teaching strategies. Members of Generation X are typically familiar with being educated and trained by people from earlier generations. The cultural gap between the generations reflects the diverse life experiences of the individuals in each generation. One of the major challenges facing instructors is the need to bridge the generation gap.

Generation X learners typically have the following characteristics:

- Independent problem solvers and self-starters. These individuals want support and feedback, but they do not want to be controlled.
- Technologically literate. Gen Xers are familiar with computer technology and prefer the quick access of the Internet, CD-ROMs, and the World Wide Web as their sources for locating information.
- Conditioned to expect immediate gratification. Gen Xers crave stimulation and expect immediate answers and feedback.
- Skeptical of society and its institutions. Members of this generation want their work to be meaningful to them. They want to know why they must learn something before they will take the time to learn the subject.
- Lifelong learners. Gen Xers do not expect to grow old working for the same company, so they view their job

environments as places to grow. They seek continuing education and training opportunities; if they do not find them, they will look for new jobs where they can get what they want.

- Ambitious. Members of Generation X often seek employment at technology start-ups, start their own small businesses, and even adopt worthy causes.
- Fearless. Adversity—far from discouraging members of this generation—has given them a harder, even ruthless edge. Many believe, "I have to take what I can get in this world because no one is going to give me anything" (Brown, 1997).

In the classroom as well as in the workplace, Gen Xers often clash with baby-boomer or traditionalist instructors. As an instructor, you may be the baby boomer trying to teach the Gen Xer. The common value that both generations share is the high priority placed on learning and developing new skills. This point poses a challenge to instructors, who must continually raise the bar in the classroom and perhaps adapt to cross-generational training. To accommodate the unique characteristics of Generation X learners, instructors need to step outside the box and promote learning that has applications in the school, work, and community settings that are part of the Gen Xers' experience.

◼ Generation Y

Generation Y (also known as the Millennial Generation or Dot-comers) is the name given to those people who were born immediately after Generation X. There is no consensus as to the exact range of birth years that are included in Generation Y, or whether this term is geographically specific. The only consensus appears to be that Generation Y includes those individuals born after Generation X.

The "Y" in Generation Y refers to the namelessness of a generation that has only recently come into an awareness of its existence as a separate group. Members of this generation typically feel dwarfed and overshadowed by the baby boomers and Generation X. Nevertheless, members of Generation Y are known for holding wide-ranging opinions on political and social issues, religion, marriage, gay rights, and behavior **FIGURE 3-4** . This lack of a common perspective may have occurred because these individuals have not encountered a personal situation where their actions or reactions have forced them to consciously choose sides on issues and maintain consistent positions. Most Gen Yers are more tolerant of alternative lifestyles and unconventional gender roles. At the same time, Generation Y tends to be more spiritual and religious than the generation that includes their parents, and disagreement on social issues between the more liberal and more conservative members of this generation is commonplace.

Generation Y learners typically have the following characteristics:

- Generation Yers are often considered young, smart, and brash.
- While they want to work, they do not want their work to be their entire life.
- The members of Generation Y have been nurtured and programmed with a slew of activities since they were

FIGURE 3-4 Generation Y, or Millennials, are often considered young, smart, and brash. Many members of Generation Y have been nurtured with a constant flow of activities since they were toddlers, making them high-performance and high-maintenance multitaskers.

toddlers, meaning they are both high performers and high maintenance (Armour, 2005).
- Generation Y are considered multitaskers, meaning they can focus on multiple things at once.

Learning in virtual classrooms and virtual labs is ideally suited to Generation Y, as is teaching by technical professionals, because these adult learners can interact with others and get hands-on experience. Virtual labs suit the "click on it and see what happens" experiential approach to learning and have the added benefit that the learner remains in a safe environment.

The Gen Yers in the classroom present a challenge to the fire service instructor who is either a baby boomer or a Gen Xer. The fire service instructor needs to have a general understanding of members of this generation and support their learning styles and needs. Instructors need to understand that success in teaching this generation requires gaining their trust by being a good example and "walking the walk" in what instructors say and do.

◼ Generation Z

Generation Z (also called Generation Next) will bring their own set of values and learning styles with them, and the instructors who teach and lead them will need to adjust and adapt to their needs as well. Many Generation Z students have taken online classes, have had computer access most of their academic life,

VOICES
OF EXPERIENCE

When I teach a class, I am always amazed at the differences in background and experience levels among the students. In a single class I have had students ranging from 16 to 70 years of age; students who dropped out of school after the 8th grade and students who have completed doctoral programs; homeless students and students with significant financial resources; students who work the night shift before coming to class and students who will be going to work after class is over.

It is my responsibility to determine how students will best grasp the material and then give them the opportunity to learn in that way.

When I first began teaching, I never realized how much all these factors would influence my classroom. I expected to go to class, present the material, and students who were paying attention would grasp the concepts. I quickly learned that was not the case…not even close.

What I found was that some students picked up the material fairly quickly, while others struggled with the basics. One of my biggest surprises came when I found people who were very intelligent with cognitive material but who struggled significantly to get the knowledge from their head to their hands. On the other hand, I had students who had a difficult time reading and understanding the cognitive work, but who were quick to pick up the psychomotor skills. My students did not fit the mold that I thought they would.

As an instructor I am always challenged to find ways to reach of all of my students. It begins with making sure they are motivated to learn. For some, this is as simple as my being enthusiastic about the material. For others I may have to dig deeper to help them understand why the material is important to them. Sometimes students are motivated by the fact that I have set high expectations for them, challenging them to be the best they can be. Once they are motivated, then they are ready to learn.

It is my responsibility to determine how students will best grasp the material and then give them the opportunity to learn in that way. It may mean using a variety of teaching methods in the classroom as well as setting them up with a study partner, providing extra study guides for outside of class, giving them skill sheets that allow them to practice at their station, or suggesting additional resources that may benefit them.

In some cases, I have learned new techniques from my students. I always ask them to share with me how they remember things or ways they have found to grasp difficult skills. Many of the things they share have come in handy with students in other classes.

As an instructor, it is important to remember it's not about me…it's about the students. Our students cannot fit into a neat little box. They don't look the same and they don't learn the same. Their backgrounds, hobbies, jobs, and interests all play a role in making them the person who has entered my classroom.

The challenge for us as instructors is to help them be successful by learning about their individual characteristics, by using a variety of methods of instruction, and most importantly by caring enough to help each student reach his or her full potential.

Diane May
Instructor Trainer
Maryland Fire and Rescue Institute
College Park, Maryland

and have based their learning on quick and very visually-oriented types of learning. They may prefer self-study or directed study opportunities as they can work around their personal schedules.

Some disconnects in learning occur when a student who has learned in high-tech environments is suddenly dropped into a fire service classroom that is a traditional lecture-based format. The Generation Z student may easily become bored and not engaged unless the instructor is able to motivate them using many application questions and more hands-on methods. Instructors may use more directed study applications and preload the lesson plan content by providing resources and information ahead of time to the students.

Whereas a chalk board and overhead projector were common to a Generation X student, a smart board and flat panel would be more recognizable to the Generation Z student. This underlines the importance of knowing your audience characteristics and adapting the learning environment to them to increase retention and understanding of class material. As more and more Generation Z students enter the work force and the fire and emergency medical services field, instructors will have to adapt the learning environment to varied generations of students who will all bring different learning backgrounds and different ways that learning is retained and put to use in the future.

Methods of Instruction

The method of instruction is the process of teaching material to students. You can also identify it as how your lesson plan will be given to your students. Many different and diverse methods of instruction exist, and each applies to different types of materials as well as different types of students. Each of the various student demographics and categories discussed earlier in this chapter will respond differently to the material according to how they are motivated and the method(s) of instruction used. If you are an Instructor I, the method of instruction will be identified on your lesson plan by the lesson plan developer. If you are an Instructor II, you are the person who decides which method of instruction is most appropriate to deliver the material in the lesson plan. The *Lesson Plans* chapter discusses formatting of a lesson plan and placeholders in the lesson plan that developers can use to identify the methods of instruction they have determined to be most appropriate for the material. The *Lesson Plans* chapter also discusses the parts of a lesson plan and gives guidance on how lessons are constructed, considering the method of instruction to be used.

Determining which method of instruction should be used to deliver information is accomplished based on several factors:

- The experience and educational level of the students
- The classroom setting or training environment
- The objectives of the material to be taught
- The number of students to be taught

As an instructor at either the Instructor I or II level, knowing and understanding the different methods of instruction is a fundamental skill that you should master. Based on this fundamental knowledge, when a lesson plan states that Instructor I personnel are to deliver a "lecture," they understand the method of instruction to be used. For Instructor II personnel, as they develop a lesson plan they will select the most appro-

priate and effective method to deliver the material in the lesson plan they have created. Any level of instructor should be able to adapt a lesson plan to the learning environment as well as to the audience who will receive the instruction so as to make it the most effective. All instructors will perform an audience analysis by looking at the demographics of the audience, including age, learning level, previous experience, and other factors, and then adapt the lesson plan accordingly.

Methods of instruction vary among disciplines, but for the fire service the following techniques are most common and effective in teaching new and in-service personnel:

- **Lecture.** This method is the most common and is used often in the fire service. It consists of the instructor being the primary source of information and delivering the course material while using various multimedia tools **FIGURE 3-5**. As part of your Instructor I job duties, you will be required to present an illustrative lecture in a practice teaching environment to a group of fellow students. An illustrative lecture allows the instructor to use a visual aid such as a PowerPoint® slide to "illustrate" key points, which is important to effective communication and is probably the most common and familiar method of fire service instruction.

- **Demonstration/skill drill.** This method is used when teaching skills or hands-on learning. This type of learning is used when the instructor demonstrates the use of a new tool or demonstrates how to perform a specific task-oriented skill **FIGURE 3-6**. Several variations in the sequence of how these types of drills should progress are possible, especially when learning a new skill. You might view the training as a crawl–walk–run sequence or engage in an instructor-led demonstration—that is, you might demonstrate the skill at full speed for the students, then slowly explain each step and stop at each step point, then perform it one more time at full speed while explaining the step. This approach is also known as a psychomotor lesson or hands-on method of instruction. As part of the Instructor 1 job duty area, you may be required to present a demonstrative

FIGURE 3-5 The lecture method of instruction consists of the instructor being the primary source of information and delivering the course material while using various multimedia tools.

FIGURE 3-6   Skill drills allow the student to learn by actively participating and physically using the new tool or practicing the new skill.

FIGURE 3-7   Discussion is used to facilitate the learning process by having a two-way open forum between instructor and student.

presentation where a student will perform a task from a prepared skill sheet as part of the student instructor's practice presentation.

Most fire fighters and other adult learners respond very well to hands-on training. This method gives each student the opportunity to learn by actively participating and physically using the new tool or practicing the new skill. While the student engages in hands-on application during this method of learning, the instructor needs to monitor student safety. This method of instruction references a skill drill and requires the instructor to ensure that personal protective equipment (PPE) is in use and proper student-to-instructor ratios are maintained to ensure student and instructor safety. The *Safety in Training* chapter discusses appropriate levels of safety considerations that must be applied to this method of instruction.

■ **Discussion.** The discussion method is used to facilitate the learning process by having a two-way open forum, from instructor to student and from student to student FIGURE 3-7 . This method fosters interaction among the instructor and the students; the instructor's role in this method is more of a facilitator or monitor to ensure the discussion stays on track. Time keeping will be an essential element when this method of instruction is employed. Discussion is typically used as the method of instruction when consensus is needed or when new ideas are necessary to explore higher levels of learning and application. In most cases, this method requires students to have some prerequisite level of topic knowledge.

Based on the material to be covered and the make-up of the class, most instructors use a combination of these methods while delivering course material. For example, instructors might deliver course material via lecture and then engage in a discussion to determine the level of understanding from

the students. When teaching a new skill or use of a tool, the instructor might begin with a lecture and then demonstrate how to use the tool; the class might then move to the training ground to practice using the new tool.

■ Enhanced Instructional Methods

Instructors have other tools at their disposal to enhance the learning process (further discussed in the *Lesson Plans* chapter) and increase the retention of the material being taught. These tools include the use of role play, labs/simulations, case studies, and assignments to be done after class, such as handouts or assignments in a course workbook.

■ **Role-play.** Students become part of the lesson by assuming roles and "acting out" a situation or scenario meant to reinforce material from the lesson. Many leadership and management classes feature a senior instructor role-playing a problem employee or angry citizen to allow students to practice handling confrontations.

■ **Labs/simulations.** Students are placed in a "real life" situation, albeit in a controlled environment. A table-top fire-ground lab where the student can run a fire scene is an example of such a tool. Placing students under pressure in a controlled scenario strengthens information learned in the classroom. Many simulation programs now exist to simulate actual fires and incidents that create a very realistic learning environment FIGURE 3-8 . A simple online search of fire scene simulations will reveal multiple sources of simulation-based programs. The chapter titled *The Learning Process* emphasizes the use of these types of technology labs in today's training environment.

■ **Case studies.** Instructors often take advantage of examples from the real world to teach material. Using an incident report that was generated after a line-of-duty death and allowing students to read the report, discuss the material, and then determine ways to apply the

Courtesy of Orland Fire Protection District

FIGURE 3-8 Many new simulation programs now exist to simulate actual fires and incidents that create a very realistic learning environment. Pictured here is a Command Training Center using the Blue Card Certification program.

knowledge is an effective way to work with a case study. NIOSH fire fighter fatality reports provide tremendous learning opportunities in many types of training sessions. The Charleston, South Carolina, furniture store fire investigation, for example, has become a staple in case study presentations discussing fire behavior, risk assessment, and tactical options. Most fire fighters can recall a presentation that included case studies when such scenarios relate to the lesson being presented, especially if the scenarios are from incidents they have heard of. All fire fighters should be aware of the significant fire fighter fatality investigations that have shaped our profession (such as the Hackensack, New Jersey, car dealership fire; the Worcester, Massachusetts, cold storage warehouse fire; and the Phoenix, Arizona, supermarket fire), as they have greatly influenced modern firefighting knowledge.

- **Out-of-class assignments.** Assignments given to students to be completed outside of class are another effective method of helping students learn. Many courses require pre-course work to be completed prior to class; the instructor might then give assignments in a workbook to be completed during the course. If the instructor does use this tool as part of the coursework, the instructor needs to make sure that these assignments are evaluated and that feedback is provided to the students. Failure to follow through with these measures will give the impression that the material was not important to learn.

Three other methods of instruction are becoming more prevalent in today's fire service: online learning, independent study, and blended learning. These methods of learning are gaining popularity for many reasons (such as budget constraints or distance to a training location) and are being used in many disciplines for new or continuing education training.

Online learning is conducted on the Internet and takes the form of students entering a "classroom" that is hosted by an organization or campus. Delivery methods for the classroom consist of either a "synchronous" or "asynchronous" delivery method. In synchronous learning, the entire class meets at the same time and the instructor delivers the course material in real time by voice and/or video. Students interact with the instructor and class verbally or with instant messaging. In asynchronous learning, in contrast, the course material is provided in the online classroom and students enter and exit it on their own schedule. Classroom discussion in an asynchronous class is facilitated by use of discussion boards where the instructor will post discussion questions or assignments and students are required to post responses to the instructor or to classmates and complete assignments that students then upload to the instructor.

Independent study has been used for many years (primarily in the military) to help students complete course material at distant locations. It entails registering for a course, receiving course material in the mail, completing assignments, and submitting them back to an instructor for evaluation. This method is one example of distance learning for those who might not have access to the Internet.

A third method of instruction combines online/independent study with a face-to-face meeting with the instructor. This blended learning format allows for students to learn material on their own at remote locations. Then, at certain benchmarks within the curriculum, the class gathers to complete the hands-on portion of the training. The chapters *The Learning Environment* and *Technology in Training* further discuss technological resources, distance delivery, and creative ways for instructors to enhance their presentations.

All of these methods of instruction are dependent on the instructor knowing how to use each method and applying them while delivering the material as outlined on the lesson plan.

Lesson Presentation Skills and Techniques

No matter which methods of instruction you use or which generations you teach, the following communications techniques will help you improve the delivery of your information to your class. (See also the *Communication Skills* chapter.)

Dress professionally. Students should see you as a professional and recognize you instantly as a subject-matter expert. Even though we like to believe that we don't "judge a book by its cover," you will be sized up by your students. A professional appearance lays the groundwork for making a good first impression.

Welcome each student. Your first impression begins the moment that students enter the classroom. Shake students' hands as they arrive, introduce yourself, and have a warm smile on your face—this will immediately convey that you are a likable, approachable person who cares about your students **FIGURE 3-9** .

Call students by name. Making every student feel important is essential in creating motivation and maintaining attention. Calling students by their names is one of the most effective methods of demonstrating to them that they are valued.

FIGURE 3-9 A good way to make a positive impression is to greet your students as they enter the classroom.

Although it may not be possible in large, lecture-style courses that meet only once, most fire service courses are small enough to allow you to learn the students' names. Many of us struggle to learn names, but a variety of methods have been developed that can help you with this task. One of the most basic strategies is to use name tents or cards with the students' names. Another technique is to keep a seating chart in front of you. Name association may also work well. Each method has its own advantages and disadvantages, but finding an effective method will go a long way toward helping you manage your class.

Use eye contact. In our culture, it is important to make eye contact with people to whom you are speaking. Making eye contact with each fire fighter for a few moments in the course of your presentation will help students feel that you are communicating directly with them. Resist the temptation to look at the wall, at the clock, or over students' heads. Lack of eye contact makes people think you are not willing to be straight with them or that you are not confident in what you are saying. In fact, it is probably one of the most disturbing mannerisms you can display.

Use your normal speaking voice. Sound interested by being interested in the subject and in what students have to say. As a fire service instructor, you have a strong motivation to build a strong relationship with your class. If you are committed to the importance of what you are teaching, then that commitment will be revealed in your voice and by the enthusiasm that you display for the topic. Remember that enthusiasm is contagious—let it show!

Speak with respect. You are teaching adults, and it is important that they perceive that they are respected as adults. When you are respectful to them, they will be respectful to you.

Use pauses when you speak. A short pause can be used to signal an important point or a transition in your lecture. A pause can also indicate the conclusion of one topic and the beginning of another. It allows you the opportunity to view your students and visually check for comments, interest levels, or questions.

Use information appropriately. It is important to gauge your students' level of understanding correctly. For example,

Fire Fighter I students will have a different understanding of information than seasoned fire officers. Make sure you speak at a knowledge and vocabulary level that students will understand and explain concepts or information on the same level. You may have to spend time ensuring that the level of training or education is appropriate for all participants in the course.

Use gestures sparingly, for emphasis. Avoid meaningless gestures—for example, jingling keys or change in your pocket, playing with a pencil, or tapping mindlessly on a podium or desk. Try to avoid playing with things, chewing on the ends of pencils, or repeatedly tapping your foot. Hide those tempting items if playing with them has become an unconscious habit for you. Eventually, these habits will become intrusive—and distracting to your students—and students will be placing bets on whether you will swallow the pencil or how many times you will tap your foot in a minute.

Use posture and movement to signal a transition of topic or point of view. Make your movements conscious and meaningful, and try to avoid pointless, nervous pacing. Think about where you want to move and when.

Practice positive mannerisms. Well-placed mannerisms can increase your effectiveness in presentation of your lectures, and appropriate gestures can add to the verbal messages. Vary your position and movement, because your ability to change the pace of the lesson will increase involvement. Your body language and facial expression can give a fire fighter support and encouragement even when you are not saying anything aloud.

Choose your teaching aids appropriately. The technology that you have at your fingertips in the classroom is simply a way to communicate with students and reinforce information. Eraser boards and paper pads have some unique advantages in communicating, as they can be used to convey a variety of information during a single lesson. If you use these tools, make sure that your writing is legible. These teaching aids also provide a visual reinforcement of your verbal message. PowerPoint® presentations, charts, posters, models, slides, films, and DVDs are other means of communicating information to students and can reinforce information being presented verbally. Technology changes rapidly, and your wise use of it can help enrich your classroom and assist you in teaching. Even so, you should not rely on technology alone to teach: It can break, so always have a backup plan.

Know how to use your chosen teaching aids in the classroom. Your communication with your students is interrupted when you do not handle technology smoothly. Each time you are guilty of mishandling your teaching aids, you will weaken the impact of the lesson by sending an unintended message ("I don't know what I'm doing").

Transition techniques during presentations. While presenting your lesson plan, make sure that you make smooth transitions between your media and verbal communication techniques. Distributing handouts before you want to have the students read them is an example of using media (nonprojected materials/handouts) at the wrong time. If you want to hold students' attention, then use media in the correct manner and use transition techniques that enhance the learning process rather than distract from it.

Transition techniques also are used during a presentation to move from one topic to the next. Many students get lost during a lecture when the instructor moves on and the students do not realize or understand the instructor has changed topics. To help students follow a presentation, instructors need to learn transition techniques. Five effective techniques are delineation, words/phrases, pictures, voice, and body language.

- **Delineation.** This is the simplest way to make a transition. As an instructor, you tell your class that you will be covering four points and then begin by stating "point number one," and then move to point two, point three, and point four, and then summarize the material.
- **Words/phrases.** Use phrases such as "the next point we need to discuss," or summarize the topic and then use a phrase like "the next step in the process...."
- **Pictures.** Because so many of today's lessons use PowerPoint® presentations, an effective tool to use within such a presentation is the insertion of historic or unique pictures between points or topics. This visual signal will allow you to pause and switch topics, and the students will quickly pick up on the fact that the topic has changed.
- **Voice.** Use your voice by changing your tone or pitch, by using different volumes to emphasize change from one idea to another, or by inserting an effective pause. Pausing will draw students' attention to you and will develop anticipation about what is coming next.
- **Body language.** Effective use of body language will assist you in moving from one topic to the next. Moving from the podium to one side of the classroom with a change in voice will clearly inform students that the topic has changed or is very important.

All of these techniques or methods will allow you to make transitions during a presentation, but they need to be practiced to be used effectively. Some instructors will develop all of these transition techniques; others will not. The key is to develop transition techniques that work for you as the instructor.

Eye contact is an important tool for engaging students and demonstrating an interest in them. Avoid overuse of eye contact, however, because it might be misconstrued as conveying your sexual interest in a student.

■ Effective Communication

Effective communication is the key to ensuring that the students learn—that they receive and remember the information you intended to transmit to them. As a general guideline, consider these findings noted by educator Edgar Dale in 1969:

- We remember 10 percent of what we read.
- We remember 20 percent of what we hear.
- We remember 30 percent of what we see.
- We remember 50 percent of what we see and hear.
- We remember 70 percent of what we say.
- We remember 90 percent of what we say while doing.

It is important to choose the right kind(s) of communication when you are preparing for class. It is also important to use the selected communication methods effectively. Understanding how the communication process works will help you develop lesson plans or adapt lesson plans based on your students' learning styles. For example, if your students are visual learners, then you may need to include more visual presentations. (See the *Communication Skills* chapter for a discussion of the elements and role of the communication process.)

Questioning Techniques

An important part of the learning process is to assess the learning process during class. One simple method used over and over is to ask students questions from the material that is being taught from the lesson plan. Many lesson plans will build in questions for the instructor to ask the students to determine whether the students are picking up the new information and understanding what is being taught.

When using questions during the teaching process, it is important to ask the correct question at the correct time. Again, many lesson plans will identify the appropriate time to ask questions to support what has just been taught. Another appropriate place to seek student feedback is at the end of a lesson during a summary activity. There are times during a presentation at which an instructor can use a question to stimulate discussion, to bring class focus to a new topic, or to break up side discussions in the classroom. When used correctly, a question can also be used to challenge a disruptive student positively and bring him or her back into the classroom.

Numerous types of questions can be used during the learning process, including the following:

- **Direct questions.** Direct questions are those beginning with "who, what, where, when, why, or how." They are meant to elicit specific information and limit discussion. This type of question is used when the instructor is seeking an answer from a specific student and is often effective in dealing with a student who is being disruptive or whose attention seems to be somewhere besides the classroom. It is important to exercise caution with this type of questioning, as it could be perceived as singling out a student.
- **Rhetorical questions.** This type of question is usually asked to stimulate discussion or to prepare the learners' minds for a new topic. Often a rhetorical question does not have an answer, or it may have multiple correct answers.

- **Open-ended questions.** This type of question could be answered with varying correct answers based on the question and the situation presented.
- **Closed-ended questions.** This type of question seeks a specific answer based on the question asked.

Other types of question formats could be added to this list, but these are the most common forms of questions used within a classroom during a presentation. Knowing which question to ask and when to ask the question is very important, but dealing with a student's response to a question is just as important. Instructors are there to teach students and to foster a positive learning environment, so responding to a student's answer is critical to maintaining a good learning environment. If the student gives the correct response, the instructor can build on the response and move on with the lesson. But what happens when a student's response or answer to a question is incorrect? Effective instructors should use the answer that was given and highlight the positive and downplay any negatives. Instructors who make light of a student's incorrect response can destroy a student's desire to learn and dampen the learning environment. Students are in a classroom to learn; if they respond incorrectly, then the instructor has an opportunity to correct the information, reinforce the positive, and help the student learn new material.

FIGURE 3-10 You are responsible for providing an appropriate learning environment to all of your students.

> ### Teaching Tip
>
> One of the instructor's most important responsibilities is to maintain a positive learning environment where students can feel comfortable with the instructor and their peers and feel free of judgment.

Disruptive and Unsafe Behaviors in the Classroom

Disruptive students come in many shapes and sizes, and with many attitudes. Likewise, your strategy for handling the problematic situation will vary depending on the type of behavior. The most important thing to remember is that you are in charge of the classroom. If you do not control your classroom, you may have some serious problems. You may be perceived by your students, peers, and administration as being a less competent instructor. Disruption in the classroom can significantly reduce or limit the amount of learning time for students **FIGURE 3-10**.

> ### Teaching Tip
>
> When preparing your lesson plans, remember the following points:
> - Adult learning should be problem centered.
> - Learning must be experienced and centered on meaningful goals.
> - Fire fighters need feedback regarding their progress toward their goal.
> - Adults learn at different speeds and in different ways.

It can also lead to unsafe and negative learning that could have serious implications when students apply what they have learned in the field. You are responsible for providing a safe learning environment and proper, safe instruction to all students.

Disruptive students come in many guises:

- The *Monopolizer* has a tendency to take control of classroom discussions and dominate class time. To deal with this type of student, make certain to call on other students during classroom discussions. You can also call a break or, when a class break occurs, pull the Monopolizer aside and discuss your concerns about his or her "over-participation" during the class. Instructors have been successful positioning themselves between the Monopolizer and other students as a physical barrier in an effort to engage other students. More direct approaches may be to ask an overhead question: "Anyone other than 'John' have an idea how this term applies to fire department operations?"
- The *Historian* has some experience and wants to make sure that everyone is aware of it. Real Historians have the benefit of years of experience and share their experiences to help illustrate an instructor's lesson; they focus on helping the instructor and other students. By contrast, Artificial Historians focus on helping themselves and interject their opinions to show other students how much they know. Real Historians can be disruptive if their stories become distracting for the other students, in which case you should carefully guide the discussion back to the main topic. Another way to redirect Real Historians is by pairing them with struggling students. Artificial Historians are almost always distracting. Assertively redirecting the discussion back to the lesson plan generally corrects this issue. If not, take the student aside and discuss the need for the class to stay focused on the lesson plan. Treat your adult students as adults—with respect.
- The *Daydreamer* is preoccupied with anything other than the lesson. He or she may have been forced to attend the training class or may be participating as part of a job assignment. Many times this student is

focusing on life outside of the classroom. Daydreamers may not directly disrupt other students, but are more of a distraction to you, the instructor. Try to draw such students into the class by using direct questioning techniques and making maximum eye contact.

- The *Expert*, also called the *Know-It-All*, is a student who may become confrontational or monopolize the class in an attempt to show off his or her brain power and perceived expertise, often as a measure of competitiveness. The Expert will provide up-to-date information, factoids, applications, and case information about the topic and can easily subvert your position as the classroom authority. You must corral the Expert and may even have to discuss his or her contributions to the class during a private break.

- The *Class Clown* is the student who seeks attention by acting out or making jokes about course content, the instructor, or other students in the class. Most often the reason for this type of behavior is to draw attention to the individual. Two effective methods for dealing with this type of behavior are to ignore the Class Clown or to engage him or her with a direct question. Sometimes a stern look will deflate this type of student. Another option is, during a break, to take a moment to engage the Class Clown in conversation and to get to know him or her better and seek the individual's help in setting an example for others in the class.

- Another type of student who can be a challenge in the classroom is the *Gifted Learner*. This type of student is usually not disruptive, but becomes a challenge to the instructor because he or she seems to be disengaged in the learning process. These students might appear to be shy or daydreaming, but in reality they are usually intelligent and are "bored" with the material being presented because they have read the material already or have good working knowledge of it. An effective method to manage this type of student behavior is to have such students assist you as an instructor during the course.

Some behaviors are clearly inappropriate in the classroom because they are unsafe or illegal. The *Legal Issues* chapter describes many of these issues, including harassment (sexual or otherwise) and discrimination. Other worrisome behaviors may include actions or words that are threatening or offensive, or that create risks to students, instructors, or property. These behaviors must be dealt with immediately and without hesitation.

Theory Into Practice

Have students identify the components of communications and demonstrate those communication skills in short, individual presentations. Students should be allowed time to prepare prior to making their presentations. Skill sheets developed from the previous activities could be used to have students provide feedback to presenters. The completed skill sheets might then be given to students to enhance their personal learning.

Behavioral problems in the classroom arise for many reasons, ranging from very complex to very simple. For example, the student who falls asleep in class may have been on a working fire during all or part of the previous night and is just plain tired. This issue is relatively easy to identify and is not a persistent problem. By contrast, the student who falls asleep in class every day may have a medical disorder that causes him or her to be unusually drowsy. Learn more about human behavior, behavioral problems, and their causes to become a more effective fire service instructor.

Behavioral problems displayed by the disruptive student may be rooted in other problems as well. As the fire service instructor, you may be put in the position where you must deal with behaviors that are the result of situations outside the instructional environment. These problems may be as simple as hunger or thirst, illness, or medical needs. On other occasions, disruptive students may be experiencing personal problems. Your first priority is to create a positive learning environment for the whole class so all behavioral expressions can be addressed. Your second priority is to ensure the existence of a positive learning environment for the individual student. If you are able to determine the reason for the behavioral issue, and it is one that you can address, this will allow the individual student to learn as well.

Other behaviors that may prove disruptive to the class include calling out, asking irrelevant questions, and giving excessive examples. In these cases, students may be seeking attention from the class or from you. Some students may seek to gain power by being argumentative, lying, displaying a temper, or not following directions. These students need to be dealt with quickly and directly before their behavior gets out of hand.

Some of these problem behaviors can be easily addressed. The hungry student may need to take a break so he or she can eat. The bored student may need a more stimulating lesson to motivate him or her to learn. No matter what the problematic student behavior is, understanding the sources of some basic behaviors will help you control the classroom environment.

As the fire service instructor, you have the opportunity to make positive behavioral changes possible for your students. Here are some simple ways to do so:

- *Lead by example.* You set the tone for the class through your own behavior; in doing so, you set a standard for the class. Act as a professional.
- *Take steps to prevent undesired behaviors.* By preventing the behavior from occurring, you will have changed it before it happens and disrupts the class. For example, in your opening discussion with the class, you can outline the rules that govern the classroom and tell students which kind of behavior is (and is not) expected of them (Bear, 1990).
- *Know what to expect.* By putting the rules in writing, you will have anticipated what the behaviors may be. When formulating these policies, it is important to include the consequences for their violation and to let the students know what rights they have in the classroom situation.
- *Stick to the rules.* Do not contradict the facility's or administration's rules.

- *Communicate.* Talk to problem-causing students and explain that it is unfair for their behavior to continue to disrupt the class.
- *Reward good and appropriate behavior.* This can be done through special assignments or may be as simple as a pat on the back—in front of the class, of course (Walker & Shea, 1991).
- *Act like a professional,* which includes being prepared, approachable, and good natured.
- *Do not either overreact or underreact.* Some issues may bother some people and not bother others. Conversely, problems left unaddressed seldom resolve themselves.

When all else fails, know when it is *time to clean house.* If a student continues to be disruptive after you have attempted to speak with him or her one-on-one and used the chain of command to resolve disruptive behavior, the time might come when the instructor requests that the student leave the class. This is an extreme step and one that should be used only when all other efforts have failed to correct disruptive or inappropriate behavior.

Transitioning Between Methods of Instruction

The Instructor I should acquire skills to effectively utilize problem-solving techniques, to facilitate and lead conferences, and to employ discussion methods during presentations. These techniques are frequently used to conduct small-group sessions where participants have advanced knowledge and experience in the subject matter and the goal is to reach a group solution to a problem or issue. When students become engaged in the learning process and actively participate in a class, the instructor can shift from a lecture format to a guided discussion. The same effect can occur when a student has adapted to a basic skill level in a hands-on training session, allowing the student to perform simultaneous actions and progressive skills.

Teaching Tip

In small groups, students should be encouraged to discuss what influences them to learn, including identifying their personal agendas and goals for the class. It certainly would be appropriate to have students present their group's results to their class. This exercise is a good way to encourage discussion and for you to learn more about the students.

Safety Tip

Always be aware of what is going on around you. As the instructor, it is your responsibility to ensure the safety of your class. Modifications to exits, changes in the environment, and unusual activities or commotion in the area should prompt a careful investigation to keep your learning environment safe. Items out of place on the floor, people not sitting with all of the legs of the chair on the floor, or furniture that is unstable might seem benign, but are actually dangers that can cause injury to the students in the classroom setting. Use common sense and practice safety.

Fire Service Instructor II

Understanding Lesson Planning

All students have special needs in the classroom that may require you to adjust some part of the lesson plan at some juncture. When you develop your plan, it should be detailed and complete enough to achieve the identified goals and desired outcomes, yet flexible enough that it can be adjusted to meet students' needs on the fly. Be prepared to repeat information in different ways or to perform demonstrations more than once in different ways. Make sure that you follow a detailed, sequential plan when demonstrating a skill; explain and outline the information to be explained; and discuss the results, including key questions to guide the discussion. These simple, yet flexible components will enable you to develop an effective, yet concise lesson plan. The *Lesson Plans* chapter details this process.

As an Instructor II, you will have to adjust certain elements of a lesson plan to accommodate students with special needs. If a student is not able to read well, for example, your content of the lesson plan will remain the same, but the teaching method may have to be adjusted to include audio recording permission and more graphic illustrations of your objectives. As stated earlier, you may have to repeat information while using different visual aids, varying voice and tone inflections, or even breaking class sessions down to smaller units to address individual students' needs.

Ask yourself three basic questions when reviewing and planning for the delivery of a lesson plan:

1. *What are the goals of the lesson?* These goals are the purpose, aims, and rationale for the class period. You and your students will be working to achieve these goals.
2. *How will you achieve those goals?* Identify the objective of the lesson and focus on what the students will do to acquire the knowledge and skills of the lesson.
3. *How will you know when those goals are achieved?* Through some kind of assessment, you need to ensure that the students have learned, at a minimum, the desired information. Be sure to provide students with an opportunity to practice what you will be evaluating.

Training
BULLETIN

Jones & Bartlett Fire District

Training Division
5 Wall Street, Burlington, MA, 01803
Phone 978-443-5000 Fax 978-443-8000
www.fire.jbpub.com

Instant Applications: Methods of Instruction

Drill Assignment

Apply the chapter content to your department's operation, training division, and your personal experiences to complete the following questions and activities.

Objective

Upon completion of the instant applications, fire service instructor students will exhibit decision making and application of job performance requirements of the fire service instructor using the text, class discussion, and their own personal experiences.

Suggested Drill Applications

1. Identify the most effective and least effective training sessions you have attended. Determine which methods of instruction were appropriate or inappropriate and may have influenced your impressions of those training sessions.

2. Review a list of upcoming training sessions scheduled for your department and determine the methods of instruction that should be used to achieve the best outcomes for them.

3. Rearrange your typical classroom seating arrangement to accommodate various methods of discussion, such as a lecture, demonstration, or group discussion.

4. Review the Incident Report in this chapter and be prepared to discuss your analysis of the incident from a training perspective and as an instructor who wishes to use the report as a training tool.

Incident Report

Hollandale, Minnesota—1987

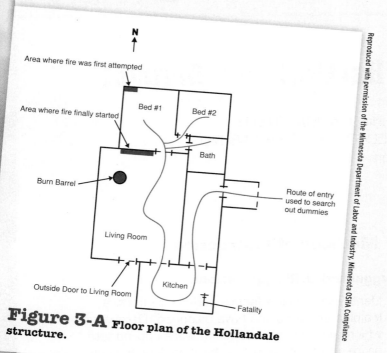

Figure 3-A Floor plan of the Hollandale structure.

Area where fire was first attempted

Area where fire finally started

Bed #1

Bed #2

Bath

Burn Barrel

Route of entry used to search out dummies

Living Room

Outside Door to Living Room

Kitchen

Fatality

Figure 3-B Photo of rear of house in the Hollandale fire. The kitchen is below the area with the flat roof. Note the minor amount of smoke and heat damage above door.

Search and rescue drills had been conducted for some time in a small two-bedroom house without problems. Leaves burning in a 35-gallon (132.5-liter) drum were used to create smoke. The drum had a metal cover to reduce the danger of fire spread, and the drum was situated above the floor. Caution was exercised to prevent fire from spreading from the container. Two fire apparatus were on scene with 1400 gallons (5299 liters) of water on one of the fire units and 1000 gallons (3785 liters) in a portable tank. The participants carried a charged 1½" (38.1 mm) hose line.

After the interior search evolutions, the intent was to destroy the house by burning it down. Four crew members, wearing full protective clothing, equipped with breathing apparatus, and protected by a charged 1½" (38.1 mm) hose line, entered the structure using the east entry that had been used during the drill **(Figure 3-A)**. They attempted to light a fire in the back bedroom but were unable to get it to ignite, most likely due to the dampness from the water. They moved to the living room and splashed #2 fuel oil on the wall and carpet. Two of the fire fighters moved to the kitchen while another used a propane torch to ignite the fire in the living room. The entire room flashed, and a third fire fighter crawled out, thinking his partner was with him. Three of the fire fighters crawled to the outside. The fourth made it to the kitchen and stopped, where he apparently removed his SCBA mask to call for help and then reentered the fire area and was killed. His Nomex hood, propane torch, and fuel can were later found in the kitchen. The three fire fighters who had exited the house reported the emergency and entry was immediately forced into the front entrance. The fire was quickly extinguished, and the victim was found dead.

It was uncovered in the investigation that the living room paneling was made of a light-weight combustible material, found in mobile homes at the time. The smoke barrel with live fire may have preheated the combustibles in the living room **(Figure 3-B)**. The report noted that the NFPA 1403 standard should have been followed. Investigators believed that if the deceased had remained in the kitchen, he most likely would have survived.

The fire department was cited by the state OSHA for failure to "examine the building more closely, failure to brief the participants on the building's layout, failure to have a building evacuation plan, and failure to have a qualified instructor on the scene." This was the first fire fighter fatality after the release of NFPA 1403.

Post-Incident Analysis: Hollandale, Minnesota

NFPA 1403 (2007 Edition) Noncompliant

Combustible wall paneling and carpeting contributed to rapid fire spread (4.2.17)

Flammable and combustible liquids used (4.3.6)

Flashover and fire spread unexpectedly (4.3.9)

No direct egress from living room (4.2.24.5)

No formal evacuation plan including evacuation signal (4.4.10)

No incident safety officer (4.4.1)

No interior hose line (4.4.6)

No EMS on scene (4.4.11)

No written emergency plan (4.4.10)

Participants not informed of emergency plans and procedures (4.4.10)

No preburn plan (4.2.25.2)

No walk-through performed with students and instructors (4.2.25.4)

Wrap-Up

Chief Concepts

- Understanding how adult learning takes place, which factors affect motivation, and how different generations approach the educational process will help you become a better instructor.
- Motivation affects how students practice, what they observe, and what they do—all of which are critical to learning. As a fire service instructor, you need to understand how learning and motivation work together to make a great fire fighter. Meeting the needs of the students is the key to motivating them and helping them to learn.
- Adult learners can be divided into three types: learner oriented, goal oriented, and activity oriented. Each type is associated with different motivational issues, which in turn affects how learning takes place.
- Learning is an interactive process in which the adult learner encounters and reacts to a specific learning environment. That learning environment encompasses all aspects of the adult learner's surroundings.
- Effective learning requires instructors to step outside the realm of personal experiences and into the world of the learner.
- Students' learning requirements differ because their experiences and development have differed.
- Adult students need realistic, practical, and knowledgeable lessons that deliver information in a variety of ways to accommodate their differing learning styles.
- When you can communicate to students that you know the information, and that you understand their unique needs and motivations, you will find that it is much easier to achieve the goal of teaching the information and having the students learn.

- Determining which method of instruction should be used to deliver information is accomplished using several factors:
  - The experience and educational level of the students
  - The classroom setting or training environment
  - The objectives of the material to be taught
  - The number of students to be taught
- Methods of instruction vary among disciplines, but for the fire service the following are most common and effective in teaching new and in-service personnel:
  - Lecture
  - Demonstration/skill drill
  - Discussion
  - Role play
  - Labs/simulations
  - Case studies
  - Out-of-class assignments
  - Online learning, independent study, and blended learning
- Five effective transition techniques are delineation, words/phrases, pictures, voice, and body language.
- An important part of the learning process is to assess the knowledge gained during class. This is most commonly done by asking the students questions about the material you have presented. Several types of questions may be used:
  - Direct questions
  - Rhetorical questions
  - Open-ended questions
  - Closed-ended questions
- Disruption in the classroom can significantly reduce or limit the amount of learning time for students. It can also lead to unsafe and negative learning that could have serious implications when students apply what they have learned in the field. As the fire service instructor, you are responsible for providing a safe

learning environment and proper, safe instruction to all students.

- When you develop your lesson plan, it should be detailed and complete enough to achieve the identified goals and desired outcomes, yet flexible enough that it can be adjusted to meet students' needs on the fly.
- Ask yourself three basic questions when reviewing and planning for the delivery of a lesson plan:
  - What are the goals of the lesson?
  - How will you achieve those goals?
  - How will you know when those goals are achieved?

Hot Terms

Adult learning The integration of new information into the values, beliefs, and behaviors of adults.

Andragogy The identification of characteristics associated with adult learning.

Asynchronous learning An online course format in which the instructor provides material, lectures, tests, and assignments that can be accessed at any time. Students are given a time frame during which they need to connect to the course and complete assignments.

Audience analysis The determination of characteristics common to a group of people; it can be used to choose the best instructional approach.

Baby boomers The generation born after World War II (1946–1964).

Blended learning An instruction method that combines online/independent study with face-to-face meetings with the instructor.

Generation X People born after the baby boomers; they are today's adult learners.

Generation Y People born immediately after Generation X.

Generation Z People born immediately after Generation Y, starting in the late 1990s.

Independent study A method of distance learning in which students order the course materials, complete them, and then return them to the instructor.

Learning style The way in which an individual prefers to learn.

Method of instruction The process of teaching material to students.

Motivation The activator or energizer for an activity or behavior.

Motivational factors States of the person that are relatively temporary and reversible and that tend to energize or activate the behavior of the individual.

Pedagogy The art and science of teaching children.

Synchronous learning An online class format that requires students and instructors to be online at the same time. Lectures, discussions, and presentations occur at a specific hour and students must be online at that time to participate and receive credit for attendance.

References

Armour, S. (2005). http://www.usatoday.com/money/workplace/2005-11-06-gen-y_x.htm.

Bear, G. G. (1990). Models and techniques that focus on prevention. In A. Thomas & J. Grimes (Eds.), *Best practices in school psychology* (p. 652). Silver Spring, MD: National Association of School Psychologists.

Bender, D. (1999). *Unique characteristics of adult learning.* www2 .bw.wdu/~dbender/csu/adultlearn.html.

Brown, G. L. (1997). *New learning strategies for Generation X.* ERIC Clearinghouse. http://www.ed.gov/databases/ERIC.

Butler, G., & McManus, F. (1998). *Psychology.* Oxford, UK: Oxford University Press.

Draper, J. (1993). *The craft of teaching adults.* Toronto: Culture Concepts.

Elias, J. L., & Merriam, S. (1980). *Philosophical foundations of adult education.* Malabar, FL: Krieger.

Houle, C. (1961). *The inquiring mind.* Madison, WI: University of Wisconsin Press.

Knowles, M. (1984). *The adult learner: A neglected species* (3rd ed.). Houston, TX Gulf Publishing.

Knowles, M. (1990). *The adult learner: A neglected species*, rev. ed. Houston, TX: Gulf Publishing.

McClincy, W. D. (1995). *Instructional methods in emergency services.* Englewood Cliffs, NJ: Brady Prentice Hall.

McClincy, W. D. (2002). *Instructional methods in emergency services* (2nd ed.). Englewood Cliffs, NJ: Brady Prentice Hall.

Merriam, S. B., & Caffarela, R. S. (1999). *Learning in adulthood.* San Francisco, CA: Jossey-Bass.

National Fire Protection Association. (2012). *NFPA 1041: Standard for Fire Service Instructor Professional Qualifications.* Quincy, MA: National Fire Protection Association.

National Fire Protection Association. (2007). *NFPA 1403: Standard on Live Fire Training Evolutions.* Quincy, MA: National Fire Protection Association.

Walker, J. E., & Shea, T. M. (1991). *Behavior management: A practical approach for educators* (5th ed.). New York: Macmillan.

Wlodkowski, R. J. (1999). *Enhancing adult motivation to learn: A comprehensive guide for teaching all adults*, rev. ed. San Francisco, CA: Jossey-Bass/Pfeiffer.

A class of students including rookies, veterans, and officers has assembled in your department to train on new methods of fire-fighting based on current research on tactics and strategies. You know that the rookies are receptive, as they have just learned about this topic in their academy, but the veterans and officers may be resistive due to their years of experience using more traditional methods. You believe that the new methods will be safer and more effective, and the department administration is soliciting input on how best to implement the research, but is still undecided whether the SOPs will be updated to reflect it.

1. Which generation label would most likely represent the most experienced fire fighters?
   A. X
   B. Y
   C. Z
   D. Baby Boomers

2. A student in the class challenges the material being taught and takes control of most of the classroom discussions. What category of disruptive student would he be in?
   A. Historian
   B. Expert
   C. Day Dreamer
   D. Monopolizer

3. To effectively communicate to a class of students, one of the keys to ensure that your students receive and remember the information you give them would be the use of which technique in your presentation?
   A. Let them read handouts and text material on their own.
   B. Show them video or still photos of the examples you are using.
   C. Lecture to them emphasizing the key points continually throughout the presentation.
   D. Have them say what they are doing, while they are doing it.

4. The most experienced fire fighters and officers in this class may best respond to the learning process if you do which of the following?
   A. Begin with what they already know and move into what they will learn as new information.
   B. Begin with the new information and work backwards to the information that they already know.
   C. Open with question and answer sessions about the new material before you present your lesson plan.
   D. The content is more important than the act of being involved in the process.

5. A key in developing the lesson plan for this class would be:
   A. Adults learn at the same pace and same manner.
   B. Adults prefer to be told what to do and what they will learn.
   C. Adult learning should be problem centered.
   D. Identify who the disruptive students will be and address the content to their needs and concerns.

The Learning Process

Fire Service Instructor I

Knowledge Objectives

After studying this chapter, you will be able to:

- Describe the laws and principles of learning (NFPA 4.3.2 NFPA 4.4.3). (pp 79–81)
- Identify the three types of learning domains. (pp 85–87)
- Define learning styles and discuss the effects of the various learning styles on the classroom (NFPA 4.4.5). (pp 89–91)
- Describe learning disabilities and methods of dealing with learning disabilities and disruptive behavior in adult learners (NFPA 4.4.5). (pp 91–93)

Skills Objectives

After studying this chapter, you will be able to:

- Analyze student learning styles and preferences. (pp 89–91)
- Demonstrate methods of dealing with learning disabilities. (NFPA 4.4.5). (p 92)

Fire Service Instructor II

Knowledge Objectives

There are no knowledge objectives for Fire Service Instructor II students.

Skills Objectives

There are no skills objectives for Fire Service Instructor II students.

Fire Service Instructor III

Knowledge Objectives

There are no knowledge objectives for Fire Service Instructor III students.

Skills Objectives

There are no skills objectives for Fire Service Instructor III students.

You Are the Fire Service Instructor

At a recent fire instructor conference, a great deal of discussion focused on the new generation of fire service students who are entering the fire service of today. They come from more diverse backgrounds and in many ways are more "tech savvy" than their predecessors. In the opinion of many conference attendees, these new students are more of a challenge to teach. You find yourself both agreeing and disagreeing with the various debate participants, because you have seen some of these different behaviors in your own younger fire fighters.

Upon returning to your department, you engage in a discussion with other instructors about what you heard and learned during the various workshops, and you bring up the debate about the new generation of fire fighters. Quite unexpectedly, one of the senior members of the assembled group of instructors makes the following statement: "Doesn't matter what generation enters the fire service, because the principles of learning are constant and our job is to understand these principles and do our jobs of teaching whoever the department hires and to teach them to stay alive." Everyone in the group is left to ponder this statement as the group breaks up and returns to work.

1. What are the principles of learning, and how do you use them properly?
2. How do you teach and develop each domain of learning?
3. How can you help all students be successful in your classes?

Introduction

Learning is a change in a person's ability to behave in certain ways. This change can be traced to two key factors—past experience with the subject (e.g., in the field) and practice (e.g., training in the classroom). Learning can occur both formally (inside the classroom) and informally (around the dinner table) (Connick, 1997). Formal learning does not occur by accident—it is the direct result of a program designed by an instructor (Butler & McManus, 1998). An adult learner may intentionally set out to learn by taking classes or by reading about a subject. He or she may also gather information through the experience of living that changes the learner's behavior. Informal learning occurs spontaneously and continually changes the adult learner's behavior. Ideally, learning is created through the blending of individual curiosity, reflection, and adaptation (Stewart, 2003).

Learning takes time and patience. As a fire service instructor, you have the opportunity to influence fire fighters positively at all levels of the fire department **FIGURE 4-1**. As a leader in the fire department, you have a responsibility to teach information, to hone skills, and to promote positive values and motivation in your students. A wise fire service leader once said, "Classroom learning plus street experience equals wisdom." As an instructor, you have a major impact on the first part of this equation.

All levels of fire service instructors will learn that each student in your class has different ways that they understand

FIGURE 4-1 As a fire service instructor, you have the opportunity to positively influence fire fighters at all levels of the fire department.

information. As you participate in this course, you are experiencing this phenomenon by reading text and reviewing charts, graphs, photographs, and different text fonts, sizes, and colors. Each of these elements allows different learning styles to be experienced. Other characteristics are demonstrated as your instructor presents course materials.

Fire Service Instructor I

The Laws and Principles of Learning

Numerous theorists, including Edward L. Thorndike and his contemporaries Edwin R. Gutherie, Clark L. Hull, and Neal E. Miller, have studied learning and developed ideas that have become recognized as the basic laws or principles of learning. According to Thorndike, there are six laws of learning: readiness, exercise, effect, primacy, recency, and intensity.

1. *The law of readiness*: A person can learn when he or she is physically, mentally, and emotionally ready to respond to instruction.

2. *The law of exercise*: Learning is an active process that exercises both the mind and the body. Through this process, the learner develops an adequate response to instruction and is able to master the learning through repetition. For example, a fire fighter can master ropes by practicing tying knots every night for a month. The opposite is also true: a skill that is not exercised after it is developed will be lost. In some areas, this concept is referred to as the *principle of disuse*.

3. *The law of effect*: Learning is most effective when it is accompanied by or results in a feeling of satisfaction, pleasantness, or reward (internal or external) for the student—for example, when the student receives an A on a pop quiz.

4. *The law of primacy*: This principle states that the first method or way a concept is taught will be the way the student learns and remembers it. As the saying goes, "First impressions are lasting impressions"—and when it comes to learning, this holds true. The concern here is that if a student learns a new skill incorrectly, then the instructor has to help the student "unlearn" and correct the knowledge.

5. *The law of recency*: Practice makes perfect, and the more recent the practice, the more effective the performance of the new skill or behavior. Running drills on new skills will reinforce and perfect training in fire fighters. This law is why we retrain fire fighters and review information that is not used on a daily basis. If a skill is not used, it goes away, just as a muscle that is not exercised loses its strength over time.

6. *The law of intensity*: This principle states that the more of the senses that are stimulated, the more likely the student will be to change his or her behavior—which is what learning is all about. A lesson that moves from lecture to real-world videos to hands-on application of the skill will increase the student's ability to learn. The more the material is presented in a real-world manner, the stronger the learning effect (Thorndike, 1932).

■ Use of Senses

Adult learners use the five senses—hearing, seeing, touching, smelling, and tasting—to obtain information **FIGURE 4-2**. The type and number of senses used to learn determines how learning occurs and what is remembered. Of the five senses, seeing (e.g., a visual demonstration of a skill) is generally the most effective means of learning, followed by hearing (e.g., listening to a lecture), smelling (e.g., smelling smoke), touching (e.g., handling fire equipment), and tasting (rarely used in training fire fighters, especially due to hazardous materials and other safety concerns). As an instructor, the more senses you can engage during the learning process, the greater the students' ability to retain the information that was presented. Effective lesson plans will address this issue by allowing time for questions, hands-on training, and other methods that bring all of the senses together.

■ The Behaviorist and Cognitive Perspectives

Two schools of thought have developed to explain how behavior evolves: behaviorist and cognitive. According to the behaviorist perspective, learning is a relatively permanent change in behavior that arises from experience. Not all changes in behavior reflect classroom learning. Many changes in behavior take place owing to a person's maturation and physical changes as he or she ages (Rathus, 1999).

According to the cognitive perspective (mental), learning is an intellectual process in which experience contributes to relatively permanent changes in the way individuals mentally represent their environment. Cognitive theories stress the use of mental capabilities to change behavior. Cognition is defined as the acquisition of knowledge through

FIGURE 4-2 Adult learners use their senses to obtain information.

the use of perception, encounters with ideas, and obtaining of experiences (Lefrancois, 1996). Thus cognitive learning is demonstrated by recall of knowledge and use of intellectual skills. It entails the comprehension of information, organization of ideas, analysis and synthesis of data, application of knowledge, choice of alternatives in problem solving, and evaluation of ideas or actions. According to this perspective, learning is an internal process that is demonstrated externally by behavioral changes.

■ Competency-Based Learning Principles

Learning that is intended to create or improve professional behavior is based on performance, rather than on content. Such competency-based learning is usually tied to skills or hands-on training. Not surprisingly, much of what is accomplished on the fire ground occurs through hands-on activities that require proficiency with skills—for example, the ability to force entry on a door. Competency-based learning is based on what the adult learner will do or perform.

Competency-based learning originates with the identification and verification of the needed competencies to perform particular skills. These competencies are actually job requirements, and ideally they will have been identified through a job analysis. For the fire service, this job task analysis has been done by the National Fire Protection Association (NFPA) via its professional qualification standards, which are all written in job performance requirement (JPR) format.

To ensure the success of a competency-based course, it is essential to follow a standardized process. This process ensures that each successive step has the proper foundation. Without the proper foundation, what may look like an excellent course might not address the real skills issues of the fire department. The design of a competency-based course includes the following requirements:

- Competencies exist.
- Job analysis is performed.
- Standards are identified and set.
- Course goals are identified and written.
- Lesson objectives are identified and written.
- Competency-based instruction is delivered.

The most critical facet of competency-based learning is the notion that the skills must be mastered by the adult learners for each of the competencies. To ensure that this requirement is met, students must be given enough time and appropriate instruction to both learn and master the skill. Both the instructor and the students must understand that with mastery comes the responsibility of maintaining a particular performance level. Thus learning objectives, instruction, and the evaluation of the skills must be tailored to performance in the field, not just to the acquisition of abstract knowledge that is out of context.

A competency-based learning process must be flexible enough to be adapted to the needs of all students. Each student inevitably learns at his or her own pace and in his or her own individual fashion. Some students may take longer than others to learn the skill, so you must take each individual's

FIGURE 4-3 Immediate feedback needs to be given on the training ground to ensure that each student masters the skill.

preferred learning style into account when teaching skills. Also, appropriate and immediate feedback needs to be provided to students to allow them to achieve mastery of the skill, thereby applying the principles of primacy, recency, and intensity **FIGURE 4-3**.

Teaching Tip

Instructors need to constantly think about what might help their students learn. The creation of an environment that addresses the physical, emotional, and cognitive needs of all students provides for a successful learning experience. To achieve this kind of supportive environment, make sure that your classroom accommodates differences in students' learning styles.

The use of teams can be an especially helpful strategy in the classroom setting. It allows students to work with people who differ from them in terms of how they learn, think, or respond.

■ Forced Learning

There is an important fact of which you need to be aware: Adult learners cannot be forced to learn. Unfortunately, too many adult learners attend training programs or engage in educational situations only because they are required to do so or because they are motivated by the wrong reason. Understanding the dynamics of the fire service, there always seem to be new continuing education credits to be accumulated for recertification or some new required certification to be obtained. Mandating learning often results in a lack of motivation related to classroom activities and a poor attitude toward the learning process in general, both of which present quite a challenge to the fire service instructor who is tasked with teaching the course.

The presence of students who are mandated to attend a class or training session should alert you to the possibility that the lesson plans might need to be altered. For example, if you are assigned to teach a cardiopulmonary resuscitation (CPR) recertification course to paramedic students, then you might look for ways to involve students in both the teaching process and the learning process. The creation of such dual roles will help capture their attention and enhance the learning process.

place. Such positive reinforcement can serve to increase the motivation of students who are forced into your classroom. You may need to acknowledge the areas of special interest of reluctant students to guide them through the learning process and to keep them engaged in your class. Knowing your students and their areas of interest allows you to capitalize on their knowledge base and become a more effective fire service instructor.

Ethics Tip

One method that is routinely used to encourage learning from forced participants is tapping into the participants' knowledge and letting them teach from their experiences. Is it ethical to have someone other than the assigned fire service instructor present the instructional material? Would it be ethical for a college business professor to have a student with equal business experience teach other students in an effort to maintain that student's interest in the class?

Like all ethical dilemmas, both positions on this issue have pros and cons. Having an otherwise disengaged student present instructional material will increase the learning for that particular student. Also, other students often learn much from the student presenter, with whom they can identify more easily. However, if the other students do not receive the necessary instruction, or if there is the perception that they may not receive proper instruction, this practice can be detrimental. Having the assigned fire service instructor teach the material ensures that the appropriate level of quality is provided and that the correct information is being presented to the students.

When an experienced student is forced to attend your class and you are thinking about taking advantage of that student's value as a peer-teacher, consider the material, the experience of the student, the reputation of the experienced student among his or her classmates, the student's attitude, and the reaction of the other students to a "guest" presenter from their own class. Most fire service instructors would agree that the student should not handle the majority of the teaching because students have paid to benefit from the instruction of the professional fire service instructor; at the same time, allowing the experienced student to assist in teaching small components of the lesson might be acceptable to both the class and the experienced student.

Teaching Tip

When teaching adults, you are really more of a facilitator than a traditional teacher. Here are some core qualities that will help you become a good facilitator:

- Be genuine in your relationships with adult learners, rather than consistently adhering to the traditional role as "teacher." Show that you really care about the students and their success in the classroom.
- Accept and trust in the adult learner as a person of worth. Provide positive reinforcement and engage the learner with respect and value. Remember— people learn more from individuals whom they trust than from individuals whom they mistrust. Establishing a mutual climate of trust is important in fostering learning.
- Provide nonjudgmental understanding of adult learners' perspectives. Respond with empathy to both their intellectual and emotional perspectives. Remember that people are more open to learning when they are respected and feel supported than when they feel judged or threatened.
- Establish a climate for learning. Openness and authenticity are essential for maintaining this supportive atmosphere. People will be more likely to examine and embrace new ideas in an open and authentic environment.

Learning Skills for Adult Learners

As a fire service instructor, you may find yourself teaching adult learners who have not been in a classroom setting since high school or college or rookie school. Study habits that these students developed while in primary school, secondary school, or college could have become rusty over time. Perhaps the most important study skill to develop as a student (and instructor) is to have the right attitude toward learning and studying. A positive attitude toward learning will help you learn new material and enjoy the classroom environment.

To assist your students in becoming effective learners, you can share study skills that enhance what they learn during classroom and personal study time. Study tips can be viewed as belonging to one of three categories: classroom, personal study, and test preparation.

Areas of Interest

All adults have certain areas of interest. Tapping into these areas may help reinforce the desire to learn. The source of this interest is less important than its mere presence, because all students need to have encouragement if learning is to take

■ Classroom Study Tips

In the classroom environment, students can learn to develop the following skills to help them in the learning process:

- **Ask questions.** If students do not understand the material being presented by the instructor, they should be encouraged to ask questions. Students should not wait for someone else to ask a question or feel that they are the only student who does not understand. In most situations, other students have the same question.
- **Take notes.** A good study technique is to develop an outline format for notes that follows the syllabus or outline given by the instructor.
- **Participate in class.** Participate in classroom activities and discussions and group activities.
- **Turn off your phone.** If students want to get the most out of classroom instruction, they should turn off their cell phones and other technological distractions and listen to the instructor.
- **Connect with peers.** As appropriate, students should get the names and phone numbers of other students in the class to contact for notes or information in case of a missed class.

■ Personal Study Time

Outside the classroom, students will need to study on their own. This is the time when most adult learners really learn the new material and develop effective learning skills and habits, including the following:

- **Time management.** Adult learners must learn to budget their time. This can be done by setting aside a predetermined period of time each day to study, and sticking with that schedule.
- **Place to study.** If studying in the fire house after hours, the student should find a location to study that is free of distractions and will allow him or her to focus on the material to be learned.
- **Prioritize material.** A student should review what needs to be studied and determine what is the highest priority and what needs to be covered first.
- **Review class notes.** To aid the student in focusing personal study time, he or she should review notes from class.
- **Reading skills.** One technique that will help students retain what they have read is to take brief notes of the material read. These notes can be used later to help in test preparation.
- **Study groups.** Depending on the class format and the material being learned, study groups can be an effective method for learning new material. These groups can be formal groups assigned by the instructor or informal groups that are organized by students who desire to study together.

■ Test Preparation Skills

Testing and course evaluation during instruction are commonly used methods for instructors to determine whether students have learned the material presented in class. Testing can take the form of quizzes or questioning by the instructor. Many courses require a comprehensive test to be administered at the end of the course to assess the students' learning. The following techniques can be used to help students prepare for a test:

- **Participate in study groups.** Study groups are effective methods for students to quiz one another and help one another in test preparation.
- **Do not cram.** Preparing for a course exam should begin on the first day of class, not hours before the test by cramming. Students should not wait until the last hours before a test to study; this method does not work.
- **Use mnemonic devices.** Fire service members and paramedics have long used mnemonic tools to help them remember information, such as RECEO-VS or SAMPLE.
- **Take a break.** One skill that students need to learn is knowing when to take a 10-minute break from studying and refresh yourself—let your mind stop thinking, get some fresh air, and then go back to work.

Teaching Tip

To be a successful fire service instructor, you must truly know yourself. Some of us are excellent speakers, whereas others are effective writers. What are your strengths? What do you need to work on? Taking a personal inventory assessment such as the Myers–Briggs Type Indicator can clue you in to behaviors that you may need to improve.

Safety Tip

Sometimes you will find that the fire fighters in your department are ready and willing to learn and experience training as a group. When it comes down to individual fire fighters, however, blocks to learning may spring up. For example, a fire fighter may be tired and unable to concentrate because he worked an incident all night. When basic needs are not met, learning becomes a second priority, which can in turn create a safety issue. A student may perform an unsafe act due to his or her less-than-optimal physical or mental state, or a student may miss information (because of inattention) that he or she must know to be able to perform at actual incidents.

JOB PERFORMANCE REQUIREMENTS (JPRS)
in action

Every student in each class you instruct learns at a different rate with different levels of comprehension and understanding. Likewise, each student responds differently to the methods of instruction that are used in a class. An experienced fire service instructor knows how to adapt a lesson plan while the presentation is under way. The need for such flexibility may be one of the more difficult concepts for new fire service instructors to understand. The goal of any presentation is to improve the student's understanding of the objectives. To do your job correctly, you must have a working knowledge of the learning process. Your understanding of these factors is critical to the success of your entire class.

Instructor I

The Instructor I delivers a prepared lesson plan using a variety of instructional methods. The appropriate method to be used might be identified after review of the content of the lesson plan, and perhaps even more so by the analysis of the audience that will participate in the training. Student-centered delivery with the students involved in the learning will normally make the longest-lasting impressions on the students' learning.

Instructor II

When developing objectives and lesson plan content, the Instructor II must be aware of the various domains of learning and the levels of comprehension intended for the students. The objectives must be written to the appropriate level and applications in the lesson plan must be developed to help students apply what they have learned. Such exercises check the students' understanding of course content.

Instructor III

The Instructor III develops major course goals and program direction. Understanding how adults learn and are motivated will improve the quality of a course, as the Instructor III can create more effective teaching strategies and student applications.

JPRs at Work

Review your assigned lesson plan and determine the best methods of presenting the material based on the objectives and evaluations developed for the class. Adjust the lesson plan delivery based on the students' understanding and progress through the material.

JPRs at Work

NFPA does not identify JPRs that relate to this chapter. Nevertheless, it is important to understand how students learn and the most effective ways of instructing them so that you can be the most effective instructor possible.

JPRs at Work

NFPA does not identify JPRs that relate to this chapter. Understanding the learning process, however, will allow the Instructor III to develop course goals and learning objectives that are effective with different audiences.

Bridging the Gap Among Instructor I, Instructor II, and Instructor III

Often a coffee-table discussion between instructors at different levels can assist in determining the best methods of delivering a prepared lesson plan. Instructors at levels II and III will typically have more experience in the development and delivery of training and should mentor the Instructor I in ways to engage the students in the learning process so as to achieve better outcomes. Select appropriate methods of instruction by understanding the learning process and recognizing how adults are motivated to learn.

Maslow's Hierarchy of Needs

Abraham H. Maslow was a psychologist who researched mental health and human potential. He is most often associated with the concepts of a hierarchy of needs, self-actualization, and peak experiences. Maslow's initial work in 1943 was followed up in 1954 by his development of what is commonly referred to today as the "hierarchy of needs." Maslow took a unique approach in conducting his research, called "biographical analysis," wherein he reviewed the biographies and writings of 21 people he identified as being "self-actualized." His subjects included such major figures as Albert Einstein, Jane Addams, Eleanor Roosevelt, and Frederick Douglass. Based on his research, Maslow determined that certain needs should be satisfied to enable a person to move forward toward self-actualization. He perceived human needs as being arranged like a ladder or a pyramid FIGURE 4-4. According to Maslow, unfulfilled needs at the lower levels on the ladder will inhibit a person from climbing to the next step. As he pointed out, a person who is dying of thirst quickly forgets the thirst when deprived of oxygen.

Understanding the basic needs of people can facilitate an appreciation of which basic needs your students have. An instructor should keep these needs in mind when creating, adapting, or modifying lessons and the classroom dynamic.

■ Level One: Physiological Needs

The most basic human needs, at the bottom of the ladder, are physical—air, water, food, and shelter. It is hard to be creative or productive when you are exhausted and hungry. For the fire service instructor, this means staying alert for conditions where the students are hungry, exhausted, dehydrated, too hot, or too cold. Although some of these conditions may arise in a long classroom session, they are more likely to occur in off-site training, during which reasonable breaks must be given for food, water, and rest (see *The Learning Environment* chapter for more information). An example of a basic need that is too often overlooked is restroom facilities at a remote training site.

Sometimes fire fighters have to work at the extremes of discomfort and personal inconvenience during emergency activities, but these situations should be the exceptions and not a normal way of getting the job done. Making a fire company train or work without a rest and rehydration period will fail to meet their most basic needs FIGURE 4-5.

■ Level Two: Safety, Security, and Order

Safety, security, and order needs are next on the ladder of needs. Safety is obviously a primary concern in the fire service, although a fire fighter's perceptions of safety are probably quite different from those of a typical office or factory worker. If students feel that their instructor is leading them in an unsafe manner or that a particular policy or practice in the fire department is exposing them to an avoidable risk, safety is likely to become a very significant issue.

The concept of security is more closely associated with maintaining employment or status within the organization. This need could be as subtle as a fear that a personality clash with a particular instructor will impede a fire fighter's professional development. An individual who feels threatened may do what is necessary to survive, but will not feel highly motivated to advance the fire department's objectives.

The need for security and order could become a factor when a fire department is undergoing a significant reorganization, such as a change in a standard procedure that the fire fighter was accustomed to over a long period of time. If a department adopts a new procedure or operational structure, the fire fighter may feel apprehension over new training responsibilities, evaluations, and scrutiny by upper levels of management. Any type of change in the established order is likely to arouse insecurities. The anticipation of significant changes in the organization can easily impede the fire fighter's creativity and motivation.

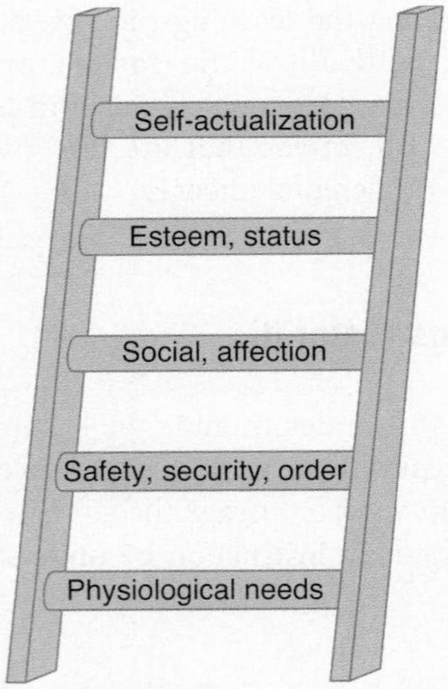

FIGURE 4-4   Maslow's hierarchy of needs.

Self-actualization

Esteem, status

Social, affection

Safety, security, order

Physiological needs

FIGURE 4-5   Fire fighters need periods of rest and rehydration to meet their most basic physiological needs.

■ Level Three: Social Needs and Affection

Social needs and affection, as identified by Maslow, are the psychological or social needs that are related to belonging to a group and feeling acceptance by the group. Group acceptance and belonging are particularly strong forces within the fire service. The majority of fire fighters have some type of identification on their personal vehicles that visibly declares their local fire department affiliation and membership in the fire service at large. Transferring a fire fighter away from a "good" assignment can be a strong demotivator, just as suspending a volunteer from responding to calls can have a significant impact on him or her at the social and affection levels. Fire fighters who are temporarily or permanently assigned to a company other than the one they have worked with and for over a long period of time will feel the same impact. For example, they must learn how they will fit into the formal and informal rank structures and work groups in the new station. In fact, many newly placed but experienced members feel like they do not fit into the company until they can begin to socialize with the new crew members.

Make your class its own community of social acceptance by acknowledging that all students present are to become the future of the fire service. This group acceptance leads the class to a greater cohesion even when many of its members come from differing backgrounds and departments.

■ Level Four: Esteem and Status

Promotions, certifications, and special awards are all symbols that relate to the student's esteem and status level **FIGURE 4-6** . For example, membership in an elite fire department technical unit is often an indicator of advanced qualifications or achievements.

FIGURE 4-6 An instructor can positively affect a fire fighter or company's esteem and status by making efforts to recognize new certifications or training milestones.

Many fire fighters are willing to invest money and off-duty time to compete for a high-esteem or high-status position. For example, fire fighters may spend months preparing for a fire officer promotional exam or taking courses to qualify for a rescue company. Most elite fire units have more applicants than available positions. As an instructor, keep these motivational factors in mind. Every student is in your class for a reason, and it is your responsibility to guide them toward meeting the course objectives, and in doing so getting another step closer to their own goals.

As an instructor you need to be aware of how self-esteem plays out in the classroom environment and the importance of protecting your students from negative comments both from you as the instructor and from their fellow students. The fire service is known for joking in a "brotherly" fashion to foster camaraderie, but as the instructor you need to be careful not to allow this type of joking to cross the line where it affects a student's self-esteem.

■ Level Five: Self-Actualization

Peak experiences are profound moments of love, understanding, happiness, or rapture, when a person feels more whole, alive, self-sufficient, and yet a part of the world—more aware of truth, justice, harmony, and goodness. Self-actualizing people have many such peak experiences. These individuals tend to focus on problems outside of themselves and have a clear sense of what is true. Self-actualization in the learning environment is seen when students finally complete a skill or task they have been struggling with and can turn their attention to the next challenge. Graduation day at a recruit academy is an ultimate moment and point of self-actualization for those students, when they have completed a hard training process and receive their badges as fire fighters.

Learning Domains

In 1956, Benjamin Bloom and his research team identified three types of <u>learning domains</u>—that is, categories in which learning takes place **FIGURE 4-7** :

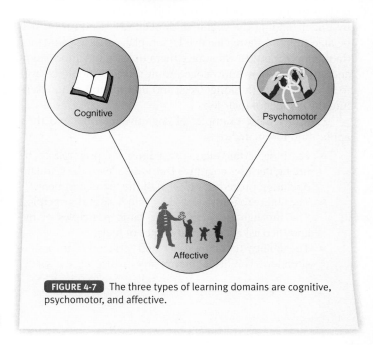

FIGURE 4-7 The three types of learning domains are cognitive, psychomotor, and affective.

- Cognitive domain: knowledge
- Psychomotor domain: physical use of knowledge
- Affective domain: attitudes, emotions, or values

Today, this classification system is widely known as Bloom's Taxonomy of the Domains of Learning. Bloom's Taxonomy has had a widespread influence in the field of learning research. Researchers have since developed new strategies of classifications and identified additional learning domains to build upon Bloom's original taxonomy.

Although Bloom identified three general learning domains, the learning process seldom takes place solely in one domain. Additionally, within each of the domains, the levels of learning build one upon the other. These levels, which have been expanded and refined by subsequent researchers, serve as the building blocks of the learning process. Within each level are found three basic sublevels: knowledge, application, and problem solving. Being aware of these levels and Bloom's three original learning domains will assist you in developing lesson objectives and course material to address a specific level of learning.

■ Cognitive Learning

The most commonly understood of the learning domains is cognitive learning, which results from instruction. Bloom's Taxonomy distinguishes six levels of cognitive learning:

- Knowledge: remembering knowledge acquired in the past
- Comprehension: understanding the meaning of the information
- Application: using the information
- Analysis: breaking the information into parts to help understand all of the information
- Synthesis: integrating the information as a whole
- Evaluation: using standards and criteria to judge the value of the information

Each of these levels builds on the previous one. The most basic is the knowledge level; the highest is the evaluation level. By understanding the cognitive domain of learning, you will be able to build more effective lesson plans and use students' thinking processes to encourage them to learn the information and apply that new knowledge. Adult learners should be able to fit the abstract thinking, facts, and information you present into the context of their jobs.

Consider these examples of cognitive learning and their application for fire fighters:

- Learning terms, facts, procedures, or principles: The fire fighter uses cognitive learning to learn the standard operating procedures (SOPs) of the fire department.
- The ability to translate and explain facts and principles: The fire fighter is taught the basic principles of fire and then asked to identify types of fuels.
- The ability to apply facts and principles to a new situation: The fire fighter is able to size up the scene,

formulate a plan based on that analysis, and evaluate all possible outcomes.

■ Psychomotor Learning

For fire fighters, psychomotor learning is the most common learning domain. The term *psychomotor* refers to the use of the brain and senses (*psycho*) to tell the body what to do, and the use of the muscles (*motor*) to tell the body how to move. The psychomotor learning domain involves the ability to physically manipulate an object or move the body to accomplish a task or perform a skill. This kind of learning is also referred to as kinesthetic learning.

As in the cognitive domain, the learning phases in the psychomotor domain build progressively one upon the other. Building on Bloom's work, Dave (1970) added the following six levels to the psychomotor domain model:

- Observation: watching the skill or activity being performed
- Imitation: copying the skill or activity in a step-by-step manner
- Manipulation: performing the skill based on instruction
- Precision: performing the skill or activity until it becomes habit
- Articulation: combining multiple skills together
- Naturalization: performing multiple skills correctly all the time

Each level of learning in Dave's psychomotor domain model builds upon the previous level: observation → imitation → manipulation → precision → articulation → naturalization. By understanding that the physical use of knowledge is the basis for students learning to perform the skill, you will be able to develop a lesson plan that involves demonstrating the skill to students, who then practice until they are able to perform that skill flawlessly. As students progress from one level to the next within the psychomotor domain, they may experiment with new ways to perform the skill. This trial-and-error process is completely natural as students learn how to adapt the skill and become more familiar with the activity.

Awareness of the following aspects of human behavior will assist you in teaching the psychomotor skill and developing an effective lesson plan:

- Gross body movement: the ability to move the arms, legs, and shoulders in a coordinated, controlled manner
- Fine motor control: the ability to use hands, fingers, hand–eye coordination, and hearing
- Verbal behaviors: the use of sound to communicate
- Nonverbal behaviors: the use of facial expressions and body gestures to communicate

In your delivery of lesson plans, ideally you will identify how a student should perform a skill or activity by carrying out a task analysis and identifying the steps involved in performing the activity. Here are some examples of psychomotor learning and its application:

- Comprehensive skill approach: The skill is demonstrated by the instructor and the fire fighter watches.
- Step-by-step approach: The fire fighter performs the skill in a precise step 1, step 2 manner until the entire skill activity is accomplished.
- Repetition of the skill or activity until it becomes habitual and is correct: The fire fighter performs the skill until it becomes second nature and he or she executes the skill flawlessly. The instructor then presents "what if" questions to help the student determine when to perform the skill.
- Combination of multiple skills: The fire fighter is able to adapt to changing situations and defend his or her choice of skills for the activity.
- Performing multiple skills correctly all the time: The fire fighter can perform the skill and it is truly a natural activity regardless of the environment or circumstance in which the skill is performed.

For students to learn skills or activities properly, they need to practice, practice, practice **FIGURE 4-8**. As the brain processes the attempts at performing the skill, it also learns to avoid what does not work or what was in error.

■ Affective Learning

Affective learning is the feeling or attitude domain. It focuses on those characteristics that make each person unique—that is, an individual's preferences, perceptions, and values. Many of these characteristics will have evolved in individual fire fighters over long periods of time, so attitudes may not change immediately after the introduction of a new concept. Instead, learning within this domain progresses from simple awareness, to acceptance, to internalization, and finally to acting out the attitude. Bloom's work with Masia and Krathwohl led to the development of the affective domain and identified the following five learning levels:

- Receiving: becoming aware of the skill or concept
- Responding: acknowledging the implications of the skill or concept and altering behavior accordingly
- Valuing: internalizing the skill or concept and having it become part of everyday life
- Organizing: comparing and contrasting skills or concepts
- Characterizing: adopting and personalizing skills or concepts

Each level of affective learning in the Bloom's Taxonomy Handbook II builds on the previous level: receiving → responding → valuing → organizing → characterizing. Because much of this learning takes place inside the mind, it is important to look for subtle changes in students to confirm that affective learning is occurring. Here are some examples of affective learning:

- Acquisition of new values: The fire fighter is willing to learn the information and attends class regularly.
- Acknowledgment of the concept: The fire fighter studies for tests and participates in class activities.
- Internalization of values: The fire fighter volunteers to participate in an extra-credit activity.
- Internalization of the organization of information: The fire fighter decides to pursue continuing education because the additional training can keep him or her abreast of new developments and technologies.
- Full adoption of the new values: The fire fighter decides to pursue a degree in fire service management.

The affective domain is difficult to measure, but it should not be overlooked. The instructor must be observant to identify a student's reaction to the learning situation and changes in his or her value system. Another way of looking at the affective domain is as the marriage of cognitive knowledge with the psychomotor skills and their application in the real world by the student.

FIGURE 4-8 For students to learn skills or activities properly, they need to practice.

Teaching Tip

Exercises that encourage students to compare two events involving similar problems demonstrate how different values may be applied to similar situations as well as how values may differ from one situation to another. For example, ask students to consider two different room and content fires: one that involves entrapment of children in their home, and one that involves entrapment of adults in a crack house. How do students perceive each of these scenarios?

VOICES
OF EXPERIENCE

When my career began as a fire fighter, there was excitement to start the fire academy. I was eager to learn new skills and face new challenges, yet nervously anxious to perform the skills and concepts I had been taught. It was evident that some of the most skilled fire fighters were also those involved in teaching the trade. Making the journey through my career I wanted to become an instructor someday.

Often the first way a student learns a task is the lasting way.

I was fortunate enough to find myself in an instructor role later in my career, taking an interest in fire behavior and the process of burning. There was an opportunity to teach fire academy students and current department members in fire behavior. With a few other instructors from the department, I completed a flashover instructor course. We also received training in the operation of our flashover simulator and burn building use.

After we completed the instructor course, we began teaching students. The course was a two-part course—part classroom instruction and part hands-on skills. Using the live fire simulator (flashover box) for the hands-on portion of the class, it was common practice to place the instructor at the front of the observation area near the fire box, and a student on a hoseline nozzle next to the instructor. Students would then rotate into the nozzle position throughout the training evolution, with the instructor giving direction to the student on the nozzle for when to apply short bursts of water to the fire box area, using the term "pencil." Having the student on the nozzle gave the opportunity for them to observe fire growth and conditions.

Department fire officers began noticing a trend in recruits at actual fire scenes. Newer fire fighters were using short bursts of water for fire control instead of applying continuous water in an aggressive interior fire attack. The short bursts would knock down the fire but not remove the heat to extinguish the fire completely. The fire academy instructors looked at the current course curriculum in an attempt to identify areas for improvement. The instructors realized the first exposure students had to live fire conditions was in the fire simulator where they were shown the penciling technique to control fire conditions. Instructors also identified an area for improvement in fire attack simulations in the burn building prop. Because it took more time to re-light a fire and build smoke conditions if the fire was completely put out, students were told to apply only a little water to knock down the fire. Often the first way a student learns a task is the lasting way.

It is easy to forget how stressful it can be for the student in their first real fire conditions. Realizing why students were demonstrating incorrect behaviors allowed the instructors to re-evaluate the teaching methods and deliver training for realistic conditions, which led to direct improvement in safety and proficiency. We should always strive to train the way we operate.

Benjamin I Andersen
Engineer
Loveland Fire Rescue Authority
Loveland, Colorado

Learning Styles

■ What Is a Learning Style?

When preparing for a class, you must ask, "What is learning?" as well as "How does a particular individual learn?" To answer these questions, you need to recognize that each student has a different learning style—that is, a way in which he or she prefers to learn. During the internal process of learning, a person encounters new information, analyzes it, and rejects or adopts it. Each and every adult learner has a distinct and consistently preferred way of perceiving, organizing, and retaining information. Learning styles are habitual modes of processing information; they show an individual's predisposition to adopt a particular learning strategy, regardless of the specific demands of the class. To use the terminology introduced by Bloom's Taxonomy, learning styles comprise characteristic cognitive, psychomotor, and affective behaviors.

The process of learning directly relates to the learning environment and interactions that are influenced by an individual's preferred learning style. These interactions are the central elements of the learning process and show a wide variation in terms of their pattern, learning style, and quality.

Additional information on how students learn and retain new information and knowledge was provided by Jean Piaget, a biologist from Switzerland who developed his learning theory based on his observations of children and how they learn. Piaget's theory (Piaget & Inhelder, 1958) states that children learn "through motor actions, develop intelligence by using their natural intuition, develop cognitively through the use of logic and finally develop the ability to think abstractly." While Piaget's theories apply to adult learners in many ways, most of all they help us see how to help a student learn by going back to basics.

Although learning is an invisible mental and emotional process, the results of the learning process are frequently visible in that they may be traced back to certain experiences (e.g., a training session) and are evident in behaviors (e.g., performance of a skill).

■ Effects of Learning Styles on Training and Educational Programs

If you teach in a manner that favors a student's less preferred learning style, you need to be aware that the student's discomfort level may interfere with learning. Therefore, it is beneficial to recognize and understand all student learning styles. This knowledge comes from knowing your students. If you are new to the organization or are serving as a guest speaker, then discovering students' preferred learning styles could present quite a challenge. Nevertheless, members of a profession or group often show a decided tendency toward certain learning styles. Part of your presentation preparation might involve contacting the agency ahead of time and gathering specific information about the audience's demographics, their learning backgrounds, and their experience levels. Of course, it is inappropriate to make assumptions about a group simply based on a single group identifier. For example, if you were assigned to teach a group of senior citizens in an assisted living center and assumed that all of them would be in wheelchairs, you would be very surprised to learn that most of them cook, play golf, and lead very active lives.

The ability to learn is not solely dependent on your catering to the student's preferred learning style. Most adult learners will adapt their learning styles and use nonpreferred learning styles if necessary. For example, a study by Mel Silberman in 1998 revealed that in every group of 30 learners, 22 are able to learn effectively as long as the instructor provides a blend of visual, auditory, and kinesthetic activity. The remaining eight learners have a much stronger preference for a certain learning style and may struggle to understand the information unless special care is taken to present the material in their preferred learning style.

To see how this works in practice, consider a group of fire fighters in a class on fire-ground support operations. Although the majority of the class wants to start placing ladders immediately and practicing ventilation, a minority prefer to watch a video and discuss the skills before actually practicing the operations **FIGURE 4-9** . Both groups want to learn, but the process is different for each. Always take everyone's learning style preferences into account.

FIGURE 4-9 Both fire fighters want to learn how to raise the ladder; they simply have different learning styles.

A learning style is a characteristic indicator of how a student learns, likes to learn, and learns most effectively. Given the importance of understanding the fire service and those who work in it, you can imagine how your appropriate use of this knowledge might affect fire fighters' training and educational opportunities. An understanding of the preferred learning styles of fire service personnel is a valuable tool for fire service instructors and curriculum designers alike. Identifying any patterns of learning that are common among fire-fighting personnel may serve to improve instructional methods and curriculum content for educational courses and training geared toward fire fighters. In short, a thorough understanding of the audience's preferred learning style makes for better curriculum developers, trainers, and fire service instructors. As in any field, the development of a well-balanced program encourages learning by optimizing the learning styles of the learners and enhancing their learning experiences.

By understanding learning styles, you will gain some basic understanding of the strengths and weaknesses of the average fire service student. This information is essential given the trend toward more continuing education for fire personnel. A myriad of new education and training programs for the fire service are being developed to meet the demands of the industry, increase skills, and elevate the level of professionalism. These courses are designed to include not only the cognitive domain for recall, but also application of new skills, problem solving, and psychomotor skills.

To accommodate this expanded scope of learning, course designers and fire service instructors need to have an even deeper understanding of their audience. By appreciating learning style preferences and determining how those learning styles form patterns based on the audience, you should be able to increase the amount of learning that actually occurs in your classroom.

Today's training and education for the fire service must change to meet the needs and wants of the new generation of fire fighters as well as veteran fire fighters. The increasing diversity of fire personnel—both in terms of backgrounds and learning styles—presents an especially thorny challenge for today's fire service instructors, who must strive to deliver increasingly complex information to a broader range of students. Traditionally, the fire service focused on the physical activities associated with extinguishing fires and providing rescue services. In the past, training was the key to developing skills. In the last 15 years, however, the scope of the fire service has expanded dramatically to include responding to terrorism, dealing with chemical and biological warfare threats, and managing new medical technology. These more sophisticated demands require similarly more sophisticated training that goes beyond hands-on activities to encompass education in explosives, chemistry, and computers—and, likewise, new methods of instruction and learning techniques.

Although much speculation has centered on the preferred learning styles of fire service personnel, little scientific research has been done to support or investigate that speculation. Three decades of studies directed toward the learning styles of non-fire service personnel has convinced hundreds of administrators and educators of the effectiveness of teaching by first identifying, and then complementing, how each person begins to concentrate on, process, internalize, and retain new and difficult information and skills. Once the learning styles of students have been identified, you can select the best teaching approach.

■ Measurement of Learning Styles

Generally speaking, the researchers who are focused on learning styles are divided into two distinct camps: those supporting a narrower view that emphasizes the cognitive learning domain, and those supporting a broader view that encompasses other learning-related factors, such as motivation and personal preferences (Biggs, 1993). In the former group, psychologist and educational theorist David Kolb suggested that individuals are likely to feel most comfortable in one of four learning modes: using either abstract or concrete perception and either active or reflective processing. His Learning Style Inventory (LSI) instrument, which is designed to determine a person's placement along these dimensions, is a model of cognitive processing—that is, how we process learning in the brain (Kolb, 1995).

In general, distinctions in three or four different learning styles are well accepted as more or less prototypes of learning styles. A variety of instruments have been developed to measure an individual's affinity for particular styles (Semeijn & van der Velden, 1999). With any of these instruments, caution needs to be taken to ensure that what is being inventoried is learning style, rather than personality type.

Some researchers have focused on the visual, auditory, and kinesthetic (VAK) characteristics of learning styles, on the assumption that we all seem to have a learning style preference based on sensory intake of information. Adult learners use all three of these sensory functions to receive information, but one or more of these receiving styles is normally dominant and hence filters the information received. The most common form of information exchange is speech, which arrives in the adult learner's ear and is considered to be an auditory means of learning. With the visual characteristic, information arrives in the form of graphs, charts, pictures, color and layouts, maps, or patterns. An adult learner using the kinesthetic characteristic would include the senses of touch, hearing, smell, taste, and sight; such learners want concrete, multisensory experiences as they learn. The VAK learning style research focuses on the *how* of learning; it does not concern itself with the *why* of learning styles.

Even though the eyes take in all visual information, information is perceived differently owing to particular aspects of that information, which in turn leads to subtle differences in students' learning styles based on the visual component of the VAK model (Fleming & Mills, 1992). For example, sometimes the information is largely composed of printed words, from which some students appear to get a greater or lesser degree of understanding; at other times the information consists of mostly graphic elements. Neil Fleming (1995), a teacher in New Zealand, suggested a modification to the basic VAK scheme in which the visual characteristic is divided into iconic (symbolic) and textual characteristics. Iconic characteristic visualizers learn more effectively when they are exposed to diagrams, symbols, and other graphic matter. Textual characteristic visualizers prefer the printed word, which includes both reading and writing.

The VARK Preferences instrument is a learning styles inventory designed to help students identify how they prefer to learn in terms of the previously described visual, auditory, read/write, and kinesthetic characteristics (Fleming, 2001). Developed in 1997 by Fleming at Lincoln University in New Zealand, VARK contains 13 questions whose answers provide students with an indication of their personal learning preferences.

The learning preferences identified via the VARK Preferences questionnaire indicate the ways the adult learner wants to take in or give out information in a learning context. The four categories of learning preferences seem to reflect the experiences of the students when they are taking in or giving out information fairly accurately. The categories overlap to some extent, but Fleming has defined the following dimensions:

- **Visual (V).** This perceptual mode emphasizes a preference for the depiction of information in charts, graphs, flow charts, and all of the symbolic arrows, circles, hierarchies, and other devices that instructors use to represent information that could have been presented in words. This definition does not include the use of television, videos, films, or computers. (Most of those media are considered primarily aural and kinesthetic because of their presentation of sound and reality.)
- **Aural (A).** This perceptual mode describes a preference for information that is heard. Students with this preference report that they learn best from lectures, tutorials, tapes, and talking to other students.
- **Read/write (R).** This preference is for information displayed as words. Many academics have a strong preference for this modality.
- **Kinesthetic (K).** This modality includes a preference for experience and practice with the information, such as hands-on training or use of a simulator. Although such an approach may invoke other modalities, the key is that the student is connected to reality through experience, example, practice, or simulation.

A fifth category (multimodal) was added to the VARK Preferences questionnaire when it was found that the majority of adult learners actually have multiple preferences for learning styles. Some multimodal students may need to process information in more than one mode to learn effectively (Fleming, 1992).

VARK is structured specifically to have practical implications—namely, to improve learning and teaching. The VARK Preferences inventory, which is available for free on the Web, has been widely accepted as both practical and thought provoking (Fleming, 1992). Its results can be used in both course design and classroom activities, as students often find their scores highly provocative. In particular, you may be able to design lesson plans and present information better if you have an idea of the learning styles of your students. It is also beneficial to take the VARK questionnaire yourself, because your teaching style often matches your preferred learning style—which may not match the preferred learning styles of your students. It is important to focus on students' learning styles, employ a particular method of instruction, and not neglect the needs of adult learners who are different from most of their classmates or who require an alternative method of instruction.

Teaching Tip

The VARK questionnaire can be downloaded from the Internet (www.vark-learn.com), filled out by students, and graded by both the students and you. The results could be used as a point of discussion to learn about preferred learning styles. Completion and scoring of this instrument is a self-defining and directed activity.

Learning Disabilities

As a fire service instructor, you may encounter students with a wide variety of learning disabilities. The Americans with Disabilities Act (ADA—federal legislation that was originally passed in 1990) classifies learning disabilities into these major categories:

- Reading disabilities range from the inability to understand the meaning of words, to the inability to read or comprehend, to dyslexia. This disability could become apparent in class if a student is asked to read a section of course material aloud or to summarize a paragraph from a textbook.
- Some students lack the ability to write, to spell, or to place words together to complete a sentence. This type of disability, known as dysphasia, might present itself when a student submits work for a course and the work is poorly written.
- Other students might suffer from dyscalculia, or difficulty with math and related subjects. Students with this disability might struggle on a hydraulics course or test.
- Dyspraxia is the inability to display physical coordination of motor skills. This condition could be observed on the training ground as a student's inability to complete a task such as climbing a ladder.
- A disorder that affects children but can be carried into adulthood is known as attention-deficit/hyperactivity disorder (ADHD). A person with ADHD has a chronic level of inattention and an impulsive hyperactivity that affects his or her ability to function on a daily basis. This disorder is a documented condition identified by the Centers for Disease Control and Prevention (CDC). Although it can be treated, there is no known cure for ADHD.
- Other types of disabilities that could affect the learning process include color blindness and poor vision, even if it is just difficulty in seeing the screen in a classroom.
- Some adult learners may have poor hearing and cannot properly hear material presented in the classroom. Hearing loss is a common problem that occurs with aging, so instructors may often teach older fire fighters who are dealing with being unable to hear.

- English as a second language (ESL) is a challenge that some organizations are facing that can affect the learning environment and create a challenge to the instructor.

Knowing and understanding each of these learning disabilities will help you teach. Unfortunately, in most cases you will not know that students have these problems until a class has already begun. If you are a guest lecturer, you will not get the opportunity to know your students until after the first few hours of instruction. Once you have identified students with some type of learning disability, then you need to adjust your teaching as necessary. There are many ways to help students in your class who are struggling to learn. For example, a few positive words in private or suggestions on a written assignment might help a struggling student understand a concept that is just beyond his or her reach. Students who have learning disabilities are not unintelligent or unable to learn; they simply learn in different ways. They usually have average to high intelligence, but may perform poorly on tests because of their disabilities if accommodations are not made for their special needs.

A student who is a hesitator may be suffering from a learning disability. Such a student is often shy, reluctant, at a loss for words, and quiet. Although he or she may know the material and have much to offer, the individual's shyness, fear, or lack of confidence may keep him or her from participating in class activities or answering questions. You can engage this student by asking nonthreatening questions and offering encouragement. You need to let a hesitator know that his or her contributions are worthwhile and important. Every student's participation in the classroom is important, so provide reassurance to this type of student as necessary.

Making accommodations for students with learning disabilities is largely a fact-based situation. Under the ADA, reasonable accommodations must be made for students with disabilities; however, accommodations are not necessary if they create an undue hardship to the employer or agency or pose a direct threat to the safety and security of others. Information and general guidance for determining when an accommodation causes an undue hardship or creates a direct threat can be found in the *Legal Issues* chapter.

■ Instructing Students with Disabilities

Instructors can employ a variety of strategies to assist students with learning disabilities. For example, breaking the material into smaller steps or components will allow a student to comprehend the material more easily. Supplying regular, positive feedback in a manner that does not single out the individual is very effective in encouraging a student with a disability to learn. Use of diagrams, graphics, and pictures to support a lecture is effective in tying learning material together.

A new teaching technique that has been developed over the past few years is called scaffolding. Scaffolding is intended to assist students with learning disabilities, but could be used with any student. Think of metal scaffolding that surrounds a building as it is constructed, serving as a support system. As the structure takes shape and is able to stand on its own, the scaffolding is removed until we see the final building.

When working with a student who has a learning disability, the instructor who is using scaffolding establishes a support system when presenting the new material. The instructor gives the student the content, motivation, and foundation of the new material. In building and establishing the scaffolding, the instructor can use some of the following strategies:

- Activate the student's prior knowledge with questions or ask for others' experiences.
- Offer the student a situation, involving the subject being taught, that gives the student a chance to think and give a response from his or her own experiences.
- Break complex course material into easier pieces that will allow a student learning "successes" in understanding the new material.
- Show a student an example of the desired outcome before he or she completes the task.
- Allow a student to "brainstorm" or think out loud while seeking a solution to questions or problems.
- Teach the student mnemonic techniques to help in memorizing material or procedures.
- Teach key vocabulary terms before reading to help improve what is learned while reading.

When using this technique, you offer students support to learn the new material and then slowly or strategically remove the scaffolding to allow the students to stand on their own. For learning to progress, scaffolds should be removed gradually as instruction continues, so that students will eventually be able to demonstrate comprehension independently.

Disruptive Students

Other types of students may pose challenges as well. For example, some students, for one reason or another, feel the need to establish their presence in the class by acting out in other ways (also discussed in the *Methods of Instruction* chapter).

- The class clown feels the need to make comments or jokes about the course material or perhaps others in the classroom.
- The class know-it-all has been there and done that already, regardless of the topic.
- The gifted learner is very familiar with the course material or has the ability to read quickly. This individual becomes bored because he or she is so far ahead of the rest of the class.

Most often, these types of students have other underlying issues that cause them to seek out attention in a negative way. A variety of methods may be used to deal with these types of students, but in general the most effective method is to minimize their impact on the class by not giving into the same type of behavior. Most often a stern look is enough to get a class clown to stop cracking jokes **FIGURE 4-10**. If this measure fails to quell the disruptive behavior, then engage the student with a question that is meant to bring him or her into the class discussion. If this technique does not work, then at the next break a one-on-one discussion might be appropriate. The last step in dealing with a student who continues to disrupt

FIGURE 4-10 Occasionally, an instructor will have to deal with inappropriate behavior. How might you deal with this situation?

the learning environment would be to ask him or her to leave the class. In using this form of progressive discipline, the goal is to bring the student back into the classroom in a positive way that does not embarrass either the student or you as the instructor.

Ethics Tip

Is it ethical to explain an unfamiliar term to a student during an exam? For example, suppose a student is asked to "select a type of ritual" and one of the answers is "tenure." If the student does not know the term "tenure," is it ethical to explain it?

This dilemma occurs because both instructor and student are trying to achieve a positive result. Not allowing any help during an exam achieves the goal of a level playing field for all students in the class. At the same time, the goal of learning is to see what the student knows. If the point of the exam is to see if the student knows what a ritual is, and not what tenure is, explaining the term *tenure* would allow for the exam to determine whether the student understands the term *ritual*.

Ethical dilemmas have no easy answer. Using sound judgment and reasoning is required to determine the appropriate action.

The student who is bored or quiet may be displaying that behavior because of a particular circumstance. Sources of this problem may include a lack of interest in the subject, an objection to being forced to take the class, unfamiliarity with the terminology in the class, boredom, and long and technical lectures. This student may just drift off mentally and refuse to participate in the classroom or become disruptive in other ways (see the *Methods of Instruction* chapter). As the fire

service instructor, you need to keep all students interested and focused on the material, so make your lectures interactive and concise.

Other students may not be interested in the topic. They may be in the class simply because they have to be. These students lack energy and attention, and you should try to determine whether this is a regular problem. If so, these students may have a disability or other problem that you may need to discover and address through counseling, tutoring, or other means.

A student who is a slow learner may be the most difficult for you to deal with. Such an individual has trouble keeping up and may not understand some or all of the material. One technique for handling this situation is to encourage input from other students—hearing the information differently may help. During a break, have a respectful one-on-one talk with the student to ensure understanding and comprehension.

Many students have some kind of impediment to learning, and most of those students have found unique ways to compensate for these obstacles. We all have times when our attention drifts or our energy is low, but if this failure to engage in the learning process is a recurring event it needs to be addressed because it will interfere with long-term learning. Some students may need help in the form of tutoring or individualized instruction, for example. These options should be discussed with students who have learning disabilities, and together you and the student should design a learning plan.

Theory Into Practice

To identify which kinds of learners—visual, auditory, or kinesthetic—are present in your classroom, try the following exercise:

Tell students to sit quietly for five minutes—no reading, no talking, and no moving around. Observe what the students are doing. You may want to jot down some notes about each student so you can share your observations with them at the end of the five minutes.

- The visual learner will look around and absorb or read everything he or she sees.
- The auditory learner will mumble or make some kind of sound.
- The kinesthetic learner will fidget, may get out something to doodle on, or slump comfortably in his or her seat.

This may not be the most accurate or fun activity, but your observations will provide some valuable information about your students. You can provide feedback to the students about their behavior, and over time you should be able to see how accurate your observations are.

Training BULLETIN

Jones & Bartlett Fire District

Training Division
5 Wall Street, Burlington, MA, 01803
Phone 978-443-5000 Fax 978-443-8000
www.fire.jbpub.com

Instant Applications: The Learning Process

Drill Assignment

Apply the chapter content to your department's operation, training division, and your personal experiences to complete the following questions and activities.

Objective

Upon completion of the instant applications, fire service instructor students will exhibit decision making and application of job performance requirements of the fire service instructor using the text, class discussion, and their own personal experiences.

Suggested Drill Applications

1. Review the laws and principles of learning, and identify an instructional method that incorporates each of these laws and principles.

2. Make a list of the students you will have in upcoming classes. Based on their knowledge, experience, and other instructional factors, identify the ways each of these students might learn and determine how you can make adaptations appropriate for the entire class.

3. Review how you would try to increase your students' understanding of basic material by adjusting the lesson plan to meet individual needs.

4. Review the Incident Report in this chapter and be prepared to discuss your analysis of the incident from a training perspective and as an instructor who wishes to use the report as a training tool.

Incident Report

Baltimore, Maryland—2007

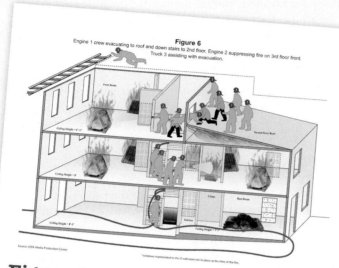

Figure 4-A Baltimore floor plan. Note fire locations.

The Baltimore City Fire Department was conducting a live fire training exercise in a row house following the department's Fire Fighter I program. The program had not been successfully completed by all participants, and some of the recruits did not successfully complete the physical performance requirements during training. Some of the recruits had also never participated in interior live fire training evolutions.

The involved building was a three-story row house, and was one of a series of very narrow row houses, only 11 ft 4 in. (3.5 m) wide. It was a 1200 ft² (111.5 m²) unit, built as a single family home. It had been vacant for almost seven years, and condemned for three years. Each unit had separate exterior walls and the side walls touched on one or two sides of the adjacent unit(s). The units were trapezoidal in shape, and this was an end unit **(Figure 4-A)**.

The live fire exercise consisted of multiple fires on all floors. The investigation indicates that approximately 11 bales of excelsior and at least 10 wooden pallets were used for the fires. Pallets were propped against the walls in several locations, with excelsior underneath. Excelsior was also placed in openings in the ceilings and in walls. In the backroom on the first floor, an automotive tire, two full-size mattresses, one twin-size mattress, one foam rubber chair, tree branches, and other debris were piled up and burned. The exact number of fires set could not be definitely determined, but there were at least nine separate fires.

It appears that the ignition of the fires was not coordinated with the preparation of students entering. Several students were not ready when the fires were ignited, and that delay allowed the fires to burn unimpeded while crews made final preparations to enter.

Five separate crews were expected to operate simultaneously as engine and truck companies. A sixth crew was designated as the rapid intervention crew (RIC), but they were not briefed and did not have a hose line.

Continued...

Incident Report Continued...

Baltimore, Maryland—2007

An adjunct instructor and four students of the first crew, designated as Engine 1, entered through the front door. All of the students had personal alert safety system (PASS) devices, but the adjunct instructor did not, nor did he have a portable radio. Per the direction of the instructor-in-charge, Engine 1 proceeded to the third floor and expected a second hose line to follow. They advanced an initially uncharged 1¾-in. (44-mm) hose line to the centrally located stairs on the first floor. There is disagreement as to when the line was actually charged.

Another crew of recruits was to advance a hose line to the second floor from the rear door. When they entered the rear of the house, they encountered the large pile of debris burning, which inhibited their ingress and egress. The fire was spreading across the ceiling, and they extinguished the fire before making a delayed advance to the second floor. As with the first crew, the four students had PASS devices, but the adjunct instructor did not, nor did he have a portable radio.

The Engine 1 crew advanced to the second floor and encountered heavy fire conditions. Although directed to go to the third floor, the instructor determined it was necessary to knock down the fire before proceeding upstairs to the third floor. Two of the crew's members stayed on the landing between the second and third floors to pull hose, as the remainder of the crew advanced upstairs. Conditions on the third floor became too hot, and the instructor did not have a radio to advise command of their situation or of the interior conditions. The instructor lifted himself up and climbed out of a high window onto the back roof of the second floor. The first student was able to lift her upper body out of the window. The instructor then grabbed her SCBA straps to pull her out the rest of the way onto the second floor roof. She was then hoisted to the third floor roof by Truck 3, where she told them that her crew needed help.

The other student was now at the window as the instructor attempted to pull her through, using her SCBA harness as he had done with the other fire fighter. She was initially able to talk to the instructor to tell him that she could not help, and that she was burning up. It appears that she still had her SCBA face mask on at this time. The instructor lost his grip, and the student fell back into the room, landing on her feet. When he was able to get hold of her again, she was still conscious but her mask was partially dislodged or removed. Her face had visibly started to blister. The instructor did not have a radio, but yelled for help. He lost his grip on her a second time, and the student fell back into the room. Shortly after regaining his grip on her for the third time, she became unresponsive. At this point, the crew jumped about 6 ft down from the third floor roof to help pull the student out, but they were unable to do so. With additional assistance, they were eventually able to help lift her onto the roof, where a student used a portable radio to advise of the fire fighter down.

Assistance was requested from the on-duty battalion, engine, and truck units that had come by to watch. The engine company engaged the fire, while the truck company entered the structure to ensure that the students were all out.

Two members of the on-duty truck company climbed the aerial ladder to the roof; they assisted in placing the victim in a Stokes basket and carried her down the ladder. An advanced life support ambulance initiated advanced life support and transported her to a shock trauma center, where her care continued until she was pronounced dead.

Post-Incident Analysis: Baltimore, Maryland

NFPA 1403 (2007 Edition) Noncompliant

Note: Near-total lack of compliance with NFPA 1403. 50 issues considered to be violations of NFPA 1403 by investigative team.

Flashover and fire spread unexpected (4.3.7)

Walk-through was not performed (4.2.25.4)

Multiple fires on different floors (4.4.15)

Post-Incident Analysis: Baltimore, Maryland

NFPA 1403 (2007 Edition) Noncompliant

Excessive fuel loading (*4.3.5*)

Fire was beyond the training and experience of the students to participate in live burn exercises in an acquired structure (*4.1.1*)

No incident safety officer (*4.4.1*)

Noncompliant gear on victim (*4.4.18.2*)

Instructors without PASS* (*4.4.18.5*)

Instructors not equipped with portable radios (*4.4.9*)

Adjunct instructors had little to no prior instructional experience (*4.5.1*)

NFPA 1403 (2007 Edition) Compliant

22 students, 11 instructors (*4.5.2*)

EMS (ALS transport) on standby on scene (*4.4.11*)

Other Contributing Factors

Students not informed of emergency plans/procedures

Rapid intervention crew staffed by students and unequipped or prepared

Wrap-Up

Chief Concepts

- As a fire service instructor, you are a leader and have a responsibility to teach information, to hone skills, and to promote positive values and motivation in your students.
- According to Edward L. Thorndike, there are six laws of learning: readiness, exercise, effect, primacy, recency, and intensity.
- Of the five senses, seeing (e.g., a visual demonstration of a skill) is generally the most effective means of learning, followed by hearing (e.g., listening to a lecture), smelling (e.g., smelling smoke), touching (e.g., handling fire equipment), and tasting (rarely used in training fire fighters).
- Two schools of thought have developed to explain how behavior evolves: behaviorist and cognitive.
  - According to the behaviorist perspective, learning is a relatively permanent change in behavior that arises from experience.
  - According to the cognitive perspective, learning is an intellectual process in which experience contributes to relatively permanent changes in the way individuals mentally represent their environment.
- Learning that is intended to create or improve professional behavior is based on performance, rather than on content. Such competency-based learning is usually tied to skills or hands-on training.
- Mandating learning often results in a lack of motivation in classroom activities and a poor attitude toward the learning process in general, which present quite a challenge to the fire service instructor who is tasked with teaching the course.

- To assist your students in becoming effective learners, you can share study skills that enhance what they learn during classroom and personal study time. Study tips can be viewed in three categories: classroom, personal study, and test preparation.
- Abraham Maslow determined that certain needs should be satisfied to enable a person to move forward toward self-actualization:
  - Physiological needs
  - Safety, security, and order
  - Social needs and affection
  - Esteem and status
  - Self-actualization
- Benjamin Bloom and his research team identified three types of learning domains:
  - Cognitive domain (knowledge)
  - Psychomotor domain (physical use of knowledge)
  - Affective domain (attitudes, emotions, or values)
- Each and every adult learner has a distinct and consistently preferred way of perceiving, organizing, and retaining information.
- The ability to learn is not solely dependent on use of the student's preferred learning style. Most adult learners will adapt their learning styles and use nonpreferred learning styles if necessary.
- Researchers focused on learning styles are divided into two distinct camps: those emphasizing the cognitive learning domain and those highlighting other learning-related factors, such as motivation and personal preferences.
- Once you have identified students with some type of learning disability, then you need to adjust your teaching as necessary to ensure their learning. There are many ways to help students in your class who are struggling to learn.
- Most often, disruptive students have other underlying issues that cause them to seek out attention in a negative way.

Hot Terms

Affective domain The domain of learning that affects attitudes, emotions, or values. It may be associated with a student's perspective or belief being changed as a result of training in this domain.

Attention-deficit/hyperactivity disorder (ADHD) A disorder in which a person has a chronic level of inattention and an impulsive hyperactivity that affects daily functions.

Behaviorist perspective The theory that learning is a relatively permanent change in behavior that arises from experience.

Bloom's Taxonomy A classification of the different objectives and skills that educators set for students (learning objectives).

Cognitive domain The domain of learning that effects a change in knowledge. It is most often associated with learning new information.

Cognitive perspective An intellectual process by which experience contributes to relatively permanent changes (learning). It may be associated by learning by experience.

Competency-based learning Learning that is intended to create or improve professional competencies.

Dyscalculia A learning disability in which students have difficulty with math and related subjects.

Dyslexia A learning disability in which students have difficulty reading due to an inability to interpret spatial relationships and integrate visual information.

Dysphasia A learning disability in which students lack the ability to write, spell, or place words together to complete a sentence.

Dyspraxia Lack of physical coordination with motor skills.

Kinesthetic learning Learning that is based on doing or experiencing the information that is being taught.

Learning A relatively permanent change in behavior potential that is traceable to experience and practice.

Learning domains Categories that describe how learning takes place—specifically, the cognitive, psychomotor, and affective domains.

Learning style The way in which an individual prefers to learn.

Psychomotor domain The domain of learning that requires the physical use of knowledge. It represents the ability to physically manipulate an object or move the body to accomplish a task or use a skill. This domain is most often associated with hands-on training or drills.

VARK Preferences A tool that measures a person's learning preferences along visual, aural, read/write, and kinesthetic sensory modalities.

Visual, auditory, and kinesthetic (VAK) characteristics Learning styles based on the idea that we all have a learning style preference based on sensory intake of information (visual, auditory, and kinesthetic).

References

Biggs, J. (1993). What do inventories of student's learning processes really measure? A theoretical review and clarification. *British Journal of Educational Psychology, 63,* 3–19.

Bloom, B. S. (1984). *Taxonomy of educational objectives.* Boston, MA: Allyn and Bacon/Pearson Education.

Butler, G., & McManus, F. (1998). *Psychology.* Oxford, UK: Oxford University Press.

Connick, G. (1997, Fall Winter). Beyond a place called school. *NLII Viewpoint.*

Dave, R. H. (1970). *Developing and writing behavioral objectives.* Tucson, AZ: Educational Innovators Press

Felder, R. (1996). Matters of style. *ASEE Prism, 6*(4), 18–23. www2ncsu.edu/unity/lockers/users/f/felder/LS-Prism.htm

Fleming, N. (1995). I'm different: Not dumb. Modes of presentation (V.A.R.K.) in the tertiary classroom. *Research and Development in Higher Education, Proceedings of the 1995 Annual Conference of the Higher Education and Research Development Society of Australasia (HERDSA), 18,* 308–313.

Fleming, N. (2001). A guide to learning styles: V.A.R.K. www.vark-learn.com

Fleming, N. (2005). A guide to learning styles: V.A.R.K. www.vark-learn.com/english/page.asp?p=whatsnew

Fleming, N. D., & Mills, C. (1992). Not another inventory, rather a catalyst for reflection. *To Improve the Academy, 11*, 137–155.

Goodman, P. (1962). *The community of scholars*. New York: Random House.

Hoetmer, G. (2000). *Fire services today: Managing a changing role and mission*. Washington, DC: International City/County Management Associations.

International Fire Service Training Association. (1994) *Fire service instructor* (5th ed.). Stillwater, OK: Fire Protection Publications.

Kolb, D. (1995). *Experiential learning: Experience as the source of learning and development*. Englewood Cliffs, NJ: Prentice-Hall.

LeFever, M. (1995). *Learning styles: Reaching everyone God gave you to teach*. Colorado Springs, CO: David C. Cook Publishing.

Lefrancois, G. (1996). *The lifespan* (5th ed.). Belmont, CA: Wadsworth.

National Fire Protection Association. (2012). *NFPA 1041: Standard for Fire Service Instructor Professional Qualifications*. Quincy, MA: National Fire Protection Association.

National Fire Protection Association. (2007). *NFPA 1403: Standard on Live Fire Training Evolutions*. Quincy, MA: National Fire Protection Association.

Piaget, J., & Inhelder, B. (1958). *The growth of logical thinking from childhood to adolescence*. London: Routledge.

Rathus, S. (1999). *Psychology in the new millennium* (7th ed.). Fort Worth, TX: Harcourt Brace College.

Semeijn, J., & van der Velden, R. (1999). *Aspects of learning style and labor market entry: An explorative study*. Maastricht: Maastricht University.

Silberman, M. (1998). *Active training* (2nd ed.). San Francisco, CA: Jossey-Bass/Pfeiffer.

Stewart, D. (2003). Computer and technology skills. http://www.ncwiseowl.org/Kscope/techknowpark/Kiosk/index.html

Thorndike, E. (1932). *The fundamentals of learning*. New York: Teachers College Press.

As the fire service instructor, you have been assigned to the training center. For the next several weeks, you will be teaching a skills recertification class. As you begin to plan your instruction materials, you think about the diverse group of people who will be in the classroom and wonder how the information you will be presenting can best be shared with them.

1. Which of the following questions would help you prepare for this class?
   A. On which day of the week will the class be given?
   B. Which shifts are the students from?
   C. Which type of learning or relearning will be going on?
   D. What time will lunch be?

2. How should the information be presented to the students?
   A. You should present the information in the way that you prefer to learn.
   B. You should pick one way to present the information and stick with it throughout the course.
   C. It does not matter how the information is presented.
   D. You should provide critical thinking opportunities that appeal to a variety of student learning styles and preferences.

3. What is the purpose of evaluating students' skills performance?
   A. To provide students with feedback that will help them master the skill
   B. To give students the opportunity to get a passing grade
   C. To provide an exercise to fill up the allotted class time
   D. To give students the opportunity to show how good they are

4. Which of the following outcomes demonstrates that learning occurred?
   A. The student passes a written exam.
   B. The student passes a practical exam.
   C. The student correctly answers questions that he or she could not answer prior to the class.
   D. The student answers all of the answers on the exam correctly.

5. If you give the VARK Preferences questionnaire at the beginning of the class, you will be able to determine
   A. Whether learning has occurred.
   B. What the preferred learning styles of the students are.
   C. How effective your teaching style will be.
   D. How effective your teaching style was for the students.

6. According to Bloom, changes in attitudes would occur in the _____ learning domain.
   A. affective
   B. psychomotor
   C. cognitive
   D. behavioral

7. According to the behaviorist, perspective changes in behavior occurs due to:
   A. teaching.
   B. intellectual process.
   C. experience.
   D. conscious thought.

Communication Skills

Fire Service Instructor I

Knowledge Objectives

After studying this chapter, you will be able to:

- Identify and describe the elements of the communication process. (NFPA 4.4.3) (pp 105–106)
- Describe the role of communication in the learning process. (pp 105–107, 109–110, 114–116)
- Compare and describe the different types and styles of communication. (pp 106–107, 109–110, 112–114)

Skills Objectives

After studying this chapter, you will be able to:

- Demonstrate effective oral communication techniques. (NFPA 4.4.3) (pp 107, 109)
- Demonstrate effective written communication techniques. (NFPA 4.4.3) (pp 110, 112–114)
- Demonstrate the ability to use various communication styles in the classroom. (pp 105–107, 109–110, 114–116)

Fire Service Instructor II

Knowledge Objectives

There are no knowledge objectives for Fire Service Instructor II students.

Skills Objectives

There are no skills objectives for Fire Service Instructor II students.

Courtesy of Joshua Bauer, Immokalee Fire Department

Fire Service Instructor III

Knowledge Objectives

There are no knowledge objectives for Fire Service Instructor III students.

Skills Objectives

There are no skills objectives for Fire Service Instructor III students.

You Are the Fire Service Instructor

You are about to begin a class for your department's weekly training session that will conclude with a written examination covering some of the many terms and definitions in the lesson plan. As you review the content of the class, you are aware of the different levels of experience and education of your audience and concerned that some of the material will be difficult for some members to apply, while others could become bored with the material if you attempt to lower the performance level for those who might have difficulty. The class content is important to the audience members' safety during an incident, and you want to do a good job in presenting the material.

Your ability to speak and present the material will be put to the test, and you will need to be able to pick up on cues from the audience to gauge how well the lesson plan is being understood. You will need to use differing levels of pitch and volume and attempt to engage the students throughout the lesson plan delivery.

1. What are the communication barriers that you must overcome?
2. How will you communicate effectively with a diverse group?
3. How will your ability to communicate enhance the students' learning?

Introduction

Communication is the most important element of the learning process. Communication in and of itself is a process that you must master to become an effective fire service instructor. When you have mastered communication skills, the environment in which you are teaching becomes less of a factor in the learning process. Even if the environment is less than perfect, strong communication skills will enable you to be an effective and dynamic fire service instructor. Your goal is to make sure that your students leave the classroom with a greater knowledge base than they had when they came in.

According to professional educator Leah Davies (2001):

Being able to communicate is vital to being an effective educator. Communication not only conveys information, but it encourages effort, modifies attitudes, and stimulates thinking. Without it, stereotypes develop, messages become distorted and learning is stifled.

As a fire service instructor, it is important for you to be an effective communicator. One of the keys to the learning process is the conveyance of information. When you demonstrate effective communication skills, you can take the learning process to even greater heights by encouraging effort, modifying attitudes, and stimulating thinking.

This chapter addresses both the spoken and written communication processes as well as the environment's effect on communication. As mentioned in *The Learning Process* chapter, students have different learning styles and use different cues when learning. Knowing how to access these diverse cues and take advantage of them to enhance the learning experience will make you a more effective fire service instructor.

The fire service classroom encompasses so much more than four walls, tables and chairs, and a dry erase board. It can also include the back of an engine after the call, a burn building, and even a kitchen table at the station surrounded by members who have just completed an emergency response. No matter what the environment, the end objective is the same: that the student learns and that the student's safety and ability to do his or her job improve. The fire service instructor must have excellent communication skills and must be able to read the students' comprehension of the training through verbal and nonverbal cues.

Teaching Tip

Although the fire service is certainly steeped in tradition, the means of communication used by fire fighters have changed dramatically over the years. The first fire officers used speaking trumpets to communicate to the members of their department at a fire scene. Today we have more modern solutions for the transmission of messages, but the message itself remains the critical piece of the communication puzzle.

Fire Service Instructor I

The Basic Communication Process

The basic communication process consists of five elements: the sender, the message, the medium, the receiver, and feedback **FIGURE 5-1**. Surrounding these elements is the environment. The five elements of the communication process act like the links of a chain. For the chain to function properly, all of the links must be attached to one another. If any one of the links breaks, then the chain falls apart.

■ Sender

In the communications chain of the fire service classroom, the sender is you, the fire service instructor. As the sender, you must know which style of communication to use. The choice of the style of communication is based on many factors, including your comfort level. If you are more comfortable using oral (verbal) communication, then you are likely to use that style most often. Many other methods of instruction are discussed in the *Methods of Instruction* chapter that center on the sender and how he or she delivers training material.

■ Message

The next link in the communication chain is the message. What, as the sender, are you trying to communicate to your students? This may sound simple, but it is actually the most complex part of the communication process. For students to understand your message, they must hear it, put it into their own terms, and then ensure that their terms match your terms. Consider the command, "Put out that fire now." A student might hear this message as "Get a hose and spray some water right now," when the message really was "Get a fire extinguisher and suppress those sparks."

■ Medium

How the message is conveyed is as important as the message itself. The means of conveying a message constitutes the medium. For example, your tone of voice, volume, speed of voice, and other qualities all convey information. If you give a lecture about a new piece of personal protective equipment (PPE) while sitting behind a desk, fiddling with a paper clip, and speaking in a very relaxed manner, you will convey to the

students the message, "This new piece of PPE really isn't that important," even though the message you are stating verbally might be "This piece of equipment will save your life."

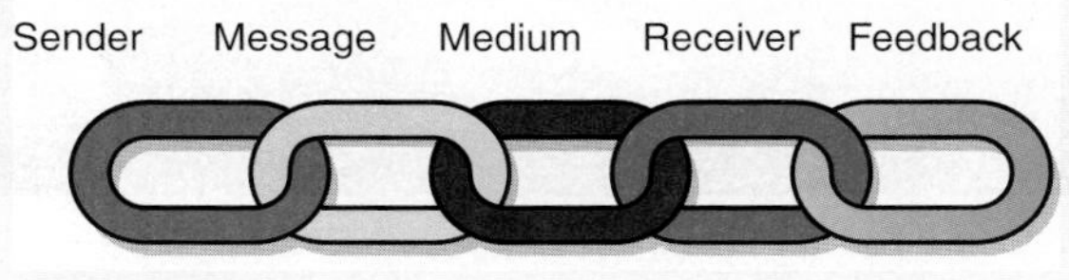

■ Receiver

The next link in the communications chain is the receiver. In the learning environment, this link is the student. The receiver plays an important, active role in the communication chain; that is, the student needs to listen actively and maintain focus on your message. At the start of each training session, remind students of their role in the communications chain and continue to keep them engaged throughout the learning process. Fulfilling this role means leaving all distractions at the door and focusing solely on the information you present. Ensure that all students understand their active role in the communications process.

> **Teaching Tip**
>
> A student's previous experience and knowledge base play a major role in determining whether the student is able to receive your message. Adapt your style and presentation for your students. If you "talk down" to students or if the presentation goes over their heads, the message is not in a form that they can receive.

■ Feedback

Feedback is the link that completes the communication process. It is often considered the most important step in the communication process because it allows the sender to determine whether the receiver understood the message. Feedback can take the form of either verbal or nonverbal response. An example of the verbal feedback process is when orders are given over the radio. Consider this example:

> **Incident Commander:** Engine 2, take a line to Division C and provide exposure protection.
> **Engine 2:** Engine 2 copies—a line to Division C for exposures.

An example of nonverbal feedback would be blank stares from students, indicating that they do not understand the message. Collectively asking a classroom of students if they understand is not the best method of determining whether the message was properly received. The natural response is "yes," because no student wants to admit that he or she does not understand. Asking a student to describe your message in his or her own words is a much better way to make sure that the student understands your message, and it will also reinforce the learning process for the student. If the student's description is incorrect, gently explain that it's not quite right and ask for other students to offer the information.

Sender   Message   Medium   Receiver   Feedback

FIGURE 5-1 The five elements of the communication process act like the links of a chain.

Face-to-face conversation is the most effective means of conveying many types of information. With this approach, the sender and the receiver can engage in a two-way exchange of information that incorporates body language, facial expressions, tone of voice, and inflection. In contrast, when verbal communication must be conducted over radio or telephone, supplementary expressions are sacrificed. When discussing communication in the fire service, particularly within the incident command system, face-to-face communication is considered the preferred method. A face-to-face exchange is required if there is a transfer of command so the sender and the receiver can pick up on any nonverbal indicators of confusion.

Another type of feedback may be apparent when you are giving feedback on a research paper. Writing notes in the margins of the paper allows the student to review and process your comments carefully, as opposed to just saying, "Nice job," and handing the student the paper.

■ The Environment

Surrounding all of the links of the communication chain is the environment. The environment—which includes physical, social, and environmental factors—can dramatically affect the communication process.

The physical environment is the room or area in which communication is taking place. It could be a classroom, the training ground, or even cyberspace. The physical environment can either inhibit or enhance the learning process. Teaching in a classroom with radios blaring and phones ringing directly outside, for example, will inhibit the communication process. When students cannot hear or see you, your message has a limited chance of being understood. Whether you are working in a classroom or on a busy training ground, make sure that all of your students can see and hear you.

The social environment is the context in which the communication takes place. If the class is not enthusiastic about the message, the message will have a limited chance of being received. Mandatory training usually has this effect. You must make sure that your students want or need to receive your message—this is the art of instruction. Show your students *why* they need your message. Instructors must also expect the students to ask *why* and must not react by getting defensive. This challenges the instructor to find additional methods of explaining the material. Although students may ask *why* for different reasons, keep in mind that they have been engaged in what you were saying and participating in the preparation step of the learning process.

The final component is the environmental factor. The receiver must first feel that his or her needs are being met. An environment that is physically difficult for its inhabitants to

tolerate distracts students, for example. If the room is too cold or too hot, then your message will not be received properly. See the chapter titled *The Learning Environment* for more information.

Nonverbal Communication

According to Charles R. Swindoll, an American writer and clergyman, "Life is 10 percent what happens to you and 90 percent how you react to it." The same can be said about nonverbal communication: Communication is 10 percent what you say and 90 percent how you say it.

Nonverbal communication is a difficult communication skill to control because you generally do not watch yourself in the mirror all day. This type of communication includes the tone of your voice, your eye movement, your posture, your hand gestures, and your facial expressions. For example, eye contact can show sincerity. If you can look a student in the eye, you are demonstrating trustworthiness and caring. If you cannot maintain eye contact, you lessen the feeling of trust. You can also use your eyes to elicit a reaction: If you continue to look at a student, then that student is compelled to focus on you.

Standing straight and tall is a sign of confidence. You should work on your posture for this reason alone **FIGURE 5-2**.

A.

B.

FIGURE 5-2   **A.** Good posture is critical for good communication. **B.** Poor posture is unprofessional.

In addition to showing confidence, such a stance helps maintain your back in good health during long days of instruction.

Hand gestures can express either aggression or enthusiasm. Rapid hand movements or repeated movements of the hands to the face can be signs of deception or uncertainty.

■ Active and Passive Listening

As an instructor, you must be well versed in both active and passive listening. Active listening is the process of hearing and understanding the communication sent and demonstrating that you are listening and have understood the message. It requires you to keep your mouth closed and your ears wide open. As part of active listening, it is also very important that you hear the entire message. All too often, instructors formulate an answer before the student finishes asking the question. If you find yourself jumping in with an answer, try restating the student's question before answering. "Why do we use foam on a chemical fire? We use foam on a chemical fire because …"

Passive listening is listening with your eyes and your senses without reacting to the message. Observe the student's body language and facial expressions **FIGURE 5-3** . Nervous movements can be indicative of misunderstanding or

FIGURE 5-3 Pay attention to what your students are telling you with their body language.

confusion, whereas a relaxed brow and alert eyes can indicate comprehension and interest.

Verbal Communication

Verbal or oral communication is more than just talking. Many people talk, yet never actually communicate. Great communicators, by contrast, use many tools to convey their message. Watch and listen to the speeches of John F. Kennedy and Martin Luther King, Jr.—both were highly effective communicators, inspiring their followers to take action. At his inauguration in 1961, Kennedy said, "And so, my fellow Americans, ask not what your country can do for you; ask what you can do for your country." The inflection in his voice added emphasis to his words and reinforced his message. In his "I Have a Dream" speech at the Lincoln Memorial in 1963, King used personalization so that the entire audience would put themselves in his shoes for a moment.

Many factors, including language and tone, affect verbal communication. The volume of your speech is important. By lowering or raising your voice, you can emphasize a point. You can also use volume to gain the attention of your students. A loud, forceful tone will let the class know that you are very serious. Thus, when a student faces imminent danger, your voice must be loud and forceful.

Likewise, the importance of speaking softly cannot be overstated. In some cases, a simple whisper in a student's ear will let the student know that he or she needs to pay closer attention because you are watching. When speaking to a student privately, maintain both a level of decorum and a softer tone to your voice. In addition, a softer voice is less threatening and does not put the student on the defensive. If the student is defensive, he or she is less likely to receive your message.

Bringing your voice to a loud crescendo and then making it softer can serve to emphasize your message. This technique is especially effective when you are using compare-and-contrast strategies. For example, during a discussion of tone of voice with fire officers, you might want to compare a dominant tone

Safety Tip

Carefully monitor what you are communicating nonverbally. Unknowingly, you may communicate through your body language or your attire that students should take unnecessary risks on the training ground.

Teaching Tip

One of the best ways to develop your communication skills is straightforward: Practice them. One method is to have another fire service instructor sit through your presentation and evaluate you. This evaluation should include how you communicated your message, which nonverbal cues you gave, and how your tone of voice affected the delivery of your message. Through such a peer evaluation, you can see the student's perspective of you.

Another technique is to make a video of your presentation. Watch the recording and look for any nonverbal cues that do not match your message. Many of the nonverbal cues you project become more obvious when you play the recording back at a faster than normal speed. Finally, close your eyes and listen to your voice at normal speed. Is your voice really sending your intended message?

JOB PERFORMANCE REQUIREMENTS (JPRS)
in action

Being able to communicate effectively is key to being a great fire service instructor. The most engaging fire service instructors are truly great communicators. A great communicator uses many senses beyond just hearing (e.g., his or her speaking). Good listeners complement the learning process by participating in the course delivery. When an instructor builds a relationship with his or her students, the bond forged improves the chances that class objectives will be achieved. Your communication skills are critical to the whole learning process because they tie together instruction methods and learning methods. The more you teach, the more you will appreciate this linkage—and the more you will understand why you must continually improve your communication skills throughout your instructional career.

Instructor I

The Instructor I is assigned the duties of presenting (using communication skills) information from a prepared lesson plan. Understanding the elements in the communication process and knowing how to interpret formal and informal communication signals from your students will allow you to present better training sessions. Your communication skills are always being observed by your students, so continual improvement of those skills must be a professional development goal.

Instructor II

Always keep in mind the methods of instruction and the learning process when developing lesson plans for the Instructor I to deliver. Awareness of various instructors' communication skills may influence who is assigned to teach certain topics or which type of lesson plan is developed to be presented. New instructors should be monitored and advised about ways to improve their communication skills.

Instructor III

Writing lesson plans and producing media for others to use in a training session are part of the curriculum development process. The Instructor III must be aware of the end user's ability to utilize materials he or she prepares. Always think of how someone else will communicate with the material you prepare. Can it be adapted and adjusted by other instructors?

JPRs at Work

You must understand the elements of the communication process if your presentation is to be effective. Know how to engage your students using a variety of communication skills throughout the training process.

JPRs at Work

NFPA does not identify any JPRs for this chapter. However, the Instructor II must have knowledge of how instructors present material and the best methods for delivery. Some lesson plans will require instructors with specific communication skill levels.

JPRs at Work

NFPA does not identify any JPRs for this chapter. However, the Instructor III must have knowledge of how to prepare written reports and have the ability to review written narratives or training records and other training documentation processes.

Bridging the Gap Among Instructor I, Instructor II, and Instructor III

Work together to identify better communication skills. Experienced instructors should always audit new instructors' training sessions to see if they can give any pointers about ways to eliminate distractions or poor communications, or better ways of getting the objectives across. The Instructor I should attempt to use a variety of communication methods and be able to interpret the communication process throughout the delivery of a lesson plan. The Instructor II can help in this area by suggesting methods of instruction and providing input into the delivery methods being used. The Instructor III will analyze reports, training records, and other communications prepared as part of the overall training program management function. Written communication skills are a core function of many Instructor III responsibilities.

versus a supportive tone. You would speak loud and forcibly to show dominance, and then speak calmly and softly to show support. The same technique can also be used when describing an escalating event. For example, when describing a flashover, you would speak at normal volume to describe the fire building. As the fire builds, your volume would increase until the flashover occurs, with your voice booming.

The learning environment is also a factor when it comes to voice volume. If you are instructing outside or in a noisy environment, the volume of your voice may have to increase to project it over any <u>ambient noise</u>. Of course, sometimes ambient noise is chattering students. In those cases, you should use volume to gain your students' attention and silence.

Projecting your voice is important. If you are stationary or do not have the ability to walk around, be aware of the need to make your voice carry to the back of the room to each student FIGURE 5-4 . This feat can be accomplished by using a microphone. If a microphone is not available, use your voice as a tool to convey your message. If you are more mobile, you may actually be behind or among the students, so be sure to project your voice so that everyone in the class can hear you.

FIGURE 5-4 Project your voice and maintain good eye contact and posture while working in large classrooms.

Teaching Tip

Use words that are familiar to your audience and explain new terms thoroughly. Mentioning the problems of fighting a fire in a "group home," for instance, may not be effective if the students do not understand what a group home is.

Theory Into Practice

The difference between instructors can sometimes be as simple as the level of enthusiasm projected through the voice.

Enthusiasm breeds enthusiasm. It creates interest both in the topic and in you. Focus on teaching the courses that fascinate you—this will add passion to your presentation. For those topics that may not be as exciting, research why the topic is important. Understanding the underlying value of the topic will help you make it more interesting for the students.

Dynamic use of voice is one of the most effective methods of holding students' attention. Knowing when and how to change your voice is a skill learned through practice. Practice making a *conscious* effort to really use your voice to convey your message until it becomes automatic.

The use of appropriate language is critical. Avoid language that does not put you or your department in the best light. Ask yourself if you would use the same language around the family dinner table. Once spoken, words cannot be taken back. Exercise restraint and caution when speaking. Always be on guard with your speech: As a fire service instructor, you are held to a higher standard. Many students have expectations of what a fire service instructor should say and do, so never let your guard down.

The tone of your voice can express more than your words. The purpose of tone is to express emotion. Your tone should always be positive, while expressing the passion of conviction. If your words say that you believe in something but your tone expresses apathy, the students will hear apathy, no matter what your words say. To see how this effect works, watch an actor on stage. He may be having a bad day, but when he steps on stage, he portrays good humor and joy. As a fire service instructor, you must put your personal feelings aside and use your tone of voice to express the message properly.

■ Language

The fire service has its own unique language. We use terms that mean something completely different to the general public. For example, when we say "company," we mean a group of personnel assigned to perform a task; by contrast, the general public defines "company" as a business. Within the fire service, terminology can change from department to department and from region to region. A *rescue* in one department is called an *ambulance* in another. It is imperative that you use the terminology of your own department when discussing equipment, apparatus, or procedures.

Ethics Tip

Fire fighters routinely use language that they would never use in front of their grandmothers. Is it ethical to use that same language when instructing?

While your first instinct may be to communicate directly in the students' everyday language, it is inappropriate to do so in the classroom. As a fire service instructor, you are judged by your actions. Always take the high road.

Teaching Tip

The fire service classroom can be as simple as a room with chairs, tables, and any number of audiovisual devices to enhance the learning experience, or it can be as complex as a state-of-the-art "techno-heaven" with multimedia consoles that respond to your requests with only a touch of a button. Sometimes the classroom may be a bunkroom that serves as a multipurpose room with an overhead projector and slide projector, but it can also be the back of a fire engine or the base of the training tower. Many facilities have "dirty" classrooms—that is, designated areas allowing fire fighters in full turnout gear to discuss the training session before or after it occurs.

The purpose of the classroom is to facilitate the learning process. Consequently, the physical location is less important than the communication process. The back of an engine can be a great learning environment if the proper communication skills are used. Many company officers take time after a call to discuss how the call went. This mini-critique can be of great value to the company, even without the traditional classroom or the audiovisual devices; only good communication is necessary for this kind of education.

Written Communications

Written communication is needed to document both routine and extraordinary fire department activities. It establishes institutional history and is the foundation of any activity that the fire department wishes to accomplish. As a senior chief explained, "If it is not written down, it did not happen." The fire service instructor is responsible for many types of written communications, including narratives, which are included in training records, and more comprehensive training reports, such as annual reports of training activities. Although e-mail has replaced many formal written letters or other communications, a written communication serves as a permanent record of an event or happening. Each agency should specify guidelines that outline which types of written documentation are necessary for specific occurrences, such as reports of injuries and documentation of training issues.

■ Reading Levels

Most newspapers in the United States are written at approximately a fourth-grade reading level to allow for understanding by the general public. The newspaper does not intend to talk down to its readers who have a higher level of education; instead, it is simply ensuring that the majority of readers can easily understand the information presented.

Do you know the reading levels of all of your students? Most fire fighters are required to have a high school education or a GED, so the expectation is that all students can read and write at the high school level. Most educational textbooks are written at a reading level that the majority of fire fighters should understand. Technical editors and select fire service personnel review textbooks to establish their readability before they are printed and bound.

Most word processing programs can perform an evaluation of the document's reading level based on the Flesch–Kincaid Readability index. With this index, a higher number indicates that the document is easier to read. Therefore, the higher Flesch–Kincaid number, the lower the reading level. The index is based on the average number of words per sentence and the number of syllables per 100 words. As a fire service instructor, you must be aware of the length of your sentences and the complexity of the words you use. Always keep your students' reading levels in mind when writing.

■ Writing Formats

It is also important to know which format you should use when writing. In general, when writing to individuals within your organization, you can use a memo (short for "memorandum"). Such a document is designed for in-house communications only. For example, a memo from the chief of your training division to all fire service instructors might inform them of an upcoming training drill. Written communications intended for readers outside the organization should appear in letter format, using departmental letterhead, and should be signed by the sender. A letter would be the appropriate form of communication to Mr. Jones confirming a preincident planning appointment at his coffee shop, for example.

The Five W's of Writing

The basic rule of thumb for writing is to include the "five W's":

- Who
- What
- Where
- When
- Why

Add an "H"—*How*—to expand the discussion further.

To: Lavalle Fire Department
Cc: Chief Michael Deforge
From: Officer Archie Reed
Date: May 23
Subject: Live Burn on July 10

On July 10, the Lavalle Fire Department will conduct a live burn at 1015 Willow Way. Fourteen students from the Fire Suppression Course will be present. Students will participate in supervised training using medium-diameter hose lines during defensive operations.

Students and Fire Service Instructors Thomas, Kelly, Hinkler, and Andrews will assemble at the South Street Fire Station at 10:00 a.m. to be transported to the live burn structure. Safety procedures will be reviewed at the site prior to the live burn. In addition to the students and the fire service instructors, Engine Company 3 will be on site.

If you are not a student in the Fire Suppression Course, an assigned fire service instructor, or a member of Engine Company 3, you may not be on the live burn site without written permission. Please see Officer Archie Reed for a permission request form.

FIGURE 5-5 A sample memo using the five W's of writing.

VOICES
OF EXPERIENCE

It has been my very good fortune to travel around the world for the express purpose of communicating with diverse groups of people. Most of the time, I was presenting some subject related to one of the many public safety disciplines. Consequently, my ability to communicate that subject would invariably affect the audience's reaction to the material and to me.

For a project in Tianjin, China, I was provided with an interpreter and warned that my students would speak very little English. So, I initiated a communication strategy that focused on my interpreter—a very sweet-natured and demure lady. Subsequently, we spent some time together and got to know each other. I learned her skill level and personal strengths. She learned my subject material and sense of humor. By sharing family photos and stories, we personally bonded and actually became a pretty competent team.

I told the class that I was a very entertaining speaker, and if they didn't enjoy me... it was the interpreter's fault.

She translated my PowerPoint® slides into Mandarin so the students could relate to the visual information in their own language. We also considered the effect translation would have on my spoken words. Basically, everything is stated twice… once in English and once in Mandarin. We would therefore be able to cover only half of the information each day that I might have presented to an English-speaking class. Finally, we were ready to begin the project.

I started the course with a brief bio that focused on where I came from and included photos from the USA that I felt might be interesting to this particular audience. I also told the class that I was a very entertaining speaker, and if they didn't enjoy me… it was the interpreter's fault. This garnered a respectable chuckle from the crowd and we were off to the races.

Each night, a different group of students would take my interpreter, my wife, and me out for the evening. Those times proved to be immensely rewarding as we ate, laughed, and sang karaoke together. Also, I noticed that my interaction with the students improved proportionately to the total number of karaoke songs logged each week. Many in the class told me that they never knew learning could be so much fun. Meanwhile, I was learning that communication is really just finding/inventing ways of sharing yourself with others. If they reciprocate, as they often do, the experience improves exponentially.

By the time we moved onto Beijing for a different course, we were already exhausted, but we wouldn't change a thing. Sharing yourself with other people is one of those precious exercises that make life worthwhile. Please don't miss the opportunity to communicate with your own corner of the world.

Kevin L. Hammons
President, Revelations, Inc.
Franktown, Colorado

By including these points in your writing, you will ensure that you address the information necessary for readers to understand your message properly. For example, suppose you need to write a memo about a live burn that will take place at a donated structure on the outskirts of town **FIGURE 5-5**. You want your students and additional fire service instructors to know where and when to meet on the live burn day. You also want the entire fire department to know when the live burn is taking place, who is scheduled to be on site, and whether permission must be granted if additional fire fighters want to be on the live burn site.

The memo in **FIGURE 5-5** contains all five W's (plus H) of writing:

- Who: Lavalle Fire Department
- What: Live burn training exercise
- Where: 1015 Willow Way
- When: July 10 at 10:00 am
- Why: Permission must be granted to be on the live burn site
- How: Fire fighters must see Officer Archie Reed for a permission request form

The Rules of Writing

Good, clear, concise writing is critical to getting your message across. Fortunately, it is easy to improve your writing: If you follow the rules of good writing in step-by-step fashion and practice for 15 minutes each day, you can become an effective written communicator. As a fire fighter, you train on a regular basis to maintain your firefighting skills. Similarly, as a fire service instructor, you need to practice your writing skills on a regular basis. This step involves actually writing and allowing fellow fire service instructors to read your writing to ensure that your message is coming through loud and clear.

The rules of good writing begin with the paragraph. Each paragraph begins with an introduction—a sentence or two that states what you plan to discuss in the paragraph. Then comes the main body of the paragraph—the frame. Like the frame of a structure, the frame of the paragraph is the support that holds everything together. The conclusion reviews what was discussed in the paragraph and reinforces the frame.

For example:

The fire officer must first consider the intended audience for the report before writing. For example, a fire officer may be assigned to prepare a study that proposes closing three companies and using quints that can respond as either engine or ladder companies. If the report is intended for internal use, the technical information would use normal fire department terminology. Conversely, if the report is intended for the fire chief to deliver to the city council, with copies going to the news media, many of the terms and concepts would need to be explained in simple terms for the general public. Always consider who your audience is before writing a report.

To practice constructing a solid paragraph, consider writing a two- to three-paragraph summary after every class. What was your message for the class? Did your students receive it?

Once you have mastered the building of a paragraph, you are on your way to constructing a complete written structure—anything from a two-paragraph memo, to a full-page letter, to a 10-page report for the city council. The paragraph is your basic building block of writing. All good writing has an introduction, a frame, and a conclusion. In an eight-paragraph report that is intended for the city council and explains why the fire department's training budget should be increased by 10 percent, for example, the report would be broken down in the following way:

- Paragraphs 1 and 2: Introduction. These paragraphs give an overview of why the fire department needs to increase the training budget.
- Paragraphs 3 through 7: Frame. These paragraphs discuss in depth the specific reasons why the training budget should be increased. For example, the city might have grown by 20 percent in three years and needs more fire fighters to provide good service.
- Paragraph 8: Conclusion. This paragraph summarizes the arguments for increasing the fire department's training budget.

Style

Most of the writing you do as a fire fighter and as a fire service instructor will be technical in nature. In such a case, your sole objective is to convey information—not to express emotion or to inspire passionate feelings. Writing in the fire service should not be boring or uninspiring, but it should remain true to the premise that you write to impart facts.

Once words are printed, they are difficult to retract. Read an editorial column in a newspaper to see how writing can affect people's thoughts and emotions. Words can be construed in different ways by different people, so it is extremely important to choose your words carefully. The meaning of words can change based on geographic location. A person's individual experiences also can change the meaning of a word. Know your audience before you begin to write. Your goal is to make certain that your message is clearly and properly conveyed to every reader.

Theory Into Practice

Choose your words carefully when writing. Make every effort to ensure that alternative interpretations cannot occur. Failure to do so can mislead students and ultimately lead to undesired performances.

Informal Communications

Informal communications include internal memos, e-mails, instant messages, and messages transmitted via mobile data terminals. Informal reports have a short life and are not archived as permanent records. Instead, they are used primarily to record or transmit information that will not be needed for reference in the future. Be aware that some agencies, as public agencies, are required to keep archived records of even informal records, and such communications may be subject to freedom of information requests (see the *Legal Issues* chapter for more information).

Some informal reports are retained for a period of time and may become a part of a formal report or investigation. For example, a written memo could document an informal conversation between a fire instructor and a fire fighter regarding a training drill attendance issue. The memo might note that the student or crew member was warned that corrective action is required by a particular date. If the issue is corrected, the warning memo is discarded after 12 months. If the fire fighter continues to demonstrate poor attendance, the memo becomes part of the formal documentation of progressive discipline.

In recent years, the storage and recall of e-mails has been a topic of much discussion. Policies will vary from agency to agency, so know your department's policy about e-mails, instant messages, and related informal communication. Ignorance will not usually work as a defense in a court of law, so know your agency's policy regarding the use of this type of informal communication.

Formal Communications

A formal communication is an official fire department document printed on business stationery with the fire department letterhead. If it is a letter or report intended for someone outside the fire department, the document is usually signed by the fire chief or a designated staff officer to establish that it is an official communication. Subordinates often prepare these documents and submit them for an administrative review to check the grammar and clarity. A fire chief or the staff officer who is designated to sign the document performs a final review before it is transmitted.

The fire department maintains a permanent copy of all formal reports and official correspondence from the fire chief and senior staff officers. Formal reports are usually archived.

Standard Operating Procedures

Standard operating procedures (SOPs) are written organizational directives that establish or prescribe specific operational or administrative methods to be followed routinely for the performance of designated operations or actions. SOPs are intended to provide a standard and consistent response to emergency incidents as well as personnel supervisory actions and administrative tasks. They serve as a prime reference source for promotional exams and departmental training.

SOPs are formal, permanent documents that are published in a standard format, signed by the fire chief, and widely distributed. They remain in effect permanently or until they are rescinded or amended. Many fire departments conduct a periodic review of all SOPs so they can revise, update, or eliminate any that are outdated or are no longer applicable. Any changes in the SOPs must also be approved by the fire chief. SOPs should always be referenced as part of a training session, especially when demonstrating skills and ability.

General Orders

General orders are formal documents that address a specific subject, policy, condition, or situation. They are usually signed by the fire chief and can be in effect for various periods, from a few days to permanently. Many departments use general orders to announce promotions and personnel transfers. Copies of the general orders should be made available for reference at all fire stations.

Announcements or Communications

The formal organization may use announcements, information bulletins, newsletters, Web sites, or other methods to share additional information with fire department members. These methods are generally used to distribute short-term and non-essential information that is of interest.

Legal Correspondence

Fire departments are often asked to produce copies of documents or reports for legal purposes. Instructors must always remember that training records are also legal documents and, as such, are available for scrutiny through the legal system. Sometimes the fire department is directly involved in the legal action, whereas in other cases the action involves other parties but relates to a situation in which the fire department responded or had some involvement. It is not unusual for a fire department to receive a subpoena—a legal order demanding all of the documentation that is on file pertaining to a particular incident or person. Gathering such documents is a task that can require extensive time and effort. In any situation for which reports or documentation is requested, consult with legal counsel.

The fire fighter or officer who prepared a report may be called on, sometimes years later, to sign an affidavit or appear in court to testify that the information provided in the report is complete and accurate. He or she might also have to respond to an interrogatory, which is a series of written questions asked by someone from an opposing party. The fire department must provide written answers to the interrogatories, under oath, and produce any associated documentation. If the initial incident report or other documents provide an accurate, factual, objective, complete, and clear presentation of the facts, the response to the interrogatory is often a simple affirmation of the written records. The task of dealing with the interrogatory can become much more complex and embarrassing if the original documents are vague, incomplete, false, or missing.

Reports

Reports should be accurate and present the necessary information in an understandable format. All reports should be proofread before being submitted. The purpose of the report might be to brief the reader, or it might provide a systematic analysis of an issue. If the fire chief wants only to be briefed on a new piece of equipment and the fire officer delivers a fully researched presentation, both parties would be frustrated.

Some reports are prepared on a regular schedule, such as daily, weekly, monthly, or yearly. Other types of reports are prepared only in response to specific occurrences or when requested. To create a useful report, you must understand the specific information that is needed and provide it in a manner that is easily interpreted.

Some reports are presented orally; others are prepared electronically and entered into computer systems. Some reports are formal, whereas others are informal. The most common form of reporting is verbal communication from one individual to another, either face-to-face or via a telephone or radio. To be effective, the transfer of information must be clear and concise, using terminology that is appropriate for the receiver.

Monthly Activity and Training Report

The monthly activity and training report documents the company's activity during the preceding month. These reports typically include the number of emergency responses, training activities, inspections, public education events, and station visits that were conducted during the previous month. Some monthly reports include details such as a list of the number of feet of hose used and the number of ladders deployed during the month. The number of training hours completed, members who receive certifications or attend classes, and any number of training-related projects or meetings should be documented in a regular format to highlight the role and importance that training plays in an organization.

In many cases, the officer delegates the preparation of routine reports that do not involve personnel actions or supervisory responsibilities to subordinates. In such a case, the officer is still responsible for checking and signing the report before it is submitted.

Some fire companies or municipal agencies post a version of their monthly reports on their public website. These reports often include digital pictures of the incidents and the people involved.

Special events or unusual situations that occurred during the month might also be included in the report. For example, the monthly report might note that the company provided standby coverage for a presidential visit or participated in a local parade.

Infrequent and Special Reports

Infrequent reports usually require a fire instructor's personal attention to ensure that the report's information is complete and concise. Such reports include the following types of documents:

- Fire fighter injury report
- Student complaints
- Property damage or liability-event report
- Vehicle accident or equipment damage report
- New equipment or procedure evaluation
- Suggestions to improve fire department operation
- Response to a grievance or complaint
- Fire fighter work improvement plans
- Research report or recommendation

A chronological statement of events is a detailed account of activities, such as a narrative report of the actions taken at an incident or accident. A recommendation report is a document that suggests a particular action or decision.

Some reports incorporate both a chronological section and a recommendation section. For example, the fire fighter line-of-duty death investigations produced by the National Institute for Occupational Safety and Health (NIOSH) include a chronological report of the event, followed by a series of recommendations that could prevent the occurrence of a similar situation.

Writing a Decision Document or Report

The goal of a decision document is to provide enough information and persuasion so the intended individual or body accepts your recommendation. This type of report includes recommendations for employee recognition or formal discipline, for a new or improved procedure, or for adoption of a new device. A decision document usually includes the following elements:

- **Statement of problem or issue:** One or two sentences.
- **Background:** Brief description of how this became a problem or an issue.
- **Restrictions:** Outline of the restrictions affecting the decision. Factors such as federal laws, state regulations, local ordinances, budget or staff restrictions, and union contracts are all restrictions that could affect a decision.
- **Options:** Where appropriate, provide more than one option and the rationale behind each option. In most cases, one option is to do nothing.
- **Recommendation:** Explain why the recommended option is the best decision. The recommendation should be based on considerations that would make sense to the decision maker. If the recommendation is going to a political body and involves a budget decision, it should be expressed in terms of lower cost, higher level of service, or reduced liability. The impact of the decision should be quantified as accurately as possible, with an explanation of how much money it will save, which new levels of service will be provided, or how much the liability will be reduced.
- **Next action:** The report should clearly state the action that should be taken to implement the recommendation. Some recommendations may require changes in departmental policy or budget. Others could involve an application for grant funds or a request for a change in state or federal legislation.

Audience Analysis

Many types of speeches exist, including persuasive, informative, and special speeches. A keynote speech is generally a persuasive speech. The purpose of such a speech is to motivate and inspire the audience to move forward with the message given. Most presentations for a fire service instructor involve informative speeches, or lectures. As a good fire service instructor, you need to determine how your message needs to be communicated before determining the appropriate format for your presentation. You also need to evaluate your audience to determine how they will best receive your message.

At the most basic level of firefighting, the lecture is usually the most appropriate presentation method. Because students in these types of classes have very little experience to draw upon, it is best to use an informative lecture to communicate concepts and objectives. You need to pass the information to the student in a very direct manner. It will be repetitive in nature—some would say almost rote memorization.

In speaking to more experienced fire fighters, the lecture is more persuasive in nature. With these students, you may be taking a skill learned in basic training and discussing it in finer detail at a higher level. This type of class requires students to be convinced about the importance of the message; otherwise, they might tune out the lecture because they mistakenly believe that they already know everything possible about the skill.

The third type of audience encountered by fire instructors is made up of experienced fire fighters and fire officers. These students tend to require discussions rather than informative lectures. Discussions engage these experienced students and encourage them to incorporate their knowledge and experiences into the learning process. This concept, which is sometimes referred to as a "tailboard chat," is used with more experienced fire fighters because it takes their existing knowledge and experience and melds it to a new concept **FIGURE 5-6**. By examining and reexamining the way we do things, and by discussing experiences and options, new ideas and new ways of doing things are developed. This sort of exchange can be very exciting and rewarding because you can witness an idea developing in front of you. In fact, many of the tools used in the fire service today evolved out of such discussions, when fire fighters saw a problem or a situation that needed to be improved and made adaptations to rectify the need. This style of instruction is sometimes referred to as facilitation.

FIGURE 5-6 The tailboard chat can provide an opportunity for a quick training exercise or to analyze the recent response.

Alternative communication methods should also be considered. For example, role playing can often reinforce learning objectives. In role playing, you can take the position of someone with whom the student interacts. This method requires you to have an idea of how a student will respond in this situation and to be prepared for that response. You should be prepared for several potential responses from the student and be able to lead the student toward the learning objective. At the end of the role playing, you should explain the learning objective of the exercise.

Another possible communication method consists of group exercises. With this method, you use the collaborative efforts of the group to benefit all members of the group. Such a technique exploits the natural tendency of fire fighters to work in groups. With group work, you can evaluate whether students can work collaboratively to solve a problem. In most cases, this is the way a fire company solves problems in the field. It is imperative that you listen to the conversations in the groups. Walk around to each group and stand near or actually sit down with the group and listen. You can see if the group is moving in the right direction or needs to be redirected. If necessary, help the group along by saying, "Have you considered…." Steering the discussion is not equivalent to giving students the answer, but rather assists them in remaining focused on the message. Be sure to solicit outcomes from each group involved in the exercise; adult learners will want to have their work discussed and will not appreciate "busy work" during a class.

Your communication style will be situational, such that you may have to use several different styles during the same presentation. When teaching a class on fire extinguishers, for example, you might begin with a lecture style and then move into a psychomotor session. When giving the lecture, you are simply presenting facts: There is no pressure and no sense of danger. When you move into the psychomotor session, however, it may be necessary to speak authoritatively to ensure that the students understand the issues of safety.

Issues of safety will inevitably affect your communication style. When you explain to new recruits that they must stay down in a fire situation, they generally nod their heads as you explain the science of thermal layering. When you get on the fire ground, be more forceful in your tone and action to keep your students focused on safety.

When an important objective needs to be reinforced, you can highlight it by changing your communication style. For example, by adding a group exercise in the middle of a lecture, you highlight the learning objective.

Time is another factor when considering communication styles. If students spend two hours straight sitting and listening to lecture, retention of the material can diminish. Nothing is worse for a student than to have a monotone instructor lulling him or her to sleep with an endless lecture. Retention of material is increased by adding activities. Lesson plans are discussed in the *Lesson Plans* chapter; activities or chances for student practice or participation are incorporated in the application step of a lesson plan. Engaging the student in the learning will pay big dividends. For example, if a student has to speak and do something, the information will be retained longer and more completely. This consideration is especially important after a meal. The more activities you can provide to students after a lunch break, the more likely they are to retain the information. The cone of learning illustrates the importance of communication skills and how students are able to remember information **FIGURE 5-7**. The more active the student is in the learning, the more effective and real the learning is to them.

Teaching Tip

When having students work in teams, it is essential to monitor each student carefully to ensure that he or she masters the material. Without close monitoring during group activities, one student may not be able to perform the task when others are not there to provide assistance.

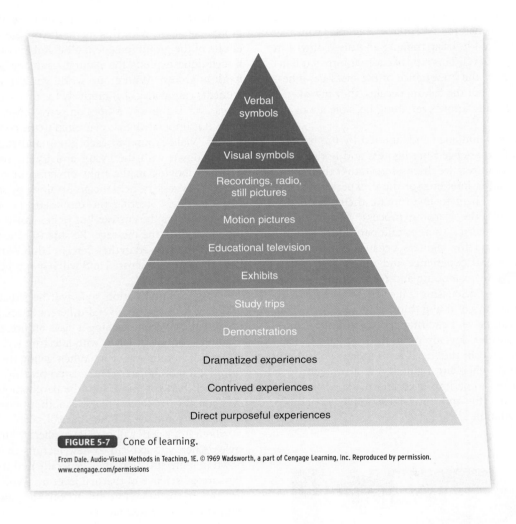

FIGURE 5-7 Cone of learning.

From Dale. Audio-Visual Methods in Teaching, 1E. © 1969 Wadsworth, a part of Cengage Learning, Inc. Reproduced by permission. www.cengage.com/permissions

Jones & Bartlett Fire District

Training Division
5 Wall Street, Burlington, MA, 01803
Phone 978-443-5000   Fax 978-443-8000
www.fire.jbpub.com

Instant Applications: Communication Skills

Drill Assignment

Apply the chapter content to your department's operation, the training division, and your personal experiences to complete the following questions and activities.

Objective

Upon completion of the instant applications, fire service instructor students will exhibit decision making and application of job performance requirements of the fire service instructor using the text, class discussion, and their own personal experiences.

Suggested Drill Applications

1. Make a list of common distracters that you have witnessed when being taught by your own instructors. How did these distracters impair the communication process? How could they have been avoided by using a different communication method?

2. Which great communicator traits can you relate to?

3. Watch a political candidate speak to an audience. Identify which communication skills the candidate uses to sway public opinion in an effort to gain votes. How can you translate that information to your own training delivery?

4. Watch news broadcasts and identify methods used to communicate the day's news. Which stories held your interest? Why? Why did you select the channel? Are there lessons you can learn from your selection?

5. Review the Incident Report in this chapter and be prepared to discuss your analysis of the incident from a training perspective and as an instructor who wishes to use the report as a training tool.

Incident Report

Poinciana, Florida—2002

Just 10 months after the incident in Lairdsville, and less than three weeks after the sentencing of the Lairdsville chief officer, a live fire training session trapped two fire fighters in Poinciana, Florida, near Kissimmee.

The 1600 ft² (148.6 m²) cement block house had three bedrooms, one of which was converted from a one-car garage with the large door removed, the wall blocked up, and a window installed **(Figure 5-A)**. The room had block walls,

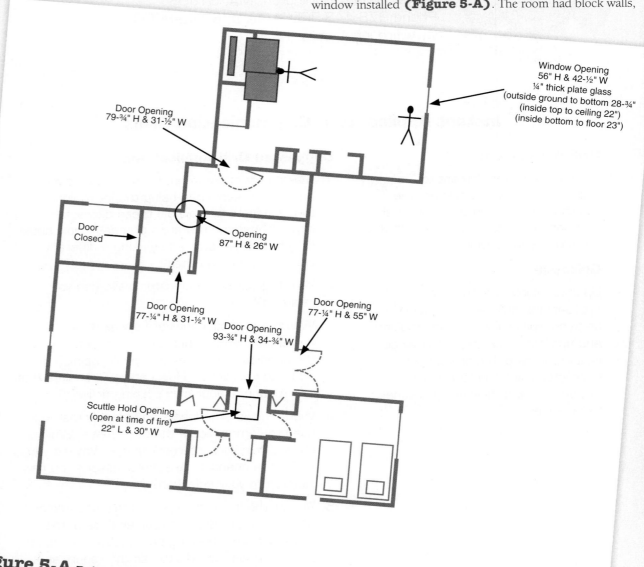

Window Opening
56" H & 42-½" W
¼" thick plate glass
(outside ground to bottom 28-¾"
(inside top to ceiling 22")
(inside bottom to floor 23")

Door Opening
79-¾" H & 31-½" W

Door Closed

Opening
87" H & 26" W

Door Opening
77-¼" H & 31-½" W

Door Opening
93-¾" H & 34-¾" W

Door Opening
77-¼" H & 55" W

Scuttle Hold Opening
(open at time of fire)
22" L & 30" W

N

Figure 5-A Poinciana floor plan.
Courtesy of the Florida State Fire Marshal

Figure 5-B Two consecutive offset turns and this narrow hallway made it difficult for fire fighters to maneuver.

Figure 5-C Interior of fire room in Poinciana.

one door, and a fixed ¼" (6.4-mm) thick, commercial-grade glass window. The exit from the room was through two consecutive offset turns, through a 26" (66 cm) opening, then through two small rooms, and out another door into the dining room **(Figure 5-B)**. A breezeway between the old garage and the main house had also been enclosed. Unlike many acquired structures, this house was in good condition and was part of an entire neighborhood being razed for a new campus. Several area fire departments were involved in a series of training events, and the structures available included other houses, as well as a motel.

All of the instructors were experienced in the fire service and had previously worked together on training fires. Safety crews were not briefed at the beginning of this exercise because of their past involvement. The training officer walked all of the participants through the structure and explained the safety aspects and goals of the training evolution, which were to conduct a search and rescue with an actual fire burning and find a mannequin in bunker gear hidden somewhere in the house and remove it. The first search team consisted of an experienced lieutenant and a trained recruit (state certified Fire Fighter II), who were to search without a hose line. Two suppression teams with 1¾" (44 mm) hose lines were in position in the house with four interior safety officers broken into two teams.

The fire was started in the converted bedroom near the only doorway to the fire room **(Figure 5-C)**. Two piles, practically vertical with pallets, wood scraps, and hay, were almost adjacent to each other, one inside and one outside the open closet. After the fire started, and with the instructor-in-charge's agreement, a foam mattress from one of the other bedrooms was added to the pile.

The search and rescue (SAR) team entered the structure at the front (east) door with a suppression team following. The suppression team stopped in the small room located between the dining room and the bedroom, where the training fire was located. The search team continued into the burning bedroom, encountering deteriorating conditions with high heat and no visibility due to heavy smoke.

Continued...

Incident Report Continued...

Poinciana, Florida—2002

On the exterior, a second suppression team waited at the front doorway, with two more fire fighters assigned as the rapid intervention crew (RIC), and a third, uncharged 1¾" (44 mm) hose line also was available outside. One fire fighter was stationed on the exterior waiting for orders to ventilate.

With near zero visibility and increasing heat conditions, two of the interior safety officers monitoring the activities of the fire room area later stated that they heard the lieutenant of the search team ask his partner if he had searched the entire room. They heard the answer "yes." Shortly thereafter, one of the safety officers yelled into the fire room asking if the search team was out. Although someone answered "yes," no one knows who replied. The interior safety officer assumed that he had missed the search team's exit from the fire room, as there were several fire fighters present by the dining room. He began to search the rest of the structure in an attempt to find them.

The instructor-in-charge ordered the front window of the fire room to be broken. After the window was broken out, heavy black smoke followed, which ignited very quickly with flames forcibly venting from the window. The suppression crew closest to the fire room applied water in short bursts, but increasing heat and steam forced the safety officers and the suppression crew to back out. Both safety officers were forced to exit the structure, after receiving burn injuries.

The second suppression team was ordered to replace the first team and engage the fire. The second team and an interior safety officer entered the fire room and extinguished the fire.

During this time the instructor-in-charge called the missing fire fighters several times and received no answer.

While the second suppression team was overhauling the fire area, they found a body in fire fighter bunker gear facedown on the floor. Both of the suppression team fire fighters initially thought it was the rescue mannequin, not realizing it was the lieutenant of the SAR team.

After no radio response from the SAR team, the instructor-in-charge ordered a personnel accountability report (PAR) and ordered the RIC to enter the structure and find the SAR team.

The suppression team that found the lieutenant dragged him to the front window of the fire room, and removed him to the outside. The missing fire fighter was also found inside that window, and he too was removed to the outside. The entire event, from the time the SAR crew entered to when the first of the two fire fighters were located, was under 14 minutes. The two fire fighters died despite working in teams with experienced instructors, two interior staffed hose lines, two interior safety teams, an RIC with its own hose line, a participant walk-though, and other safeguards in place.

The National Institute for Occupational Safety and Health (NIOSH) investigated the fire. Two separate investigations were conducted by the state fire marshal: one for criminal violations, and an administrative investigation of state training codes. No criminal charges were filed; however, the findings included the following:

- "All of the participants stated…they did not have any concerns regarding the conditions of the fire inside the structure and it appeared to them as normal fire behavior."

- Although NFPA 1403 was already required in state code, it was under the environmental laws and so the law enforcement department of the state fire marshal's office did not have the authority to enforce it.

- The fire was started in a room with too much fuel for the size of the room, including a foam mattress that was added after ignition. Flashover was precipitated by too high a fuel load and inadequate ventilation.

- National Institute of Standards and Technology (NIST) determined that the mattress was contributory to considerable smoke production; however in testing, the room flashed over in roughly the same time without the mattress present.

Post-Incident Analysis

NFPA 1403 (2007 Edition) Noncompliant

No written preburn plan prepared (4.2.25.2)

Accountability was not maintained at the point of entry to the fire room (4.5.6)

Safety crews did not have specific assignments or emergency procedures (4.2.25)

Fire started near the only doorway in and out of the fire room (4.4.16)

Excessive fuel loading (4.3.4)

No communications plan in place (4.4.9)

NFPA 1403 (2007 Edition) Compliant

Experienced instructors had worked together on training fires before (4.5.1)

Primary and secondary suppression crews inside with hose lines (4.4.6.2)

A walk-through of the structure was conducted (4.2.25.4)

Crews were alerted of the possibility of a victim (4.2.25.1)

Structure in good condition (4.2.1)

Other Issues

Radio communications less than optimal

Egress limited by offset turns, a 26" (66 cm) opening, and multiple rooms and turns

Wrap-Up

Chief Concepts

- Communication is the most important element of the learning process.
- The basic communication process consists of five elements: the sender, the message, the medium, the receiver, and feedback.
- Be aware of the message that you are sending to students nonverbally through your body language.
- Your communication style must be adapted to match the situation. You may have to use several different styles during a presentation.
- As an instructor, you must be well versed in both active and passive listening.
- Many factors, including language and tone, affect verbal communication.
- Written communication is needed to document both routine and extraordinary fire department activities. It establishes institutional history and serves as the foundation of any activity that the fire department wishes to accomplish.
- The most common form of reporting is verbal communication from one individual to another, either face-to-face or via a telephone or radio. To be effective, the transfer of information must be clear and concise, using terminology that is appropriate for the receiver.
- Reports should be accurate and present the necessary information in an understandable format. The purpose of the report might be to brief the reader, or it might provide a systematic analysis of an issue.
- Most presentations for a fire service instructor involve informative speeches, or lectures. As a good fire service instructor, you must determine how your message needs to be communicated before selecting the appropriate format for your presentation.

Hot Terms

Active listening The process of hearing and understanding the communication sent; demonstrating that you are listening and have understood the message.

Ambient noise The general level of background sound.

Communication process The process of conveying an intended message from the sender to the receiver and getting feedback to ensure accuracy.

Feedback The fifth and final link of the communication chain. Feedback allows the sender (the instructor) to determine whether the receiver (the student) understood the message.

Formal communication An official fire department communication. The letter or report is presented on stationery with the fire department letterhead and generally is signed by a chief officer or headquarters staff member.

General orders Short-term documents signed by the fire chief that remain in force for a period of days to one year or more.

Informal communications Internal memos, e-mails, instant messages, and computer-aided dispatch/mobile data terminal messages. Informal reports have a short life and are not archived as permanent records.

Interrogatory A series of formal written questions sent to the opposing side of a legal argument. The opposition must provide written answers under oath.

Medium The third link of the communication chain, which describes how you convey the message.

Message The second and most complex link of the communication chain; it describes what you are trying to convey to your students.

Passive listening Listening with your eyes and other senses without reacting to the message.

Receiver The fourth link of the communication chain. In the fire service classroom, the receiver is the student.

Recommendation report A decision document prepared by a fire officer for the senior staff. The goal is support for a decision or an action.

Sender The first link of the communication chain. In the fire service classroom, the sender is the instructor.

References

Davies, L. (2001). Effective communication. Kelly Bear. http://www.kellybear.com/TeacherArticles/TeacherTip15.html

Kennedy, John F. (1961). Inaugural address. http://www.jfklibrary.org/Asset-Viewer/BqXIEM9F4024ntFl7SVAjA.aspx

King, Martin Luther Jr. (1963). Martin Luther King's speech: "I have a dream"—the full text. http://abcnews.go.com/Politics/martin-luther-kings-speech-dream-full-text/story?id=14358231&page=2

National Fire Protection Association. (2012). *NFPA 1041: Standard for Fire Service Instructor Professional Qualifications*. Quincy, MA: National Fire Protection Association.

National Fire Protection Association. (2007). *NFPA 1403: Standard on Live Fire Training Evolutions*. Quincy, MA: National Fire Protection Association.

Swindoll, Charles R. (2003). *Strengthening your grip*. Cengage.

You are assigned to teach an in-service training class for members of your own crew and station. Your anxiety increases as you begin to think of the possible outcomes and things that could go wrong with the presentation. You might be teaching students with more seniority than you, you are probably teaching a topic everyone should already know, and your crew members may have different levels of participation or may even engage in classroom antics during the class.

As you prepare, you remember some of the keys to effective instruction: Be enthusiastic and use good communication skills. You search for the latest information in the trade publications, you review departmental policies and procedures, and you go over the equipment. As you develop your teaching outline, you should try to make the class material relevant to current department operations and tie your presentation into calls or other case studies. Focus your attention on creative ways to cover the basics as you teach information that your students might not otherwise know and attempt to engage all levels of experience into the class.

1. Which of the following is not a form of communication?
   A. Verbal
   B. Nonverbal
   C. Written
   D. Audiovisual

2. Which of the following is *not* part of the communication process?
   A. Sender
   B. Receiver
   C. Message
   D. Objectives

3. Imparting or interchange of thoughts or opinions by speech, writing, or signs is called:
   A. verbal communication.
   B. written communication.
   C. nonverbal communication.
   D. communication.

4. Enthusiasm is which type of communication?
   A. Verbal communication
   B. Written communication
   C. Nonverbal communication
   D. Communication

5. Which type of communication are departmental policy and procedure manuals?
   A. Verbal communication
   B. Written communication
   C. Nonverbal communication
   D. Communication

6. What is the primary reason for adding an activity into your presentation?
   A. To increase retention of information
   B. To evaluate instructor delivery
   C. To fill the time requirements
   D. To rest your voice

7. During your presentation, a student asks a question to clarify what you said. Which part of the communications process is this follow-up?
   A. Sender
   B. Receiver
   C. Message
   D. Feedback

Lesson Plans

Fire Service Instructor I

Knowledge Objectives

After studying this chapter, you will be able to:

- Identify and describe the components of learning objectives. (pp 127–129)
- Identify and describe the parts of a lesson plan. (NFPA 4.3.2 , NFPA 4.4.3) (pp 129–132)
- Describe the four-step method of instruction. (pp 132, 134)
- Describe the instructional preparation process. (NFPA 4.2.2 , NFPA 4.3.2 , NFPA 4.3.3) (pp 134–136)
- Describe the lesson plan adaptation process for the Fire Service Instructor I. (NFPA 4.3 , NFPA 4.3.1 , NFPA 4.4.4) (pp 138–139)

Skills Objectives

After studying this chapter, you will be able to:

- Demonstrate the four-step method of instruction. (pp 132, 134)
- Review a lesson plan and identify the adaptations needed. (NFPA 4.3.3) (pp 138–139)
- Adapt a lesson plan so that it both meets the needs of the students and ensures that learning objectives are met. (NFPA 4.3.3) (pp 138–139)

Fire Service Instructor II

Knowledge Objectives

After studying this chapter, you will be able to:

- Describe how a Fire Service Instructor II creates a lesson plan. (NFPA 5.3 , NFPA 5.3.1) (pp 140–149)
- Describe how a Fire Service Instructor II modifies a lesson plan. (NFPA 5.3.3) (p 149)

Skills Objectives

After studying this chapter, you will be able to:

- Create a lesson plan that includes learning objectives, a lesson outline, instructional materials, instructional aids, and an evaluation plan. (NFPA 5.3.2) (pp 140–149)
- Modify a lesson plan so that it both meets the needs of the students and ensures that all learning objectives are met. (NFPA 5.3.3) (p 149)

Fire Service Instructor III

Knowledge Objectives

There are no knowledge objectives for Fire Service Instructor III students.

Skills Objectives

There are no skills objectives for Fire Service Instructor III students.

A recent seminar you attended has motivated you to present to your department some information you learned at the keynote presentation and workshops. You have a brief handout of key points and terms, a pad full of written notes that you scribbled down, and links to various Web sites that the instructors provided. As part of your research, you note that the instructor from one of the workshops has made his lesson plan available to class participants for use in passing along this information. You have only two days to gather this information and review the material from the Web site before you will conduct the training session.

1. What are the responsibilities of an Instructor I in adapting a lesson plan for an audience?
2. How will you determine the outcomes of the lesson plan provided by the instructor?
3. Which parts of the lesson plan provided will contain the content of the class material? How will you use each part of the lesson plan in your presentation?

Introduction

When most people think about the job of a fire service instructor, they picture the actual delivery of a presentation in front of the classroom. Although lectures are an important aspect of instruction, they are not the only part of the job. Most fire service instructors spend many hours planning and preparing for a class before students ever arrive in the classroom. There are many details to address when planning a class:

- What are the expected outcomes (objectives) of the training session?
- How much time will the class take?
- How many students will attend the class?
- Are there student prerequisites required to understand the objectives?
- Which training aids and equipment will be needed?
- In what order will the instructional material be presented?

- Will the delivery schedule be affected by availability of specific resources?

All of these questions and more are answered during the planning and preparation for the class. This information is compiled into a document called a lesson plan. A lesson plan is a detailed guide used by the fire service instructor for preparing and delivering instruction to students. A fire service instructor who uses a well-prepared and thorough lesson plan to organize and prepare for class greatly increases the odds of ensuring quality student learning. A Fire Service Instructor I uses a lesson plan that is already developed, usually by an instructor who is certified as an Instructor II or higher. The Instructor II has received training in how to develop his or her own lesson plan and may be responsible for developing all parts of the lesson plan, including objectives, lesson outline, suggested student activities, methods of evaluation, and many other components of a properly crafted class session.

Fire Service Instructor I

Why Use a Lesson Plan?

Many fire fighters who are assigned to instruct a class have a lot of emergency scene experience and may have even participated in group discussions on department operations, personnel evaluations, or the budget process. However, working from prepared lesson plan materials is a process with which many fire fighters have no experience, as they may have participated as students and not in delivering information *to* students.

Most people without experience in the field of education do not understand the importance of a lesson plan. Attempting to deliver instruction without a lesson plan is like driving in a foreign country without a map FIGURE 6-1 . The goal in both situations is to reach your intended destination. In a lesson plan, the learning objectives are the intended destination. Without a map (the lesson plan), you most likely will not reach the destination. If an instructor attempts to shoot from the hip without a prepared lesson plan that details the expected outcomes, content may be skipped, safety points may be omitted,

FIGURE 6-1 Attempting to deliver instruction without a lesson plan is like driving in a foreign country without a map. Don't waste valuable class time searching for directions.

and inconsistency between deliveries will occur. Also, without a lesson plan that contains learning objectives, you may not even know what the destination for the class is. In other words, if you do not have clearly written learning objectives for your class and a plan for how to achieve them, there is a high probability that you will not be successful.

Written lesson plans also ensure consistency of training throughout the various companies of a large fire department, or when a class is taught multiple times, especially by different fire service instructors. In such cases, a common lesson plan ensures that all students receive the same information. Lesson plans are also used to document what was taught in a class. When the class needs to be taught again in the future, the new fire service instructor will be able to refer to the existing lesson plans and achieve the same learning objectives.

Learning Objectives

All instructional planning begins by identifying the desired outcomes. What do you want the students to know or be able to do by the end of class? These desired outcomes are called objectives. A learning objective is defined as a goal that is achieved through the attainment of a skill, knowledge, or both, and that can be observed or measured. Sometimes these learning objectives are referred to as performance outcomes or behavioral outcomes, for a simple reason: If students are able to achieve the learning objectives of a lesson, they will achieve the desired outcome of the class. Effective instructors always start their presentations by discussing and reviewing the objectives of the presentation with the students.

A terminal objective is a broader outcome that requires the learner to have a specific set of skills or knowledge after a learning process. An enabling objective is an intermediate objective and is usually part of a series of objectives that direct instructors on what they need to instruct and what the learners will learn to accomplish the terminal objective. Consider the enabling objectives to be the steps that allow you to reach the top floor—that is, the terminal objective. An example of how terminal and enabling objectives are developed from JPRs is provided later in the chapter.

■ Understanding the Components of Learning Objectives

Many different methods may be used for writing learning objectives. One method commonly employed in the fire service is the ABCD method, where the acronym stands for **A**udience (Who?), **B**ehavior (What?), **C**ondition (How or using what?), and **D**egree (How well?). (Learning objectives do not always need to be written in that order, however.) The ABCD method was introduced in the book *Instructional Media and the New Technologies of Education* written by Robert Heinich, Michael Molenda, and James D. Russell (Macmillan, 1996).

Audience
The audience of the learning objective describes who the students are. Are your students experienced fire fighters or new recruits? Fire service learning objectives often use terms such as *fire fighter trainees, cadets, fire officers,* or *students* to describe the

audience. Some lesson plans will contain multiple objectives to meet the course goal. In this case, it is acceptable to reference the audience once at the beginning of the objective listing, as long as all of the objectives relate to the same audience members. If the objective is written correctly, the audience will be evident in the structure of the objective.

Behavior

Once the students have been identified, typically the behavior is listed next. The behavior must be an observable and measurable action. A common error in writing learning objectives is using words such as "know" or "understand" for the behavior. Is there really a method for determining whether someone understands something? It is preferable to use words such as "state," "describe," or "identify" as part of learning objectives—these are actions that you can see and measure. It is much easier to evaluate the ability of a student to identify the parts of a portable fire extinguisher than to evaluate how well the student understands the parts of a portable fire extinguisher. The importance of the behavior portion of the objective will be discussed later in the chapter.

The behavior may identify the type of presentation or class that will be conducted. Words such as *describe* or *state* in the objective imply that the student will know something, whereas words such as *demonstrate* or *perform* indicate that the student will be able to do something. This is where the terms *cognitive* or *psychomotor* objectives are applied in a properly formatted objective. Instructors should blend presentation styles to enhance the learning environment whenever possible. Appealing to multiple senses and allowing for many application opportunities enhances the learning.

Condition

The condition describes the situation in which the student will perform the behavior. Items that are often listed as conditions include specific equipment or resources given to the student, personal protective clothing or safety items that must be used when performing the behavior, and the physical location or circumstances for performing the behavior. For example, the following phrases in an objective specify the condition:

- "… in full protective equipment, including self-contained breathing apparatus."
- "… using the water from a static source, such as a pond or pool."

Degree

The degree is the last part of the learning objective; it indicates how well the student is expected to perform the behavior in the listed conditions. With what percentage of completion is the student expected to perform the behavior? Total mastery of a skill would require 100 percent completion—this means perfection, following every step on the skill sheet. For example, a student performing a ladder raise needs to complete all of the steps on the skill checklist to raise the ladder safely. Skipping any step (failing the skill) may result in a serious injury, damage to the ladder, or death. In contrast, many times knowledge-based learning objectives are expected to be learned to the degree stated in the passing rate for written

exams, such as 70 percent or 80 percent. Another degree that is frequently used is a time limit, which can be included in learning objectives dealing with both knowledge and skills.

Using the ABCD Method

Strictly speaking, well-written learning objectives should contain all four elements of the ABCD method. Nevertheless, learning objectives are often shortened because one or more of the elements are assumed to be known. If a lesson plan is identified as being used for teaching potential fire service instructors, for example, every single objective may not need to start with "the fire service instructor trainee." The audience component of the ABCD method may be listed once, at the top of all the objectives, or not listed at all.

The same principle applies to the condition component. If it is understood that a class requires all skills to be performed in full personal protective gear, it may not be necessary to list this condition in each individual learning objective. It is also common to omit the degree component, as many learning objectives are written with the understanding that the degree will be determined by the testing method. If the required passing grade for class written exams is 80 percent, it is assumed that knowledge learning objectives will be performed to that degree. Similarly, if the skill learning objectives for a class are required to be performed perfectly, a 100 percent degree for those learning objectives can be assumed.

Learning objectives should be shortened in this way only when the assumptions for the missing components are clearly stated elsewhere in the lesson plan. Of course, a learning objective is unlikely to omit the behavior component, because it is the backbone of the learning objective. Some curricula do require the ABCD to be stated each time, in sequence, and in complete form. For enabling objectives, a shortened form may sometimes be permitted; otherwise, many of the components in the series of objectives would simply be restated over and over again, which may confuse or frustrate a learner trying to identify course expectations.

ABCD learning objectives do not need to contain all of the parts in the ABCD order. Consider the following example:

> In full protective equipment including SCBA, two fire fighter trainees will carry a 24-foot extension ladder 100 feet and then perform a flat raise to a second-floor window in less than one minute and thirty seconds.

Here the audience is "the fire fighter trainees." The behaviors are "carry a 24-foot extension ladder" and "perform a flat raise." Both carrying and raising are observable and measurable actions. The conditions are "full protective equipment including SCBA," "100 feet," and "to a second-floor window;" they describe the circumstances for carrying and raising the ladder. The degree is "less than one minute and thirty seconds." The fire fighter trainees must demonstrate the ability to perform these behaviors to the proper degree to meet this learning objective successfully.

The ABCD method is a clear and appropriate way to address objectives. Inclusion of all four of its elements is essential in the construction of the terminal objective, which is the main idea for the lesson. This objective should contain all the components to inform the students what will be taught, the method of evaluation, and the resources consulted for the information presented. Subsequent (enabling) objectives may assume certain points previously stated in the main objective, such as audience, degree, and references, as long as these points are clarified in the main objective. Each enabling objective allows the student to meet the intent or goal of the terminal objective.

Parts of a Lesson Plan

Many different styles and formats for lesson plans exist. No matter which lesson plan format is used, however, certain components should always be included. Each of these components is necessary for you to understand and follow a lesson plan **FIGURE 6-2**.

Instructor Guide
Lesson Plan

Lesson Title: Use of Fire Extinguishers ← **Lesson Title**
Level of Instruction: Firefighter I ← **Level of Instruction**

Method of Instruction: Demonstration

Learning Objective: The student shall demonstrate the ability to extinguish a Class A fire with ← **Learning Objective**
a stored-pressure water-type fire extinguisher. (NFPA 1001, 5.3.16)

References: Fundamentals of Firefighter Skills, 3rd Edition, Chapter 8 ← **References**

Time: 50 Minutes

Materials Needed: Portable water extinguishers, Class A combustible burn materials, Skills ← **Instructional Materials**
checklist, suitable area for hands-on demonstration, assigned PPE for skill **Needed**

Slides: 73–78*

Step #1 Lesson Preparation:
- Fire extinguishers are first line of defense on incipient fires
- Civilians use for containment until FD arrives
- Must match extinguisher class with fire class
- FD personnel can use in certain situations, may limit water damage
- Review of fire behavior and fuel classifications
- Discuss types of extinguishers on apparatus
Demonstrate methods for operation

Step #2 Presentation	**Step # 3 Application**

A. Fire extinguishers should be simple to operate.
  1. An individual with only basic training should be able to use most fire extinguishers safely and effectively.
  2. Every portable extinguisher should be labeled with printed operating instructions.
  3. There are six basic steps in extinguishing a fire with a portable fire extinguisher. They are:
    a. Locate the fire extinguisher.
    b. Select the proper classification of extinguisher.
    c. Transport the extinguisher to the location of the fire.
    d. Activate the extinguisher to release the extinguishing agent.
    e. Apply the extinguishing agent to the fire for maximum effect.
    f. Ensure your personal safety by having an exit route.
  4. Although these steps are not complicated, practice and training are essential for effective fire suppression.
  5. Tests have shown that the effective use of Class B portable fire extinguishers depends heavily on user training and expertise.
    a. A trained expert can extinguish a fire up to twice as large as a non-expert can, using the same extinguisher.
  6. As a fire fighter, you should be able to operate any fire extinguisher that you might be required to use, whether it is carried on your fire apparatus, hanging on the wall of your firehouse, or placed in some other location.
B. Knowing the exact locations of extinguishers can save valuable time in an emergency.
  1. Fire fighters should know what types of fire extinguishers are carried on department apparatus and where each type of extinguisher is located.
  2. You should also know where fire extinguishers are located in and around the fire station and other work places.
  3. You should have at least one fire extinguisher in your home and another in your personal vehicle and you should know exactly where they are located.
C. It is important to be able to select the proper extinguisher.
  1. This requires an understanding of the classification and rating system for fire extinguishers.
  2. Knowing the different types of agents, how they work, the ratings of the fire extinguishers carried on your fire apparatus, and which extinguisher is appropriate for a particular fire situation is also important.
	Slides 7–10
	
	Ask students to locate closest extinguisher to training area
	
	Review rating systems handout— Have students complete work book activity page #389

(Continues)

(Continued)

3. Fire fighters should be able to assess a fire quickly, determine if the fire can be controlled by an extinguisher, and identify the appropriate extinguisher.
   a. Using an extinguisher with an insufficient rating may not completely extinguish the fire, which can place the operator in danger of being burned or otherwise injured.
   b. If the fire is too large for the extinguisher, you will have to consider other options such as obtaining additional extinguishers or making sure that a charged hose line is ready to provide back-up.
4. Fire fighters should also be able to determine the most appropriate type of fire extinguisher to place in a given area, based on the types of fires that could occur and the hazards that are present.
   a. In some cases, one type of extinguisher might be preferred over another.

What happens if wrong type or size extinguisher is used?

D. The best method of transporting a hand-held portable fire extinguisher depends on the size, weight, and design of the extinguisher.
  1. Hand-held portable fire extinguishers can weigh as little as 1 lb to as much as 50 lb.
  2. Extinguishers with a fixed nozzle should be carried in the favored or stronger hand.
    a. This enables the operator to depress the trigger and direct the discharge easily.
  3. Extinguishers that have a hose between the trigger and the nozzle should be carried in the weaker or less-favored hand so that the favored hand can grip and aim the nozzle.
  4. Heavier extinguishers may have to be carried as close as possible to the fire and placed upright on the ground.
    a. The operator can depress the trigger with one hand, while holding the nozzle and directing the stream with the other hand.
  5. Transporting a fire extinguisher will be practiced in Skill Drill 8-1.

Display available types of extinguishers

E. Activating a fire extinguisher to apply the extinguishing agent is a single operation in four steps.
  1. The P-A-S-S acronym is a helpful way to remember these steps:
    a. Pull the safety pin.
    b. Aim the nozzle at the base of the flames.
    c. Squeeze the trigger to discharge the agent.
    d. Sweep the nozzle across the base of the flames.
  2. Most fire extinguishers have very simple operation systems.
  3. Practice discharging different types of extinguishers in training situations to build confidence in your ability to use them properly and effectively.
  4. When using a fire extinguisher, always approach the fire with an exit behind you.
    a. If the fire suddenly expands or the extinguisher fails to control it, you must have a planned escape route.
    b. Never let the fire get between you and a safe exit. After suppressing a fire, do not turn your back on it.
  5. Always watch and be prepared for a rekindle until the fire has been fully overhauled.
  6. As a fire fighter, you should wear your personal protective clothing and use appropriate personal protective equipment (PPE).
  7. If you must enter an enclosed area where an extinguisher has been discharged, wear full PPE and use SCBA.
    a. The atmosphere within the enclosed area will probably contain a mixture of combustion products and extinguishing agents.

Have students demonstrate steps using empty extinguisher

Complete skills sheet #7–9 for each student

Review PPE required for extinguisher use

F. The oxygen content within the space may be dangerously depleted.

Discuss hazards of extinguishing agents

Step #4 Evaluation:
1. Each student will properly extinguish a Class A combustible fire using a stored-pressure type water extinguisher. (Skill Sheet x-1)
2. Each student will return extinguisher to service. (Skill Sheet x-2)

Lesson Summary: ← **Lesson Summary**
- Classifications of fire extinguishers
- Ratings of fire extinguishers
- Types of extinguishers and agents
- Operation of each type of fire extinguishers
- Demonstration of Class A fire extinguishment using a stored pressure water extinguisher

Assignments: ← **Assignment(s)**
1. Read Chapter 8 prior to next class.
2. Complete "You are the Firefighter" activity for Chapter 8 and be prepared to discuss your answers.

FIGURE 6-2 The components of a lesson plan.

■ Lesson Title or Topic

The lesson title or topic describes what the lesson plan is about. For example, a lesson title may be "Portable Fire Extinguishers" or "Fire Personnel Management." Just by the lesson title, you should be able to determine whether a particular lesson plan contains information about the topic you are planning to teach. A title page may be used to highlight or preview the content of the lesson plan package **FIGURE 6-3**. It may serve as a summary of all of the contents and help prepare the instructor for the class.

Des Plaines Fire Department
Division of Training   Lesson Title

Training Date: _____   Location of Training: _____

FireHouse Code: _____   Safety Plan Required? Y ☐   N ☐

Topic:	Instructor(s):
Teaching Method(s):	Time allotted:
Handouts:	
AV needs:	Teaching resources:
Level of Instruction:	Evaluation Method:
NFPA JPR's:	Equipment needed:

LEARNING OBJECTIVES: *Upon completion of the class and study questions, each participant will independently do the following with a degree of accuracy that meets or exceeds the standards established for their scope of practice:*

General class activities for student application

| Safety Briefing | SAFETY RED FLAGS |
| 1. | ALL STOPS: |

FIGURE 6-3 Sample cover sheet for a lesson plan.

■ Level of Instruction

It is important for a lesson plan to identify the <u>level of instruction</u>, because your students must be able to understand the instructional material. Just as an elementary school teacher would not use a lesson plan developed for high school students, so you as a fire service instructor must ensure that the lesson plan is written at an appropriate level for your students. Often the level of instruction in the fire service corresponds with National Fire Protection Association (NFPA) standards for professional qualifications. If you are teaching new recruits or cadets, you would use lesson plans that are designated as having a Fire Fighter I or II level of instruction. If you are teaching fire service professional development classes, you may use lesson plans that are specified as having a Fire Officer I or II level of instruction. Another method of indicating the level of instruction is by labeling the lesson plan with terms such as "beginner," "intermediate," or "advanced." No matter which method is used to indicate the level of instruction, you should ensure that the material contained in a lesson plan is at the appropriate level for your students.

Another component of the level of instruction is the identification of any prerequisites. A prerequisite is a condition that must be met before the student is permitted to receive further instruction. Often, a prerequisite is another class. For example, a Fire Service Administration class would be a prerequisite for taking an Advanced Fire Service Administration class. A certification or rank may also be a prerequisite. Before being allowed to receive training on driving an aerial apparatus, for example, the department may require a student to hold the rank of a Driver and possess Driver/Operator—Pumper certification.

Safety Tip

You should ensure that the proper prerequisites are met by each of your students. Failure to do so may mean that a student performs tasks that he or she is not qualified or prepared to perform.

Behavioral Objectives, Performance Objectives, and Learning Outcomes

As mentioned earlier in this chapter, learning objectives are essential to the lesson plan. All lesson plans must have learning objectives. Many methods for determining and listing learning objectives are available. The specific method used to write the learning objectives is not as important as ensuring that you understand the learning objectives for the lesson plan that you must present to your students. The Fire Service Instructor II will use JPRs to develop the learning objectives in the ABCD format, and the Instructor III will write course objectives.

Instructional Materials Needed

Most lesson plans require some type of instructional materials to be used in the delivery of the lesson plan. Instructional materials are tools designed to help you present the lesson plan to your students. For instance, audiovisual aids are the type of instructional material most frequently listed in a lesson plan—that is, a lesson plan may require the use of a video, DVD, or computer. Other commonly listed instructional materials include handouts, pictures, diagrams, and models. Also, instructional materials may be used to indicate whether additional supplies are necessary to deliver the lesson plan. For example, a preincident planning lesson plan may list paper, pencils, and rulers as the instructional materials needed. The actual equipment students will be using or operating could be

shown and passed through the classroom portion of the lesson (e.g., nozzles, hand tools, PPE).

Lesson Outline

The lesson outline is the main body of the lesson plan. This element is discussed in detail later in this chapter. The lesson outline comprises four main elements: preparation, presentation, application, and evaluation. Each area fulfills a specific purpose in the delivery of instruction. Instructors who engage their students effectively use each part to ensure that the objectives of the lesson plan are properly met to the level intended.

References/Resources

Lesson plans often simply contain an outline of the information that must be understood to deliver the learning objectives. Fire service instructors who are not experts in a subject may need to refer to additional references or resources to obtain further information on these topics. The references/resources section may contain names of books, Web sites, or even names of experts who may be contacted for further information. By citing references in the lesson plan, the validity of the lesson plan can also be verified.

Lesson Summary

The lesson summary simply summarizes the lesson plan. It reviews and reinforces the main points of the lesson plan. The lesson summary plays an important role in the overall lesson, allowing the instructor to enhance the application step by asking summary questions on key objective and lesson points. You may view this as the instructor asking the student, "What did I just teach you or show you?"

Assignment

Lesson plans often contain an assignment, such as a homework-type exercise that will allow the student to further explore or apply the material presented in the lesson plan. Be prepared to explain the assignment, its due date, the method for submitting the assignment, and the grading criteria to be used.

The Four-Step Method of Instruction

While reviewing and preparing for class with your lesson plan, the four-step method of instruction is the primary process used to relate the material contained in the lesson plan to the students TABLE 6-1. These four steps are found within the lesson plan. The four-step method of instruction is the method of instruction most commonly used in the fire service.

Step 1: Preparation

The preparation step is the first phase in the four-step method of instruction. The preparation step—also called the motivation step—prepares or motivates students to learn. When beginning instruction, you should provide information to students that explains why they will benefit from the class. Adult learners need to understand what they will get out of attending the class, because very few adults have time to waste in sitting through a presentation that will not directly benefit them.

JOB PERFORMANCE REQUIREMENTS (JPRS)
in action

The lesson plan is the tool used by a fire service instructor to conduct a training session. It is as essential as personal protective equipment (PPE) is to a fire fighter. The lesson plan details the information necessary to present the training session, which includes everything from the title of the class to the assignment for the next training session. In between are the resources needed, the behavioral objectives, the content outline, and the various teaching applications used to complete the training. As a fire service instructor, you must review and practice the delivery of the lesson plan, check the materials needed for the class, and be ready to present the materials. Using the lesson plan, you must present a structured training session by taking advantage of appropriate methods of instruction to engage the students and use a variety of communication skills to complete the learning objectives.

Instructor I

The Instructor I will teach from a prepared lesson plan using appropriate methods of delivery and communication skills to ensure that the learning process is effective. The instructor must understand each component of the lesson plan. It may be necessary to adapt the lesson plan to the needs and abilities of the audience and the teaching environment.

Instructor II

The Instructor II will prepare the lesson plan components and determine the expected outcomes of the training session. The four-step method of instruction should be defined within the lesson plan and all instructional requirements outlined for the presenter's use.

Instructor III

The Instructor III will conduct a training needs assessment and develop a curriculum to meet the training need, including course goals and evaluation strategies.

JPRs at Work

Present prepared lesson plans by using various methods of instruction that allow for achievement of the instructional objectives. Adapt the lesson plan based on student needs and specific conditions.

JPRs at Work

Create and modify existing lesson plans to satisfy the student needs, JPRs, and objectives developed for the training session.

JPRs at Work

There are no JPRs at this instructor level.

Bridging the Gap Among Instructor I, Instructor II, and Instructor III

A partnership must exist between the developer of the lesson plan and the instructors who will deliver that lesson plan. In many cases, these individuals may be the same person. In other cases, such as in large departments, you may never know who wrote the lesson plan. Your skill in developing a lesson plan that another instructor can use—perhaps long after it was developed—is therefore important. If another person must deliver your lesson plan, however, you must be sure that all components of the lesson plan are clear and concise and that the material and instructional methods match the needs of the students. Your communications skills and knowledge of the learning process will be used at both Instructor I and Instructor II levels in the development and delivery of this content. The Instructor III develops curricula based on a training needs assessment and provides direction to instructors at the other levels on the goal or outcome of the material they are developing or delivering.

Table 6-1	The Four-Step Method of Instruction
Step in the Instructional Process	**Instructor Action**
1. Preparation	The instructor prepares the students to learn by identifying the importance of the topic, stating the intended outcomes, and noting the relevance of the topic to the student. This should not be confused with the time used by the instructor to prepare course material and review lesson plan content.
2. Presentation	The presentation of content is usually organized in an outline form that supports an understanding of the learning objectives. Varied presentation techniques should be used to keep the students' interest and to maximize learning.
3. Application	The instructor applies the presentation material as it relates to students' understanding. Often the fire service instructor will ask questions of the students or ask students to practice the skill being taught. This step also allows the instructor to do a progress check on the students and to identify any learning issues before moving on to other class content.
4. Evaluation	The students' understanding is evaluated through written exams or in a practical skill session.

The benefit of a class can be explained in many ways:

- The class may count toward required hours of training.
- The class may provide a desired certification.
- The class may increase students' knowledge of a subject.

Whatever the benefit may be, you should explain it thoroughly during the preparation step. In a lesson plan, the preparation section usually consists of a paragraph or a bulleted list describing the rationale for the class. During the preparation step, the Fire Service Instructor I needs to grab students' attention and prepare them to learn when the instructor begins presenting the prepared lesson plan. Adult learners like to learn quickly how the class material will affect them: Will it make them safer or more knowledgeable about their job? Will it improve their efficiency on the fire ground or make the students better leaders? This step in the process is critical as a motivational tool to encourage the students to participate in the learning process, not just to sit in on the class as a way to pass the time. The Fire Service Instructor II, while developing the lesson plan, will include suggested preparation points, including safety- and survival-related information, local examples, and explanations of how the material will help improve students' ability to do their job.

■ Step 2: Presentation

The presentation step is the second step in the four-step method of instruction; it comprises the actual presentation of the lesson plan. During this step, you lecture, lead discussions, use audiovisual aids, answer student questions, and perform other techniques to present the lesson plan. (The *Methods of Instruction* chapter discusses the various methods of instruction used during this step.) In a lesson plan, the presentation section normally contains an outline of the information to be presented. It may also contain notes indicating when to use teaching aids, when to take breaks, or where to obtain more information.

■ Step 3: Application

The application step, which is the third step in the four-step method of instruction, is the most important step because it is during this phase that students apply the knowledge presented in class. The root word *apply* is very important to illustrate what the instructor needs to let the students do: apply what

they are learning to examples or actual hands-on practice. Normally, this is where learning occurs, as students practice skills, perhaps make mistakes, and retry skills as necessary. You should provide direction and support as each student performs this step. You must also ensure that all safety rules are followed as students engage in new behaviors.

In a lesson plan, the application section usually lists the activities or assignments that the student will perform. In the fire service, the application section of a psychomotor objective lesson often requires the use of skill sheets for evaluation purposes. The experienced fire service instructor uses the application step to make sure that each student is progressing along with the lesson plan. This step also allows students to participate actively and to remain engaged in the learning process.

■ Step 4: Evaluation

The evaluation step is the final step of the four-step method of instruction. It ensures that students have correctly acquired the knowledge and skills presented in the lesson plan. The evaluation may, for example, take the form of a written test or a skill performance test. No matter which method of evaluation is used, the student must demonstrate competency without assistance. In a lesson plan, the evaluation section indicates the type of evaluation method and the procedures for performing the evaluation.

Instructional Preparation

Once you have a lesson plan, the instructional preparation begins. Which materials are needed for the class? Which audiovisual equipment will be used? Where will the class be conducted? How much time will be needed? These and many other questions must be answered during instructional preparation. The information contained in the lesson plan should be used as a guide for instructional preparation.

■ Student Preparation

Students should prepare for instruction by coming to a class prepared and ready to learn. Certain classroom or drill-ground rules may exist that prohibit bringing cell phones, newspapers, or other reading materials into the learning environment. The instructor should monitor the preparedness of the students as they come to the classroom or drill ground, and may enhance their readiness to learn by providing class information and objectives ahead of time. The expectations and outcomes may be better if the students take the time to prepare before the class. Bringing textbooks, notebooks, and writing supplies is another important part of the student's preparation.

■ Organizational Skills

Taking the information from a lesson plan and transforming it into a well-planned class takes good organizational skills. First, you should organize the class planning timeline **FIGURE 6-4**. Identify the time available for you to plan and prepare for the class. The time available for preparation is usually the amount of time from the point when the lesson plan is identified until the day when the class is scheduled to be taught. Identify the milestones that must be accomplished as part of this timeline. Depending on the lesson plan, milestones may include obtaining audiovisual equipment, purchasing materials, reserving a classroom, or previewing audiovisual aids.

■ Procuring Instructional Materials and Equipment

Most classes take advantage of instructional materials or equipment. The method of obtaining these instructional materials and equipment differs from fire department to fire department.

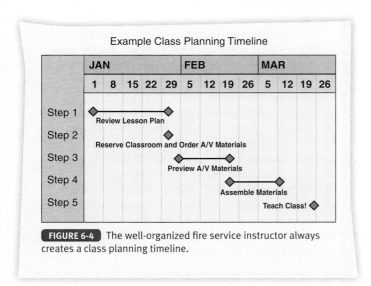

FIGURE 6-4 The well-organized fire service instructor always creates a class planning timeline.

A common method of procuring materials is for the fire service instructor to contact the person in the fire department who is responsible for purchasing training materials, such as a training officer or someone assigned to the training division. You may be required to provide a list of needed materials to the training officer. Often, this list of materials must be submitted well in advance of when the class is scheduled to begin, due to the purchasing requirements of your agency. The key is to know your agency's time frame and to work within it to allow you to follow policy. The training officer, in turn, compiles the materials either by purchasing new materials or by securing materials already available at the training division. The training officer contacts the fire service instructor when all class-related materials are available.

Another common method for procuring class equipment is the equipment checkout process, which is typically managed by the fire department's training division. For example, if you need a multimedia projector for a class, you would submit a request for the projector in which you indicate the date and time when the projector is needed. The training division would then reserve the projector for you. On the day of the class, the projector would be available for you. Depending on the organizational procedures, you might be required to pick up the projector and set it up, or the training division might set up the projector at the class location for you. Regardless of how you obtain your projector, you should make sure you understand how to use it and how to troubleshoot any problems before class begins.

■ Preparing to Instruct

The most important part of the instruction's preparation is preparing for actual delivery of the lesson plan in the classroom or on the drill ground. If you obtain the necessary materials, equipment, apparatus, drill tower, or classroom, but you do not prepare to deliver the lesson plan, the class will not be successful. You should be thoroughly familiar with the information contained in the lesson plan, which may require you to consult the references listed in the lesson plan and research the topic further. If the lesson plan includes a computer presentation, then practice using this technology to deliver the presentation.

No matter which method of instructional delivery is used, you should always rehearse your presentation before delivering it to a training session full of students. A class is destined for failure if you are seeing the presentation material for the first time in front of the class. Successful fire service instructors have a sound understanding of the information that they are delivering and can adapt to the particular needs of their class because they always know what is coming next.

■ Adapting Versus Modifying a Lesson Plan

One of the most important—yet confusing—distinctions between a Fire Service Instructor I and a Fire Service Instructor II is the Instructor II's ability to modify a lesson plan. A lesson plan is a guide or a roadmap for delivering instruction, but it is rarely implemented exactly as written. To understand what can and cannot be changed by each level of fire service instructor, let us review what the JPRs for NFPA 1041, *Standard for Fire Service Instructor Professional Qualifications*, say about modifying and adapting a lesson plan.

Fire Service Instructor I

4.3.2 Review instructional materials, given the materials for a specific topic, target audience, and learning environment, so that elements of the lesson plan, learning environment, and resources that need adaptation are identified.

4.3.3 Adapt a prepared lesson plan, given course materials and an assignment, so that the needs of the student and the objectives of the lesson plan are achieved.

The Fire Service Instructor I should not alter the content or the lesson objectives. Prior to the beginning of the class, the Instructor I should be able to evaluate local conditions, assess facilities for appropriateness, meet local standard operating procedures (SOPs), and recognize students' limitations. He or she should be able to change the method of instruction and course materials to meet the needs of the students and accommodate their individual learning styles, including making adaptations as necessary due to the learning environment, audience, capability of facilities, and types of equipment available.

Fire Service Instructor II

5.3.3 Modify an existing lesson plan, given a topic, audience characteristics, and a lesson plan, so that the JPRs or learning objectives for the topic are addressed and the plan includes learning objectives, a lesson outline, course materials, instructional aids, and an evaluation plan.

To clearly understand the difference between adapting and modifying, you must understand the proper definitions of these terms:

- Modify: to make basic or fundamental changes
- Adapt: to make fit (as for a specific use or situation)

Put simply, a Fire Service Instructor II can make basic or fundamental changes to the lesson plan, whereas a Fire Service Instructor I can adapt only to local conditions without altering objectives. Fundamental changes include changing the performance outcomes, rewriting the learning objectives, modifying the content of the lesson, and so on.

What can a Fire Service Instructor I do? He or she can make the lesson plan fit the situation and conditions. These conditions include the facility, the local SOPs, the environment, limitations of the students, and other local factors.

NFPA 1041 specifically states that a Fire Service Instructor I may modify the method of instruction and course materials to meet the needs of the students and accommodate the individual fire service instructor's style. Here are a few real-life examples:

- A Fire Service Instructor I may change a lesson plan's method of instruction from lecture to discussion if he or she determines that the latter method would be a better presentation format because of the students' level of knowledge.
- A Fire Service Instructor I may adapt the classroom setting if the facility cannot meet the seating arrangement listed in the lesson plan.

- A Fire Service Instructor I may adapt the number of fire fighters performing an evolution in a lesson plan from three to four to meet local staffing SOP requirements.
- A Fire Service Instructor I *cannot* modify a lesson plan learning objective that states a fire fighter must raise a 24-foot extension ladder because he or she feels the task is too difficult for one fire fighter.
- A Fire Service Instructor I *cannot* change the JPR of developing a budget in a Fire Officer lesson plan because he or she does not feel comfortable teaching that subject.

As with all other positions within the fire service, it is important that fire service instructors perform only those actions within their level of training. As a Fire Service Instructor I, you must recognize what you can and cannot do. Acting

Teaching Tip

All instructors routinely adapt and modify courses. While a Fire Service Instructor I may believe that a curriculum needs to be modified, those changes should be made only by a Fire Service Instructor II. This process will ensure that development of the curriculum is done correctly and that coverage of the lesson objectives is not reduced.

Theory Into Practice

Lesson plans must remain dynamic in both the short term and the long term. In the short term, you should understand when it is appropriate to adjust a lesson plan during its delivery based on students' learning styles, changing conditions, timing considerations, and students' progress. In the long term, you should provide input to your supervisor regarding the success of the delivery. If problems occurred or improvements are needed, report this feedback as well.

One critical component of lesson plan adaptability is the break times. Break times are built into course schedules and may need to be adjusted to reflect the training environment or to accommodate other classes being conducted at the same time.

If you make adjustments to the delivery of a lesson plan, it is critical that you ensure that all learning objectives are still covered. For example, many times activities must be scheduled around the activities of other courses. Scheduling use of resources in the field with other instructors can help reduce conflicts. The program coordinator should be advised if an instructor intends to move portions of the program around so that the coordinator can ensure the change does not affect other programs and shared resources.

VOICES
OF EXPERIENCE

As a 16-year-old junior fire fighter in my hometown, I was baffled by all of the possibilities that one could achieve in the fire service. All the thrills of going on calls and learning the ropes were at times quite overwhelming. I learned early on how to distinguish a good fire instructor from those who simply blew smoke. I did my time and soon became interested in the teaching aspect of the fire business. I had taken numerous classes on a variety of fire subjects, as well as English and writing classes at Penn State University. The time had come to go after state instructor certification in Pennsylvania.

The program to become state certified as a fire service instructor at the time consisted of attending a 40 Instructor Methodology class and passing a written test at the state fire academy. Upon successful completion of that week-long session and passing the written test, I was granted

> **"You won't appreciate having an updated lesson plan to teach a class until you actually need it."**

state certification as an Instructor I. I later took an Instructor II class and was certified to that level. After apprenticing in several classes, I learned different teaching styles from different instructors. Most instructors were using canned lesson plans, which were okay but left much to the imagination.

As time went on and I began to develop my own programs, I researched and selected a lesson plan format that suited me and met all the requirements of a professional plan. Like many instructors, occasionally I had to teach a class with no lesson plan, commonly known as "winging it." These classes never went quite right. It was like driving somewhere new without a map or GPS. I also was more aware of how many instructors could not develop a quality lesson plan.

Being taught to write a lesson plan that anyone could pick up and teach a class has always stuck with me. That is the true test of a well-written plan. I have developed many, many lesson plans, but one in particular stands out. It was the one I had to write for a 15-minute presentation I had to do in front of a review board as part of the interview process for the ultimate position of my career. After writing many possible lesson plans, I finally was satisfied with my final product and went with it. I felt I had the road map that I needed to make my presentation successful. That one experience truly made me fully understand the importance of having a well-written lesson plan when instructing.

Outdated lesson plans have always been a thorn in my side. It is one thing to develop a good lesson plan, but it is another thing to keep the lesson plan current. As techniques, strategies, and technology change, so must the teaching outline for an affected subject. You won't appreciate having an updated lesson plan to teach a class until you actually need it.

In my opinion, the most important component of being a successful instructor is mastering the ability to develop and *use* quality lesson plans. Teaching is so much easier when you have a lesson plan to refer to when necessary. I can truly state that my success and longevity in the instructing field can be largely attributed to the ability to write and use good lesson plans. Couple the lesson plan with all the other necessary attributes of a good instructor, and you will be a success story of your own.

F. Tom Hand
Fire Training Coordinator (Retired)
Mesa Fire & Medical Department
Mesa, Arizona

outside your scope of training may lead to legal liability. If you are ever unsure whether you have sufficient knowledge or skills to teach a topic, you should discuss this issue with your superior.

Reviewing Instructional Materials for Adaptation

There are many ways for a Fire Service Instructor I to obtain a lesson plan: fire service Web sites, commercially published curriculum packages, the National Fire Academy, the instructor's fire department's training library, or other fire departments. No matter which method is used, the lesson plan must be reviewed and any areas that need adaptation must be identified. This is true even for lesson plans that were originally developed within your own fire department. Over time, standards and procedures change, so that a lesson plan that was completely correct for your department when it was created may be out-of-date in just a few months.

After obtaining a lesson plan, you must review the entire lesson plan and determine whether adaptations are needed to make it usable for your class. As part of the class planning and preparation process, lesson plan adaptations must be scheduled and completed before you deliver the presentation to the class. A lesson plan might need adaptations for many reasons, such as differences related to the learning environment, the audience, the capability of facilities, and the types of equipment available. Always be prepared to adapt the lesson plan to accommodate last-minute classroom or equipment changes. While these occurrences should be rare, they do happen.

Instructors may find they need to adapt a lesson plan in situations where all members of the audience do not come from one department, as in a regional or academy delivery. If you are teaching from a prepared lesson plan within your department only, the audience factors, prerequisite knowledge, and abilities are likely already known, so you may have to adapt the lesson plan only if it is from an outside source, or to tailor it to the available delivery time.

Evaluating Local Conditions

The main focus when adapting a lesson plan is to make minor adjustments so it fits your local conditions and your students' needs. To accomplish this, you must first be familiar with your audience. The Fire Service Instructor I should contemplate the following questions when reviewing a lesson plan for any adaptations needed to accommodate the intended audience:

- Which organizational policies and procedures apply to the lesson plan?
- What is the current level of knowledge and ability of your students?
- Which types of tools and equipment will your students use in performing the skills within the lesson plan?

The second area pertaining to the local conditions that must be considered is you, the Fire Service Instructor I:

- What is your experience level and ability? Can you identify resources and materials to improve your knowledge and background? If not, be prepared to ask for assistance from another instructor.
- How familiar are you with the topic that will be taught? Teaching material you are not familiar with can pose severe safety and credibility hazards. If you are not familiar with the topic, do not attempt to fake it, as wrong or incomplete information could be dangerous.
- What is your teaching style?

The answers to these types of questions will allow you to adapt the lesson plan so that you deliver the lesson in the most effective way given your own abilities.

Evaluating Facilities for Appropriateness

You should also review and adapt the lesson plan based on the facilities that will be used when delivering the class. Several factors—for example, the equipment available, student seating, classroom size, lighting, and environmental noise—must be considered as part of this evaluation. For example, a lesson plan may call for students to sit at tables that have been moved into a U-shaped arrangement. However, if the local classroom has desks that are fixed to the floor, you will not be able to arrange the seating as indicated in the lesson plan. The lesson plan would then need to be adapted to meet the conditions of the facility and the seating arrangement changed accordingly. You should make this adaptation, keeping in mind the reason for the indicated seating arrangement.

Meeting Local SOPs

A lesson plan must be reviewed to ensure that it meets and follows local SOPs. This is one of the most important

Teaching Tip

The National Fire Academy maintains a Web site called TRADE's Virtual TRADEing Post, which enables fire service instructors to share non-copyrighted information, PowerPoint® presentations, lesson plans, and training programs or other downloadable materials. The training information is provided free of charge with the understanding that you must give credit to the department or agency that developed it.

This Web site is part of the Training Resources and Data Exchange (TRADE) program, which is a network designed to foster the exchange of fire-related training information and resources among federal, state, and local levels of government.

considerations when adapting a lesson plan. You should never teach information that contradicts a SOP. Not only would this lesson be confusing for the students, but it would also create a liability for you. If a student were to be injured or killed while performing a skill in violation of a SOP, you would be held responsible. At a minimum, you would be disciplined within the organization—but it is also possible that you might be held legally responsible in either criminal or civil court.

When reviewing a lesson plan, make note of the SOPs that may cover the topics in the lesson plan. After completely reviewing the lesson plan, research the SOPs and ensure that no conflicts exist. If your research turns up conflicting information, you should adapt the lesson plan to meet the local SOPs as long as it does not change the lesson objectives. If you are not familiar with your local SOPs, contact someone within the department who can assist you with ensuring that the lesson plan is consistent with local SOPs.

■ Evaluating Limitations of Students

The lesson plan should also be reviewed in light of student limitations and adapted to accommodate those limitations if possible. The lesson plan should be at the appropriate educational level for the students, and the prerequisite knowledge and skills should be verified. For example, if you were reviewing a lesson plan to teach an advanced hazardous materials monitoring class, students should have already undergone basic hazardous materials training. If you were training new fire fighters and reviewing this lesson plan, you may not be able to adapt it. Instead, you would most likely have to require students to undergo additional training before the lesson plan would be appropriate to present to them.

Safety Tip

When adapting a lesson plan, closely evaluate the revised plan's safety implications. It is all too easy to omit important safety information that was previously included in the lesson plan or to include information that may create a safety issue when combined with other material.

■ Adapting the Method of Instruction

The method of instruction is the one area that a Fire Service Instructor I may readily modify. Such a modification may be needed to allow you to deliver the lesson plan effectively, but it should not change the learning objectives. For example, you may not be comfortable using the discussion method to deliver a class on fire service sexual harassment as indicated

Ethics Tip

Fire service instructors regularly adapt material to meet their departmental needs and to improve the curriculum. Is it ethical to alter material to reflect your personal opinions when those opinions run counter to the traditional way of thinking? The solution to this dilemma is not as simple as it may seem. Imagine where the fire service would be today if only a few years ago bold fire service instructors did not step up to the plate and refuse to present material that was not based on safe practices. "Doing the right thing" is what ethical decisions are all about—but there is a fuzzy line between "the right thing" and "my way is the only right way."

For example, changing a course by eliminating the use of fog nozzles because you believe that these nozzles lead to hand burns is as dangerous as another fire service instructor eliminating the use of smooth-bore nozzles from a lesson plan. The reality is that students must understand the appropriate use, benefits, and dangers of each type of nozzle. What might seem like a simple adaptation could have serious consequences for a student who is not trained thoroughly and properly to department standards.

by the lesson plan. Instead, you might modify the lesson plan and change the method of instruction to lecture. This would allow the same information to be taught, just in a different format, and the same learning objectives would still be achieved.

■ Accommodating Instructor Style

In addition to ensuring that the method of instruction best suits your abilities, lesson plans may be adapted to accommodate your personal style. A lesson plan often reflects the style of the fire service instructor who wrote it. When reviewing and adapting a lesson plan, consider whether the lesson plan—and especially the presentation section—fits your own style. For example, a lesson plan may call for a humorous activity designed to establish a relationship between the instructor and the students. If you are teaching a military-style academy class, this may not be the best style, so you may need to adapt the presentation accordingly.

■ Meeting the Needs of the Students

All adaptations should be done with one purpose in mind—namely, meeting the needs of the students. As with all lesson plans, the main goal is to provide instruction that allows students to obtain knowledge or skills. This goal should be verified after you review and adapt a lesson plan.

Fire Service Instructor II

Creating a Lesson Plan

The Fire Service Instructor II is responsible for creating lesson plans. Depending on the subject, this task can take anywhere from several hours to several weeks. Regardless of the size of the lesson plan, the ultimate goal is to create a document that any fire service instructor can use to teach the subject and ensure that students achieve the learning objectives.

Many fire departments have lesson plan templates for the Fire Service Instructor II to use as a starting point. Such a standard format makes it easier for all fire service instructors in the department to understand the lesson plan and ensures consistency in training. If readily accessible and available, it may be easy for the instructor to access one style of lesson plan provided by a publisher and use that as the template for the training program. Consistency may be achieved if a variety of instructors all use the same lesson plan format to write or teach from. The lesson plans made available with this textbook to the instructor as part of the instructor's toolkit would be an example of a prepared lesson plan template that is written in Microsoft Word and is easily reused as a standard format. The instructor's toolkit and online companion Web site offer a sample lesson plan template (intentionally left blank) for you to use in the future.

A simple way of approaching the development of an outline that will be the basis of your lesson plan is to use a basic *introduction*, *body*, and *conclusion* sequence to organize your thoughts.

Safety Tip

Once a lesson plan is modified, go back and confirm that all learning objectives are met.

■ Achieving Job Performance Requirements

The first step of lesson plan development is to determine the learning objectives. What are students expected to achieve as a result of taking the class? Many times this desired outcome is obvious, because you are teaching a class to prepare students to perform a certain job or skill. For example, if you were to develop a lesson plan for a class to train fire fighters to drive a fire engine, you would start by listing the JPRs for a fire engine driver. On many other occasions, however, the learning objectives are not that clear.

It is very difficult to develop a lesson plan when the learning objectives are not clearly stated. Many fire service instructors have been in the unhappy position of being told to teach a certain class, such as one dealing with workplace diversity or safety, without clear direction on the intended learning objectives. Although the person requesting the class may have a general idea of what the class is intended to accomplish, he or she might not know the specific learning objectives that the Fire Service Instructor II needs to develop a lesson plan.

For example, the fire chief may want to improve fire fighter safety through training. Unless given specific learning objectives, the Fire Service Instructor II cannot develop a lesson plan to "improve fire fighter safety." When placed in this position, the Fire Service Instructor II should attempt to clarify the fire chief's vision of improving fire fighter safety. Would he like all fire fighters to understand the chain of events that leads to an accident and to know how to break that chain so that an accident is avoided? A learning objective can be written to achieve that goal. Or does the fire chief expect all fire fighters to don their structural firefighting protective equipment properly within a time limit? A learning objective can be written to achieve that goal, too. Whenever you are asked to develop a lesson plan for a class, start by clarifying the intended outcome of the class with the person who requests the class.

■ Learning Objectives

Once the Fire Service Instructor II has a clear outcome for a class, he or she should work backward to develop the learning objectives for the class. As described earlier in this chapter, learning objectives can be written utilizing the ABCD method TABLE 6-2 .

Audience

The audience should describe the students who will take the class. If the lesson plan is being developed specifically for a certain audience, the learning objectives should be written to indicate that fact. For example, if a Fire Service Instructor II is writing a lesson plan for a driver training class, the audience would be described as "the driver trainee" or "the driver candidate." Both of these terms indicate that the audience consists

Table 6-2	Learning Objectives Using the ABCD Method	
Method Step	**Question**	**Example**
Audience	Who	The fire fighter
Behavior	Will do or know what	Will perform a one-person ladder raise
Condition	Using which equipment or resources	Using a two-section, 24-foot ground ladder
Degree	How well or to which level	So that the proper climbing angle is set and all safety precautions are observed

of individuals who are learning to be drivers. If the audience is not specifically known or if it includes students from mixed backgrounds, the audience part of the learning objective could be written more generically, such as "the fire fighter" or even "the student."

Behavior

As described earlier, the behavior part of the learning objective should be specified using a clearly measurable action word, which allows the evaluation of the student's achievement of the learning objective. Another important consideration is the level to which a student will achieve the learning objective. This level is most often determined using Bloom's Taxonomy (*Taxonomy of Educational Objectives*, 1956), a method to identify levels of learning within the cognitive domain (discussed further in the chapter *The Learning Process*) **TABLE 6-3**.

Cognitive Domain Objectives

For the fire service, the most commonly used levels when developing *cognitive* learning objectives are (in order from simplest to most difficult) knowledge, comprehension, and application:

- *Knowledge* is simply remembering facts, definitions, numbers, and other specific items.
- *Comprehension* is displayed when students clarify or summarize important points.
- *Application* is the ability to solve problems or apply the information learned to situations.

Higher levels of application and understanding occur when the learning objectives are written at these levels:

- *Analysis* is the ability to break down information into components and the ability to know which components affect understanding.

- *Synthesis* is the ability to reassemble information in a new manner using abstract thinking and creativity.
- *Evaluation* is the ability to make judgments, critiques, or appraisals of the information.

In 2001, Dr. David Karthwohl and several other educational theorists modified the original Bloom's Taxonomy of Educational Objectives to expand its scope **FIGURE 6-5**. Several levels were revised:

- Knowledge became Remembering
- Comprehension became Understanding
- Application became Applying
- Analysis became Analyzing
- Evaluation became a level 5 task
- Synthesis was moved to the highest level and became Creating

A Fire Service Instructor II must determine which level within the cognitive domain is the appropriate level for the student to achieve for the lesson plan. For example, if an Instructor II is developing objectives for a class on portable extinguishers, the following objectives could be written for each basic level:

- Knowledge: "The fire fighter trainee will identify the four steps of the PASS method of portable extinguisher application." For this objective, the student simply needs to memorize and repeat back the four steps of pull, aim, squeeze, and sweep. Achievement of this very simple knowledge-based objective is easily evaluated with a multiple-choice or fill-in-the-blank question.
- Comprehension: "The fire fighter trainee will explain the advantages and disadvantages of using a dry-chemical extinguisher for a Class A fire." This objective

Table 6-3	Bloom's Taxonomy of Educational Objectives		
	Definition	**Example**	**Sample Verbs**
Knowledge	Ability to assemble terms or facts, memorize material, and recall information	Identify the wraparound short backboard device.	List, Define, Recall, State, Recite
Comprehension	Ability to use knowledge and interpret or translate the information (i.e., understand the meaning of information)	Recognize situations where the wraparound short backboard is indicated.	Explain, Summarize, Describe, Restate, Interpret
Application	Ability to use knowledge and comprehension to apply information to new situations	Illustrate situations of poor application of the wraparound short backboard.	Solve, Illustrate, Apply, Put into practice
Analysis	Ability to break down information into components and determine how each component affects understanding (i.e., looking at elements and the relationships)	Analyze situations when a wraparound short backboard is or is not indicated.	Organize, Analyze, Compare, Contrast
Synthesis	Ability to reassemble the information in a new manner using abstract thinking and creativity	Describe a scenario that illustrates other uses of the wraparound short backboard device.	Design, Hypothesize, Discuss, Devise
Evaluation	Ability to make judgments, critiques, or appraisals of the information	Following the completion of the scenario, evaluate the use of the wraparound short backboard device on an entrapped vehicle crash victim.	Evaluate, Judge, Defend, Justify

COGNITIVE DOMAIN	PSYCHOMOTOR DOMAIN	AFFECTIVE DOMAIN
KNOWLEDGE Recognition and recall of facts and specifics EXAMPLES: Define, Describe, List, State	**IMITATION** Observes skills and attempts to repeat them EXAMPLES: Assemble, build, connect, couple, repeat	**RECEIVING** Listening passively; attending to EXAMPLES: Ask, name
COMPREHENSION Interprets, translates, summarizes, or paraphrases given information EXAMPLES: Convert, infer, rewrite	**MANIPULATION** Performs skills by instruction rather than observation EXAMPLES: Transmit, arrange, recreate	**RESPONDING** Complies to given expectation; shows interest EXAMPLES: Answer, recite
APPLICATION Processes information in a situation different from original learning context EXAMPLES: Demonstrate, relate, produce	**PRECISION** Reproduces a skill with accuracy, proportion, and exactness; usually performed independently of original sources EXAMPLES: Modify, demonstrate	**VALUE** Displays behavior consistent with single belief or attitude; unforced compliance EXAMPLES: Complete, explain, justify
ANALYSIS Separates whole into parts; clarifies relationships among elements EXAMPLES: Diagram, outline, illustrate	**ARTICULATION** Combines more than one skill in sequence with harmony and consistency EXAMPLES: Combine, coordinate, develop, modify	**ORGANIZING** Committed to set of values as displayed by behavior EXAMPLES: Integrate, adhere
SYNTHESIS Combines elements to form new entity from original one EXAMPLES: Compile, compose, design	**NATURALIZATION** Completes one or more skills with with ease; requires limited physical or mental exertions EXAMPLES: Design, specify, manage	**CHARACTERIZING** Behavior consistent with values internalized EXAMPLES: Qualify, modify, perform
EVALUATION Makes decisions, judges, or selects based on criteria and rationale EXAMPLES: Compare, contrast, justify, summarize		

FIGURE 6-5 Revised matrix for the new taxonomy.

requires the student first to identify the advantages and disadvantages of a dry-chemical extinguisher, and then to select and summarize those that apply to use of such an extinguisher on a Class A fire. This higher-level objective may be evaluated by a multiple-choice question but is better evaluated with a short answer-type question.

- Application: "The fire fighter trainee, given a portable fire extinguisher scenario, shall identify the correct type of extinguisher and demonstrate the method for using it to extinguish the fire." This is the highest level of objective because it requires the student to recall several pieces of information and apply them correctly based on the situation. This type of objective is often evaluated with scenario-based questions that may be answered with multiple-choice items or short answers.

Psychomotor Domain Objectives

For the fire service, the most commonly used levels when developing *psychomotor* learning objectives are (in order from simplest to most difficult) imitation, manipulation, precision, articulation, and naturalization:

- *Imitation* is when the student observes the skill and attempts to repeat it.
- *Manipulation* is performing a skill based on instruction rather than on observation.
- *Precision* is performing a single skill or task correctly.

- *Articulation* is when the student combines multiple skills.
- *Naturalization* is when the student performs multiple skills correctly all of the time.

Action verbs associated with the psychomotor domain include the following:

- Demonstrate
- Practice
- Apply
- Perform
- Display
- Show
- Assemble

Affective Domain Objectives

The *affective* domain deals with an individual's expressed interests, ambitions, and values. These kinds of emotional behaviors are essential to the overall learning experience but may not be readily visible during initial learning. The taxonomy of the affective domain identifies five levels of understanding:

- *Receiving* is paying attention and displaying a willingness to learn.
- *Responding* is displaying an acknowledged behavior within the learning experience and participating when given an opportunity.
- *Valuing* is showing active involvement, passion, or commitment toward a topic.
- *Organization* is accepting a new value as one's own and setting a specific goal.
- *Characterization* is comparing and contrasting one's own values to others and using the new value.

Action verbs associated with affective domain objectives include the following examples:

- Accept
- Participate
- Share
- Judge

- Attempt
- Challenge

Condition

The condition(s) should describe the situation in which the student will perform the behavior. Specific equipment or resources must be listed in the objective; these are the "givens" necessary to demonstrate the skill or knowledge. For example, the phrases "in full protective clothing, including SCBA" and "using water from a static source such as a pool or pond" would tell both the instructor and the student what needs to be present to complete the behavior. You might notice that many skill sheets have a specific area inside their template that lists the equipment needed for part of the skill sheet. Before teaching the lesson, the instructor must check whether all equipment or resources needed for the performance of the objective are in working order and whether the student can properly operate all equipment or resources.

Degree

The degree may also be known as the standard because it describes how well the behavior must be performed. Both the student and the evaluator need to know the criteria against which the student is being measured. Percentage scores, "without errors," and "within a designated time" are all examples of the degree of performance that objectives should contain. A reference to a skill sheet during manipulative performance will guide both the student and instructor in how to approach proper completion and evaluation of the skill.

There is no one correct format for determining which level or how many learning objectives should be written for a lesson plan TABLE 6-4 . Typically, a lesson plan will contain knowledge-based (cognitive) learning objectives to ensure that students learn all of the facts and definitions within the class. Comprehension or cognitive objectives are then used to ensure that students can summarize or clarify the material. Psychomotor objectives are used to ensure that the student can actually perform a task identified in the objective and presented in the

Table 6-4		Examples of Formatted Objectives by Domains and Levels				
Duty Area	*Domain Level*	**Cognitive**	*Domain Level*	**Psychomotor**	*Domain Level*	**Affective**
Fire fighter	Knowledge	A fire fighter trainee, given a written exam, will identify the tools needed to perform vertical ventilation on a peaked roof according to J&B Fundamentals of Fire Fighter Skills 3rd edition.	Manipulation	The fire fighter, given a K-950 rotary saw and an assortment of blades, will follow manufacturer's directions to change the wood blade to a metal blade.	Receiving	Basic fire fighter academy students will name the safety precautions to be taken when using a ground ladder during vertical ventilation.
Officer	Synthesis	Company officers will be given video clips of five different fire scenes and will explain the type of ventilation required based on smoke conditions according to department SOPs.	Articulation	As the officer of a crew with three experienced fire fighters, and given PPE, ladders, and tools, conduct vertical ventilation on a pitched roof so that all barriers are removed and structural integrity is not compromised within four minutes.	Organizing	The officer of a crew assigned to ventilation at a structure fire will display adherence to all safety and risk management requirements for the entire crew, according to department policy.

Courtesy of Bryant Krizik

lesson plan. Objectives pertaining to the third domain—that is, affective objectives—are often listed with the psychomotor objectives and identify "when" to perform the task. These learning domains are discussed in more detail in the chapter *The Learning Process*.

■ Converting Job Performance Requirements into Learning Objectives

Often, a Fire Service Instructor II needs to develop learning objectives to meet job performance requirements (JPRs) listed in a NFPA professional qualification standard. A JPR describes

a specific job task, lists the items necessary to complete that task, and defines measurable or observable outcomes and evaluation areas for the specific task. Matching of learning objectives to JPRs occurs when a lesson plan is being developed to meet the professional qualifications for a position such as Fire Officer, Fire Instructor, or Fire Fighter.

The JPRs listed in the NFPA standards of professional qualifications are not learning objectives per se, but learning objectives can be created based on the JPRs. Each NFPA professional qualification standard has an annex section that explains the process of converting a JPR into an instructional objective, including examples of how to do so **FIGURE 6-6**. By following this format,

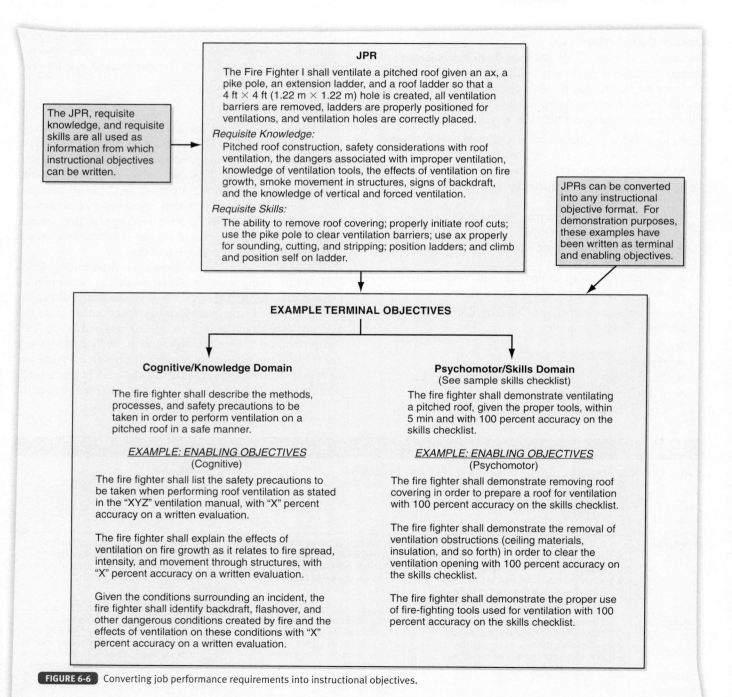

JPR

The Fire Fighter I shall ventilate a pitched roof given an ax, a pike pole, an extension ladder, and a roof ladder so that a 4 ft × 4 ft (1.22 m × 1.22 m) hole is created, all ventilation barriers are removed, ladders are properly positioned for ventilations, and ventilation holes are correctly placed.

Requisite Knowledge:
Pitched roof construction, safety considerations with roof ventilation, the dangers associated with improper ventilation, knowledge of ventilation tools, the effects of ventilation on fire growth, smoke movement in structures, signs of backdraft, and the knowledge of vertical and forced ventilation.

Requisite Skills:
The ability to remove roof covering; properly initiate roof cuts; use the pike pole to clear ventilation barriers; use ax properly for sounding, cutting, and stripping; position ladders; and climb and position self on ladder.

The JPR, requisite knowledge, and requisite skills are all used as information from which instructional objectives can be written.

JPRs can be converted into any instructional objective format. For demonstration purposes, these examples have been written as terminal and enabling objectives.

EXAMPLE TERMINAL OBJECTIVES

Cognitive/Knowledge Domain

The fire fighter shall describe the methods, processes, and safety precautions to be taken in order to perform ventilation on a pitched roof in a safe manner.

EXAMPLE: ENABLING OBJECTIVES
(Cognitive)

The fire fighter shall list the safety precautions to be taken when performing roof ventilation as stated in the "XYZ" ventilation manual, with "X" percent accuracy on a written evaluation.

The fire fighter shall explain the effects of ventilation on fire growth as it relates to fire spread, intensity, and movement through structures, with "X" percent accuracy on a written evaluation.

Given the conditions surrounding an incident, the fire fighter shall identify backdraft, flashover, and other dangerous conditions created by fire and the effects of ventilation on these conditions with "X" percent accuracy on a written evaluation.

Psychomotor/Skills Domain
(See sample skills checklist)

The fire fighter shall demonstrate ventilating a pitched roof, given the proper tools, within 5 min and with 100 percent accuracy on the skills checklist.

EXAMPLE: ENABLING OBJECTIVES
(Psychomotor)

The fire fighter shall demonstrate removing roof covering in order to prepare a roof for ventilation with 100 percent accuracy on the skills checklist.

The fire fighter shall demonstrate the removal of ventilation obstructions (ceiling materials, insulation, and so forth) in order to clear the ventilation opening with 100 percent accuracy on the skills checklist.

The fire fighter shall demonstrate the proper use of fire-fighting tools used for ventilation with 100 percent accuracy on the skills checklist.

FIGURE 6-6 Converting job performance requirements into instructional objectives.

a Fire Service Instructor II is able to develop learning objectives for a lesson plan to meet the professional qualifications for NFPA standards. This process includes the breaking down of a JPR into a terminal objective and several enabling objectives, including cognitive and psychomotor objectives. Further into the development process, cognitive objectives are written into a lesson plan and include evaluation tools such as test questions, whereas psychomotor objectives are broken down into task steps and made into skill sheets. An Instructor III will use the JPRs to help write course objectives for larger curricula.

Safety Tip

When writing objectives in a psychomotor format, be sure to address any safety issues required to meet the objective.

■ Lesson Outline

After determining the performance outcomes and writing the learning objectives for the lesson plan, the next step for the Fire Service Instructor II is to develop the lesson outline **FIGURE 6-7**

. The lesson outline is the main body of the lesson plan and is the major component of the presentation step in the four-step method of instruction.

One method for creating a lesson outline involves brainstorming the topics to be covered and then arranging them in a logical order. Begin listing all of the information that needs to be taught to achieve the learning objectives. Which terms do students need to learn? Which concepts must be presented? Which skills need to be practiced? Which stories or real-life examples would demonstrate the need to learn this material?

Once you have listed all of the topics that should be covered in the lesson outline, organize them into presentation and application sections. Arrange the listed topics you will lecture on in a logical and orderly fashion in the presentation section. It may be practical to structure a lesson outline by progressing from the known and working toward the unknown. This is particularly useful in a large course where the instructor aims to ensure that the topics previously covered have been mastered by the students. This establishes a relationship among topics in the larger curriculum. For example, you will need to instruct students on building construction and ladders before you can focus on vertical ventilation. Topics should be presented in order starting from the basic and then moving on to

Pre-Lecture (Preparation Step)
I. You Are the Fire Fighter

Time: 5 Minutes
Small Group Activity/Discussion

Use this activity to motivate students to learn the knowledge and skills needed to understand the history of the fire service and how it functions today.

Purpose
To allow students an opportunity to explore the significance and concerns associated with the history and present operation of the fire service.

Instructor Directions
1. Direct students to read the "You Are the Fire Fighter" scenario found in the beginning of Chapter 1.
2. You may assign students to a partner or a group. Direct them to review the discussion questions at the end of the scenario and prepare a response to each question. Facilitate a class dialogue centered on the discussion questions.
3. You may also assign this as an individual activity and ask students to turn in their comments on a separate piece of paper.

Lecture (Presentation Step)
I. Introduction

Time: 5 Minutes
Slides: 1–6
Level: Fire Fighter I
Lecture/Discussion

A. Training to become a fire fighter is not easy.
   1. The work is physically and mentally challenging.
   2. Firefighting is more complex than most people imagine.
B. Fire fighter training will expand your understanding of fire suppression.
   1. The new fire fighter must understand the roots of the fire service, how it has developed, and the fire service "culture" in order to excel.
   2. This course equips fire fighters to continue a centuries-old tradition of preserving lives and property threatened by fire.

(Continues)

(Continued)

II. Fire Fighter Guidelines

Time: 5 Minutes
Slide: 7
Level: Fire Fighter I
Lecture/Discussion

A. Be safe.
   1. Safety should always be uppermost in your mind.
B. Follow orders.
   1. If you follow orders, you will become a dependable member of the department.
C. Work as a team.
   1. Firefighting requires the coordinated efforts of each department member.
D. Think!
   1. Lives will depend on the choices you make.
E. Follow the golden rule.
   1. Treat each person, patient, or victim as an important person.

III. Fire Fighter Qualifications

Time: 30 Minutes
Slides: 8–10
Level: Fire Fighter I
Lecture/Discussion

A. Age requirements
   1. Most career fire departments require that candidates be between the ages of 18 and 21.
B. Education requirements
   1. Most career fire departments require a minimum of a high school diploma or equivalent.
C. Medical requirements
   1. Medical evaluations are often required before training can begin.
   2. Medical requirements for fire fighters are specified in NFPA 1582, *Standard on Comprehensive Operational Medical Program for Fire Departments*.
D. Physical fitness requirements
   1. Physical fitness requirements are established to ensure that fire fighters have the strength and stamina needed to perform the tasks associated with firefighting and emergency operations.
E. Emergency medical requirements
   1. Many departments require fire fighters to become certified at the first responder, Emergency Medical Technician (EMT)–Basic, or higher levels.

IV. Roles and Responsibilities of the Fire Fighter I and Fire Fighter II

Time: 30 Minutes
Slides: 11–17
Level: Fire Fighter I and II
Lecture/Discussion

A. The roles and responsibilities for Fire Fighter I include:
   1. Don and doff personal protective equipment properly.
   2. Hoist hand tools using appropriate ropes and knots.
   3. Understand and correctly apply appropriate communication protocols.
   4. Use self-contained breathing apparatus (SCBA).
   5. Respond on apparatus to an emergency scene.
   6. Force entry into a structure.
   7. Exit a hazardous area safely as a team.
   8. Set up ground ladders safely and correctly.
   9. Attack a passenger vehicle fire, an exterior Class A fire, and an interior structure fire.
   10. Conduct search and rescue in a structure.
   11. Perform ventilation of an involved structure.
   12. Overhaul a fire scene.
   13. Conserve property with salvage tools and equipment.
   14. Connect a fire department engine to a water supply.
   15. Extinguish incipient Class A, Class B, and Class C fires.
   16. Illuminate an emergency scene.
   17. Turn off utilities.

(Continued)

18. Perform fire safety surveys.
19. Clean and maintain equipment.
20. Present fire safety information to station visitors, community groups, or schools.

B. Additional roles and responsibilities for Fire Fighter II include:
1. Coordinate an interior attack line team.
2. Extinguish an ignitable liquid fire.
3. Control a flammable gas cylinder fire.
4. Protect evidence of fire cause and origin.
5. Assess and disentangle victims from motor vehicle accidents.
6. Assist special rescue team operations.
7. Perform annual service tests on fire hose.
8. Test the operability of and flow from a fire hydrant.
9. Fire fighters must also be prepared to assist visitors to the fire station and use the opportunity to discuss additional fire safety information.

V. Summary

Time: 5 Minutes
Slides: 51–53
Level: Fire Fighter I
Lecture/Discussion

A. Remember the five guidelines: Be safe, follow orders, work as a team, think, and follow the golden rule.
B. Fire fighter qualifications consider age, education, medical and physical fitness, and emergency medical certifications.
C. The roles and responsibilities of Fire Fighter I and Fire Fighter II vary.

Post-Lecture
I. Wrap-Up Activities (Application Step)

Time: 40 Minutes
Small Group Activity/Individual Activity/Discussion

A. Fire Fighter in Action
This activity is designed to assist the student in gaining a further understanding of the roles and responsibilities of the Fire Fighter I and II. The activity incorporates both critical thinking and the application of fire fighter knowledge.

Purpose
This activity allows students an opportunity to analyze a firefighting scenario and develop responses to critical thinking questions.

Instructor Directions
1. Direct students to read the "Fire Fighter in Action" scenario located in the Wrap-Up section at the end of Chapter 1.
2. Direct students to read and individually answer the quiz questions at the end of the scenario. Allow approximately 10 minutes for this part of the activity. Facilitate a class review and dialogue of the answers, allowing students to correct responses as needed. Use the answers noted below to assist in building this review. Allow approximately 10 minutes for this part of the activity.
3. You may also assign these as individual activities and ask students to turn in their comments on a separate piece of paper.
4. Direct students to read the "Near Miss Report." Conduct a discussion that allows for feedback on this report. Allow 10–15 minutes for this activity.

Answers to Multiple Choice Questions
1. A, D
2. B
3. B
4. D

B. Technology Resources
This activity requires students to have access to the Internet. This may be accomplished through personal access, employer access, or through a local educational institution. Some community colleges, universities, or adult education centers may have classrooms with Internet capability that will allow for this activity to be completed in class. Check out local access points and encourage students to complete this activity as part of their ongoing reinforcement of firefighting knowledge and skills.

Purpose
To provide students an opportunity to reinforce chapter material through use of online Internet activities.

Instructor Directions
1. Use the Internet and go to **www.Fire.jbpub.com**. Follow the directions on the Web site to access the exercises for Chapter 1.
2. Review the chapter activities and take note of desired or correct student responses.
3. As time allows, conduct an in-class review of the Internet activities and provide feedback to students as needed.
4. Be sure to check the Web site before assigning these activities, as specific chapter-related activities may change from time to time.

(Continues)

(Continued)

II. Lesson Review (Evaluation Step)

Time: 15 Minutes
Discussion
Note: Facilitate the review of this lesson's major topics using the review questions as direct questions or overhead transparencies. Answers are found throughout this lesson plan.

A. Name some of the physical fitness requirements established for firefighters.
B. What requirements do the firefighter qualifications focus on?
C. How do the roles and responsibilities of the Fire Fighter I and II differ?

III. Assignments

Time: 5 Minutes
Lecture

A. Advise students to review materials for a quiz (determine date/time).
B. Direct students to read the next chapter in *Fundamentals of Fire Fighter Skills* as listed in your syllabus (or reading assignment sheet) to prepare for the next class session.

FIGURE 6-7 A sample lesson outline.

the more complex. Ensure that the topics flow together and that the presentation does not contain any gaps that might confuse a student. If you identify a gap, you may need to create a new topic to bridge it.

In the application section of the lesson, list the topics that require students to apply the information learned in the presentation section. Most often the topics in the application section will be activities or skills practice. If the lesson does not include actual hands-on activities, the application should at least consist of discussion points for you to talk about with the students to ensure the information in the lecture was learned and can be applied.

Many lesson outlines utilize a two-column format. The first column contains the actual outline of the material to be taught. If this lesson outline is to be used by experienced fire service instructors, a simple outline of the material may suffice. For less experienced fire service instructors (or to ensure consistency among multiple instructors), the outline may be more detailed. The second column of the lesson outline contains comments or suggestions intended to help a fire service instructor understand and present the lesson outline. It is also a good practice to indicate in the second column which learning objectives are being achieved during the presentation or application sections. This information is especially helpful when you are developing a lesson plan to teach an established curriculum that uses a numbering system to identify learning objectives.

Instructional Materials

Once the lesson outline is developed, all instructional materials needed to deliver instruction should be identified and listed in the lesson plan. This list should be specific so that the exact instructional aid can be identified. For example, if the lesson plan is a fire safety lesson for children that incorporates a DVD as an instructional aid, just listing "fire safety video" in the lesson plan does not provide enough information for the

fire service instructor. Instead, give the exact title of the video, such as "*Sparky Says: Join My Fire Safety Club*, by the NFPA." This information will allow any fire service instructor who uses the lesson plan to obtain the correct instructional aid. Instructional materials may range from handouts to projectors to the hoses used during a skills practice.

Often the inclusion of one instructional aid creates a need for more instructional materials. For example, if a lesson plan lists a DVD as an instructional aid, the instructional materials would need to be revised to include a DVD player and projector. Ask the following types of questions to determine precisely what you need:

- Are additional informational resources needed to present the learning objectives to students—for example, a handout describing your department's SOPs?
- Are supplies needed to make props or demonstrations?
- Is equipment needed for the activities or skills practice?
- Is equipment needed to ensure student safety?

■ Evaluation Plan

The evaluation plan is the final part of the lesson plan. Each part of the evaluation plan should be directly tied to one or more learning objectives. When writing this component of the lesson plan, simply describe the evaluation plan—do not provide the actual evaluation. In other words, the lesson plan could indicate that the evaluation plan is a 50-question multiple-choice test, but it should not list the actual test questions. The test questions should be a separate document that is securely kept and made available only as needed to fire service instructors. When the evaluation plan lists skills performance tests, these documents should be included with the instructional materials and distributed to students so they can prepare for skills testing. This step is covered in the chapter *Evaluating the Learning Process*.

Theory Into Practice

A set of clear learning objectives, a thorough lesson outline, and a method of ensuring that the learning objectives are met should form the backbone of every course taught in the fire service. Proper construction of learning objectives will ensure that the course meets the identified needs of the students. Conversely, when they are not supported with clear learning objectives, courses may stray from their intended purpose. To ensure development of clear learning objectives, use the ABCD method during the creation process.

The lesson outline is a necessity to ensure that you cover the material required to meet the learning objectives. This outline should flow logically to assist in student comprehension. Brainstorming will help guide your thought process in covering all relevant topics. Arranging topics in a logical sequence requires an understanding of the learning objectives. Once it is completed, run your lesson outline by a colleague to see if he or she can follow your logical sequence. If your colleague is confused, then your students are also likely to be confused.

The last step to ensure success is to create an evaluation process to confirm that the learning objectives have been met. Without an evaluation process, there is no way to measure the objectives or to guarantee that a student has met the goals of the class.

Teaching Tip

If you are creating a lesson plan, the evaluation should directly link back to the initial learning objectives outlined at the beginning of class.

Modifying a Lesson Plan

Unlike a Fire Service I, who can only *adapt* lesson plans (as discussed earlier in this chapter), a Fire Service Instructor II may *modify* lesson plans. Modifying a lesson plan occurs when the Instructor II makes fundamental changes, such as revising the learning objectives. When these kinds of substantial changes to a lesson plan are made, the lesson plan should be completely revised, following the step-by-step process used to develop the original lesson plan. To ensure that the lesson plan is written to meet a new learning objective, follow through each step of the lesson plan development process and make the necessary changes in all sections of the lesson plan.

When modifying a lesson plan, always obtain necessary approval from the authority having jurisdiction. Even though a Fire Service Instructor II has the training to modify learning objectives, the change typically must be approved by a curriculum committee, a training officer, or the fire chief. Similarly, any lesson plan modification must comply with all agency policies and procedures. If a reference used to develop the lesson plan is updated, such as a department SOP or an NFPA standard, make sure that the reference cited in the lesson plan is current.

After modifying a lesson plan, retain a copy of the original lesson plan. This original must be kept to document the classes that were taught from that lesson plan. It can also be referred to when making future lesson plan modifications.

The fire service instructor greatly improves his or her ability to deliver training information to students by using a standard lesson plan format that incorporates the four-step method of instruction. Consistency and accuracy of information must be relayed to varied audiences, and use of a lesson plan template can facilitate this process. Moreover, in the event of unexpected emergency runs or other breaks that may occur during instruction, having a well-organized lesson plan allows for you to pick up where you left off. Fellow instructors can use the same lesson plan and achieve similar outcomes. The lesson plan can be compared to an incident action plan, in that it identifies expected outcomes of a training session, clarifies resources available or needed, and provides a step-by-step measurable set of instruction material that brings a training session to a successful outcome. Existing or published lesson plans should be reviewed and modified to reflect your department procedures and practices. Utilization of fire service references and NFPA JPRs also provide content validity to the material being taught. Using a standard form for instruction ensures that the instructor covers many legal and ethical concerns relating to the delivery of training in the modern fire service.

Jones & Bartlett Fire District

Training Division
5 Wall Street, Burlington, MA, 01803
Phone 978-443-5000 Fax 978-443-8000
www.fire.jbpub.com

Instant Applications: Lesson Plans

Drill Assignment

Apply the chapter content to your department's operation, its training division, and your personal experiences to complete the following questions and activities.

Objective

Upon completion of the instant applications, fire service instructor students will exhibit decision making and application of job performance requirements of the fire service instructor using the text, class discussion, and their own personal experiences.

Suggested Drill Applications

1. Using a sample lesson plan included in supplemental course material, identify the components of the lesson plan.

2. Using the same sample lesson plan, adjust the lesson plan based on the needs of different audiences.

3. Analyze an existing lesson plan from your department. Are the components complete and accurate?

4. Review the Incident Report in this chapter and be prepared to discuss your analysis of the incident from a training perspective and as an instructor who wishes to use the report as a training tool.

Incident Report

Milford, Michigan—1987

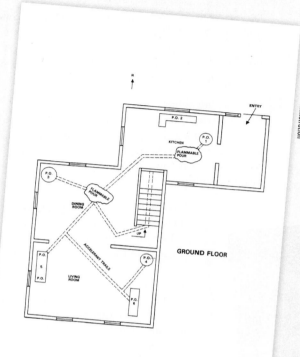

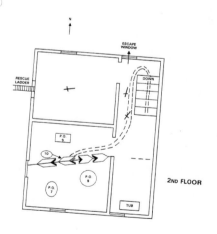

Figure 6A Floor plan of involved Milford structure.

Four days after the Hollandale, Michigan fire, a training session that was supposed to teach fire fighters to recognize physical evidence of arson fires trapped six fire fighters on the second floor of a 120-year-old house **(Figure 6A)**. The wood-framed house had low-density ceiling tiles with lightweight wood paneling on the first floor. There were numerous holes in the walls and ceilings.

Various arson scenarios were set up in each of the rooms on both floors using furniture, clothing, and other items. Both flammable and combustible fluids were used.

Multiple hose lines were charged and ready on the outside of the structure. Portable water tanks were set up and filled with tenders, prepared for a water shuttle operation.

Fire fighters toured the building to see the scenarios before ignition. Initial ignition efforts failed, and several windows on the second floor were broken to improve ventilation. A fire set on a couch in the southwest corner of the living room using flammable or combustible liquids was openly burning and had breached the exterior wall and entered the attic space.

One fire fighter was already inside trying to ignite the fires on the second floor. Four fire fighters were directed to enter the house, along with the assistant chief. Without a hose line, these fire fighters passed multiple burning fire sets that were producing little heat or smoke. Most likely, they did not see the fire in the living room or realize it had spread to the attic. They met with the assistant chief and another fire fighter on the second floor to ignite a fire in one of the bedrooms. At this time, the other upstairs bedroom was already burning. The assistant chief directed them to exit the house as fire conditions rapidly intensified. The escape route down the stairs to the first floor was cut off and, under very adverse conditions, the fire

Continued...

Incident Report Continued...

Milford, Michigan—1987

fighters were able to locate a window on the second floor for egress **(Figure 6B)**. Three of them were able to exit through the window and onto the first floor roof below. Reportedly, the SCBA face piece was melting off the last fire fighter able to exit.

Outside, fire fighters observed the change in conditions and the rescue of the assistant chief and the two other fire fighters. They initiated suppression and rescue operations. Ladders were raised to the second-floor windows in the now almost fully involved house.

The first trapped fire fighter was located and removed in approximately 10 minutes. The others were all located on the second floor shortly thereafter. Per the NFPA report, "One of the fire fighters who was able to escape did not know that he would be part of an interior training until he was instructed to 'suit up,' and even then he was unsure of his specific assignment." Three fire fighters died in the fire, which was the first multiple-death training incident in the United States since the release of NFPA 1403 in 1986.

Reproduced with permission from NFPA, © 1987, National Fire Protection Association.

Figure 6B Stairs to the second floor of the Milford house show significant fire involvement that blocked escape from second floor.

Postincident Analysis

NFPA 1403 (2007 Edition) Noncompliant

Combustible wall paneling and ceiling tiles contributed to rapid fire spread *(4.2.17) (4.2.10.5)*

Flammable and combustible liquids used *(4.3.6)*

Multiple simultaneous fires on two floors *(4.4.15)*

No interior hose line *(4.4.6)*

Numerous holes in walls and ceilings allowed fire spread *(4.2.10.4)*

Interior stairs only exit other than windows (Note: a violation due to the limited normal means of egress, which may have precluded the use of the upper floor without additional provisions put in place) *(4.2.12.1) (4.2.13)*

Flashover and fire spread unexpected *(4.3.9)*

Fire chief claimed no knowledge of NFPA 1403 *(4.5.4)*

Wrap-Up

Chief Concepts

- A fire service instructor who uses a well-prepared and thorough lesson plan to organize and prepare for class greatly increases the odds of ensuring quality student learning.
- All instructional planning begins by identifying the desired outcomes, called objectives.
- In the ABCD method of writing learning objectives, ABCD stands for **A**udience (Who?), **B**ehavior (What?), **C**ondition (How or using what?), and **D**egree (How well?).
- A lesson plan includes the following parts:
  - Lesson title or topic
  - Level of instruction
  - Behavioral objectives, performance objectives, and learning outcomes
  - Instructional materials needed
  - Lesson outline
  - References/resources
  - Lesson summary
  - Assignment
- While reviewing and preparing for class with a lesson plan, the four-step method of instruction is the primary process used to relate the material contained in the lesson plan to the students:
  - Preparation
  - Presentation
  - Application
  - Evaluation
- Preparing for instruction is very important. You may need to spend several hours preparing to teach a class, including reviewing the lesson plan, reserving classrooms and instructional aids, and purchasing materials.
- A Fire Service Instructor I can use a lesson plan to teach a class and may adapt the lesson plan to the local needs of the class.
- A Fire Service Instructor II can create a new lesson plan to teach a class and may modify an existing lesson plan.
- Over time, standards and procedures change, so that a lesson plan that was completely correct for your department when it was created may be out-of-date in just a few months.
- The main focus when adapting a lesson plan is to make minor adjustments so it fits your local conditions and your students' needs. To accomplish this goal, you must be familiar with your audience.
- You should review and adapt the lesson plan based on the facilities that will be used when delivering the class.
- A lesson plan must be reviewed to ensure that it meets and follows local SOPs. After completely reviewing the lesson plan, research the SOPs and ensure that no conflicts exist.
- The lesson plan should be reviewed based on student limitations and adapted to accommodate those limitations if possible.
- Reviewing and adapting a lesson plan should be a formal process. You should document in writing which adaptations have been made.
- The method of instruction is the one area that a Fire Service Instructor I may readily alter. Such a change may be needed to allow you to deliver the lesson plan effectively, but it should not change the learning objectives.
- When reviewing and adapting a lesson plan, consider whether the lesson plan—and especially the presentation section—fits your personal style.

- All adaptations should be done with one purpose in mind—namely, meeting the needs of the students.
- When creating a lesson plan, a Fire Service Instructor II should ensure that the lesson plan is complete and clearly understandable so that any other fire service instructor can use it.
- Developing lesson plans includes the following steps:
  - Achievement of job performance requirements
  - Learning objectives
  - Conversion of job performance requirements into learning objectives
  - Lesson outline
  - Evaluation plan
- When modifying a lesson plan, always obtain necessary approval from the authority having jurisdiction. Even though a Fire Service Instructor II has the training to modify learning objectives, the change typically must be approved by a curriculum committee, a training officer, or the fire chief.

Hot Terms

ABCD method Process for writing lesson plan objectives that includes four components: audience, behavior, condition, and degree.

Adapt To make fit (as for a specific use or situation).

Application step The third step of the four-step method of instruction, in which the student applies the information learned during the presentation step.

Assignment The part of the lesson plan that provides the student with opportunities for additional application or exploration of the lesson topic, often in the form of homework that is completed outside of the classroom.

Audience Who the students are.

Behavior An observable and measurable action for the student to complete.

Condition The situation in which the student will perform the behavior.

Degree The last part of a learning objective, which indicates how well the student is expected to perform the behavior in the listed conditions.

Enabling objective An intermediate learning objective, usually part of a series that directs the instructor on what he or she needs to instruct and what the learner will learn to accomplish the terminal objective.

Evaluation step The fourth step of the four-step method of instruction, in which the student is evaluated by the instructor.

Four-step method of instruction The most commonly used method of instruction in the fire service. The four steps are preparation, presentation, application, and evaluation.

Job performance requirement (JPR) A statement that describes a specific job task, lists the items necessary to complete that task, and defines measurable or observable outcomes and evaluation areas for the specific task.

Learning objective A goal that is achieved through the attainment of a skill, knowledge, or both, and that can be measured or observed.

Lesson outline The main body of the lesson plan; a chronological listing of the information presented in the lesson plan.

Lesson plan A detailed guide used by an instructor for preparing and delivering instruction.

Lesson summary The part of the lesson plan that briefly reviews the information from the presentation and application sections.

Lesson title or topic The part of the lesson plan that indicates the name or main subject of the lesson plan.

Level of instruction The part of the lesson plan that indicates the difficulty or appropriateness of the lesson for students.

Modify To make basic or fundamental changes.

Preparation step The first step of the four-step method of instruction, in which the instructor prepares to deliver the class and provides motivation for the students.

Prerequisite A condition that must be met before a student is allowed to receive the instruction contained within a lesson plan—often a certification, rank, or attendance of another class.

Presentation step The second step of the four-step method of instruction, in which the instructor delivers the class to the students.

Terminal objective A broader outcome that requires the learner to have a specific set of skills or knowledge after a learning process.

References

Bloom B. S. (1956). *Taxonomy of Educational Objectives, Handbook I: The Cognitive Domain*. New York: David McKay Company.

Heinich, Robert, Michael Molenda, and James D. Russell. (1998). *Instructional Media and the New Technologies of Education*. 6th ed. Upper Saddle River, NJ: Prentice Hall College Division.

National Fire Protection Association. (2012). *NFPA 1041: Standard for Fire Service Instructor Professional Qualifications*. Quincy, MA: National Fire Protection Association.

National Fire Protection Association. (2007). *NFPA 1403: Standard on Live Fire Training Evolutions*. Quincy, MA: National Fire Protection Association.

U.S. Fire Administration/FEMA. (2012). TRADE. http://www.usfa.fema.gov/nfa/trade/index.shtm. Accessed October 17, 2012.

You are a Fire Service Instructor I who has been asked to teach an SCBA class to your department's new-recruit class. The captain in charge of the training academy provides you with the lesson plan that was used during the last class. He asks you to review the lesson plan and let him know if you need anything before you teach the class in two weeks.

1. Which statement best describes your next action?
   A. Safely store the lesson plan away until the day of the class.
   B. Begin creating your own lesson plan and compare it to the one you were given.
   C. Review the lesson plan you were given and develop a timeline to prepare for the class.
   D. Tell the captain that a Fire Service Instructor I cannot teach this class.

2. As you review the SCBA lesson plan, you notice that some of the learning objectives are no longer needed because of an equipment change. As a Fire Service Instructor I, what should you do?
   A. Delete the unnecessary objectives from the lesson plan.
   B. Notify the captain, so a Fire Service Instructor II can modify the lesson plan.
   C. Teach the learning objectives anyway because they are in the lesson plan.
   D. Rewrite the learning objectives so they apply to the new equipment.

3. The last class contained 20 recruits. The class you will teach will have 40 students. As a Fire Service Instructor I, you can adapt the lesson plan to accommodate the additional students.
   A. True
   B. False

4. Your department has a standard operating procedure for the use and maintenance of SCBA. Which of the following statements is true concerning the SOP and the lesson plan?
   A. The lesson plan should never contradict the SOP.
   B. The lesson plan may reference the SOP but you do not need to teach it.
   C. The lesson plan should not include the SOP because students will learn it later.
   D. The lesson plan should cover textbook material only, not SOPs.

5. As a Fire Service Instructor I preparing to teach this class, which of the following issues would you normally be responsible for?
   A. Selecting the type of SCBA for your department
   B. Establishing a budget for the class
   C. Writing the exam questions
   D. Reviewing and preparing audio/visual aids

6. Why would a less experienced fire service instructor have a more detailed lesson outline?
   A. A less experienced fire service instructor should not have a more detailed lesson outline because it will be distracting.
   B. A more detailed lesson plan will allow the fire service instructor to cover areas that are not part of the learning objectives.
   C. The fire service instructor will have more basic knowledge about the topic, so less information is required in the lesson outline.
   D. All lesson outlines should be the same regardless of the fire service instructor's experience.

7. When reviewing a lesson plan, should you consider your personal style of presentation and adapt the plan to meet your style?
   A. This is acceptable because the original fire service instructor will have incorporated his or her own personal style when developing the lesson plan.
   B. This is acceptable because the original fire service instructor will have ensured that no personal style is reflected in the lesson plan.
   C. This is not acceptable because the material should be about the students and not the fire service instructor.
   D. This is not acceptable because no adaptations should ever be made by a Fire Service Instructor I.

The Learning Environment

Fire Service Instructor I

Knowledge Objectives

After studying this chapter, you will be able to:

- Describe the effect of demographics on the learning environment. (pp 169–170)
- Describe how to adapt the learning environment to suit the needs of your students. (pp 163–167, 169–170)
- Describe how to create a learning environment that facilitates the learning process. (**NFPA 4.4.2**) (pp 163–167)
- Describe the effect of the audience and the venue on the learning process. (pp 161, 163–167)

Skills Objectives

After studying this chapter, you will be able to:

- Analyze the learning environment according to the students' needs and the learning objectives. (pp 163–167)
- Organize the learning environment by arranging it to meet the requirements of the material to be presented. (**NFPA 4.3.2** **NFPA 4.4.2**). (pp 163–167)
- Present material to a diverse audience following acceptable presentation methods and techniques (**NFPA 4.3.2**). (pp 161, 169–170)

Fire Service Instructor II

Knowledge Objectives

After studying this chapter, you will be able to:

- Describe how demographics of the learning environment affect development of a lesson plan. (**NFPA 5.4.2**) (pp 161, 163, 169–170)
- Describe how to evaluate the learning environment when developing a lesson plan. (**NFPA 5.4.2**) (pp 163–167)
- Identify the impact of the learning environment as part of the lesson development process. (**NFPA 5.4.2**) (pp 161, 163–167, 169–170)

Skills Objectives

After studying this chapter, you will be able to:

- Present material to a diverse audience following acceptable presentation methods and techniques. (**NFPA 5.4.2**) (pp 161, 169–170)

Fire Service Instructor III

Knowledge Objectives

There are no knowledge objectives for Fire Service Instructor III students.

Skills Objectives

There are no skills objectives for Fire Service Instructor III students.

You Are the Fire Service Instructor

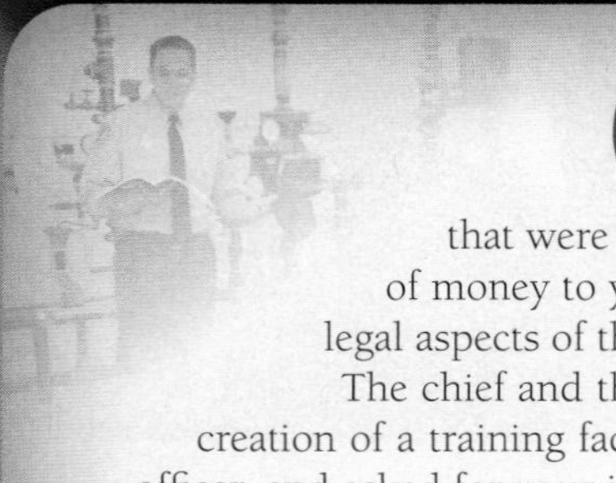

On a recent call, your agency responded to and assisted with an incident at a major chemical company's complex. The incident was mitigated without any injury or major damage to the facility. The company was extremely grateful to your agency for the resources and support that were provided during the incident, and in an effort to show its appreciation, it offered to donate a sum of money to your agency. The chief met with the city manager, the city council, and an attorney to examine all legal aspects of the proposed donation, and it was determined to be an acceptable gift.

The chief and the department command staff discussed several options for using the funds and decided that the creation of a training facility would be the best use of the money. The chief has approached you, as the training division officer, and asked for your input into the creation of a classroom and basic fire fighter training facility.

1. Based on your agency, which type of classroom would you suggest be built?
2. Which basic equipment and resources would you list and prioritize for your new classroom to make it a state-of-the-art training room?
3. Which basic training props would you recommend to your chief to meet the needs of your agency?

Introduction

There is an old saying: When two people are together, one is always the leader. In training, whenever two fire fighters are together, one is always the fire service instructor. The setting is not important, the time is not important, and the subject is not important—what *is* important is that you use the environment to ensure that learning takes place.

The learning environment is not always a classroom, because learning takes place in many different locations. In many company officer training classes, the concept of tailboard chats is encouraged. Many fire officers report that a great deal is learned sitting around the kitchen table in a firehouse or standing around a charred table in a kitchen that has just been overhauled **FIGURE 7-1**. The place is not as important as recognizing the opportunity to share a message or thought and to deliver a piece of knowledge with others. The instructor should be well versed in training in a wide array of environments and situations. Training might start in the classroom lecture or a demonstration by the instructor. From the classroom,

the demonstration moves to the training ground, where the students get to apply hands-on training to further develop the skill set. Skill sets become building blocks that develop into scenarios, and scenarios move to larger-scale evolutions that are based on training as close to reality as possible. This process encompasses a variety of learning environments that require the instructor to know how to teach in each venue.

In many agencies, the learning environment has moved well beyond the traditional classroom. Many agencies are now using online training for their personnel for either initial or continuing education needs. As an instructor, the time may come when you will present a lesson in an online environment, which is different than the traditional classroom but is now considered standard for many agencies.

FIGURE 7-1 Learning can take place almost anywhere, and in some cases can be even more effective in informal settings than in a classroom or on a drillground.

Teaching Tip

The learning environment for an online course is different because you do not have a classroom to prepare or set up. However, depending on the type of online learning platform, you can have a big impact on how your virtual classroom is set up and how the students feel in that environment. See the *Technology Based Instruction* chapter for more information on online courses.

To be an effective fire service instructor, you need a solid understanding of the benefits and weaknesses of potential learning environments and must be able to select the best environment for presenting your message. For example, you would not demonstrate a hose drill in the computer room. The key point to remember is that learning occurs in many different places. Assessing your learning environment is an important skill that you as an instructor will need to develop.

Because learning can take place in many locations—such as a traditional classroom, fire-ground training facilities, or an acquired structure—you as the instructor need to be constantly aware of your environment. First and foremost, you need to ensure the safety of your students, yourself, and others involved in the training. Eliminating all safety hazards and concerns should be priority number one. After evaluating the learning environment for safety, you next need to address the environment to ensure that learning can take place. Learning to control distractions, environmental conditions such as room temperature and lighting, and seating arrangements is part of being an instructor. Honing these skills will enable you to be effective in presenting the lesson objectives and training your fire fighters.

> ### Teaching Tip
>
> To be effective in teaching, you must know your audience, the message you are trying to convey, your best delivery method, and the ideal atmosphere in which to deliver your message. This information is not always easy to come by. You may not have knowledge of exactly with whom you are going to speak or even the context in which you will be presenting. You must be able to think on your feet and be prepared to take the presentation in the direction it needs to go so that the learning objectives are met. Remember—a good fire service instructor always has a backup plan.

Fire Service Instructor I and II

Audience and Department Culture

■ Audience

Knowing your audience is critical to your presentation's success. Depending on whether you are presenting at a conference with attendees from multiple jurisdictions or presenting locally to your own department, the manner in which you share the material with your audience will change. When presenting to a diverse audience, your presentation may need to be more generic. Avoid using specifics, such as the exact procedures of any one department. Consider speaking in generalities or giving examples instead of describing mandates or specific procedures. Taking this kind of broader perspective will allow each audience member to adapt your message for his or her own department. For example, when speaking about a procedure, give examples of several procedures used in fire departments across the country. Even when you are presenting to one department, be careful not to mistake instruction for department policy.

You may also need to adapt your presentation style to better match the learning environment. As your audience and your space increase in size, your presentation loses its intimacy. If you typically encourage a lot of student interaction, you may have to adapt your style to deliver a more traditional lecture because the larger room may not allow for personal student interaction.

> ### Teaching Tip
>
> Avoid the phrase, "If I were you, I would do it this way." The best fire service instructors help lead students into making the best decisions for their own departments based on the knowledge that the instructors provide.

■ Department Culture

When making a presentation or teaching a class, you must also be aware of culture. Here we speak not of customs associated with a particular national origin, but rather of departmental culture. Every department has its own culture. It is what makes fire fighters different, yet exactly the same. It is based on the mission of each and every fire department in the world. Departmental culture specifies what we will do, what we can do, and what we will not do. When you instruct students, you must put aside your personal culture. If you are brought in to teach, you must be able to separate facts from opinions.

Your misgivings about a technique do not automatically make that approach a bad idea. For example, some fire fighters might insist that positive-pressure ventilation causes a building to burn down; they may even show examples and argue that this technique introduces air into a hostile environment. These people may have had unpleasant experiences with positive-pressure ventilation, but that does not mean the technique is at fault. The misapplication or misuse of any technique will end poorly. This is the information a student needs—not personal opinions.

You must listen to the culture in which you are instructing and adapt your training accordingly. Every fire department says it is in existence to save lives. Likewise, every department says it is in existence to protect property. But how will the members of that culture save lives and property? To what extent are they willing to risk their own lives? These are the questions a departmental culture must answer.

As fire fighters, a major portion of our job is risk management. Risk management must be viewed within the context of the fire department's culture so that you can place the correct emphasis on the decision-making process. For example, although you may not agree with the decision to go offensive, you must respect and acknowledge the department's decision.

JOB PERFORMANCE REQUIREMENTS (JPRS)
in action

Where the student learns reflects on *how* the student learns. Establishing the proper learning environment enhances and complements your instructional skills. Whether it is the classroom or a training tower, this environment needs to be safe; with limited distractions; and arranged so that all students can participate, view the demonstration, see you, and be an active part of the learning process. Experienced fire service instructors often suggest that the best way to accomplish this goal is to put yourself physically where your students will be during the session. Sit in their chairs, stand where they will view the demonstration, and identify any barriers to effective communication and learning.

Instructor I

The Instructor I will organize the actual setting for the training session so that all participants are able to take an active role in the learning process. He or she can create a safe learning area during practical skills training and enhance student participation in classroom settings through seating arrangements and proper lighting.

Instructor II

The Instructor II will create lesson plans and evaluations in a manner that will take advantage of the learning areas available to the instructor. Suggestions for classroom arrangements, student-to-instructor ratios, and other delivery tips should be included in the instructor instructions for each training session.

Instructor III

The Instructor III should be aware of the eventual delivery of the curriculum he or she develops. Consider multiple delivery configurations and learning styles when you consider your course goals and objectives.

JPRs at Work

Set up your learning area so that all types of learning styles and all methods of instruction that will be used for the training can be maximized. Be able to adapt training materials to the learning environment so that all students can see, hear, and participate in the learning session.

JPRs at Work

No direct JPRs are identified for the Instructor II for this area. Nevertheless, the Instructor II must be fully aware of where the lesson plan will be presented and how the selected instructor will present the information.

JPRs at Work

No direct JPRs are identified for the Instructor III for this area. Nevertheless, the Instructor III must be fully aware of where the lesson plan will be presented and how the selected instructor will present the information.

 Bridging the Gap Among Instructor I, Instructor II, and Instructor III

If instructors can influence where the training session will take place, they can take into account the intended methods of instruction, the various aspects of the learning process, and the communications skills of the instructor who will conduct the session in making their recommendations for the class's location. Consider the final delivery of the lesson plan by the instructor in the design phase of this process, with the goal being to create the best possible learning environment for the student. Learning environments may include both formal and informal settings, ranging from the classroom to the apparatus tailboard. Instructors should take advantage of any opportunity to train when it arises.

Prerequisites

Some fire service instructors use programs that require students to take a pre-course survey. This survey might try to determine student motivation for taking the class or the students' educational level, work experience, or previous training. Its results can also be used to adapt the lesson plan based on the students' needs.

A tool used in some classes is a pre-course test or quiz, which is used to gauge the students' knowledge of the material. After reviewing the results, you can adjust the course material accordingly so as to meet the needs of the students better. These tests or quizzes do not count toward the final grade, but they do provide an understanding of how the fire service instructor can best help each student.

Some programs require that students take prerequisite classes before attending certain advanced classes. The fact that students have completed such preliminaries ensures that the students taking your course have a certain base level of training and education. It also ensures that students do not get into a course that is beyond their experience level. Success in a class will be impaired if a student is already behind the knowledge level for the course. You must ensure that your students have the correct competencies before moving to the next level.

The Physical Environment

Learning takes place in two general settings: formal and informal. The formal environment consists of a classroom plus learning objectives, lesson plans, lesson outlines, and a final evaluation. By comparison, informal learning takes place every day in every location. Controlling the learning environment is a critical task for instructors. As an instructor you need to consider the following factors to ensure you have a positive learning environment that will facilitate the learning process:

- Classroom environment
- Classroom setup
- Safety
- Lighting
- Teamwork and self-actualization
- Minimizing distractions

If you are instructing locally, the venue could be anything from the back of a fire engine to a formal classroom in a training center. Knowing where you will present the information is important so that you can tailor your presentation style to the venue. The size of the room is one factor; the seating in the room is another. The receptivity of students to your message will vary depending on whether the instruction takes place in a formal classroom with tables and chairs or in an apparatus room where students are just standing around. A formal training center with tables and chairs automatically puts students in the learning mind frame; that is, students are more focused in this venue. Because this venue represents a more formal situation, student behavior tends to be more formal. By comparison, the more casual atmosphere of the apparatus room puts students at ease and minimizes the formality of the student–fire service instructor relationship. This opens up opportunities for the discussion to go more in the direction of the students' interests.

The concept of tailboard chats is not a new one. These discussions have been used for many years and can be an effective method of educating fire fighters. The tailboard chat is an informal gathering of fire fighters at the back of an engine where the company talks about various subjects. The tailboard is the equivalent to the water cooler in a business setting. It puts all members of the company at ease and on the same level when you instruct them there. For example, you might use the previous call as an opportunity to reinforce a learning objective. These tailboard chats can happen either back at the station or on the scene after the call is over.

Some fire service instructors use the structure that the company was just at during a call as an example and "simulate" a fire to reinforce learning objectives. The tailboard chat may begin with you saying, "If there were a fire on the third floor of this building, in the B-C corner, where would we place the engine and the truck? What would the first engine company be responsible for doing? What would the first truck company do? What if this fire was at 2:00 a.m.? At 4:00 p.m.?" This is tactics and strategy training at its best.

■ Indoor Classroom

Perhaps the most common setting for conducting a class is in a traditional classroom. Thus, when an instructor thinks of a "classroom environment," he or she probably thinks of a classroom in a building. Most often, classes start in an indoor setting and either remain there for the entire class or, once introductory material is covered, move outside to a training ground.

An important consideration in many classrooms today is the addition of new technology that affects the learning process. Most classrooms today have projectors and computers that are used for presenting lesson material. Many agencies have provided wireless Internet service to assist students with the learning process or to complete assignments online. Of course, with this technology comes the possibility that the student might be distracted by "surfing" the Internet or go to Web sites that are not class related. Issues related to classroom Internet use can be addressed with agency policies or by turning the wireless access on or off as appropriate.

Classroom Environment

The perfect environment for students is one in which all of their needs are met. In Abraham Maslow's hierarchy of needs (discussed further in *The Learning Process* chapter), the first need relates to creature comforts and physiological needs. Warmth and shelter make you feel safe in your environment. In the teaching sense, this need would be met when the classroom has enough light, the temperature is comfortable, and the seating and work space are adequate. In the case of an outdoor classroom, it may mean a place of shelter is available where students can warm up in cold weather and cool off in hot weather.

You must also provide time for breaks so that students can attend to their personal needs. An old instructor's rule states that "The brain can learn only as long as the bottom will allow it." Some instructors (and students) may consider these breaks to be a waste; nevertheless, they are necessary both for the students and for you. In addition to enhancing students' personal comfort, the informal conversations that take place during a break can actually enhance the learning process.

> ## Teaching Tip
>
> Controlling the classroom temperature is very important to the learning process. If a room is too hot or too cold, students will focus more intently on the temperature than on what is being taught. You cannot make every student happy, so you must go with the majority opinion. If possible, lean toward keeping the room cooler versus warmer. In a colder environment, students are less likely to become drowsy. Also, if the temperature is too cool, students can always put on a sweater or light jacket.

Classroom Setup

Setting up the classroom is as important to the learning process as copying handouts or creating a PowerPoint® presentation. Arranging the classroom sets the tone for the learning process. The first question you need to ask is, "Which type of learning do I want to accomplish?" If you plan to lecture to the students, then the room can be set up in rows facing a dry erase board **FIGURE 7-2**. **TABLE 7-1** lists various classroom setups, along with their advantages and disadvantages **FIGURE 7-3**.

Safety

After the basic needs are met, the next highest-priority needs are those relating to safety. In a classroom environment, the term "safety" is used frequently.

Safety in the classroom is just as important as safety on the fire ground, but the hazards are much different. Ensuring that a classroom is safe involves removing both physical hazards (e.g., the trip hazard of a cord, book-bags around students' feet, wet floors from rain or spilled coffee). Physical hazards need to be controlled to ensure a safe classroom environment.

The second type of safety provided in the learning environment is emotional safety of the students. Students need to have an environment, free of judgment or harassment, that will allow them to learn and express their thoughts and opinions. For example, students in a learning environment need to feel that if they make a mistake, there is an opportunity to learn from the mistake. In training, whether in the confines of a classroom or on the expanse of the drill ground, students must feel confident that if they make a mistake, you as the instructor will not belittle or otherwise demean them. Your task is to support students and to help them learn from their mistakes. Football coach Vince Lombardi once said that a person is not measured by how many times he is knocked down, but rather by how many times he gets back up. Your students must believe that you will help them get back up, and this sense of comfort can be provided in the right learning environment.

Lighting

In addition to the arrangement of tables and chairs, having the proper lighting for the room is important. Depending on

FIGURE 7-2 Your style of presentation will determine the best classroom setup.

Table 7-1	Classroom Setups		
Method of Instruction	**Learning Environment**	**Advantage**	**Disadvantage**
Lecture	• Auditorium, traditional, or theater	• Good choice when students are able to see the front of the room • Instructor has more control in this setting • Can be effective for demonstrating skills	• Difficult for students to see one another • Depending on how close chairs and tables are, the instructor may be very limited in moving around the classroom • Depending on the size of the room, students may have a difficult time hearing the instructor or other students
Discussion	• Small U-shaped arrangement • Hollow square • Conference table	• Students have the opportunity to see one another • Good environment for small-group interaction • Good environment for instructors to work in smaller groups with students	• Students can easily be distracted by one another • Difficult for all students to see the front of the room • Difficult for instructors to control the teaching environment
Demonstration	• Large U-shaped/horse-shoe arrangement	• Allows students to see the demonstration of skills • Allows the instructor to move about the room and interact with students	• Students can easily be distracted by one another • Difficult for all students to see the front of the room

A. Traditional.

B. U-shaped (horse shoe).

C. Conference table.

FIGURE 7-3 Sample classroom configurations.

Courtesy of L. Charles Smeby, Jr./University of Florida

Minimizing Distractions

One challenge that all fire service instructors face is the variety of distractions offered by today's communications technology. Cell phones, smart phones, pagers, and station alarms—all of these ubiquitous devices can cause students to be distracted from learning. Many of these distractions can be eliminated at the beginning of class by asking students to turn off or silence all electronic devices. If you have on-duty crew members in class, then make arrangements with them to turn down their pagers or radios.

Other distractions, such as people entering and leaving the classroom during a lecture, can be addressed before the class begins by posting a sign on the door stating that class is in session. Request that latecomers wait until a break to enter the classroom.

> **Teaching Tip**
>
> If training is to occur with on-duty crews, consider having one student monitor a portable radio with the volume turned low.

If training is taking place outdoors, then controlling the learning environment is more difficult. Noise and distractions can disrupt the learning environment, so try to conduct your training in an isolated place. Eliminate what you can and minimize distractions as much as possible. An effective tool is to keep your groups small, so that every student is engaged in the learning process with the oversight of a fire service instructor. Also, be selective regarding what has to be taught outside versus what could be taught inside. Limiting instructional time outdoors is a good method of controlling outside distractions.

> **Teaching Tip**
>
> Proper selection of the learning environment will determine the success or failure of the class in meeting the learning objectives.

Classroom Arrangement for Testing

The setup of the indoor classroom typically changes when students are required to take a test. Specifically, the classroom should be arranged to provide each student with enough space to put both the testing booklet and the answer key on a writing surface. It should also be designed to minimize the chance of wandering eyes, while simultaneously allowing the instructor to observe all students. Given that most testing is individualized, the classroom should remain quiet. If the testing is group based, you must be able to distinguish between testing noise and distractions from a group. The personal needs of students should be recognized. If a student needs to leave the room to attend to personal needs, the security of the testing process should be maintained. Because testing and evaluating the

the type of presentation that you are giving, you may need to adjust the lighting. In some cases, lights may be left on during a PowerPoint® presentation or during other projected presentations. In the ideal classroom, the lights will be split, allowing lights in the front of the room to be dimmed while lights in the back of the room remain on to allow students to take notes. Your podium should have a light source so that you can read your notes. If the classroom does not have split lighting or adjustable lighting over the students, then you need to decide which type of lighting is best for your students.

learning process are important to the fire service, having the right testing environment will ensure that students can focus on the test.

■ Outdoor Classroom

Although it might appear that the work conducted in the outdoor classroom is more important than the instruction that occurs in the indoor classroom, both are needed to ensure that students achieve the learning objectives. Your challenge is to match the environment to the learning objectives to make sure that the student is always in the correct environment in which to learn.

Outdoor classrooms differ in more respects than just their location **FIGURE 7-4**. When moving the learning process to an outdoor venue, the instructor needs to address issues such as weather, training and ambient noise, and logistical issues such as equipment needs, parking, or rehabilitation facilities. The key to training outdoors is to plan ahead and to conduct the proper pretraining assessment to ensure a safe and effective learning experience for your students.

When curriculum or training needs require an outdoor training environment, the instructor needs to be prepared to teach in this environment. These outdoor classrooms can be organized on a developed and organized training ground,

FIGURE 7-4 Outdoor locations can present a set of challenges for the fire service instructor.

which could contain a fixed burn building to teach interior fire attack, a mobile prop used to teach fire behavior, or a pond to teach drafting skills. Another type of outdoor classroom could be an off-site location such as a mountain used to teach high-angle rope rescue skills.

Outdoor locations, whether on a training ground or an off-site location, pose similar challenges and considerations for the instructor to address to facilitate the learning process. These challenges include the weather, outside noise, logistics/resources, and the ability to teach so that students can learn in the outdoor classroom.

A classic example of an outdoor learning environment that would present several challenges would be training at an acquired structure. These buildings present a great learning opportunity for students, but also create many difficulties for instructors. Following NFPA 1403, *Standard on Live Fire Training Evolutions*, will assist the instructor in correctly preparing for such a training event.

■ Communicating Outdoors

Communicating in the outdoor classroom can be a challenge to any instructor. Learning to establish an environment that will allow you to teach outdoors is just as important as it is when in a traditional classroom.

For example, when teaching in a live fire training scenario, communication can be difficult. In this environment, you first need to be cognizant of your own safety. Taking off your self-contained breathing apparatus (SCBA) mask to communicate in a smoke-filled environment not only puts you at risk, but also sends a message to the students that the use of a SCBA mask is not important. Communication in the live fire environment is essential, however, and can be accomplished by first gaining the students' attention and then by using a calm and controlled voice to communicate what you are attempting to teach.

A live fire training scenario is also an environment in which teaching is a constant process. You need to be close enough to the students to observe their behaviors and far enough back to observe the scene and allow the students to perform and learn.

Learning how to communicate in the outdoor classroom is a skill that you will need to develop and always be mindful of when outdoors. Observing your students will be critical to determining whether your voice and the content of your message are reaching everyone. Observing students' body language will be a good indicator of whether your message is getting to everyone.

> **Teaching Tip**
>
> The use of multiple senses as part of the learning process increases learning and long-term retention of material. It is a useful technique in the outdoor environment.

Safety

One point cannot be emphasized enough: Safety should be your greatest concern in the outdoor classroom. During outdoor training, the stakes become higher and the attention to detail becomes more intense. When performing live fire training, for example, the environment needs to be as safe as possible while allowing learning to take place. You hold students' lives in your hands, and students need reassurance that you take this responsibility seriously. It is important to follow standards such as NFPA 1403, among others. Having the right number of personnel, in the right positions, doing the right things, with an emphasis on safety, makes the learning environment the safest place for fire fighters to work and learn in.

The purpose of many outdoor training sessions is to teach or evaluate students' psychomotor skills; these skills are the ones that fire fighters use to protect the lives and property of citizens. For that reason, when training and evaluating students, you need to be in the most realistic situation possible while maintaining a safe environment for students. In rope rescue, this concept is called being "redundantly redundant"—that is, there is a safety for everything.

Make sure that students' psychomotor skills are practiced and completed perfectly. A skill is either acceptable or unacceptable—there is no middle ground. If a fire fighter is asked to don a full protective ensemble and forgets to put the hood up, he or she has failed in that skill. Vince Lombardi said, "Practice doesn't make perfect; perfect practice makes perfect." Another relevant saying is, "The more we sweat in training, the less we bleed in battle."

Weather

A major consideration outdoors is the weather. Obviously, outdoor classrooms are subject to the weather conditions. The physical location of your outdoor classroom can determine how you prepare for and handle the class. If you are in the South or Southwest, you may need to deal with high heat levels. If you are in the Northeast or Midwest, you may be more concerned about cold temperatures and snow. No matter where you are, the threat of rain is a consideration when an outdoor classroom will be used for instruction.

The safety of your students should always be your highest concern. Many fire departments use either a temperature/humidity (heat) index or a wind chill index as a guide when determining whether outdoor training is feasible or acceptable within department policy. NFPA 1584, *Standard on Training and Emergency Incident Rehabilitation Practices*, offers guidelines on temperature and physical exertion levels relating to how much training, personal protective equipment, and rehabilitation is required during extreme temperature and weather.

Always have a "plan B" in case inclement weather forces the class to use an alternative location that is unaffected by the weather, such as an apparatus bay floor or a large, unused warehouse. Plans to use these areas should be put into place before the training occurs, rather than decisions being made spontaneously during the training as an emergency measure. Rain is a fact of life. A small amount of rain may not be an issue during outdoor training, but torrential downpours and lightning are definitely reasons to seek an alternative environment. Warming or cooling areas and shelter from wind, rain, or other weather phenomena are necessities for your students' safety.

Safety Tip

Know your department's policies on temperature indices for training. If the temperatures reach those points, follow the policy in adapting or suspending the training activity.

Noise

Noise in the outdoor environment is also a concern during training. In most cases, you will need to project your voice so that all of the students can hear you without the use of an electronic device such as a microphone. Environmental factors such as wind can be a deterrent in allowing your voice to carry. Other distractions—such as vehicle traffic, ambient noise from the field, or other training activities going on around you—can add to the difficulty of communicating in outdoor environments. During your preplanning of the outdoor classroom, look for locations in which you can present verbal information effectively. Many training grounds have outdoor pavilions that are designed for teaching and for sharing information. Shielding yourself near a fire apparatus can help minimize noise and project your voice. It might not be possible to eliminate all outside noise, so it falls on you as an instructor to use your voice and other teaching skills to help communicate the information and knowledge you are attempting to teach.

Logistics and Resources

When the classroom moves to the outdoors, an instructor must address several other issues as well. As part of your classroom setup, ensure that all logistical arrangements have been secured prior to moving outdoors. Reviewing your lesson plan and ensuring that all required equipment and materials are in place before class are steps that are expected of you as the instructor. Knowing the location of restroom facilities and having a rehabilitation area established are just as important as having the right equipment ready for your class. At an established training ground, these resources are usually in place and have been designed into the facility. The real challenge comes when training moves to an off-site location, because you need to convey all of your resources to that site. Making arrangements before class to address the physical needs of your students for restroom facilities and rehabilitation becomes an important challenge when moving to an off-site location. The need to take required and extra equipment with you is also important to help teach so downtime is minimized if something breaks or additional equipment is needed.

Safety Tip

The importance of proper hydration cannot be overstated when students will participate in physical activities. Even professional athletes have suffered fatal cases of heat stroke when temperatures were not extreme.

VOICES
OF EXPERIENCE

In my tenure as a fire service instructor, I have had to learn the skills necessary for teaching in an adult learning environment. Many of the skills needed to teach an adult learner are found in the basics of understanding how we communicate and relate to each other in the learning environment. Adult students have the need for safety, acceptance, reassurance, and acknowledgement of achievement like any other student, but they also have life circumstances that are thrown into the mix. Remember that the adult learner's time is valuable. He or she is taking your class while trying to balance work, home, and social demands.

The student and I were able to come up with an agreeable schedule so he could catch up on the missed classroom contents.

Not every situation while teaching always goes as planned. I once had a student approach me in a panic during a classroom break. He had just learned of a family crisis and was worried about whether he should leave the class to attend to the emergency or ride out the event until the end of the class and then rush home. The student needed the class to qualify to take an officer examination in the coming weeks and there wouldn't be another offering of the class prior to the test date.

I explained to the student that he wouldn't be able to focus on any part of the instruction if he was sitting in the classroom while a crisis was happening at home. I assured him that should he resolve the situation at home and let me know if he would return to class, that we would be able to make up the missed time. Fortunately, although the emergency was serious, it was able to be handled that same evening and the student and I were able to come up with an agreeable schedule so he could catch up on the missed classroom contents.

The empathy that was given to the student was relayed to the rest of the class when he returned the following day and shared the story with his classmates. He expressed how pleased he was with the manner in which the problem had been handled. Most importantly, we were able to work together to solve his dilemma and achieve his goal of class completion. All of this was possible by knowing the permissions of my instruction and the importance of the learning environment for the benefits of the student and the class. This is an example of retaining positive instruction while managing the unexpected life events of teaching in an adult learning environment.

Jason Loeb
Hoffman Estates Fire Department
Hoffman Estates, Illinois
Illinois Fire Service Institute

Contingency Plans

You must have a plan B regardless of the environment. What do you do if the projector fails to work? What do you do if lightning begins to flash in the sky on a live fire training day? What happens if an assistant instructor does not arrive on a practical skill day? All of these contingencies need to be considered and planned for. Many times the solutions are simple—for example, keeping a spare projector bulb or a spare projector, or having a rain date for outdoor training. In any event, you as the instructor need to be able to adjust your training plan so that your training plan can continue. If an evolution or training session cannot continue due to some type of issue, many rescheduling or make-up logistics will be involved in finishing the training. Without question, safety and proper delivery will always take precedence over any of these considerations. A best practice may be to have a set of "ready-to-go" backup drills in the event that training must be stopped. You do not want to lose momentum or the opportunity to train when your department meets only a few times per month.

Another way to look at this situation is much like you would approach an emergency scene. Do you have the resources in place if something should go wrong? For example, suppose you are teaching a psychomotor drill and a trainee gets hurt. Is there an ambulance available? Who is notified? Which forms must be completed? How will the training continue if you are missing one student? On an emergency scene, fire fighters constantly plan for contingencies.

Teaching Tip

Always have a contingency plan. Things can and will go wrong, and having at least considered these issues will allow you to adapt to the changing circumstances quickly.

Teamwork and Self-Actualization

In Maslow's hierarchy of needs, Maslow speaks of the issue of belonging. A main focus of fire training is teamwork, and the learning environment you create must be one that supports and recognizes the crucial nature of teamwork. Many of the best friends you have in the fire service are likely to be people whom you met as you trained to become a fire fighter. During your training, you built a strong sense of camaraderie, which ultimately helped you in completing a very difficult task. An important part of the instruction should be providing time either before, during, or after the formal instruction to allow students to network.

The next level in Maslow's hierarchy of needs is the need for self-esteem or accomplishment in the student. You should create an environment that celebrates students' victories as they are accomplished. Status and recognition can be very powerful motivators in the learning environment. In his book *The One-Minute Manager*, Ken Blanchard talks about catching people doing things right. In a good learning environment, this idea is expanded to not only catch students doing things right, but also recognizing and celebrating that accomplishment. People want to be in an environment where they can succeed.

The highest-priority need in Maslow's hierarchy is the need for self-actualization. At this level, the student is able to synthesize material by taking multiple learning situations and putting them together to solve a problem. The need for self-actualization is satisfied in the environment where you perform live fire simulations or even live fire training. Look to challenge your students to take your message and use it to solve a problem.

Safety Tip

Failure to provide for students' sense of safety will prevent a positive learning experience.

Demographics

To deliver your message effectively, you must know who your audience is. A fire fighter is a fire fighter. There is no race, there is no male or female, and there is no ethnicity. We are no longer called "firemen"; we are called "fire fighters." As fire service instructors in the new millennium, we must realize that demographic factors do not apply just to the private sector. Demographic considerations extend beyond issues of race and national origin; in the fire service, they require looking in a more holistic manner at everything a fire fighter brings to the department.

Demographics include the age, gender, marital status, family size, and educational background of a fire fighter. As a fire service instructor, you should look at a fire fighter's demographics and ask yourself a number of questions: Which type of skills does this fire fighter bring? Could these skills help the class learn, or might the skills hold the fire fighter back in the learning process? Are you dealing with a highly educated individual or a fire fighter who has met only the basic educational requirements? Is the student an experienced fire fighter or a rookie? Does the same class include a mix of experienced fire fighters and rookies? It is this type of demographic information that you must understand and appreciate as you approach your teaching assignment.

Another demographic consideration unique to the fire service is the different types of staffing encountered among fire departments. The fire service includes career, volunteer, combination, and paid-on-call fire fighters. The learning objectives for paid-on-call or volunteer fire fighters do not differ, but the presentation to these groups does. Because volunteer and paid-on-call fire fighters generally have full-time jobs, training accommodations must be made to fit their schedules. By gearing instruction toward the demands of the students' schedules, you allow each student to focus fully on the subject at hand and not be distracted. For example, if a class is made up of all career fire fighters, the class may be six to eight weeks long, eight hours a day. By contrast, if the same class is made up of all volunteers, the class may meet Tuesday and Thursday nights for three hours and all day on Saturdays for six months. Both classes should cover the same learning objectives to the same levels of proficiency, but the differing demographics of the participants require different schedules.

Demographics in the Classroom

As a fire service instructor, you should be aware of how demographics affect the classroom. Always exercise caution and common sense in the classroom. Never make a comment that refers to someone's background, such as where they attended school, their marital status, or their ethnicity.

Gender Considerations

Gender is a hot-button issue in today's society. One change that has affected the fire service in the last few years is the abandonment of the term "fireman." Many qualified women have joined the ranks of this once male-dominated profession, so it is more appropriate to refer to students as "fire fighters." Firefighting does not, according to the fire, have a gender. Although learning objectives do not change with the gender of the student, you do need to be cognizant of students' gender. If you use the term "guys" to address your students, your students need to understand that "guys" includes both men and women. Be gender neutral when developing learning objectives. The key is to be respectful of all your students at all times. As an instructor you should present yourself with professionalism at all times.

Offensive Language, Gestures, and Dress

The use of offensive language should be avoided—period. Some people may contend that to make a point, sometimes you must swear or use slang terms; others contend that this is part of the fire service culture. This is not the case. Jokes and language of a sexual or explicit nature are unacceptable in the learning environment. As a fire service instructor, you should have a mastery of the English language. You should be able to make your point without the use of profanity. You will find that students gain a deeper respect for you when you demonstrate eloquence in the classroom.

Do not participate in improper situations. A wise fire service instructor once shared a key piece of wisdom when he said, "If you hear an off-color joke and grin, you're in." If you hear an offensive joke, show that you do not agree with what was said. It is your responsibility to stop any unsafe or inappropriate behavior immediately. It is important to take pride in your department, and one source of that pride involves showing respect to everyone.

If a student is offended by your behavior or your language, you should apologize. If you offended the student in front of the class, then it would be appropriate to apologize first to the student in private and then in front of the class. This can go a long way toward making amends. An equally important step is to establish a policy that covers acceptable behavior and language. Your organization should have a harassment policy that addresses these issues, including remedies for offenses, reporting of incidents, and conduct. The best policy is to understand the audience to whom you are speaking. Ensure that your presentation occurs in a context that is appropriate for your audience. A wise fire service instructor used to say, "Make sure that your language would be acceptable to your grandmother."

Like words, gestures and dress can sometimes be offensive. How you dress and how you move are representations of your position **FIGURE 7-5**. Keep a neutral position at all times, which allows the student to feel free to express his or her opinion and concerns in a safe environment.

FIGURE 7-5 How you present yourself is a reflection on the fire department.

Adapting the Class Based on Demographics

The information that you learn as a result of students' verbal introductions may require slight modifications to the lesson plan. Perhaps students will need to spend more time in one area of the class and less time in another. The information gained in this way can also help you decide the depths you need to go in the various areas of the presentation, perhaps allowing you to lengthen parts of the program and shorten others to meet the needs of the class. For example, if you are presenting a Disaster Preparedness Course in Arizona, you may not need to spend an entire hour on blizzards.

If a student requests that you cover a certain topic, you should not rewrite the entire course; however, you should keep the student's request in mind. If time and circumstances permit, you could briefly present the topic and then ask the student, "Did that meet your need?" By doing so, you ensure that students know you are available to instruct them on whatever they need to know so as to meet the learning objectives of the course. Bear in mind that as a fire service instructor, you must be prepared either to adapt your lesson plan or to tell the student that the topic is outside the scope of the class.

Teaching Tip

If you indicate that you will cover an area or an objective, you need to ensure that you fully address any concerns or questions about that topic. Meeting this goal requires knowing and understanding your audience. Many fire departments have target hazards or situations that are unique to their locations. For example, many communities in the Northeast include row houses, whereas these structures are rarely found in the Midwest. When asked, "How would you handle this situation?" you should be prepared to render an opinion. By anticipating what the questions will be, you will be prepared to offer an opinion because you have had time to research the question and present an intelligent response. If you do not know the answer, do not—under any circumstance—make up the answer. It is completely acceptable to say, "I do not know, but I will try to find out for you." Fire fighters respect honesty in their fire service instructors.

Jones & Bartlett Fire District

Training Division
5 Wall Street, Burlington, MA, 01803
Phone 978-443-5000   Fax 978-443-8000
www.fire.jbpub.com

Instant Applications: The Learning Environment

Drill Assignment

Apply the chapter content to your department's operation, training division, and your personal experiences to complete the following questions and activities.

Objective

Upon completion of the instant applications, fire service instructor students will exhibit decision making and application of job performance requirements of the fire service instructor using this text, class discussion, and their own personal experiences.

Suggested Drill Applications

1. Review your current classroom facilities and identify any barriers to effective learning or communications. Identify methods of eliminating or reducing these barriers.

2. Review your current practical training area and identify any barriers to effective learning or communications. Identify methods of eliminating or reducing these barriers.

3. Experiment by moving tables and chairs into different configurations to identify the best way of arranging the room for your presentation.

4. Troubleshoot the practical training area and consider alternatives to limited-view props or practical stations in the event of inclement weather or a large number of students.

5. Review the Incident Report in this chapter and be prepared to discuss your analysis of the incident from a training perspective and as an instructor who wishes to use the report as a training tool.

Incident Report

Boulder, Colorado—1982

A search and rescue drill was being held in a large 28' × 61' (8.5 × 18.6 m) shed. The fire department had divided the shed into smaller rooms with temporary partitions made of combustible materials to provide better search scenarios. The shed had a wood frame with a ceiling made of combustible, low-density fiberboard tiles. This engine company was the third to participate that day.

Although this was set to be a smoke drill only, the smoke was produced using motor crankcase oil to burn tires on the floor in each of the rooms. Smoke bombs were also used. A road flare and small amounts of gasoline mixed with cleaning solvent were used to ignite each fire.

A water supply was not established and the nearest hydrant was 1100' (335.3 m) away. A reserve engine with a 500-gallon (1892.5-liter) booster tank was used to supply water. The primary engine had a 300-gallon (1135.5-liter) booster tank, with only a charged booster hose line used by the crew participating in the training exercise. The pump operator was one of the participating crew members, and both pumpers were left unattended.

Prior to the incident during the engine company's training, there were fires in the rooms for roughly two hours. Before the last crew entered the structure, another tire was added to each pile. The three engine company members and a training officer were inside the structure when the preheated ceiling tiles ignited, trapping the four fire fighters. They became separated, and only the training officer and the company officer were able to make it out alive. The company officer received third-degree burns on over 35 percent of his body. It is important to note that the training officer in charge had never previously been directly involved with this sort of training.

Post-Incident Analysis: Boulder, Colorado

NFPA 1403 (2007 Edition) Noncompliant

NOTE: This training fire occurred before the creation of NFPA 1403.

Combustible ceiling tiles contributed to rapid fire spread (*4.2.17*)

Rubber tires and flammable and combustible liquids used (*4.3.3*)(*4.3.6*)

Multiple simultaneous fires on the floor in open containers (*4.4.15*)

No interior hose line (*4.4.6*)

No established water supply (*4.2.23*)

No incident safety officer (*4.4.1*)

Flashover and fire spread unexpected (*4.3.9*)

No written emergency plan (*4.4.10*)

Students not informed of emergency plans and procedures (*4.4.10*)

No back-up hose lines (*4.4.6.2*)

No EMS on scene (*4.4.11*)

Instructor not trained or experienced with topic (*4.5.1*)

Deviation from the plan (*4.2.25*)

No preburn plan (*4.2.25.2*)

No walk-through performed with students and instructors (*4.2.25.4*)

Wrap-Up

Chief Concepts

- Because learning can take place in many locations—such as a traditional classroom, fire-ground training facilities, or another location such as an acquired structure—you as the instructor need to be keenly aware of your environment.
- When making a presentation or teaching a class, you must also be aware of both the audience and their culture.
- Some programs require that students take prerequisite classes before attending certain advanced classes. The fact that students have completed these preliminaries ensures that the students taking your course have a certain base level of training and education.
- Establishing a learning environment is just as important as having a lesson plan. Without the proper environment, the learning process will be limited and can be ineffective.
- Matching the correct classroom arrangement to the needs of the lesson plan will allow students to learn.
- When presenting a lesson in an indoor classroom, consider the classroom environment and setup, safety, lighting, and ways to minimize distractions. Adjustments may need to be made for testing.
- When presenting a lesson in an outdoor classroom, considerations include communication, safety, weather, background noise, logistics, and resources.
- Ensuring a safe learning environment is a high priority for all fire service instructors.
- You must have a plan B when using any environment. All potential contingencies need to be considered and planned for.

- Understanding demographics extends beyond issues of race and national origin. In the fire service, taking demographics into account during instruction requires looking in a more holistic manner at everything a fire fighter brings to the department.

Hot Terms

<u>Demographics</u> Characteristics of a given population, possibly including such information as age, race, gender, education, marital status, family structure, and location of agency.

<u>Learning environment</u> A combination of the classroom's physical and emotional elements.

<u>Tailboard chat</u> An informal gathering where fire fighters discuss various issues.

References

Blanchard, Kenneth and Spencer Johnson. (1982). *The One-Minute Manager*. New York, NY: William Morrow and Company, Inc.

National Fire Protection Association. (2012). *NFPA 1041: Standard for Fire Service Instructor Professional Qualifications*. Quincy, MA: National Fire Protection Association.

National Fire Protection Association. (2012). *NFPA 1403: Standard on Live Fire Training Evolutions*. Quincy, MA: National Fire Protection Association.

National Fire Protection Association. (2007). *NFPA 1403: Standard on Live Fire Training Evolutions*. Quincy, MA: National Fire Protection Association.

National Fire Protection Association. (2008). *NFPA 1584: Standard on Training and Emergency Incident Rehabilitation Practices*. Quincy, MA: National Fire Protection Association.

You are presenting a class on leadership and motivation and have several case studies that you plan to use as part of the application step in your delivery. You envision a lot of group participation and discussion but also have some lecture material to present. Your classroom has the typical accommodations for standard course delivery.

1. Which type of seating arrangement would be best for the lecture portion of your class?
   A. A large auditorium-style setting
   B. A horse-shoe or U-shaped pattern
   C. The tailboard of an engine
   D. A circle pattern

2. Which type of seating arrangement would be best for group discussion and participation?
   A. A large auditorium style setting
   B. A horse-shoe or U-shaped pattern
   C. The tailboard of an engine
   D. A circle pattern

3. How important is the arrangement of tables and chairs to the effective participation of all of your students?
   A. Table and chair placement is very important, as it will allow for effective information exchange and participation.
   B. Table and chair placement is not that important, as it is more the students' desire to participate that will drive the flow of information.
   C. The instructor's skill in engaging the students will be more important than table and chair placement.
   D. Both A and C are correct.

Technology in Training

Fire Service Instructor I

Knowledge Objectives

After studying this chapter, you will be able to:

- Describe the types of multimedia tools available for the fire service instructor. (NFPA 4.3.3A) (NFPA 4.4.2) (pp 178–199)
- Describe the advantages and limitations of audiovisual equipment and teaching aids. (NFPA 4.4.2) (pp 189–191, 193–194, 196–198)
- Describe how to use multimedia tools. (NFPA 4.4.6) (pp 185, 187–189)
- Describe how to maintain multimedia tools. (NFPA 4.4.6) (p 197)
- Describe when to use multimedia tools in a presentation. (NFPA 4.4.7) (pp 197–199)

Skills Objectives

After studying this chapter, you will be able to:

- Demonstrate how to use multimedia tools during a classroom presentation. (NFPA 4.4.2) (pp 185, 187–189)
- Demonstrate how to clean and maintain audiovisual equipment in the field. (NFPA 4.4.6) (p 197)

Fire Service Instructor II

Knowledge Objectives

After studying this chapter, you will be able to:

- Describe the types of multimedia tools available for the fire service instructor and identify their use in a lesson plan. (NFPA 5.3.2) (pp 178–199)
- Describe how to use multimedia tools during a classroom presentation. (NFPA 5.3.2) (NFPA 5.3.3) (pp 189–191, 193–194, 196–198)
- Describe when to use multimedia tools in a presentation. (NFPA 5.3.2) (NFPA 5.3.3) (pp 197–199)

Skills Objectives

After studying this chapter, you will be able to:

- Demonstrate how to use multimedia tools during a classroom presentation. (NFPA 5.4.2) (pp 185, 187–189)

Fire Service Instructor III

Knowledge Objectives

There are no knowledge objectives for Fire Service Instructor III students.

Skills Objectives

There are no skills objectives for Fire Service Instructor III students.

You are a relatively new instructor but have already learned that the methods of instruction used can influence students' retention of material and participation in class. In reviewing the training calendar for the next training period, you notice a variety of instructional opportunities in which you can participate. One class will include primarily young fire fighters who are in the early stages of the fire academy; another class will consist of a mix of new and experienced members. Your experience with technology-based instruction is sound, and you contemplate the methods that might be used to deliver these classes. You can foresee the use of blended learning where the students complete some of the coursework online and other parts in traditional environments.

1. Which type of technology-based instruction is the younger generation of fire fighters accustomed to in their education so far?
2. Which traditional methods of instruction have the experienced members become accustomed to?
3. How does a blended learning approach maximize your ability to share information with your students? What are the downfalls of this approach?

Introduction

The use of training aids has been part of the training process, regardless of the discipline, since the beginning of humankind. Ancient warriors learned how to fight with practice swords and shields before they used real ones; commanders would draw scenarios in the sand while trainees would gather around to observe. Table-top simulators were used to run mock battles while commanders looked on and learned what was to happen before ever stepping foot on the battlefield. All of these activities are forms of training aids to assist the instructor in training and the students in learning.

Today, the training aids we use in the fire service have evolved from basic chalkboards and erasers to PowerPoint® (PPT) presentations and computer simulations run in a classroom or over the Internet. Although training aids are fundamental in assisting the fire service instructor in the learning process, they are merely "aids" to help the instructor impart the knowledge the fire fighter needs to know and understand. As an instructor, it is your responsibility to learn how and when to use these tools during your presentations and, when necessary, to be able to troubleshoot or repair them in the field. The key is to remember that these tools are there to "assist" you during your presentation—but ultimately you are the fire service instructor, and the students learn from you.

Fire Service Instructor I and II

Technology-Based Instruction

One of the most dynamic and fast-changing concepts in education and training is technology-based instruction (TBI). TBI may also be called other names, such as *e-instruction*, *Internet-based instruction*, or *distance learning*. In general, TBI is defined as presenting training and education using the Internet or a multimedia tool such as a DVD or CD-ROM. As is true for traditional methods of instruction, there are both advantages and disadvantages to TBI. TBI tends to be a student-centered instructional method because in most cases students can control the pace, the rate of delivery, and even the time at which they view lessons and information. This flexibility is highly desirable to many busy adults, fire fighters and emergency medical services (EMS) providers included. Research also supports the student-centered nature of TBI as being more effective, in that students can repeat lesson areas they are struggling with, while instructors enjoy the ability to monitor each student's participation and progress more closely.

Consider a traditional PPT presentation given by even the most dynamic instructor. Some students may disengage from the learning process, while other, more engaged and participatory students carry the load in answering questions and giving examples. Unless the instructor carefully engages and involves everyone, some students could sit through an entire presentation and receive a passing grade simply by attending the session, without truly participating in it. Consider this example: The instructor displays a slide of a house fire. The instructor asks for a student to perform a size-up and initial radio report. In a class of 25 students in a typical classroom-based delivery, chances are that only a few students will be able to participate in answering the question. Using TBI, the instructor can set up a system that shows the same slide, but then requires all students to perform the size-up and initial radio report in a narrative example. The TBI method will certainly involve all students, whereas the more traditional approach may not.

Distance Learning

Distance learning does not require computers, modems, DVD players, or any other fancy technology. Instead, it simply consists of instruction from a location "a distance away" from a training center or college/university. Such a program can be as simple as a package of printed materials sent through the mail, or as complex as Flash animation presented during a video conference. Today, the majority of distance learning appears to be accomplished through a combination of reading assignments, conferencing, video presentation, testing with proctors, and written assignments. Student evaluations are processed through an instructor.

Computer-Based Training and Learning Management Systems

Computer-based training (CBT) software provides a virtual course environment with tools for course preparation, delivery, and management. With this software, you have the tools to prepare course materials and efficiently manage day-to-day teaching tasks. Many CBT programs offer features that allow you to facilitate collaborative learning, personalize content based on students' unique needs, and positively affect learning outcomes. CBT is not limited to students in colleges and universities; it has also changed the way in which private industry and public safety agencies train their staff members.

CBT can be a powerful way of teaching. It can inform, illustrate, test, and demonstrate complex concepts simply. For example, students can view Flash animations with a voiceover that explains how a fire pump works internally. They can also click on parts of the fire pump image to see labels or learn more details. If real-world experience is desired, some systems have built-in simulators to enable students to experiment with key concepts while using different parameters. This allows students to benefit from real-world learning while ensuring that they can make mistakes without endangering personnel or equipment.

E-learning (short for "electronic learning") is a form of computer-based training—one that has enjoyed increased popularity and prevalence in recent years. Web-based learning management systems (LMS) are at the heart of the e-learning experience. The following terms will help you understand the systems that support computer-based training:

- **Learning management system (LMS).** A learning management system is a Web-based software application that allows online courseware and content to be delivered to learners, wrapped with classroom administration tools such as activity tracking, grades, communications, and calendars. An LMS is sometimes also referred to as a virtual learning environment (VLE) or learning course management system (LCMS).
- Synchronous learning. Synchronous learning happens when learners are all engaged in or receiving educational content simultaneously. An example might be a live Webcast, Webinar, or online chat session.
- Asynchronous learning. Asynchronous learning simply means that all learners are not consuming content at the same time. In a traditional face-to-face classroom, all learners would hear a lecture simultaneously. In

an asynchronous learning environment, however, the same learners might consume educational content at different times, usually at their own pace, provided they adhere to established submission deadlines set by the instructor.
- Hybrid learning. Sometimes referred to as "blended" or "distributed" learning, hybrid learning occurs when some content is available for completion online and some learning requires face-to-face instruction or demonstration.
- Courseware. Courseware is any educational content that is delivered via a computer.
- Virtual classroom. A virtual classroom is a digital environment where content can be posted and shared, and where instructors can create quizzes and assignments to be completed and tracked online via an Internet-connected computer.
- Webinar. An online meeting, the word *webinar* is the portmanteau of "Web" (denoting the online component) and "seminar." These meetings occur in real time, and usually allow for participants to share their desktop screens, Web browsers, and documents live as the meeting is occurring.
- Social learning. Social learning is an informal method of learning using technologies that enable collaborative content creation. Examples include blogs, social media (e.g., Twitter, Facebook), and wikis.
- Instructor-led training (ILT). In the world of e-learning, ILT usually refers to a scenario in which a human instructor facilitates both the in-person sections and the online component of the course. Instructors usually set deadlines for submission, create quizzes and assignments, and track student progress.
- Self-directed study. Prepackaged learning modules can be designed to allow learners to engage in self-directed study—to stop and start as they desire, progress to completion of the module, and submit a final score to determine whether they passed or failed the module, all without an instructor's involvement. An example would be the online ICS courses freely available from FEMA's Emergency Management Institute.

The popularity of e-learning has exploded in the past few years. In fact, many colleges and universities have doubled or tripled their courses offered via the Web. Some institutions even offer entire associate or bachelor's degree programs over the Internet. Recognizing this fact, many students now expect courses to be offered in this manner. Colleges, universities, fire departments, and other training organizations are all considering how best to integrate e-learning into the instructional experience.

Of course, no technology can ever replace the wisdom and experience of a seasoned fire service instructor. E-learning is not a threat to instructors, but rather a tool that can be used to address challenges posed by live training. For example, in the future, virtual reality simulators could be used instead of live burns in the early stages of training. The skill will eventually need to be performed to be truly learned, but work with simulators can increase students' comfort level with the scenarios before encountering them with all dangers present.

An LMS is a Web-based software application that acts as both a digital content warehouse and a classroom management tool. Learning objects (content) reside within the LMS and are accessible to any member of a course who has Internet access. Most LMS platforms allow instructors to upload (and thereby save and store) documents into their online classroom and then link those resources in a meaningful way for learners to consume. In this regard, the LMS acts like a repository for your content.

Uploading documents to a Web site, however, hardly provides the entire solution with regard to managing an online classroom. Just as you would do in a traditional classroom, as the instructor, you need to track *how well* your e-learning students are performing, *what* they are consuming, and *whether* relevant objectives have been met. That is where the classroom management piece of an LMS comes into play. **FIGURE 8-1** shows possible course management tools.

Typical features of all LMS platforms allow an instructor to administer their course in a variety of ways:

- Add new users or roles
- Access course editing tools
- View the grade book
- Access course files and folders
- Run learner activity reports
- Import, back up, or reset content
- Communicate with course members

Although an LMS can be used to facilitate fully online learning, in the context of fire service instruction it is critical that students demonstrate comprehension of topics as well as a capability to perform critical tasks. Given this requirement, the hybrid learning approach is preferable.

The National Fire Protection Association (NFPA) estimates that the number of fire fighters working in the United States was 1,103,300 as of 2010. Of that number, 335,150 (30%) were

career firefighters and 768,150 (70%) were volunteers. Of the volunteers, 50% were located in rural areas. When you consider the difficulties facing today's fire service instructor—shrinking budgets, departments located at a geographical distance from the nearest training facility, increased hours and curriculum requirements, and the composition of the fire service (70% volunteer, meaning they have other primary responsibilities)—a compelling argument for the inclusion of technology begins to emerge.

By combining LMS technology with the tried-and-true traditional classroom setting, you achieve a learning experience where instructors can capitalize on the following features:

- Take advantage of the power of Internet-enabled learning
- Offload low-level or routine items to the online system for completion
- Recover classroom hours that can be repurposed for hands-on instruction
- Focus their energy on critically important skills
- Track all activity (online and in-person) in a single location, accessible via the Web
- Save and export all student data in a shareable (e.g., Excel, Open Office) format
- Take the learning to the learners via the medium they are most familiar with

Interestingly, the U.S. Department of Education (DOE) released a study in 2009 in which the conclusion was drawn that those students who were exposed to both traditional and some online learning materials, in a hybrid approach, performed better than their counterparts who learned exclusively in a traditional, face-to-face setting. This study was performed using a meta-analysis, which simply means that several studies (46 in the case of the DOE study) were compared to determine meaningful patterns and relationships.

This result does not mean that traditionally based instruction is inferior to hybrid learning, nor does it infer that all students across all fields of study will automatically perform better in a hybrid learning environment. What it does suggest, however, is that with the popularity of technology in our daily lives, it stands to reason that today's learners would embrace—and therefore benefit from—the inclusion of technology during their studies.

The DOE report notwithstanding, now that you understand what an LMS *is*, we should discuss what it *is not*. An LMS is *not* a replacement for the human instructor, hands-on evolutions, or face-to-face interactions. In the world of public safety education, the LMS should be seen as an extension of the instructor—a tool to augment what you are already doing in the classroom, which is why the concept of hybrid learning is both viable and appealing. It represents an alternative to totally Internet/computer-based training. It takes the best parts of the classroom environment and partners them with computer-based elements so that the student has greater flexibility in completing assignments. Texts, assignments, and exercises can be completed online, and discussion and hands-on learning can be completed in the classroom. This creative alternative is useful in attempting to meet the needs of both the student and the fire service instructor. It can be difficult for any fire fighter—not just volunteers—to be in a classroom at the same time, five days a week.

FIGURE 8-1 Course management tools.

Choosing an LMS

Before deciding on a particular system, it is a good idea to complete a simple needs assessment. Prior to selecting an LMS, as an instructor you should ask yourself the following:

- What do I want the LMS to do for me?
- How will I support the LMS, the learners, and the content?
- How can I implement an LMS in an organization that is not currently familiar with one?
- Who will handle administrative tasks such as student enrollments?
- Do I need this system to "speak" or connect with other systems?

Deciding to use an LMS in your training program is merely the first step; from there you need to decide which system will best suit your needs. So how do you choose which system to use? There are hundreds of systems available, although a study conducted by The Campus Computing Project in 2011 identified the four most prevalent systems:

- Blackboard (51% of respondents), www.blackboard.com
- Moodle (19% of respondents), www.moodle.org
- Desire2Learn (11% of respondents), www.desire2learn.com
- Sakai (7% of respondents), www.sakaiproject.org

This list indicates the most widely used systems, not your only options. Also, before you go much further in considering a system for your own use, be advised that despite some subtle differences among the various systems available, the functionality ("What can I do with this?") is largely consistent among platforms.

First things first: You need to decide if you want to configure, maintain, and support your own system (open-source systems, such as the popular Moodle LMS, are free or low-cost options) or if you would rather pay a third party to perform those tasks for you (such as a publisher-hosted system or a proprietary solution).

If you are technologically savvy, have a Web hosting account or your own Web servers, have basic site design experience, and want to own every aspect of your own system, then an open-source option such as Moodle might just be the solution for you. Conversely, if you do not have your own IT staff, do not want to have to acquire the technical skills or equipment necessary to run and host your own LMS, or you simply do not have the time to manage all the back-end administrative tasks associated with supporting an LMS, it's probably best to purchase access to a hosted solution from a publisher or vendor (sometimes referred to as "SaaS" model or *Software as a Service*). Blackboard and Desire2Learn are two examples of this type of system.

If you like the idea of an open-source system, but would rather have it hosted and maintained by a third party, Moodle remains a popular choice. This type of LMS may be hosted by publishers (JBLearning.com, for example: http://www.jblearning.com/elearning/navigate/) or by third-party vendors (MoodleRooms, www.moodlerooms.com, and RemoteLearner, www.remote-learner.net, are options).

Understanding the *what, why,* and *how* of your decision to use an LMS will greatly reduce the time it takes to select, configure, and successfully rollout a system for your class.

Functional Items in LMS Platform

The next step is to understand the key functional items on LMS platforms:

- Creating quizzes
- Creating assignments
- Delivering digital content to your learners
- Tracking learner activity and grades
- Communicating via discussion forums

For this section, we'll use the JBLearning.com LMS, branded as *Navigate*, to illustrate these commonly encountered features of learning management systems **FIGURE 8-2**.

Creating Quizzes

All LMS platforms allow for instructors to create, distribute, and grade assessment items such as quizzes, tests, and exams **FIGURE 8-3**. Depending on which system you use, the layout and availability of options may vary; however, most platforms allow instructors to determine the when, where, and what of an assessment with ease. Additionally, banks of test questions are easily authored or imported into all the major LMS systems.

If you have doubts about deploying assessments (regardless of whether you call them quizzes, exams, or tests) online, take comfort in the knowledge that almost all LMS platforms have some level of security and control. Instructors will likely be able to set options such as question randomization, browser locking, JavaScript blocking (controls right-clicking, for example), number of attempts allowed, and time limits on student access.

When creating an online assessment, the best security is still an engaged and informed instructor. Assuming software alone will stop cheating will simply set you up for failure or disappointment. Just as you would set conditions for an assessment in a traditional classroom, so you should take advantage of options that allow you to prescribe a fair but challenging time limit and

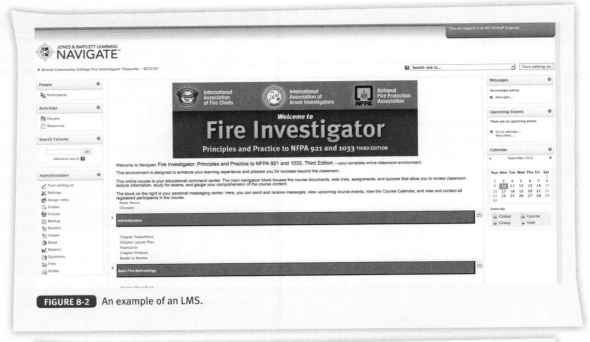

FIGURE 8-2 An example of an LMS.

FIGURE 8-3 LMS assessment options.

submission deadline for the online components of your course. You should avail yourself of question randomization options, and you should restrict the amount of feedback given at the time of submission to the basic grade received. Most of the premium systems available will allow instructors to control when feedback is viewable to students, meaning you can disable this feature when the quiz is still being taken, but automatically open it for review once no more submissions can be made.

Creating Assignments

Assignments are tasks (for example, homework) that students must turn in for a grade or completion. Most LMS platforms support a variety of assignment types to allow for different methods to interact with the system **FIGURE 8-4**.

Examples of assignment types typically found in an LMS would include the following:

- Responding to an online text-based question
- Uploading an essay or document for grading
- Uploading multiple documents
- Completing an offline (face-to-face) activity, but allowing it to be tracked in the LMS

Assignments, much like assessments and forums, are an excellent way to engage your learners, measure their comprehension of material presented, and help foster the learning environment. A good practice for adding assignments would be to keep it to a maximum of 1 or 2 assignments per week. You do not want to overwhelm your learners when such assignments are balanced against forum requirements or assessments in the same week (forums are discussed later in this section).

Another helpful tip for student success is to create both a course syllabus and a grading rubric matrix to ensure students fully understand not only what is expected of them, but also how they will be graded and the criteria for both. Fortunately, both types of documents can be easily uploaded to your main LMS classroom.

Delivery of Digital Content

Two pieces of the LMS make this type of system an especially attractive educational tool—namely, the classroom management tools available and the ability to store digital documents. Think of your LMS as being akin to a warehouse for those documents. Just like with a real warehouse, you can store a great number of items in your LMS. Be sure, however, that you know exactly how much storage (usually expressed in megabytes [MB] or gigabytes [GB]) you have been allocated by your hosting provider and whether any upload restrictions (file size) might preclude you from adding exceptionally large files.

Also be aware that documents you store in your LMS, unless you have opted to host and support your own system, will technically reside on the LMS provider's servers. You might want to inquire as to the host's retention and recovery policies, should you need this information long after a course ends.

Tracking Student Activity and Grades

Ensuring your learners are actually learning and being able to document their actions while in an online course are at the heart of what an LMS provides. Regardless of the name given to this capability (e.g., logs, activities, reports, tracking) in a system, the ability to view *who* has done *what* and *when* is a standard feature in online learning platforms.

A typical activity report will list not only who logged in but also when and from where (typically an IP address, or the unique identifier for a given computer/device on a network). In most cases, you will be able to export and save these records directly to your computer in a shareable format (e.g., Microsoft Excel).

As far as grading options, there are usually a multitude of features available to instructors. A typical grade book will appear much like a Microsoft Excel spreadsheet with columns and rows **FIGURE 8-5**. Not only does this make the experience a familiar one for most users, but it also helps facilitate the export of this data into a Microsoft Excel format for sharing or saving.

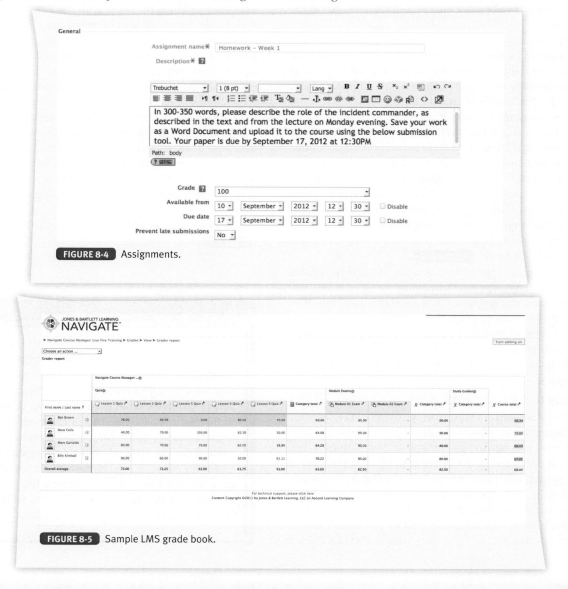

FIGURE 8-4 Assignments.

FIGURE 8-5 Sample LMS grade book.

In addition to the standard grades (whether they are presented in the form of points or a percentage earned), a number of other reports are available in most platforms. For example, a statistical item analysis of individual quiz questions can provide a wealth of granular information about the questions being asked in your assessments and the performance of your students interacting with them.

Communicating via Discussion Forums

Discussion forums should be thought of as online bulletin boards, in which users can post text-based messages asynchronously. Instructors considering the use of an LMS as an adjunct educational tool should similarly be thinking about how they might use forums in their classroom. These communication tools, if properly managed and created, can spark conversation, debate, analysis, and discussions in a manner unrivaled in a traditional classroom. Unlike in a traditional classroom where a "back row" exists, there is no such place to hide in an online course. To ensure that a student does not hide, the instructor can require all students to post comments to questions and classmates in the discussion forum. The instructor can then issue a grade based on the student's participation in the discussion process. If a student fails to participate in the discussion forum, it will have a direct effect on his or her grade for the course.

Additionally, given the diverse nature of incoming public safety trainees, a discussion forum can be leveraged for collaboration, employing a social learning approach whereby the students become instructors, sharing their knowledge and experiences for the benefit of all.

Discussion forums can vary from system to system, but all will have some form of forum available to instructors. Standard forums are ones that allow all users to see all posts, all the time. Thus, if you post material at 3:00 P.M. and your first student responds at 3:30 P.M., the next students into the forum will see both your original post and posts from the student(s) who came before them. These types of forums are great for "ice-breakers" at the beginning of a course when you want to get to know your students better and to simultaneously acclimate them to the use of a forum **FIGURE 8-6**.

Another popular type of forum is the "Q and A." In this type of forum, unlike the standard forum, your students will not be able to see the responses of their classmates until they respond to the instructor's original post. These types of forums are excellent for instructors who wish to spark interesting conversations and engage their students, yet ensure (at least initially) that the responses made by all students are their own original replies.

Limitations to CBT and LMS Use

For fire service instructors who plan to instruct via e-learning systems, consider the following issue: Do you have the time needed to fully commit to this method of instruction? When instruction is presented in the classroom, all students meet at a set time period, ask questions, and get immediate responses. With e-learning, students have the same expectations, yet the course is available on their schedule, not yours. As a consequence, you may need to check e-mail or discussion boards throughout the day and night to give your students feedback

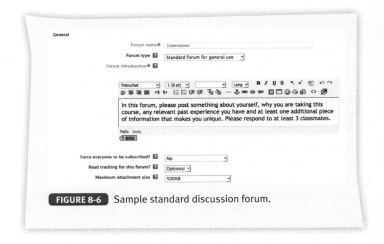

FIGURE 8-6 Sample standard discussion forum.

quickly unless you first establish how and when you will communicate with them.

Designing e-learning courses can be difficult and time consuming. Understanding the language of e-learning is just the beginning. All the rules of keeping the course interactive, challenging, and interesting to students still apply. Can you adequately address this issue with the time you have available? It can take dozens of hours to produce 1 hour of material for an e-learning system. If you are following a traditional course length of 16 hours for continuing education or 45 hours for college or university level, this demand represents a considerable time investment. The advantage is that, once the course is developed, it can be easily updated for new challenges, new information, or improved operational capabilities.

In the past, the focus has been placed on the instructor developing the e-learning course. This may not be the best use of a fire service instructor's talents, however. In some cases it might be wiser to work with one of the professional outsourcing avenues to develop an e-learning solution that is right for you.

When using an LMS, follow the do's and don'ts listed in **TABLE 8-1**.

Ethics Tip

The fire station is a natural collection of diverse individuals. Some may be good at one thing, while others are good at something else. At times, some personnel may want to take advantage of this fact by having one person take the online test and give the correct answers to the rest of the class. This is cheating. Not only is it unprofessional, but it can be life threatening. If a fire fighter does not know which fire extinguisher to use on a grease fire because a buddy helped him with the answers on a test, the grease fire will not help the fire fighter out by providing the right answer. If asked to cheat, don't. Remind students of the dangers of cheating.

Table 8-1	Do's and Don'ts for Using an LMS
Do	**Don't**
<ul><li>Determine your "must haves" for an LMS.</li><li>Determine whether you will host your own system or pay a vendor to host it.</li><li>Determine whether the budget you have available for an LMS is adequate or whether the cost will be passed to the learners.</li><li>Choose the LMS that most closely matches your specific needs.</li><li>Create a syllabus, classroom policies, and grading rubric document so learners understand the expectations and metric by which they are measured.</li><li>Take advantage of security and testing features common to most LMS platforms.</li><li>Create evenly balanced assignments, limiting assignments during weeks with assessments.</li><li>Create discussion forums, even if just to facilitate an "icebreaker" conversation. Require students to log in at specific intervals and respond to at least three classmates.</li><li>Use the LMS as a tool, an extension of yourself, as a means to take the learning to the learners.</li><li>Learn from the examples of others. Collaborate and share best practices with your colleagues. Many of the techniques used in a traditional classroom will also work online.</li></ul>	<ul><li>Expect an LMS or any software application to replace the human instructor, especially in fire service education.</li><li>Create too many assignments, assessments, forums, or other requirements for students to complete in a given time period. Balance their workload and see how it develops. A good rule of thumb:<ul><li>1–2 assignments per week</li><li>1 original discussion post per week</li><li>Students respond to at least 3 classmate posts per week</li><li>No more than 1 quiz/assessment per week</li><li>Allow for nontraditional submission hours</li></ul></li><li>Be afraid to create and deploy graded assessments, tests, or quizzes online. You will have options to select to help make your testing secure.</li><li>Underestimate the power of simplicity. A clean user interface and intuitive layout of an LMS will trump bells and whistles every day.</li></ul>

■ National Fire Academy CD and Virtual Academy

Since 1975, nearly 1.5 million students have received training through the National Fire Academy (NFA) on many different types of training platforms. NFA is continually evaluating its training methods for fire and EMS personnel. This organization offers courses via residential delivery on its campus in Emmitsburg, Maryland, but the scope of this program is obviously too small to meet the needs of the fire service community across the United States. To fill this gap, NFA offers coursework via off-campus delivery, thereby reaching students through a distance-learning system. This system includes regional delivery, direct delivery, and the National Emergency Training Center (NETC) virtual academy. The last of these options—the NETC virtual academy—leverages technology through the combination of CD distribution, simulations, and Internet-based resources.

NFA courses can be delivered right to the student's door via CD-ROMs that are mailed directly to the student. The student views the content and takes a test evaluating his or her understanding of the material presented. Upon successful completion of the test, a certificate is mailed to the student. Each course is available to the student at no cost.

In the NETC virtual academy, students register online, view course material via the Internet, and take a course examination. Upon successful completion of the course, a certificate is provided to the student. The NETC system incorporates the use of slides, audio and video presentations, and quizzes with interactive participation. Through interactive participation, for example, students can assess their comprehension of the materials before they participate in testing. In this active learning environment, students gain confidence while undergoing training in an atmosphere where any mistakes will not cause physical injury.

In addition to the NETC virtual academy, the NFA participates in an academic outreach program. If personnel cannot attend traditional college courses owing to their work hours or location, and have met certain prerequisites, they can earn a degree in fire science through independent study as part of a distance-learning program offered by seven colleges and universities across the United States. For more information on this or any other program offered by NFA, visit the organization's Web site (http://www.nfaonline.dhs.gov/).

Multimedia Applications in Instruction

A key element in most presentations is the use of a multimedia tool in some form, ranging from handouts to PPT presentations. Using multimedia tools enhances the learning process for students and can make your job easier. At the same time, multimedia tools can be a distraction if they are poorly developed or used improperly **FIGURE 8-7** . Knowing how to use multimedia tools effectively is just as important as knowing the lesson plan. When preparing to give a presentation, part of

FIGURE 8-7 The use of too many multimedia tools can be a distraction in the classroom.

VOICES
OF EXPERIENCE

Technology is a huge asset to the modern fire service instructor. We can make dazzling presentations using PowerPoint® and digital simulations that depict an evolving incident with realistic flames, smoke, and sound—all from a desktop computer. The ability to deliver content has changed as well. With the implementation of learning management systems, material can be delivered to multiple stations, rural areas, or volunteer and paid-on-call fire fighters at home. This flexibility allows the modern fire service instructor to adapt content that students will have access to outside the classroom, allowing for more hands-on training to occur during scheduled training time.

I was never more prepared to teach a course...or so I thought!

My organization currently uses a learning management system to deliver annual required training such as blood-borne pathogen refreshers, standard operating guideline reviews, documentation, and similar courses. These important and necessary courses, which involve a lot of PowerPoint® and video-based learning, can be completed at a fire fighter's own pace and time schedule. This gives the department more flexibility to deliver monthly training topics that are more hands-on. This arrangement allows our instructors to focus more of their time on teaching hands-on skills—such as hose line advancement, ventilation, and automobile extrication—rather than dedicating precious training time solely to the classroom.

However, it's important to recognize that as much as technology can benefit the fire service instructor, we must also plan for when it fails! As a relatively young instructor, I was given the opportunity to teach a course that would span multiple weekends. I spent an extraordinary amount of time polishing my PowerPoint® presentations and preparing my course syllabus. I was never more prepared to teach a course...or so I thought!

The first few classes went without incident. But right as I was in the middle of a sentence during the morning session, the multimedia projector light bulb burned out. I was only an hour into the lecture that was scheduled to last the entire day. There were no replacement bulbs or projectors to be found! I was speechless and devastated. As an instructor, I should've known to plan for the unexpected, but I took for granted that everything would be in working order. It was an embarrassment to me and equally as difficult to finish the class for the rest of the day without any type of audiovisual stimulation. I went out the next day and purchased my own projector with a spare bulb and cables so that would never happen again. Even today, I still remind myself to always prepare for the unexpected! Technology can and will fail. We must be ready with a Plan B.

Justin R. Heim, MPA, EFO
Fire Chief
Eagle Fire Department
Eagle, Wisconsin

your efforts should focus on reviewing the audiovisual portions of the lesson plan.

For some fire departments without access to burn buildings, technology is used to bridge the training gap in experiential learning. For this reason, an increasing number of fire departments are using presentation technologies in fire fighter training. On college campuses, most students now expect a professor to use technology to distribute subject matter to them. Whether it takes the form of lecture slides in front of the class, hard-copy handouts, or the Internet, technology makes it possible for students to get information faster and easier than ever before. Coincidentally, this trend makes technology more affordable for fire and emergency services agencies to utilize.

> ### Safety Tip
>
> Multimedia tools inevitably require a power supply. It is all too easy for both students and instructor to trip on power cords snaking around the classroom, even if they are taped to the floor.

Although technology provides you with tools to immerse your students in the subject material, it comes with additional burdens. To engage students with technology effectively, you face many questions. Which sorts of material should go on a PPT slide? Are there limits to the amount of text you should put on a PPT slide? How should you arrange the material for optimal viewing? Should handouts of the PPT slides be distributed before class, after class, or not at all? If you decide to distribute the material, should it be in the form of a printout, on a CD-ROM, or via a Web site? Transitions between media types should occur smoothly with minimal interruption. Visual aids should be incorporated

into your general approach to teaching in a harmonious way. There are many rules and guidelines that you can follow to ensure that you meet the goal of teaching your class effectively.

As late as the 1990s, multimedia tools used in fire fighter training included only the basics: movies, videotapes, chalkboards, easel pads, duplicated materials (handouts), audiocassettes, overhead projectors, models, cutaway demonstrators, and simulators. Good fire service instructors created their own make-shift simulators consisting of a slide projector, pictures of area buildings, a piece of glass with wooden frame, and silica sand, all on an overhead projector. When used in combination, these tools created an inexpensive fire simulator—complete with the illusion of fire and smoke. Scenarios could be changed depending on the needs of the students or at the whim of the fire service instructor.

Today, these home-made simulators have been replaced by computer, projector, and software programs that create digital images of a fire incident showing specifically the elements of smoke, fire, and other fire behavior processes overlaid on structures or local areas. With a few simple keystrokes, you can change the simulation dramatically. Enhanced through the use of digital photography, local target hazards can be imported into the simulation for a customized learning process. These multimedia tools have an enormous impact on a student's understanding of the concepts presented in the classroom.

Today's students learn differently than the students of the last couple of decades. Many members of the new generation of fire fighters grew up with instant access to the Internet, videogame systems, and instant messaging. Their need for understanding transcends mere technology: These students have grown up with the advantage of technology in their everyday lives, so their learning needs relate to understanding the materials and concepts presented—not the technology used to present it **FIGURE 8-8**. Although modern-day students need to

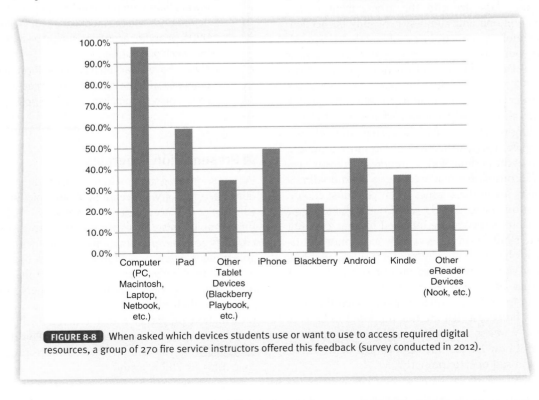

FIGURE 8-8 When asked which devices students use or want to use to access required digital resources, a group of 270 fire service instructors offered this feedback (survey conducted in 2012).

learn the same concepts as were presented to their counterparts in earlier years, the way these students learn will be enhanced by technology because it is the baseline from which they can launch into understanding of the material.

Understanding Multimedia Tools

Multimedia tools come in a variety of types and are intended for a variety of uses. For example, they include the PowerPoint® and Keynote software programs; videos; DVD presentations; graphic presentations developed by a publishing company for use with its textbooks; CBT programs such as WebCt, Blackboard, and Angel; Internet-based training; distance learning; National Fire Academy training programs; tablet computers such as iPads; digital audio players such as iPods; and use of smart phones. It might seem that it would be impossible to choose from such an abundance of multimedia tools, but each fills a specific niche based on the type of learning that is to take place.

Learn how to use multimedia tools before you apply them in the classroom. It may take a fair amount of preparation to master the tools you choose to make your presentation. Sometimes use is as simple as pressing the "play" button; at other times it is as complicated as learning a whole new "language" or software program or transitioning among various types of media. As a fire service instructor, it is up to you to learn how to use these tools correctly and incorporate them smoothly into your presentation skills. Don't fall into the trap of blaming a poor presentation on your inability to use the equipment, and don't depend on multimedia tools to instruct for you. You need to practice with these tools to make them an effective medium that enhances your existing instructional skills.

■ Audio Systems

Depending on the class size and the arrangement of the classroom, a microphone and audio system may be needed. Although this technology is not new, it is often misunderstood. Remember who your audience is. Can everyone see and hear your presentation? Do all of your students have the same hearing abilities? Just as some people don't like to wear bifocals when their vision begins to deteriorate with age, some do not want to admit that their hearing just isn't what it used to be.

When deciding which kind of audio system you will use, keep your presentation style in mind. Do you move around a lot, or do you stay at the podium with your notes? Will you have a long stretch of lecturing that may wear out a hand holding a microphone? Think about these issues when you must choose between a wireless microphone and a wired microphone setup.

A wireless microphone can be attached to your clothing and the transmitter hidden from view. This looks more natural and allows your hands to remain free for writing. Wireless microphones offer you the most freedom to move around the classroom, but they have some problems, too. Wireless systems may have an issue with frequency usage, such that you might hear voices broadcast that are not supposed to be part of your program. Wireless microphones may also have limitations in terms of how far the microphone can be from the remote receiver and still operate properly.

The greatest limitation for wired systems relates to the cable. Is it long enough for you to move around effectively? Some wired microphones are permanently attached to the podium, in which case you lose the freedom to move around the classroom. Another disadvantage of wired systems is the potential for a mishap if your feet become entangled in the wire.

Theory Into Practice

During the course of your career, you will find that the use of multimedia tools can greatly enhance your ability to convey your message. But use of technology can also come with some drawbacks. To minimize the potential pitfalls, consider these issues when using multimedia tools:

- Do not read from the PPT slides. All fire service instructors make this mistake at one time or another. Most slides are outlines of the lesson plan and serve to emphasize critical points. The only time it is acceptable to read from the PPT slide is if you are reading a quote, legal definition, standard operating procedure, or legal statement.
- Select the appropriate level of lighting for the type of multimedia equipment you are using. If you want students to take notes during the presentation, the room must be dark enough to see the PPT slides, yet light enough to write.
- There is an art to transitioning from the lecture to the multimedia technology—and practice makes perfect.
- When lecturing with PPT slides being projected behind you, do not stand in front of the projection screen. It is difficult for students to see what you are talking about if you are in the way.
- Always have your multimedia equipment set up before class and the media queued. Videotapes, DVDs, or audio recordings should be ready to go at the touch of a button.
- Always have a backup plan. What if you have problems with the computer or projector? Always have a plan B, and even a plan C if possible.

■ Presentation Programs

Presentation programs include a variety of software programs developed by software vendors. Companies such as Microsoft (the maker of PowerPoint®) and Apple (the maker of Keynote), for example, create software programs to assist customers in creating portable presentations for business, education, and government. Each presentation program has its own unique features and benefits that you can use to enhance your instruction. Some are easier to use than others; some allow for greater flexibility. For the purposes of this chapter, we refer to presentation programs with the acronym PPT, reflecting the fact that most fire departments use the PowerPoint® program **FIGURE 8-9**. If you use another software program, the basic style rules are still relevant.

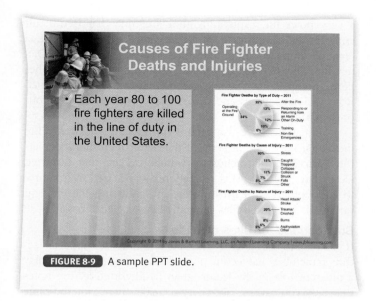

FIGURE 8-9 A sample PPT slide.

The ability to embed video and audio clips into your presentation is one advantage of using PPT, and it adds a new layer of sophistication and professionalism to your presentation. A disadvantage is the potential for over-reliance on technology as part of your instruction. If your computer doesn't work or problems crop up between the computer and projector, then you cannot use the PPT presentation. If your instruction relies solely on your PPT slides and the technology fails, then you may have to reschedule your presentation if you do not have backup technology available.

Another disadvantage associated with these kinds of presentations is the overload factor. Some PPT creators tend to include too much information, too many sounds, and too many visual effects on their slides. Keep your presentation simple and concise for greatest effect.

Designing and Using PPTs

When using PPT software or a similar program, instructors need to follow several simple rules to make it easier for students to visualize their message and avoid distractions. When you begin developing a presentation with such software, you will be asked if you wish to utilize a "design template," "blank presentation," or "auto content wizard." If you have not created a presentation in the past, the easiest way to get started is to use the wizard or select a design template. The software will then guide you through the process of making choices about background color, font, graphics, and other aspects of the presentation. After you have completed your design, the resulting presentation should look and feel professional.

Font

The words you choose impart to your audience the message you are trying to convey, but the appearance of your message can distract or maintain students' attention, thereby affecting how well they heed that message. Fonts are a critical factor when you are developing a presentation for others to read and understand. Some fonts are better choices than others for getting your message across. For instance, sans serif fonts are better choices than serif fonts for enhancing readability when the material will be displayed on screen. Serif fonts have

small embellishments or lines at the base of each letter. These embellishments make it easier to follow a line of text on the printed page, but they are a distraction on a screen. Select a sans serif font (e.g., Helvetica or Arial) instead of a serif font (e.g., Times New Roman) for your presentation.

Font size is crucial. You will find many rules for determining the proper font size for presentation settings. A general rule of thumb is to use a font size of at least 28 points for body text and 38 points for heading text.

Some people are colorblind, so avoid using color combinations that might cause these individuals difficulty, such as red text on a green background. Normally you will use light text on a dark background or dark text on a light background in projected presentations, to help with contrast, but pay attention to the strength of the image projected by the projector.

The shades of color you see on your monitor when you are developing your presentation are not necessarily the same ones you will see when projecting that presentation. Before showing your presentation to students, project it through the equipment you will use during class. You may find you need to change colors or fonts.

Keeping the right amount of blank space on a slide is important. If you find yourself wanting to reduce the font size so that you can cram more text onto the screen, consider redesigning the slide so that you have less text on it. Limit the number of words on your slides to either 7 × 7 (seven lines, no more than seven words each) or 5 × 5 (five lines, no more than five words each). Always remember the needs of your audience: Is your focus on your message or on how much you can pack onto one slide? In the latter case, considering splitting the material across more slides for greater impact in your PPT presentation. Not only does this make the text easier to read, but it also helps to keep students' interest throughout the program.

Teaching Tip

If you are using an overhead projector and transparencies, don't use dark backgrounds. They will wash out, diminishing the impact of your presentation.

Image Selection

Illustrate your PPT slides with images to engage the audience. Such images appeal to the senses. Remember the old saying, "A picture is worth a thousand words"? This is especially true for fire service personnel, because fire fighters tend to be hands-on learners. If students can relate their actions to visual images, the presentation is likely to have a greater impact.

Images can also be used to support classroom discussions. They help to move students through stages of understanding, allowing them to organize their thoughts and responses based on patterns in the image. For example, in a safety presentation, a slide might contain an image of an intersection with a car slammed into a power pole, fluids on the ground, and people milling around. Ask the students to work together to identify safety patterns and decide what they should be aware of in response to this situation. You might then present a slide that shows improved patterns. The discussion would continue, supported at each stage by a slide that exhibits the patterns identified at that stage.

Appropriate images can often be obtained from the clip art files provided with many commercial presentation programs. If you want to add or create your own images, use a digital camera, scanner, or graphics program. Cameras and scanners generally come with their own software, allowing you to save your images in the format that you prefer. If you want to create your own graphics, you will need graphics software (also known as a drawing program). Commercial programs such as Illustrator, Photoshop, Paint, and Paint It are readily available. Each has its own language and setup, so you will need to spend some time becoming familiar with the software before you attempt to edit or create images with such a program.

Sound

Sounds can be either helpful or a hindrance. Files containing sounds for use with computers are labeled with a variety of file extensions. Be sure your computer can accommodate the particular file extension (i.e., the type of file) so that the sound will perform properly. For example, files with a .mov extension are intended for use with Quicktime media programs, whereas .voc files are intended for use with Soundblaster programs. One kind of file may not work with the other program.

Depending on the source from which you obtained a sound clip, some of the material may be copyrighted. Contact the source of a clip before you include it in your presentation to find out whether you need to obtain permission for its use.

Teaching Tip

Inevitably, your preferred medium will crash at the worst possible time. Always back up your material in at least one additional location, such as a CD, thumb drive, or other storage device, to prevent a catastrophic loss.

Video and Animation

Locally produced video can add to the realism of the presentation and help hold students' interest. Students can absorb material only to the extent that their attention is engaged. If your images aren't interesting, you have lost your students.

Like sound clips, video takes up a large amount of storage space on a computer. Essentially, every frame of video is comparable to a photo with sound. Usually, video is captured at a rate of 24 to 30 frames per second. If you have a 5-minute video, that equates to approximately 7200 pictures (24 frames × 60 seconds × 5 minutes). Such a 5-minute video may consume over 100 megabytes of storage space. Most training videos are 15 to 30 minutes long, so they require considerable storage space. If you plan on editing this video, you will need to approximately double the size estimate. More information on storage space considerations can be obtained from the Internet by visiting the software developer's Web site.

Animation is similar to video in that it consists of a series of still images put together to create motion. Animation is a very powerful, captivating tool, but one that is difficult and time consuming to produce. Consider that it took a team of animators more than three years to develop the Disney film *Snow White*. The old adage "Time is money" applies here, so consider using animations that can be freely copied from various sources and incorporated seamlessly into your presentation.

Animations have also been used in reconstructions of fire incidents. For example, one of the better recreations depicts the Worcester, Massachusetts warehouse fire that occurred December 3, 1999. This recreation can be used in a course to allow others to learn how the fire grew and the incident unfolded. Animation files generally have a .gif extension.

Suggestions for Uses of Slides

If you plan to use slides to illustrate or otherwise support a lecture, remember that lecture notes displayed on your slides play a different role in the lecture than handwritten lecture notes that only you can see. Don't try to make them play the same role, or you may find students reading your slides instead of listening to you.

You can use slides in the lecture to list the major points of your presentation. Several major points might stay on the screen as you develop each of them in turn, providing a way for students to place each point within the larger context.

Also use slides to list important terms. Again, one slide with several terms may remain on the screen for some time, allowing you to refer to each term as you introduce it.

In addition, slides can be used to create prompts for group work, being projected for all groups to see as the students do their work. This approach helps students move from one stage to the next when you manually change the slide or through the use of a timer built into the slide presentation. Slides can also contain breakouts that indicate how each group will handle different aspects of a scenario.

At times, you may find that you need to use audio or video as part of your presentation. This becomes possible with a PowerPoint® presentation, thereby eliminating the need for videotape machines, audiocassette players, or other expensive audiovisual (A/V) equipment. Streamlining the technology employed in the class will not only keep you on track, but also enhance the flow of your presentation. This can be a simple process. The audio or video can be started when you click on a slide, for example, or it can be started when you click on an icon (a pictorial representation of an object used to represent documents, file folders, and software) on the slide. To work correctly, the files need to be in the correct location. If you copy the file to a different medium (CD-ROM), be aware that the program may not locate the audio/video file when you attempt to play it later.

Remember, more is not always better. Sometimes a simple, clean slide is better than one with bells and whistles such as sound and animation. Multimedia tools should *enhance* your students' understanding of the material; they are not designed to make you look cool with your techie toys.

When inserting video onto a slide, keep in mind the visual area of the slide. Several complications can arise when incorporating video into a presentation. If the image is too big, the picture will appear out of focus. If it is too small, it may be difficult to see and distracting. If the computer processor is not fast enough to show the video, the video playback will appear jerky or not play at all, or the picture may appear out of focus or with <u>ghosting</u> (faint shadows to the right of each letter or number). Check your video on the equipment you will be using for the presentation before you choose not to use a videotape machine or DVD player. Don't give up quality for convenience.

© Goyvel-Sokol Dmitry/ShutterStock, Inc.

FIGURE 8-10 A data projector can enhance your presentation.

Ethics Tip

Give credit where credit is due. If materials came from a journal or other publication, remember to credit the authors. Regardless of how you found the material or how long ago the information was developed, give credit to its author. This will assist you in presenting credible and defendable information. The fire service has had an unhappy reputation of borrowing information and calling it research; do your part to dispel this image.

■ Visual Projectors (Document Camera)

You may find it useful to refer to the textbook to illustrate your message. Instead of requiring your students to get out their books and turn to the right page, you can use a visual projector or document camera to show the material to the entire class. This type of multimedia tool can project almost any one-dimensional printed material onto a screen. The visual projector has backlighting, overhead lighting, and color capabilities that can be tied in with a computer. It combines the flexibility of the overhead projector with some of the capabilities of a high-tech device such as a liquid crystal display (LCD) projector. Major disadvantages of visual projectors include their costs and unwieldy size.

Teaching Tip

Block print should be used whenever possible when writing material that will be projected. Script is difficult to read and should not be used.

■ Data Projectors

LCD projectors and digital light processor (DLP) projectors (also known more simply as data projectors) are used with video players, DVD players, computers, and visual projectors to project documents, images, and motion pictures so that a group of students can easily view the material simultaneously. Data projectors can present material on screens ranging in size from 2 feet wide to 30 feet wide **FIGURE 8-10**. Important features to consider when selecting a data projector are the ability to show videos or DVDs directly, audio capabilities, screen resolution, and the brightness of light emitted by the projector (measured in lumens). You should also consider the price of the equipment and the price and lifespan of the bulb.

Many data projectors have become more affordable in recent years. In addition, the size of the projector is fairly compact, so these devices can be moved from facility to facility with little effort. The various pros and cons should be examined before purchasing a data projector. As with a slide projector or overhead projector, having an extra bulb available is a good idea. However, bulbs for LCD and DLP projectors are very expensive and you may not be able to replace them in the field.

You should also consider the issues that can arise in the classroom when you use a data projector. First, connections can be an issue, because the right kinds of cables must be available. Also, make sure that the cables are long enough, because short cables will hinder your setup and use of the projector.

You must learn the proper start-up procedures for your data projector as well. Generally, the data projector must be connected to the computer before the computer is turned on. Learn how to get the computer to send its output to the data projector. Some computers will automatically recognize the data projector device, but often you must press a control-function key to activate the connection. Some of your presentations may not be compatible with the data projector, and video clips that play on your computer screen may not project properly with such a device. Given these caveats, always test your presentation before class and have a plan B.

■ Videotapes

Videotapes have been around for more than 25 years. The technology standard is the VHS format. Most VHS videotapes consist of a magnetic tape encased in a plastic cassette, similar to an audio cassette tape, only larger.

Video cassette recorders (VCRs) and video cassette players (used to play VHS tapes) are very affordable. The VCR requires another component for viewing the images—either a data projector or a television. If you are using a television to view the videotape, keep in mind the size of the class and the lighting. The larger the class, the larger the screen must be. A 27-inch or larger screen is adequate for most classrooms. Also think about the audio component. Most televisions have adequate sound quality for classroom presentations. In larger classrooms where

JOB PERFORMANCE REQUIREMENTS (JPRS)
in action

After you decide on the learning environment, the next step is to identify how the lesson plan will be presented. The more of the student's senses that you can appeal to during the learning process, the more effective the training will be. Using visual aids is one way to improve a lecture presentation by appealing to the senses of sight and sound—but remember to match the equipment to your instruction style. Today's students desire the inclusion of multimedia, visual aids, and as much hands-on practice as you can deliver. Part of your continuing education responsibilities should be to keep up-to-date on the latest delivery methods and technology available for use in the classroom.

Instructor I

When delivering the lesson plan in a classroom setting, a multimedia tool will probably be used to enhance the lecture. The Instructor I must be knowledgeable both in the use of the media and in how to adapt to classroom setting limitations. All media should be selected to supplement the objectives being presented, and all of them have pros and cons for their usage. Take the time to practice their use well before your presentation and adapt them to the learning environment.

JPRs at Work

The Instructor I will organize the learning environment and operate all audiovisual equipment to present the content of the lesson plan so that all learning objectives are effectively covered. Multimedia tools include a wide range of projectable aids, handouts, Internet-based products, and DVDs or videos. The Instructor I must be skilled in using all types of media equipment available.

Instructor II

The learning environment plays a big part in the selection and use of media in your presentation. Today's training applications require a mix of multiple types of media and visual aids. Media should be designed to enhance the learning, not to replace traditional learning methods. For example, a DVD needs to be presented and discussed and made relevant by the instructor using the media.

JPRs at Work

NFPA does not identify any JPRs that relate to this chapter, although the Instructor II will have a great amount of influence in choosing which multimedia equipment is used when he or she develops a lesson plan. Whether choosing a mass-produced curriculum or designing your own, consider the skills of the instructor presenting the material and prepare backup media in the event of equipment failure.

Instructor III

At the Instructor III level, the instructor is focused on the overall course design and development process. Understanding how instructional media can impact a course is important more from the aspect of budget needs and costs.

JPRs at Work

NFPA does not identify any JPRs that relate to this chapter directly, but because the Instructor III level is usually a senior instructor with responsibility for budgets and making recommendations, understanding the content of multimedia aids is highly important to the overall success of a program.

Bridging the Gap Among Instructor I, Instructor II, and Instructor III

The Instructor I presents objectives using a lesson plan and media selected by the Instructor II. In some situations, the Instructor II may be conveying a message that the Instructor I does not completely understand. It is sometimes difficult to teach another person's original thought and work. The Instructor II must be clear in the construction of the message and the media used to present the objectives. Work together to ensure clarity in the message being presented. The focus of the Instructor III level is overall curriculum development, budget, and management of a program.

data projectors are used, however, a more sophisticated audio system may be required.

Videotapes are copyrighted material. As a consequence, they should not be copied unless you obtain permission to do so from the video production company. An alternative to using copyrighted material is to produce your own videos. The price for video cameras continues to decline, allowing some fire departments the opportunity to develop their own training materials. Videos can be produced through a variety of sources, including your local television/cable provider. You can take advantage of software programs such as Final Cut Express, Final Cut Pro, and Adobe's Premiere Elements to produce quality videos for training purposes. But remember: Students don't want to watch your home movies. Your videotapes must be of professional quality before they can be screened in the classroom. Amateur videos are no different than the overused war story.

DVDs

Like CD-ROMs, DVDs are available as either blank or pre-programmed media. Information ranging from video clips to audio clips to PPT presentations can be stored on a DVD. DVD players and burners are often standard hardware on new computers. With a DVD burner and digital media editing software, you can develop your own digital presentations and ensure that the end result is a professional product.

In addition to a DVD player (either as a separate device or in a computer), DVDs require either a television or data projector for viewing. As with use of videotapes, you must remember the audio component when showing DVDs. Depending on the size of the classroom, you may need additional external speakers.

Digital Cameras

To capture students' attention and keep them engaged in your presentation, you may want to include local examples in your multimedia presentations. If you are giving a PPT presentation on the implications of special occupancies for preincident planning, you might want to include some local examples, such as the assisted care center on Main Street or the group home on Center Street. A digital camera allows you to capture an electronic image, upload it onto your computer, and insert the image into your PPT presentation fairly easily. Many digital cameras also have video recording capability.

Digital cameras are available at every price point. You can purchase anything from a simple point-and-shoot digital model to a professional digital camera with multiple lenses. The resolution of these cameras is constantly changing and improving. Most digital cameras currently on the market offer resolution of 4 megapixels or higher, which is sufficient for use in PPTs.

The number of megapixels matters because it directly affects the image quality. To see how this works, take a newspaper from the recycling bin and look at a photo very closely with a magnifying glass. The photo is made up of tiny dots of color, called pixels. In printing, pixels (dots) are measured in units of number of dots per inch (dpi). The more pixels (or dots) per inch, the sharper the resolution.

A printing press would never consider printing an image at less than 360 dpi, but a data projector is capable of projecting at a resolution of only 72 dpi. What does this mean for you, the fire service instructor? You can use smaller-size image files in your PPT presentation than if you were going to actually print out the image. A pixel-heavy image file will just take up time and space in your PPT presentation. Save copies of these images' files at a lower resolution, thereby creating smaller files and saving space and time in your presentation. Keep the higher resolution copies on hand, however, in case you need the image for a different application. Resolution can be lowered, but it cannot be increased once the photo is taken.

Flash

Flash refers to both the Adobe Flash Player and a multimedia authoring program used to create content for Web applications, games, and movies. The Flash Player, which was developed and is distributed by Adobe Systems, is an application that is readily available in most Web browsers. It features support for vector and raster graphics, a scripting language, and bidirectional streaming of audio and video. Since its introduction in 1996, Flash has become a popular method of adding animation and interactivity to Web pages FIGURE 8-11 . Several software programs and devices are able to create or display Flash images. Flash files (traditionally called "flash movies") carry the file extension .swf. These files may be either accessed as objects on a Web page or played in a stand-alone Flash Player. Your local community college may offer classes or workshops on how to use this program. Flash is not available on certain devices, however, including many tablets (like the iPad) and mobile devices.

Scanners

While the scanner can be a helpful tool, its use does raise copyright concerns. For example, suppose you find a photo of a house fire in your local newspaper. This picture may perfectly illustrate your message about the dangers of leaving food

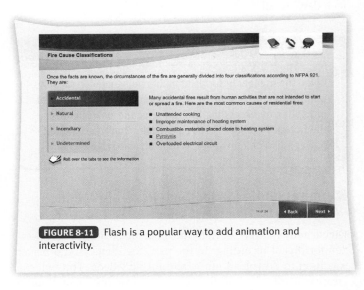

FIGURE 8-11 Flash is a popular way to add animation and interactivity.

unattended on the stove. Excited to add this image to your PPT presentation, you head over to the office scanner with your laptop, scan the newspaper picture, save the image file onto your laptop, and drop the image into your PPT presentation. If you haven't contacted the newspaper and received written permission to use the newspaper's photo in your presentation, however, you are committing a copyright violation.

■ Presentation-Ready Programs

Prepackaged software programs are a valuable resource when you are planning your instruction. For instance, Jones & Bartlett Learning provides instructors with presentation-ready programs for use on computers at the local level. This material includes PowerPoint® presentations, testbanks, handouts, and other course-specific material that will assist the instructor. Thanks to these programs, you do not need to spend valuable preparation time developing your own presentation; instead, you can simply use the professionally developed materials. Of course, you must still familiarize yourself with the presentation before class, rather than seeing it for the first time in front of the class. Professionally developed presentations are intended to be dynamic instructional tools, used to enhance students' learning experience, but will never replace the experienced instructor. No presentation-ready programs should truly be considered ready to use. Instead, instructors must review materials ahead of time and adapt them to the students who will use them. The best instructors will customize stock backgrounds with department logos and department scenes to make the material more relevant to the students.

■ Overhead Projectors

Low-tech solutions are sometimes just as effective as the newest technology. Overhead projectors, for example, have been used in training for more than 40 years and were once considered the industry standard for quality educational enhancement. The overhead projector displays materials that have been developed on transparencies and then manually placed on the projector for viewing on a projection screen. This low-tech device is often a good choice as your plan B in case your high-tech devices fail.

Safety Tip

Given the number of forms of media being used in presentations, instructors might be tempted to violate the local fire codes by exceeding the rating of the power cords or the outlet. Practice what you preach: Don't use multiple power strips or lightweight extension cords.

■ Digital Audio Players and Portable Media Devices

Digital audio players, such as the iPod, are creating a new horizon for digital recording. In the past, the digital audio player was used to download music and music videos. Through the creative work of many fire service instructors, digital audio players are becoming a mainstream technology for students at most universities. Before long, digital audio player usage within the fire service will be commonplace. These devices are so compact that students can carry them easily and download lectures or pictures that can be reviewed while traveling to class FIGURE 8-12 .

Some digital audio players allow students to use an audio record feature, similar to tape recorders. Several media outlets are actively engaged in pushing the technological edge by offering downloads to these devices featuring such notable fire service figures as Chiefs Billy Goldfeder and Rick Lasky.

Students may also find that digital audio players allow them to complete class assignments. In some courses, students may be able to use desktop audio/video editing software to create and edit projects for course assignments. The results can be uploaded to the training center's computer system for presentation, or content playback can occur immediately via classroom audiovisual equipment. Digital audio player devices can store not only images and audio, but also entire presentations created in PowerPoint®, word processing documents, or files created with other software.

As an instructor, you can benefit greatly from these devices as well. The possibilities for their use are endless: oral examinations, classroom performance and practice, group discussions,

FIGURE 8-12 Digital audio players allow students to listen to lectures and view images outside the classroom.

field interviews, interviews with content experts, classroom lectures, and discussions or presentations. The possibility of a podcast replacing a Webcast allows you to make the materials for a class more widely available, too.

As with other technologies, we must wait to see which benefits this technology will ultimately yield. As more digital systems become available and their costs are driven lower by greater demand, fire service instruction may be transformed from a system of learning confined to a specific locale to one where information can be obtained "on the fly."

Another portable media device, the personal digital assistant (PDA), effectively allows students to carry a computer in the palm of their hand. PDAs are making an entrance into educational media as well. While known by various names, PDAs can include technology such as cellular telephone capabilities; computer operating systems such as Linux, Windows, and others; e-mail; Web browsing; and much more. Their use can bring new dimensions to educational media—but it can also introduce potential problems. For example, cheating, use of calculator functions during examinations, and answering of telephone calls in the middle of coursework are all new challenges associated with PDAs. For information transfer, however, PDAs combine the benefits of digital audio players and the flexibility of mobile communications technology.

As technology adapts, improves, and advances, new opportunities will be found to augment educational media. Satellite communications, cell phones with media downloads, and text messaging—all are new frontiers for you to provide quality training to fire service personnel.

Tablet Computers

A tablet is a wireless portable computer with a touch screen that is smaller than a notebook computer but larger than a smart phone **FIGURE 8-13**. These devices either include the keyboard on the screen or allow an external keyboard to be attached to the device. Such computers can access the Internet and offer a wide range of capabilities and features such as cameras, phones, e-books, e-mail, videoconferencing, and GPS

satellite navigation. Many fire and EMS agencies have adopted these devices for fire inspections, preplanning activities, and gathering information during a medical emergency. Students can use these devices to take notes, complete assignments and e-mail them to the instructor, view course material, and conduct research during class. In some instances, instructors may send the students assignments by e-mail, which are then completed and returned via the tablet. An emerging trend in many college and academy programs is to issue tablet computers to their students that come preloaded with textbooks, online course materials, and other interactivities ready to use.

Smart Boards

A smart board is an interactive whiteboard that uses touch detection for user input, much in the same way a personal computer does with a mouse **FIGURE 8-14**. A projector is used to display a computer's video output on the interactive whiteboard, which then acts as a large touch screen. Smart boards come with pens that allow the instructor to use "digital ink" to write on the whiteboard during a presentation. While using the system, a computer records the material, which can be saved and reviewed at a later time or emailed to students.

Satellite Programming

Satellite television is available worldwide. Vendors such as Dish Network and Direct TV market such services to residential customers as a replacement for their landline cable systems. Many fire stations are equipped with satellite systems that rival those of the local and national commercial broadcast companies and that utilize a 3-meter base station. Two of the most popular satellite broadcasters for fire service personnel are the United States Fire Administration (USFA) and the Fire and Emergency Training Network (FETN). The USFA provides interactive Internet and satellite distance learning, making a wide variety of programs readily available to any community in the United States. Through the Preparedness Network (PREPnet), all programming is available to any community with access to the Internet, or a C-band or Ku-band

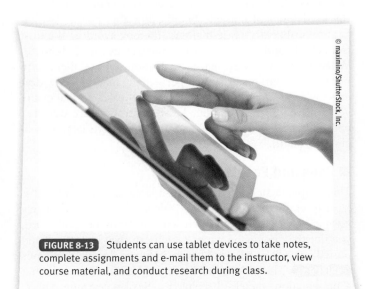
FIGURE 8-13 Students can use tablet devices to take notes, complete assignments and e-mail them to the instructor, view course material, and conduct research during class.

FIGURE 8-14 A smart board is an interactive whiteboard that uses touch detection for user input.

analog satellite dish. Vendors such as FETN, whose programs are produced by Trinity Learning Corporation, may require the recipient to obtain an additional satellite dish to take advantage of their programming.

■ Virtual Reality

Virtual reality devices allow students to interact with a computer-simulated environment. By displaying images either on a computer screen or via a headset, the goal is to immerse the student in an environment that can be controlled by the student and the programmer. This sensory experience is designed to force feedback from the student as he or she interacts with the environment. Although this technology is, in essence, a simulation, the visual, auditory, and tactile sensations create an artificial world wherein the student can make mistakes and cause injury.

The virtual environment can be designed to mimic a real-world scenario such as a burning grocery store, or it can be set up to resemble a video game. This technology is developing daily by quantum leaps, and it may become the standard means of training in the near future. Unfortunately, the currently available technology does not achieve the realistic look and feel instructors want for their students. In the future, as faster processors and more cost-effective systems emerge, instructors may be able to apply the best possible technology for the benefit of their students.

■ Simulations

Like virtual reality systems, simulators allow students to interact safely with hazards in a controlled environment. This practical application was developed by the military to train soldiers for tactical operations during peacetime. Students today have the benefit of growing up with simplified simulators. With the advent of popular video games such as *Halo* and *Doom*, many students are accustomed to being immersed in an imaginary world. Today, gaming has taken on new dimensions, as newer systems mimic the movements of the operator to accomplish a mission in the game. Simulations—no matter how simple or complex—allow students to be challenged and to learn from their mistakes.

Simulations have been developed both commercially and noncommercially. Commercially available simulators provide emergency responders with the opportunity to develop skills in command, control, coordination, and communications. Incidents faced in simulations may include structure and auto fire responses, aircraft accidents, terrorist acts, hazardous materials spills, fires, and natural disasters, with the programs being designed to test and validate students' understanding of emergency operations.

While this educational activity is not the same as live fire training, after a simulation the student walks away with concepts that can be applied and internalized long before he or she faces a live fire event. As with an actual live fire, replication of exact circumstances is difficult at best, yet with the simulator the same conditions can be repeated until the outcomes are acceptable. Simulators are no substitute for live fire training, of course.

Online certification courses such as Blue Card Hazard Zone Management utilize simulations for both instruction and evaluation purposes. Directed and distance learning opportunities can capture training opportunities with this technology and bring realism to a classroom environment.

■ Models

Many fire service instructors find the use of models helpful when teaching basic principles prior to going into the field. Models are built by the fire service instructor. To be effective, they should be built to scale or all pieces in the entire model should be on the same scale. Such models are often incorporated into the table-top exercises used for incident command training. For example, a model city may be built for the various types of scenarios that could be encountered within a jurisdiction.

Models can be used to teach basic principles before students go out into the field. A good example of this application would be a trench rescue scenario that uses a scale model to demonstrate demarcation of zones, placement of apparatus, shoring techniques, and incident management. Once the students have mastered the basic principles using the models, then they can progress to an actual trench. One disadvantage when using models is that you have to build your own models, which may prove difficult for all students to see in a large classroom.

■ Chalkboards and Erasable Boards

Chalkboards can be used to illustrate points that the instructors believe are important. Chalkboards have been around since the 1800s and have several advantages: They are very easy to use, most classrooms provide one, they are relatively inexpensive, and they can be reused.

One downside to this technology is the time spent with your back to the classroom, writing information on the board. You can alleviate this problem by arriving early and writing your points on the board prior to class.

Another issue related to chalkboards is not so easy to overcome—namely, how do you contain the large amount of chalk dust? With students who are allergic to chalk dust or have asthma, creating a healthy learning environment can be a challenge. In these cases, you should look into the availability of erasable boards. Erasable boards (whiteboards) utilize dry-erase markers, thereby avoiding the dust from chalk, and are used in the same way as the chalkboard. When using dry-erase markers, do not confuse them with permanent markers. Using a permanent marker on a dry-erase whiteboard will damage the whiteboard, so make sure you use the correct marker. Most classrooms are outfitted with either a chalkboard or, as in most modern facilities, an erasable board.

■ Pads and Easels

This simple technology comprises a large drawing pad that is used to emphasize key points. Pads and easels are relatively inexpensive tools and can be reused. Similar to the chalkboard, you illustrate points on paper, which can then be torn off and posted on walls or retrieved for use in future class presentations **FIGURE 8-15**. The pad can be developed before, during, or after a presentation, thereby becoming a flexible learning device.

FIGURE 8-15 Paper easels can be a valuable addition to your presentation.

The greatest advantage to the pad and easel is its flexible use in small-group settings. Members of each group can write their thoughts and ideas in response to the activity and post them for others to see, review, and discuss. Each group can write its responses to activities and post them for later discussion. The pads can also be used for later discussion to review information or to refer back to while clarifying a point or concept.

When using pads/easels (and chalkboards/erasable boards), legibility is very important. If you cannot write so that students can read the material, then you should have the material preprinted. A simple tip to help improve legibility is to use lined easel pads or to draw lines lightly with pencil onto the paper. Then, use the pencil to do any writing before class. During the presentation, the pencil marks are not visible but can quickly be traced to give a nice, neat visual.

■ Handouts

Today, handouts play a vital role in student comprehension and retention. Many students prefer to take away handouts prepared with key points highlighted, rather than a stack of copies of the PowerPoint® slides that they have just viewed. Environmentally conscious instructors will reduce waste by consolidating key points onto a compressed handout. Handouts remain an effective way to enhance lecture material and provide additional information. The copy machine has made it cost-effective to produce a variety of materials for students to take with them when the class is over.

Teaching Tip

Include your name and contact information on each student handout. This practice will allow students to ask questions at a later date, or allow others to contact you for more information regarding similar training. Handouts allow students to review the information on their own time and gain a more comprehensive understanding of the material.

How to Maintain Technology

With any technology, maintaining its quality is an important consideration. Some vendors charge an additional fee to repair or replace damaged materials. When purchasing multimedia tools, get the maintenance contract in writing and understand what the contract covers in terms of repair and maintenance. Although this contract may seem to be just an additional step or an extra cost, it can pay dividends when you need it. Remember—fire fighters can and do break things.

Some multimedia products require less maintenance than others. For instance, a chalkboard will not get a computer virus—but it also cannot transition from one slide to the next at the touch of a button. Be aware of the trade-offs between high technology and good old-fashioned methods. You need to decide based on a number of factors (e.g., cost, ease in transportation, audience) what is appropriate for your event or presentation.

Audiovisual equipment, like any other equipment, needs to be kept clean and in good order so that it will work when you need it. Understanding and following the manufacturer's recommended maintenance practices is the simplest way to understand what is needed for a particular device. Another important reminder is to make sure your device has the latest software, as this factor can affect its ability to "talk" with other devices.

When to Use Multimedia Presentations

Multimedia presentations allow students to see, hear, and (in some cases) touch the educational content, which enhances learning. Such applications should be used to simplify complex theories or hypotheses. Multimedia resources can be used to reinforce materials through exercises, pictures, or words that relate to the information you are presenting. They can summarize previously integrated material for use in the current program so that you do not waste a lot of time covering prerequisites. In combination with the learning styles of today's students, multimedia presentations help keep students interested as well.

Despite all of these advantages, there are some drawbacks to using multimedia in training. In particular, the failure of the media to perform as designed can put instructors at a loss. Such a failure can be as simple as the bulb on the slide projector burning out or as complex as the computer presentation not interfacing properly with the projector. As an instructor, you must be versatile enough to be able to overcome these glitches and not let failure impede what you are trying to present. If a technology failure does happen, then you must have a plan B.

Perhaps the biggest issue facing instructors who want to use multimedia is determining when it is appropriate. Multimedia resources are not always needed. Remember—the technology always needs to be appropriate for the audience and the content to be presented.

The classroom has long been known as a place where networking and information combine to create the learning experience. This milieu allows the students to learn from one another as well as from the instructor and the materials presented. At its best, this environment tends to encourage students to learn for themselves. Remember the old saying, "You can lead a horse to water, but you cannot make it drink." The same is true for students: You can present all the material in the world but, if students do not want to learn, they won't. Multimedia tools try to break down the resistance to learning by engaging the student.

Multimedia should be used whenever its application makes it easier for students to grasp the material. It does not matter if the application means bringing the emergency incident into the classroom, demonstrating a nonthreatening environment, exemplifying safety concerns, or utilizing simulation or distance education. The only reason to use multimedia tools is to help students learn.

■ Selecting the Right Applications

Selecting the right media application is critical. When used as an instructional aid, the right tool allows you to clarify the information you present to students. Essentially, the various multimedia features organize the way in which students receive the material. Diagrams, flowcharts, and graphics impart information that can be retained by students who are visual learners. The lecture or audio cues assist auditory learners. Used in combination, these methods engage students by involving multiple senses.

Emphasis alone will not guarantee that every student retains all of the information presented in a course, but it certainly makes it possible to clarify difficult concepts. Through visual representations, students can travel to places that are unreachable in real life. For example, imagine the advantage of being able to follow a drop of blood through the body to understand the circulatory system. In this case, words are not an adequate medium to get the point across, yet the desired effect is easily achieved through simulation. Likewise, students could recognize the different ways that heat is transferred by allowing them to feel the heat from a light bulb to clarify the concept of radiant heat transfer.

■ Troubleshooting Commonly Encountered Multimedia Problems

Many things can go wrong when using multimedia applications. As a consequence, instructors need to know and practice a few alternatives in case problems crop up with this technology. **TABLE 8-2** can assist with some of the more common

Table 8-2	Common Troubleshooting Procedures
Try each "solution" in order. There is no need to continue down the list if the first step solves your problem. If your problem remains unsolved, you may need more expert assistance.	
Symptom	**Solution**
Presentation projector does not turn on/off	1. Point the remote directly at the projector and try again. 2. The projector may be unplugged. 3. The remote may need new batteries.
No audio from podium	1. Make sure the audio system is turned on. 2. Make sure the microphone is turned on.
No audio from wireless microphone	1. Make sure the audio system is turned on. 2. Make sure the body pack is turned on. 3. The microphone may need new batteries.
VCR, audiocassette, or DVD will not play	1. Check to make sure the VCR, DVD, or other player is plugged in. 2. Eject and reload the media and try again.
No audio from VCR, DVD, or computer	1. Check the volume controls to make sure they are not turned down. 2. Make sure the audio system is turned on. 3. Make sure connections are made to the audio output on the device and the input on the audio system.
No light from visual projector	1. Check that the light has not been switched off on the base unit.
No image from computer/laptop	1. Check that the correct input source has been selected. 2. Make sure the computer has not been turned off. 3. Press the Alt and F5 keys simultaneously (or other Alt Function keys, depending on the computer).
Projector light will not come on	1. If you hear a fan, make sure the lens cover is off. 2. If you hear a fan and the lens cover is off, you may need to replace the bulb. 3. If you don't hear the fan, make sure the projector is plugged in.
Computer will not read media	1. Take the disk out and confirm that it was inserted properly. 2. Make sure the computer has the proper software needed for the media to work properly.

problems. The simplest things that can go wrong can often be easily fixed by taking a deep breath and thinking about the problem and potential solutions. Take the time to work with your equipment and know its limitations. Be prepared for the possibility that something may go wrong. Have a plan B, a plan C, or even a plan D.

Sometimes solving a problem is not as easy as referring to a simple chart. That is, sometimes you need to be able to diagnose the problem and verbalize it. This step is necessary to use the *help* files contained within most software programs. To find these files, look at the menu bar at the top of your screen. It typically lists options such as File, Edit, and View; at the very end of that list is Help. You can either double-click on the word Help or press the Alt and H keys simultaneously to open this script. Then select the type of help you need.

Teaching Tip

To determine whether your multimedia presentation is appropriate and useful, make sure it meets the following criteria:

- Is appropriate for the audience and topic. Is the presentation geared toward the specific group of students, is it easy to grasp the material, and is the material right for the subject matter?
- Covers the material completely and accurately.
- Is easy to use and dependable.
- Is current and up-to-date. You may need to identify trends and potential future challenges.
- Promotes expected behavior and outcomes, such as demonstrating the use of personal protective equipment (PPE) or apparatus positioning.

Training BULLETIN

Jones & Bartlett Fire District

Training Division
5 Wall Street, Burlington, MA, 01803
Phone 978-443-5000 Fax 978-443-8000
www.fire.jbpub.com

Instant Applications: Technology in Training

Drill Assignment

Apply the chapter content to your department's operation, training division, and your personal experiences to complete the following questions and activities.

Objective

Upon completion of the instant applications, fire service instructor students will exhibit decision making and application of job performance requirements of the fire service instructor using the text, class discussion, and their own personal experiences.

Suggested Drill Applications

1. Experiment with a PowerPoint® presentation. Which color schemes work best for dark classrooms? Which work best for bright classrooms? Learn how to change a PowerPoint® presentation quickly, so that you can instantly adapt to existing conditions.

2. Modify a PowerPoint® presentation by adding text, photos, or other illustrations to clarify the message.

3. Select a learning platform such as an online or distance-learning course. Discover what the benefits of that platform are for your students.

4. Review the Incident Report in this chapter and be prepared to discuss your analysis of the incident from a training perspective and as an instructor who wishes to use the report as a training tool.

Incident Report

Green Cove Springs, Florida—1990

In Green Cove Springs, Florida, a live training exercise was set up near the end of a fire fighter training course. The course led to state certification, and was taught by a state-certified fire training center operated by a local school district. Fire fighters from a municipal and county fire department joined the students for the final live exercises. A picnic for the students and their families was held before the fire and the families remained for the training fire.

The wood-framed, two-story house was almost 100 years old and had previous fire damage to it. The 24-person class with two state-certified instructors and one noncertified instructor set up four 1¾" (44.5-mm) hose lines, wyed off of two 2½" (63.5-mm) hose lines. Two additional 2½" (63.5-mm) hose lines with nozzles were positioned, with the pumper supplied by a 5" (127-mm) large-diameter hose (LDH) line. Ground ladders were set on three sides as emergency egress from the second floor. The stairs leading to the second floor were 36" (914.5 mm) wide.

An initial burn was conducted in a second-floor bedroom, measuring 12' × 19' (3.7 × 5.8 m) with 9' (2.7-m) ceilings. There were two doors to the room: an entry door and a door that connected to an adjoining room. There were two windows, one of which looked out over the front porch roof. A 1¾" (44.5-mm) hose line was brought up and into the bedroom by the first crew. The exercises upstairs were primarily for observing fire behavior, and the actual suppression exercises were to follow afterward in the rooms downstairs.

Approximately a half-gallon of diesel fuel was poured on a mattress that was lying on its side against a wall about 4' (1.2 m) from the entry door. It was ignited with a road flare by a noncertified instructor. The first instructor took the students into the room to observe the fire spread as he explained the fire's behavior. The students then knocked the fire down under the direction of the instructor. The instructor flipped the mattress over for reuse. The first crew and the instructor exited down the stairs.

Another certified instructor then led a second crew up the stairs. He swapped self-contained breathing apparatus (SCBA) with the other instructor who had just exited because there was a failure with his own. As the second crew advanced the 1¾" (44.5-mm) hose line, it lost pressure because the supplying 2½" (63.5-mm) hose line burst. The 2½" (63.5-mm) hose line was quickly changed out, and the crew continued to the fire room.

Once in the fire room, the students were met with deteriorating fire conditions. Apparently, the nozzle was pulled from a student's hands. Conditions became untenable. One student crashed headfirst through the window onto the front porch roof. Another student and the instructor exited through another bedroom window down one of the ground ladders. The last student exited though the bedroom door and down the stairs. An instructor outside ordered a safety line to enter and extinguish the fire after observing the student escape though the front window. The first, noncertified instructor now became the incident commander. He requested EMS and additional fire units.

The instructor inside apparently did not have his gloves on and received third-degree burns to both of his hands, right arm, and shoulder. He lost parts of two fingers due to the burn injuries. Two students received second-degree burns on their upper torsos, and one received third-degree burns to his back. All were flown to medical facilities by EMS helicopters.

Richard Knoff, Deputy Chief of Clay County Fire Rescue, said this of the incident:

> At first we thought there might be a problem when we saw heavy black smoke coming from the bedroom windows upstairs. I called the lead instructor who had the students inside but I did not get an answer. I tried again and then we saw a student bail out headfirst through a second-floor window, landing on the porch roof. I called for additional fire units and EMS by radio, and instructed a crew to get the fire fighter down from the porch roof. Then, I deployed the RIC [rapid intervention crew] and went in with them. We got inside and everybody was already out. Looking at the room afterwards, you could not imagine how anyone was hurt. There was no heavy charring, no significant damage to their SCBA or bunker gear. Only damage to their helmet shields.

Continued…

Incident Report Continued...

Green Cove Springs, Florida—1990

Post-incident Analysis: Green Cove Springs, Florida—1990

NFPA 1403 (2007 Edition) Noncompliant

Flashover and fire spread unexpected (4.3.9)

Fire "set" was close to primary entry/egress of fire room (4.4.16)

Student–instructor ratio was 8:1 (4.5.2)

Noncertified instructor (4.5.1)

No backup line (4.4.6.2)

NFPA 1403 (2007 Edition) Compliant

Ground ladders were set as secondary means of egress (4.2.12.1)

Chief Concepts

- Although training aids are fundamental in assisting the fire service instructor in the learning process, they are simply "aids" to help the instructor impart the knowledge the fire fighter needs to know and understand.
- One of the most dynamic and fast-changing concepts in education and training is technology-based instruction (TBI), also known as e-instruction, Internet-based instruction, or distance learning.
- The majority of distance learning is accomplished through a combination of reading assignments, conferencing, video presentation, testing with proctors, and written assignments. Student evaluations are processed through an instructor.
- Computer-based training (CBT) software enables instructors to prepare course materials and efficiently manage day-to-day teaching tasks. Many CBT programs offer features that allow you to facilitate collaborative learning, personalize content based on students' unique needs, and positively affect learning outcomes.
- A learning management system (LMS) is a Web-based software application that allows online courseware and content to be delivered to learners, wrapped with classroom administration tools such as activity tracking, grades, communications, and calendars. This type of system is also referred to as a virtual learning environment (VLE) or learning course management system (LCMS).
- A key element in most presentations is the use of a multimedia tool in some form, ranging from handouts to PowerPoint® (PPT) presentations. Using multimedia tools enhances the learning process for students and can make the instructor's job easier. At the same time, multimedia tools can be a distraction if they are poorly developed or used improperly.
- Multimedia tools come in a variety of types and uses, ranging from audio systems to Flash animation to simulations.
- Depending on the class size and the arrangement of the classroom, the instructor may need to use a microphone and audio system. When deciding which kind of audio system you will use, keep your presentation style in mind.
- Presentation programs include a variety of software programs developed by software vendors. Each presentation program has its own unique features and benefits that you can use to enhance your instruction.
- The visual projector has backlighting, overhead lighting, and color capabilities that can be tied in with a computer. It combines the flexibility of the overhead projector with some of the capabilities of a high-tech device such as a liquid crystal display (LCD) projector. This multimedia tool can project almost any one-dimensional printed material onto a screen.
- LCD projectors and digital light processor (DLP) projectors are used with video players, DVD players, computers, and visual projectors to project documents, images, and motion pictures so that a group of students can easily view the material simultaneously.
- The VCR requires another component for viewing the video cassette images—either a data projector or a television. If you are using a television to view the videotape, keep in mind the size of the class and lighting. The larger the class, the larger the screen must be.
- Like CD-ROMs, DVDs are available as either blank or pre-programmed media. Information ranging from video clips to audio clips to PPT presentations can be stored on a DVD.
- A digital camera allows you to capture an electronic image, upload it onto your computer, and insert the image into your PPT presentation fairly easily.
- Flash has become a popular method of adding animation and interactivity to Web pages. Flash files carry the file extension .swf and may be accessed as objects on a Web page or played in a stand-alone Flash Player.
- While the scanner can be a helpful tool, its use does raise copyright concerns. Obtain written permission before using copyrighted material in your presentations.
- Professionally developed presentations are intended to enhance students' learning experience, but do not replace the experienced instructor. Instructors must review materials ahead of time and adapt them to the students who will use them.
- Overhead projectors have been used in training for more than 40 years and were once considered the industry standard for quality educational enhancement.
- Digital audio players are becoming a mainstream technology for students at most universities. Students can carry these devices easily and download lectures or pictures that can be reviewed while traveling to class.
- Students can use tablet computers to take notes, complete assignments and e-mail them to the instructor, and conduct research during class.
- A smart board is an interactive whiteboard that uses touch detection for user input.
- Many fire stations are equipped with satellite programming systems that rival those of local and national commercial broadcast companies and that utilize a 3-meter base station.
- Virtual reality devices allow students to interact with a computer-simulated environment. The goal is to immerse the student in an environment that can be controlled by the student and the programmer.
- Like virtual reality systems, simulators allow students to interact safely with hazards in a controlled environment.

- Many fire service instructors find the use of models helpful when teaching basic principles prior to going into the field.
- Chalkboards, erasable boards, pads and easels, and handouts are great low-tech options to keep the students focused on the lesson.
- As with any technology, maintaining the quality of multimedia tools is important.
- Multimedia tools should be used to simplify complex theories or to assist instructors in reinforcing their message through exercises or images.
- Selecting the right media application is critical. When used as an instructional aid, the right tool allows you to clarify the information you present to students.

Hot Terms

<u>Asynchronous learning</u> Environment in which learners are not consuming content at the same time; instead, students work at their own pace, adhering to established submission deadlines set by the instructor.

<u>Courseware</u> Any educational content that is delivered via a computer.

<u>Ghosting</u> When viewing text, faint shadows that appear to the right of each letter or number.

<u>Hybrid learning</u> Learning environment in which some content is available for completion online and other learning requires face-to-face instruction or demonstration; sometimes referred to as "blended" or "distributed" learning.

<u>Icon</u> A small, pictorial, on-screen representation of an object used to indicate the existence of documents, file folders, and software.

<u>Instructor-led training (ILT)</u> Learning environment in which a human instructor facilitates both the in-person sections and the online components of coursework. Instructors usually set deadlines for submission, create quizzes and assignments, and track student progress.

<u>Learning management system (LMS)</u> A Web-based software application that allows online courseware and content to be delivered to learners, wrapped with classroom administration tools such as activity tracking, grades, communications, and calendars.

<u>Self-directed study</u> Learning environment in which learners can stop and start as they desire, progress to completion of a module, and submit a final score to determine whether they passed or failed the module, all without an instructor's involvement.

<u>Social learning</u> An informal method of learning using technologies that enable collaborative content creation. Examples include blogs, social media (e.g., Twitter, Facebook), and wikis.

<u>Synchronous learning</u> Learning environment in which learners are all engaged in or receiving educational content simultaneously. An example might be a live Webcast, Webinar, or online chat session.

<u>Technology-based instruction (TBI)</u> Training and education that uses the Internet or a multimedia tool such as a DVD or CD-ROM.

<u>Virtual classroom</u> A digital environment where content can be posted and shared, and where instructors can create quizzes and assignments to be completed and tracked online via an Internet-connected computer.

<u>Webinar</u> An online meeting, occurring in real time, which usually allows for participants to share their desktop screens, Web browsers, and documents live as the meeting is occurring.

References

Blended Learning: More Effective Than Face-to-Face. Digital Education: *Education Week. Education Week: Blogs.* N.p., n.d. Retrieved September 10, 2012. <http://blogs.edweek.org/edweek/DigitalEducation/2009/06/blended_learning_more_effectiv_1.html>.

EMI Courses and Schedules. National Preparedness Directorate National Training and Education Portal. N.p., n.d. Retrieved September 10, 2012. <http://training.fema.gov/EMICourses/>.

Evidence-Based Practices in Online Learning: A Meta-analysis and Review of Online Learning Studies. US Department of Education. N.p., n.d. Retrieved September 8, 2012. <www2.ed.gov/about/offices/list/opepd/ppss/reports.html#edtech>.

The Evidence on Online Education. *Inside Higher Ed: Higher Education News, Career Advice, Events and Jobs.* N.p., June 29, 2009. Retrieved September 8, 2012. <http://www.insidehighered.com/news/2009/06/29/online>.

Fischer, Katherine. Electronic Communication Discussion Tips. Donna Reiss, WordsWorth2 Consulting. N.p., n.d. Retrieved September 10, 2012. <http://wordsworth2.net/activelearning/ecacdiscustips.htm>.

Learning Circuits. Top 10 LMS Purchasing Mistakes (and How to Avoid Them). N.p., n.d. Retrieved September 10, 2012. <http://www.learningcircuits.org/2002/mar2002/egan.html>.

National Fire Protection Association. (2012). *NFPA 1041: Standard for Fire Service Instructor Professional Qualifications.* Quincy, MA: National Fire Protection Association.

National Fire Protection Association. (2007). *NFPA 1403: Standard on Live Fire Training Evolutions.* Quincy, MA: National Fire Protection Association.

National Fire Protection Association. Research: Fire Statistics: The U.S. Fire Service. N.p., n.d. Retrieved September 10, 2012. <http://www.nfpa.org/displayContent.asp?categoryID=955>.

10 Principles of Effective Online Teaching: Best Practices in Distance Education. Effective Teaching Strategies for the College Classroom. *Faculty Focus.* N.p., n.d. Retrieved September 10, 2012. <http://www.facultyfocus.com/free-reports/principles-of-effective-online-teaching-best-practices-in-distance-education/>.

Weaver, Skip. In Time of Budget Shortfalls, Firefighters Training Online. *Middletown Journal.* N.p., April 13, 2012. Retrieved September 10, 2012. <www.middletownjournal.com/news/news/local/in-time-of-budget-shortfalls-firefighters-training/nNSNb/>.

Things seem to be going well in your career. You have been promoted to Battalion Chief of Training. You are finally getting to apply your skills in making great presentations. Your first big assignment is to teach a refresher course on strategy and tactics. Given that only a few new topics, such as compressed-air foam, must be included in the class, you can use most of the previous curriculum. While the content is still current, however, the presentation is antiquated.

You decide to upgrade the presentation from 35-mm slides to a PowerPoint® slide show. You also want to embed some photos and video files into your presentation. You have some great footage from a recent house fire that provides a good example of how to read the smoke. You also know that dispatch has the sound files to go with it.

1. If you decide to use PowerPoint® slides, which other equipment will you need to give the presentation?
   A. An overhead projector
   B. A slide projector
   C. A data projector
   D. An iPod

2. Which file format could be associated with the video clips?
   A. .mov
   B. .tif
   C. .wav
   D. .pdf

3. Which of the following statements is true?
   A. Video files are essential for a good class.
   B. Including sound files will hamper learning.
   C. Only multimedia tools that support learning should be included.
   D. Both A and C are true.

4. If you decided to turn this course into a distance-learning course, what does that mean?
   A. It will be a computer-based course.
   B. The course will be provided "a distance away" from the training center/college/university.
   C. The course requires college credit to be assigned to it.
   D. The course cannot be computer based.

5. Which unit is used to measure the brightness of the light emitted by a data projector?
   A. Megapixels
   B. Pixels
   C. Megahertz
   D. Lumens

6. _____ are an effective means of providing students with materials to take home.
   A. Movies
   B. Handouts
   C. Cassettes
   D. Slides

Safety During the Learning Process

Fire Service Instructor I

Knowledge Objectives

After studying this chapter, you will be able to:

- Discuss the relationship between training and fire fighter safety. (pp 208–221)
- List the 16 fire fighter life-safety initiatives. (p 209)
- Describe how to ensure safety in the classroom. (NFPA 4.4.2 , NFPA 4.4.5) (pp 209–210)
- Describe how to promote and teach safety by example. (p 209)
- Describe your responsibility for student safety during training. (NFPA 4.4.2) (pp 211–213, 215–216)
- Describe the laws and standards pertaining to safety during live fire training. (pp 218–221)
- Discuss how to develop safety as part of your department's culture. (pp 219–220)

Skills Objectives

After studying this chapter, you will be able to:

- Demonstrate how to lead by example. (p 209)
- Demonstrate safety in the classroom and on the training ground. (pp 208–221)
- Demonstrate safety during the training process. (pp 208–221)

Fire Service Instructor II

Knowledge Objectives

After studying this chapter, you will be able to:

- Discuss the relationship between training and fire fighter safety. (pp 208–221)
- List the 16 fire fighter life-safety initiatives. (p 209)
- Describe how to ensure safety in the classroom. (pp 209–210)
- Describe how to promote and teach safety by example. (p 209)
- Describe your responsibility to student and other instructors during training. (NFPA 5.4.3) (pp 211–213, 215–216)
- Describe the laws and standards pertaining to safety during live fire training while supervising other instructors. (NFPA 5.4.3) (pp 218–221)
- Discuss how you as a supervisor can develop a culture of safety in your department. (pp 219–221)
- Describe your role as a supervisor in using the incident management system during all high-risk training activities. (pp 211, 218, 221) (NFPA 5.3.3)

Skills Objectives

After studying this chapter, you will be able to:

- Demonstrate how to lead by example. (p 209)
- Demonstrate safety in the classroom and on the training ground. (pp 208–221)
- Demonstrate the use of the incident management system during a high-risk training evolution. (pp 211, 218)

© Photos.com

Fire Service Instructor III

Knowledge Objective

There are no knowledge objectives for Fire Service Instructor III students.

Skills Objective

There are no skills objectives for Fire Service Instructor III students.

Recently an instructor at a neighboring training facility was injured during a training drill that left the instructor unable to return to work. During the investigation, it was determined that the instructor had failed to wear all the correct personal protective equipment (PPE) and that this failure had caused his injuries to be more severe than if he had been wearing his PPE. This incident and other related incidents on a national level have prompted your training division staff to review your agency's policies and procedures—not only those dealing with PPE, but also those pertaining to instructor conduct during training.

1. As an instructor, how do you ensure the safety of fellow instructors working with you during training?
2. How do you ensure the safety of your students during training evolutions?
3. Which policies and guidelines do you have in place to ensure the safety of your fellow instructors and the students for whom you are responsible?

Introduction

Each year in the United States, countless fire fighters are injured and, on average, 10 fire fighters are killed during training-related activities. This trend has accelerated over the past few years and is a growing concern among fire service leaders. In addition to the higher number of injuries to students during training, the number of injuries to instructors who teach fire fighters has increased.

As a fire service instructor, you must take measures to stop the preventable injuries and deaths that occur during training. Of course, for training to be effective, it has to be realistic and, therefore, a certain level of danger will exist in all training events. Live fire training tops the list of high-risk training events, but this same high risk can be found in many other training activities.

You must always place students' safety ahead of any performance expectations for the training session. This mandate includes everything from assigning safety officers to implementing rehabilitation protocols, to decreasing instructor–student ratios, to ensuring that enough trained eyes are watching out for the safety of all students. The standards, technology, and information needed to protect your personnel from training tragedies are readily available. You also need to

apply relevant checklists, standards, and control measures to your training sessions correctly to keep your students safe. Fire fighters want meaningful drills that allow them to apply their skills, knowledge, and abilities. It is your responsibility to provide this training while keeping your students safe FIGURE 9-1.

FIGURE 9-1 It is your responsibility to provide meaningful drills that allow fire fighters to apply their skills, knowledge, and abilities while keeping your students safe.

Fire Service Instructor I and II

The Sixteen Fire Fighter Life-Safety Initiatives

In 2004, NFFF brought together fire service experts and leaders to discuss ways to slow the number of line-of-duty deaths (LODDs) experienced by the fire service. Their work culminated in 16 initiatives that were given to the fire service with

the goal of reducing LODDs by 50 percent over a 10-year period TABLE 9-1. These initiatives should provide both guidance and motivation to you while you are developing and conducting any training. They highlight the importance of training as a strategy that will help reduce the number of fire fighter LODDs. As a fire service instructor, you will be specifically charged with carrying out initiative 5, which states, "Develop

Table 9-1	16 Fire Fighter Life Safety Initiatives

1. Define and advocate the need for a cultural change within the fire service relating to safety, incorporating leadership, management, supervision, accountability, and personal responsibility.

2. Enhance the personal and organizational accountability for health and safety throughout the fire service.

3. Focus greater attention on the integration of risk management with incident management at all levels, including strategic, tactical, and planning responsibilities.

4. All fire fighters must be empowered to stop unsafe practices.

5. Develop and implement national standards for training, qualifications, and certification (including regular recertification) that are equally applicable to all fire fighters based on the duties they are expected to perform.

6. Develop and implement national medical and physical fitness standards that are equally applicable to all fire fighters, based on the duties they are expected to perform.

7. Create a national research agenda and data collection system that relates to the initiatives.

8. Utilize available technology wherever it can produce higher levels of health and safety.

9. Thoroughly investigate all fire fighter fatalities, injuries, and near misses.

10. Grant programs should support the implementation of safe practices and/or mandate safe practices as an eligibility requirement.

11. National standards for emergency response policies and procedures should be developed and championed.

12. National protocols for response to violent incidents should be developed and championed.

13. Fire fighters and their families must have access to counseling and psychological support.

14. Public education must receive more resources and be championed as a critical fire and life safety program.

15. Advocacy must be strengthened for the enforcement of codes and the installation of home fire sprinklers.

16. Safety must be a primary consideration in the design of apparatus and equipment.

Courtesy of National Fallen Firefighters Foundation's Everyone Goes Home® Program

and implement national standards for training, qualifications, and certification … that are equally applicable to all fire fighters based on the duties they are expected to perform." This initiative should be applied every time you instruct a course and in every drill-ground evolution you carry out.

Leading by Example

As a fire service instructor, you will use many methods of delivery to teach. One of the most powerful methods is leading by example. Fire fighters, regardless of their rank, years on the job, or past experiences, will look to you and make mental notes on how the job *should* be done **FIGURE 9-2**. As an instructor, you play an extremely influential role in determining the performance of your students. Given this responsibility, you must ensure that your own practices and principles clearly demonstrate the safety values employed in the fire service. If you do not correct an unsafe practice during training, your students will mistakenly believe that you condone the error. You must set a positive tone and embrace safety as the number-one priority of every lesson plan delivered. Both in the classroom and on the drill ground, use the controlled environment of the training arena to point out where things can go wrong in real-world applications.

FIGURE 9-2 As the fire service instructor, you will be demonstrating how the job should be done.

Teaching Tip

Prior to class, survey the classroom or teaching environment and correct any safety concerns that may exist.

Safety in the Classroom

Safety during training begins with safety in the classroom. First, consider the location of the classroom. While you may not have complete control over where the class is held, you can still affect some things. Survey the general classroom environment. Are there inadequate fire exits or other unsafe conditions, such as remodeling in the area or icy steps leading to the classroom? If the classroom is in an engine bay, is the area subject to excessive exhaust fumes, or is there oil on the floor? Evaluate the lighting conditions in the classroom. Not all students have great vision, and either low or high lighting may create eye strain. Excessive noise can be a problem as well—not only because it interferes with learning, but also because it can cause unnecessary damage to students' hearing.

Next, review the physical arrangements of the classroom. Unsecured extension cords running throughout the classroom may trip either you or the students. Prior to each class, reroute or tape down extension cords. Also, be aware that fire stations may be the recipients of hand-me-downs from other city agencies. Often, chairs and tables that have outlived their usefulness in other areas are pressed into service in a fire department classroom and could be prone to collapsing. Make sure that each item is in good condition to prevent injuries to students.

At the beginning of class, be sure to raise safety awareness with the students. Go over the locations of the nearest fire exit and fire extinguishers. If you are teaching a class to an on-duty crew or a group of volunteer fire fighters, you should address what to do if an alarm sounds. For example, if some students in the class are on-call fire fighters, have them sit near an exit so that if they need to respond during class they can leave without interrupting the class. Explain the emergency plan to follow should a fire alarm activate or a severe weather event or natural disaster occur. It is important to emulate the behavior you teach to the general public and to remind your students that safety is paramount.

Hands-on Training Safety

Hands-on training includes all training activities conducted outside the classroom environment. Whether it consists of a simple tailboard chat on nozzles or a full-blown live fire training exercise, hands-on training is often the most powerful method of instruction for a student. Unfortunately, hands-on training also produces hazards that must be considered and mitigated. Slippery surfaces, sharp or jagged metal, and products of combustion need to be accounted for. Risk management practices need to be applied to limit, reduce, or eliminate risk exposure to students. Environmental factors—such as heat, cold, wind, rain, and storms—must be considered as well. Mother Nature can pose a significant risk to your students.

■ Personal Protective Equipment

Determine the appropriate level of personal protective equipment (PPE) after reviewing the lesson plan. Lesson plans often specify the level of PPE required. However, local standard operating procedures (SOPs) may also exist that specify the level of PPE for training. A general statement of the PPE requirements for the training session needs to be part of your presentation. As an instructor, you must always wear the PPE required for the drill being conducted. The appropriate level of PPE depends on both the potential hazards and your department's SOPs.

The best practice is always to use the same PPE in training as you would in real-world incidents. In some departments, it may be acceptable to conduct ladder drills using helmets, gloves, and boots as opposed to full fire gear and self-contained breathing apparatus (SCBA), due to the skill being taught or reinforced. This modification to the PPE policy may make the drill safer by reducing the potential for heat stress. Another option may be to begin the training session in full levels of PPE and then to reduce the PPE required after signs of heat stress appear or when the skill has been performed properly. When these adaptations are implemented, you must make it very clear to your students that the modification is only allowable in the training session and should never be employed when responding to actual emergencies. If any level of hazard exists in the session, then the highest level of PPE available should be used.

■ Rehabilitation Practices and Hands-on Training

NFPA 1584, *Standard on the Rehabilitation Process for Members During Emergency Operations and Training Exercises*, requires that rehabilitation protocols be established and practiced at training events that involve strenuous activities or prolonged operations. As a fire service instructor, you must be aware of the content of this standard and apply it to these types of training sessions. These practices further reinforce the importance of following appropriate rehabilitation practices on the fire ground FIGURE 9-3 . Training should mirror as closely as possible the actual fire-ground operation. Medical care and monitoring, hydration, and cooling/warming materials are just a few of the rehabilitation considerations that NFPA 1584 requires to be available during training events.

Not to be overlooked is the importance of instructors participating in the rehab process. Too often, instructors will

FIGURE 9-3 Rehabilitation is an important part of training.

rotate students through the rehab station but neglect to rehab themselves. In reality, participation in rehabilitation is a safety factor as well as an important way to set the proper example for your students. Instructors who remain in an immediately dangerous to life and health (IDLH) environment in full protective clothing are putting themselves at high risk both physically and mentally.

Safety Policies and Procedures for Training

As a fire service instructor, you must have a firm grasp of the safety policies mandated by your fire department. All safety policies required on the fire ground must be reinforced and practiced during training. Indeed, one of the more common causes of injuries during training is failure to follow established safety practices. The instructor has a responsibility to ensure that all policies of the department are followed during training. Policies such as establishment of an incident command system, fire-ground accountability, rehabilitation, and full use of PPE are followed on the emergency scene, and adherence to those policies must not be set aside in training. Enforce those policies in the same manner a safety officer or incident commander would at an emergency incident.

On occasion, personnel from another agency might train with you at your facility if you are conducting mutual aid training or a special course. It is important in these types of situations that all fire fighters know which policies will be followed during the training. If you are hosting the training, all visitors must follow your policies, and your visitors need to be aware of what they are expected to do and which policies they will follow.

The Fire Service Instructor III is responsible for drafting and forwarding policies relating to training through the chain of command. At the minimum, a training policy relating to training safety will include procedures for instructor qualifications, PPE for both students and instructors, safe student-to-instructor ratios, and effective rehabilitation practices during high-risk/exertion exercises. All levels of instructors must develop and deliver training that adheres to applicable safety points and must enforce a strict, zero-tolerance policy for deviations from accepted safety standards or practices FIGURE 9-4.

As an instructor, you must also be informed of and follow the manufacturer's recommendations for using equipment. Read the accompanying owner's manuals for equipment such as chainsaws and heavy hydraulic equipment. Often the limitations of these tools are assumed, and an injury occurs when the tool is used in a manner counter to its intended use. When new apparatus is delivered to your department, you should be present for the initial training by the apparatus builder. Essential information presented by the manufacturer needs to be recorded, practiced, and reinforced whenever that piece of equipment is used.

Influencing Safety Through Training

As a fire service instructor, you are vitally important in setting a good example in terms of safe fire-ground practices. Constantly remind yourself that you serve as the role model of a safe and effective fire fighter. Everything from your fitness level to the way you wear your uniform should exemplify your values related to safety and adherence to rules. The example that you set sends a far more powerful message to your students than any theories and concepts you present in a lecture or drill.

You should be acutely aware of how influential a role the fire service instructor plays, especially with new fire fighters. New fire fighters are uncertain about what is expected of them—that is, what is considered right versus what is considered wrong. With these students, whatever you do or say will be construed as the right thing. If you use improper techniques or fail to follow safety protocols while teaching, students will perceive that as the right way to do things. This faulty knowledge becomes dangerous when students leave your protection and perform on the emergency fire ground.

Your primary responsibility during a drill or training session is to ensure the safety of all students attending the drill. The actions of a fire fighter on the fire ground are the result of training. If you allow students to use unsafe practices in training, they will repeat those same actions on the fire ground, because decisions and actions on the fire ground are based on previous training.

You have many responsibilities related to the safety of your students during training, but one that stands above all others is the responsibility to have students practice safe principles and practices during training so that they remain safe on the fire ground. As part of your duties as an instructor, you must correct poor skills or bad decisions as they occur. If you witness a skill being performed incorrectly or a safety procedure not being followed but you do not correct the student's error, then you are reinforcing the incorrect behavior. This constant monitoring and correction is important with new recruits but can be more difficult with experienced personnel. Correcting unsafe or poor behavior demonstrated by a veteran fire fighter can be very challenging—but it is a challenge you cannot ignore. Correcting an unsafe skill should be the same whether the fire fighter is a raw recruit or a seasoned veteran.

■ Planning Safe Training

You should begin to address safety at the earliest stages of course development and preparation, while formulating or modifying lesson plans. Analyze and critique your training to ensure that safety measures are correctly addressed and emphasized. It is too late to address safety issues after an injury has occurred. Preventing injury becomes much easier when the hazards are identified and addressed early in the planning process.

Many of the NFPA standards provide additional information concerning safety and training and should be consulted during the planning process. NFPA 1500, *Standard on Fire Department Occupational Safety and Health Program*, would be an important document to review when planning a program of instruction. In general, almost all NFPA standards that begin with the prefix 14 relate to a training recommendation and focus on providing safe training sessions. NFPA 1584, *Standard on the Rehabilitation Process for Members During Emergency*

Training Safety Plan		

Drill Date: **Time:** **Shift:**

Drill Location:

Type of Training: *(check all that apply)*

	Fire Suppression		EMS		Technical Rescue
	Live Fire Training		Vehicle/Machinery Extrication		Hazardous Materials/WMD
	Driver Training		Other Acquired Structure Training		Water/Dive Rescue
	Apparatus Operation		Preplan Survey or Simulation		Physical Fitness Activity
	Other:				

Drill Risk Assessment: ☐ **High**   ☐ **Medium**   ☐ **Low**

Maximum Student/Instructor Ratio: ___ **to** ___   **Safety Officer Needed:** (Required on High Risk Drills)

Instructor PPE Requirements:

☐ **SCBA**	☐ **Full PPE**	☐ **Helmet**	☐ **Eye Prot**	☐ **Filter Mask**	☐ **Hearing**
☐ **Gloves**	☐ **Radio**	☐ **Lights**	☐ **High Visibility Vest**		

Drill Objective(s): *(Brief explanation of objectives)*

Description of Training: *(e.g., extricate victim from vehicle, victim search in limited visibility)*

PPE/Equipment Required for Each Participant: *(check all that apply)*

	Helmet		Personal Flotation Device
	Eye Protection		Buoyancy Compensator
	Hearing Protection		Mask/Snorkel/Fins
	Gloves (*Type*):		SCBA
	Bunker Coat		SCUBA
	Hood		Other Respiratory Protection (*Type*):
	Bunker Pants		Hazardous Materials CPC (*Type*):
	Safety Boots		Radio
	Other (*Specify*):		

Department Related SOPs or Technical References: *(List number and name)*

Hazards and Control Measures: (*Check all Hazards AND write in control measure*)

	Atmospheric (e.g., smoke, dust, low oxygen):	Sewage/Septic:
	Combustible/Flammable Environment:	Sharp Edges/Objects:
	Confined Space:	Structural:
	Electrical:	Terrain:
	Elevation:	Traffic:
	Hazardous Substances (e.g., asbestos, chemicals):	Water:
	Nighttime Conditions:	Weather:
	Other:	

Accountability: (*Check all that apply*)

	Buddy System
	Visual
	Passport
	Dive Master Control Sheet
	Other:

In Case of Emergency: (*Check all that apply*)

	Code or Signal Used:
	RIT Assigned:
	ALS Standby:

Communications:

	Radio—Primary Frequency:
	Radio—Secondary Frequency:
	Hand Signals
	Rope Line
	Lights
	Other:

Resources Assigned: (*Check all that apply AND fill in designated unit*)

	Battalion Chief(s):
	Rehab Officer/Area:
	Rescue Unit(s):
	Safety Officer:
	Specialty Unit(s):
	Suppression Unit(s):
	Other Resources/Equipment:

Rehabilitation Plan: (*Describe rehabilitation plan and guidelines*)

Job Safety Analysis

1. Identify level of required PPE for each participant.

2. List basic steps required to safely complete evolution.

3. Identify potential accidents or hazards that may occur.

4. Determine recommended safe procedures.

Safety Planning Notes: (*e.g., site plan, drawings*)

Lead Instructor:	
Signature:	Date:
Reviewed by (*print*):	
Signature:	Date:

FIGURE 9-4 Sample drill safety plan.

JOB PERFORMANCE REQUIREMENTS (JPRS)
in action

One of the primary goals of training is to improve students' knowledge and skills in a subject so as to increase their ability to do their jobs safely. Opportunities exist throughout the entire learning process to educate and practice specific safety skills. During classroom training sessions, objectives relating to safety can be added to almost every subject area. Every practical session requires diligence in each component of the training to ensure that students and instructors alike remain safe during the session.

Instructor I

The Instructor I understands the application of safety policies and practices those policies in every training session. Lessons learned through experience and case studies help in providing valuable references to the proper safety precautions taken during training. In each lesson presented, the instructor emphasizes topical safety points.

JPRs at Work

Organize and adjust the learning environment to ensure the safety of students and instructors at all times. This includes applying NFPA standards such as NFPA 1403, *Standard on Live Fire Training Evolutions*, and following established procedures for conducting hands-on training.

Instructor II

The Instructor II is responsible for the inclusion of safety-related behaviors in each lesson plan that is developed. Classroom sessions may not present obvious physical hazards, but the instructor must incorporate information on the safe applications of the objectives in the field. In practical sessions, safety rules for each evolution must be reviewed prior to the start of each session.

JPRs at Work

The Instructor II is also responsible for supervising other instructors during increased hazard exposure to ensure that safety policies are followed.

Instructor III

The Instructor III is responsible for the inclusion of safety-related behaviors in each curriculum that is developed. Classroom sessions may not present obvious physical hazards, but the instructor must incorporate information on the safe applications of the objectives in the field. In practical sessions, safety rules for each evolution must be reviewed prior to the start of each session.

JPRs at Work

The Instructor III needs to be mindful that safety-related material is present in curriculum as it is developed and taught. In addition, the Instructor III needs to be aware of how instructors presenting the material adhere to policy in regard to classroom and drill ground training.

Bridging the Gap Among Instructor I, Instructor II, and Instructor III

The Instructor I must learn how to apply safety rules to the training process. When developing skill sheets and lesson plans, the Instructor II must be clear when describing issues relating to safety performance. This will allow for easier application of the lesson plan by the Instructor I. The Instructor I should seek out opportunities to learn about the many standards that apply to training and identify ways to implement safety information in every session. Likewise, the Instructor II needs to be on constant alert for trends in training and must make sure lesson plans are updated if necessary. The Instructor III needs to ensure that all training material and instructor conduct is within policy and that safety is the highest priority in all training sessions.

Theory Into Practice

Creating a safety-conscious student must be a primary goal of virtually all training in the fire service. While there are inherent dangers in their occupation, fire fighters should constantly strive to reduce those risks through good awareness and safety practices. When determining the appropriate level of protective gear to use during training, for example, the instructor should consider the exercise's safety implications for students:

- First, is PPE necessary to ensure students' safety during the training activity? If so, full PPE must be used and any inconveniences mitigated. For example, because full PPE is needed for live fire training, be sure to provide rehabilitation in such a scenario.
- Second, consider whether *not* using full PPE will make the training safer. For example, perhaps hoods need not be worn during ladder practice because of the potential for heat stress in July.
- Third, consider whether *not* wearing full PPE will create a false perception that real scenes can be safely mitigated without its use.
- Fourth, ask yourself if you are considering reduced PPE owing to peer pressure or if this decision is *truly* in the best interest of the student.

Each of these considerations should be evaluated prior to making a determination of the level of PPE to be used. Safety cannot be an afterthought or an unwanted formality: Instead, it must be the first and foremost consideration to ensure the lives and safety of all.

Teaching Tip

Review safety policies that pertain to the drill you are practicing. Reinforce the safety policies during the drill. Also, inform students about where they can find the local safety SOPs for your department.

Operations and Training Exercises, is also a great resource that can assist in planning safe training evolutions.

An important part of the planning process is determining how many fire service instructors are needed to conduct the desired training. Two factors should be considered in making this decision: the type or risk level of the training and the skill level of the students.

Training can be classified based on risk as either high risk, medium risk, or low risk. An example of high-risk training would be live fire evolutions in an acquired structure or swiftwater rescue training. Medium-risk training would include fire extinguisher training or ladder evolutions. Low-risk training would include a classroom lecture or replacing a sprinkler head in a sprinkler lab.

The skill level of those being taught should also be considered. High-risk training of new recruits should generate different concerns than high-risk training involving veteran fire fighters.

Other areas of planning include assembling the equipment that may be needed. Sometimes outdated equipment or equipment that has been replaced by upgrades within the department is used when conducting drills. Using old or outdated equipment has long been an accepted practice in the fire service, but it has often led to injuries that could have been prevented. If the tool or apparatus does not meet the accepted performance standards, then it should not be used in training. You should also question whether using an outdated power saw, for example, is beneficial to students given that this type of saw is not used on current apparatus.

The fire service instructor is also responsible for inspecting all equipment being used in the training to ensure that it meets the required standards prior to the drill. Do not assume that the equipment is in safe and operable condition. If your department does not have the equipment available to make the training safe, then it needs to find alternative equipment—perhaps by borrowing items from a neighboring fire department or the state fire training academy. Acquisition of the appropriate equipment should be arranged prior to the drill.

As a fire service instructor, you are responsible for protecting students from physical and emotional harm in the classroom and on the drill ground. Most of this chapter focuses on means to protect students from physical injury. Nevertheless, instructors cannot overlook the psychological dangers students encounter during training both in and out of the classroom.

Teaching Tip

After development of a lesson plan, have an experienced fire service instructor or health and safety officer review the lesson plan to identify any safety concerns. The department safety officer or safety committee may be able to assist with this task.

■ Hidden Hazards During Training

During training, students are asked to perform various tasks, recall policies, or demonstrate the ability to perform. When placed in an environment of peers, most students realize that their performance is being evaluated—not only by the instructor, but also by their fellow classmates. Failure to accomplish the task or answer the question correctly can label the student as not competent or someone who cannot be trusted on the fire ground. This situation is particularly dangerous because the typical personality of fire fighters is driven by the need to succeed, regardless of the conditions or situation.

This scenario sets the stage for disaster if a fire fighter is placed in an overwhelming situation, such as being asked to conduct a simulated rescue of another fire fighter or a rescue mannequin that may be twice the candidate's size or weight. If the fire fighter is asked to perform this task while fellow classmates are watching, it forces the student to attempt the impossible. Often, the student makes a valiant attempt, only to suffer an injury in the process. You have a responsibility to construct the learning environment so that situations such as these do not arise.

Do not minimize your levels of performance or drop your benchmarks so that every student can achieve success; this is not realistic. At the same time, it is unacceptable for you to stand idly by as a student is placed into a scenario where he or she cannot perform successfully without being exposed to physical harm.

■ Overcoming Obstacles

Money may be an obstacle in the training process, because instructors often face demands to train students while using limited resources. In the past, the fire service has been creative in developing methods and practices to accomplish drills or tasks with minimal cost. For example, when an acquired structure is not available for training on overhaul, it is easy to construct a wooden simulator from scrap 2-inch by 4-inch (2 × 4) lumber and cover it with discarded gypsum board that can be obtained from a lumber supply shop, often at very little expense **FIGURE 9-5**. When foam is needed for a fire stream training session and expired or surplus foam is not available, you can make a homemade solution using dish soap liquid and water. Of course, if you are using something other than equipment or products that would normally be used in the evolution, make sure that they will not produce a safety hazard.

One item that cannot be minimized is personnel: An adequate number of personnel must be present to conduct the training safely. This is particularly true when conducting live fire training. Live fire training is the pinnacle of all firefighting training, but it is also one of the most hazardous drills. The hazard becomes even more pronounced when the drill occurs with fewer personnel than are acceptable to ensure a safe environment.

FIGURE 9-5 If an acquired structure is not available for training on overhaul, it is easy to construct a wooden simulator of scrap 2 × 4 lumber and cover it with discarded gypsum board.

Ethics Tip

Which messages are you sending to your fire fighters? Look at the equipment being used during training. Is it the best and most up-to-date equipment in the department? New fire fighters learn the culture of the organization from their first day on the department—and the equipment used for training is one method that communicates this culture. Is your department—and are you personally—communicating a culture that emphasizes safety?

When some fire departments replace front-line equipment, they send the outdated equipment that was replaced to the training division. For example, a truck company's chainsaw that does not have a chain brake might be replaced with the newest equipment, with the old saw going to the drill ground to train the new recruits. Does this practice reinforce a culture of safety for those students who may have never even used a chainsaw before? Similarly, new recruits need to build their confidence until they can determine margins of safety for themselves. Using a new ladder, for example, allows recruits to build confidence in how much safety is designed into fire service ladders.

As fire service instructors, we have an obligation to think about the future of the fire service. Specifically, we have an obligation to develop fire fighters who have internalized our concern for safety. Teach your students through words and actions that as fire fighters, we all have an obligation to see that everybody goes home.

Safety Tip

When using hands-on teaching methods, make sure an adequate number of fire service instructors are present to ensure students' safety as well as learning.

Theory Into Practice

Analysis of your audience will help you select the best method of instruction for your training session. Some training sessions will require you to demonstrate the proper techniques in a step-by-step fashion, pausing to allow for questions and to provide clarification. This method may be appropriate for entry-level or new-skills training. Other sessions will require you to review the objectives or desired outcomes of the training session along with key safety behaviors. This approach may be applied to training sessions for experienced fire fighters or when doing evaluations of skill levels.

VOICES
OF EXPERIENCE

My first class toward becoming a fire service instructor occurred as a newly appointed fire captain. At the time I was working on becoming a certified fire officer. Because the department did not have a designated training officer, the captains were assigned to do the bulk of the training. This is when I realized I was responsible for the training of staff I supervised, as well as their health and safety. This realization made me adopt a very serious mindset of personnel safety.

As I progressed through the ranks, I made it my personal goal not to become a focused point of a serious injury or line of duty death investigation or report. That goal allowed me to focus on training methods that did not allow for inappropriate behavior, activities or actions, or lack of proper personal protective equipment (PPE). During my teaching activities I have learned to stay current on the common practices of the fire service; stay current with the National Fire Protection Association's standards, and any other national, state, and local standards and department policies; and keep my curriculum current. I have run curriculum past department training committees to get their opinion on leading edge materials and how they apply to the agency.

> **"You are in a position to influence the agency's safe operations through the way you conduct your training. Take that responsibility seriously."**

The fire service is a high-risk occupation. Our training must remain current and it will involve a range of incidents from low risk–low frequency activities to high risk–low frequency activities. My goal—and it should be every instructor's goal—is to stop preventable injuries and deaths.

You, as the instructor, are in a position to influence the agency's safe operations through the way you conduct your training. Take that responsibility seriously, because fire fighters perform as they are trained. I have seen this countless times, from forgetting to wear their protective hood or gloves in a stressful situation, to not conducting a proper size-up of the incident before starting operations.

Adopt the Sixteen Fire Fighter Life-Safety Initiatives; embrace the national, state, and local standards and mandates; adhere to department policies; develop plans; review curriculum before training to insure that it is current and safe; deploy safety officers as needed; and make everyone is a safety officer for each other.

Train like your life depends on it, because it does.

Michael Ridley
Fire Chief, Retired
Wilton Fire Protection District
Wilton, CA

Live Fire Training

Perhaps one of the most hazardous but exciting portions of training involves live fire training evolutions. Safety considerations during live fire training should be the highest priority for all instructors involved with this type of high-risk training.

Due to past fire fighter fatalities involved with live fire, the NFPA developed a standard that is to be followed during all live fire training evolutions. NFPA 1403, *Standard on Live Fire Training Evolutions*, describes all aspects of the training process involving live fire training. It identifies the positions that must be staffed when conducting live fire training and the prerequisite training that students must complete before they can participate in any live fire exercises addressed by this standard. This standard applies to all types of live fire training, including acquired structures, gas-fired and non-gas-fired live fire props, and exterior live fire props. It includes the permit process, pre/post training, and documentation of the training process. NFPA 1403 must be consulted before the live burn exercise and followed to the letter when conducting live fire drills. As an instructor, you have a responsibility to ensure that all positions required by this standard are staffed at your training sessions. Do not attempt the drill with anything less than a full complement of trained and qualified staff.

Positions that must be filled during live burn activities include a designated safety officer, who has received training and background in the responsibilities of this position during live fire training. This safety officer is responsible for intervention in and control of any aspect of the operation that he or she determines to present a potential or actual danger at all live fire training scenarios. An instructor-in-charge is required to plan and coordinate all training activities and is responsible for following NFPA 1403. An additional position identified in this standard is the ignition officer, who is responsible for igniting and controlling the material being burned. The ignition officer coordinates all fire-lighting activities with the instructor-in-charge, the safety officer, and the fire control team.

The proper student-to-instructor ratio when conducting training involving live fire is recommended to be 5:1. In some training sessions involving multiple fire departments, you will need a large pool of fire service instructors to assist you and maintain the 5:1 ratio of crew members to instructors. This 5:1 ratio must be maintained throughout the evolutions and is set so that the typical span of control is not exceeded.

NFPA 1403 states that the agency conducting the live fire training is responsible for identifying the qualifications for instructors who direct live fire training, but experience has demonstrated over time that senior instructors and fire fighters should staff each of the positions as identified in the standard. In addition to being qualified, each person who carries out these functions should be properly trained.

NFPA 1403 identifies specific positions to be staffed during live fire training due to the nature of the risks involved while working around fire. This same "safety" mind-set should apply during any type of high-risk hazard training, such as high-angle, swiftwater, or confined-space training. Having an experienced instructor-in-charge with appropriate safety officer(s) and a low student-to-instructor ratio will enhance a safe training environment and promote a safety culture.

■ Other Training Considerations

In many fire-ground incidents, if a particular position is not filled, then the incident commander is responsible for completing all of the functions of that position. Likewise, in training, the instructor-in-charge must perform the duties of the safety officer and assistant fire service instructors if these positions are not filled. If company officers are available to assist with training, the responsibilities associated with these positions must be covered in depth before training.

You must pay attention to details when planning and conducting drills where students may face a risk of injury. With practical drills, many variables come into play and many things can go wrong, resulting in an injury to a student. One way in which you can minimize that possibility is to pay special attention to every detail of the drill. Mentally walking through the evolution during the planning session and drawing upon your own experiences or case reports of actual incidents can help develop your awareness of what might potentially go wrong. This review will allow you to create backup plans or, in the best case, to eliminate the hazard altogether.

When the drill includes large numbers of people whom you may not know (e.g., in disaster drills where many outside agencies are invited to participate), your preplanning skills will be put to the test. Asking for assistance from other fire service instructors or agencies that have conducted similar exercises will help you in covering all the bases during the planning phase. As the instructor-in-charge, run through each aspect of the session, always asking, "What might go wrong here?" Aspects of the training to be considered include the weather, topography, equipment being used, and steps of the task.

> ## Safety Tip
>
> When relying on assistant fire service instructors, ensure that they have the appropriate level of knowledge and skills to fulfill their roles based on your department standards.

Additional fire service instructors should be employed at training sessions where the following conditions are present:

- Extreme temperatures are expected.
- The qualifications of the students are unknown, which could require more one-on-one time.
- Large groups of people will participate.
- Training will take place over a long duration.
- Training involves complex evolutions and procedures where additional instruction and safety oversight may be valuable.

Establish a file of newspaper articles, magazine articles, near-miss reports, and NIOSH reports pertaining to fire training accidents. Refer to this file as you prepare your lesson plans to ensure that the mistakes that led to a prior tragedy are not repeated in your own training session.

Developing a Safety Culture

Fire service instructors are in a position to select personnel to assist with their training programs. Often company officers are used to help deliver the session or watch over a skill being evaluated. As an instructor, you have the responsibility to know about the people you are asking to help conduct this training. What are their qualifications? What are their values concerning fire service safety? Despite your best efforts to keep your training safe, you cannot be everywhere at all times. Thus you must rely on other fire service instructors to help supervise the drill.

Fire service instructors who are not competent in or confident with the subject being presented should not be placed in a situation where they attempt to teach that material to others. The motivation of your subordinate fire service instructors must be assessed as well. An instructor who has a propensity to show off or try to prove that he or she is superior to students is an accident waiting to happen. Often the focus of such an instructor is not on the students or their safety. Know your staff, their attitudes toward safety, the way they operate on the fire ground, their motivations for being fire service instructors, and their teaching methods. All support personnel for the drill, including those who simply assist in prop construction, should be included in this assessment, because this information is essential to conduct training safely.

■ Anticipating Problems

As part of the preparation for instruction, you need to assess potential problems and rectify them before they result in an injury. You need to train your eye to pick out these hazards on first glance. It may be easy for a company to throw up a ladder and leave it unheeled; you should be able to foresee the potential consequences of that action and take steps to make sure the hazard is addressed. Most experienced fire service instructors have this quality already. The more fire-ground experiences you have, the better you will become at recognizing the small details that might escalate into a full-scale disaster.

Your fire-ground experiences translate to the drill ground, and you should apply your firefighting experiences to every evolution you instruct. Tell your fire fighter students what can go wrong, what to watch out for, and how to be prepared for changing conditions throughout the training evolution. If a smoke generator will be used for the training session, let students know how it compares to smoke at a real incident,

Is it ethical for a fire service instructor to teach a topic without the practical experience to support it? Can a fire fighter teach leadership skills to company officers? Can an instructor without aircraft rescue firefighting (ARFF) experience teach a class solely through extensive research? Can a fire officer effectively teach other fire officers solely based on his or her own personal experiences without receiving any relevant education in the subject? You will likely have to come to grips with these questions at some point during your tenure teaching other fire fighters.

As a fire fighter, you must have confidence in yourself—you would not be successful in the fire service without it. Nevertheless, that confidence can be detrimental to you and others if you are not aware of your limitations. As a fire service instructor, you have an obligation to your students, your department, and your profession. Teaching based solely on experience without research is nothing more than a history lesson given from a single perspective. Conversely, teaching based solely on research without experience is a regurgitation of other people's opinions without any grounding in reality.

If you are asked to teach a course in an area where you have little background, you owe it to everyone to decline the request. If you are asked to teach a course in which you do not have enough interest to research the topic and obtain a broader understanding of it, you also owe it to everyone to decline the request. You have an ethical responsibility to provide your students with high-quality training. Do not fall into the trap of teaching a class in which you are not interested or experienced, hoping that your students will not notice—they will.

if the situation may be significantly different from the training experience.

Often training tragedies are not the result of a single catastrophic event, but rather represent the culmination of several small safety violations that combine to create a fire training tragedy. The instructor-in-charge may identify that an attack line of a smaller size is appropriate for the amount of fuel being used if it is advanced without delay. But hose lines can get kinked, water may be slow to reach the nozzle from the pump, and conditions may deteriorate and lead to a delayed attack that can bring the fuel load above the capacity of the slowly advancing attack line. Collectively, these elements are a prescription for disaster.

Changes in wind or weather, crew fatigue, or student motivation may also affect the evolution. Your guard must stay up until the evolution is complete. You might imagine that if just one thing goes wrong you can quickly correct it, but the escalation of events often occurs so rapidly that you cannot gain control of the one small mistake before an injury takes place. If you see a potential hazard, correct it immediately.

Safety Tip

Review the causes of accidents for the type of training you will be conducting to ensure you have taken all appropriate preventive measures.

■ Accident and Injury Investigation

Despite all of your careful preparation, sometimes an accident might happen. In such circumstances, it is important to conduct a comprehensive investigation of the incident. A key factor in any investigation is the attitude held by those conducting the investigation. The purpose of such an investigation should be to find the reasons for and contributing factors to the accident or injury—not to assign blame. If the main focus is on finding the cause of the accident, then individuals will be more honest and open to this important process.

The investigation should include the date, location, events prior to and during the incident, equipment involved, and personnel who were present when the incident occurred. This information will help the investigator determine the cause of the injury and enable the agency and training personnel to put control measures in place to prevent future occurrences. Keep all injury investigation notes, doctors' reports, and any information generated during the investigation on file. Periodically review these files for any emerging trends that might cue you to change a policy or procedure that is contributing to injuries. When types of injuries are compared over a period of time, trends in their incidence often become apparent. You should see that modifications are made to the training program based on this information.

■ Student Responsibilities for Safety During Training

Every student participating in training has a stake in ensuring his or her own personal safety. Each student is responsible for his or her actions as well as the actions of other crew members, training team members, and fellow students when it comes to addressing hazardous conditions. Students should be knowledgeable about the training to take place and should meet all of the prerequisites specified before participating in the drill. During high-risk training, this background becomes even more important because of how an unprepared student's performance can affect other students involved in the training. Several case studies have dealt with improperly prepared students contributing to the injury of fellow students.

Students should also adhere to all PPE requirements specified for the evolution and wear all other assigned safety devices or equipment. Observance of all safety instructions and safety rules throughout the training session will assist in maintaining the overall safety profile of the training session.

Each student should also be honest regarding his or her strengths and weaknesses. If a student is not prepared for the skill or evolution, he or she should notify you so that other students or instructors are not put at risk.

Legal Considerations

Fire service instructors are held accountable for planning, lack of planning, adherence to standards and codes, and decision making during the training process. The *Legal Issues* chapter describes the many types of liability that arise during training. As a fire service instructor, you have a responsibility to stay informed about all laws and standards governing fire fighter training and safety. These laws and standards change over time, so you must keep up-to-date with the most current editions. Fire service instructor networks, Web sites, and electronic bulletin boards are all popular methods of staying abreast of these ever-changing rules.

In many cases, the standard of care or best practices relating to training safety may be tied to your department's SOPs. As mentioned in earlier chapters, those SOPs must be referenced within the scope of training along with the application to the individual job description.

Another legal test that may be applied to training is the "reasonable person" concept. This standard tests your decisions against the decisions made by others with similar training and background, asking what your peers would do when faced with the same situation.

Many of the NFPA standards in the 1400 series provide guidance in the application of different types of training exercises. For example, NFPA 1451, *Standard for a Fire Service Vehicle Operations Training Program*, provides information about the development and delivery of a vehicle training program. Within that standard and others such as NFPA 1500, *Standard on Fire Department Occupational Safety and Health Program*, it is required that training on vehicle driving take place before the fire fighter is allowed to respond to emergencies as a driver. If the department does not adopt that policy as a standard practice, the failure to do so might be held against the department in a lawsuit.

Most of the safety protocols that govern fire fighter training are embedded in laws that describe broader topics, such as Occupational Safety and Health Administration (OSHA) rules on respiratory protection (or SCBA) and hazardous materials. The NFPA standard specific to conducting live fire training safely is NFPA 1403, *Standard on Live Fire Training Evolutions*, which addresses conducting live fire training in acquired structures, gas-fired live fire training structures, non-gas-fired live fire training structures, exterior prop fires, and Class B fuel fires. Many of the requirements for these exercises must also be applied to skills training.

NFPA 1403 was developed as a result of fire fighter fatalities and injuries in live fire training accidents. This standard is

Safety Tip

Strictly adhering to all recognized standards and laws for the type of training you are conducting will ensure that such training is conducted in a manner that will be safe for all participants.

almost entirely devoted to practices that must be followed to conduct live fire training safely. Although NFPA standards are not laws, you can be held accountable for adhering to them just the same. For example, a training officer was recently held criminally liable for a fire fighter fatality that occurred during live fire training because he did not follow the guidelines in NFPA 1403 when conducting the drill.

Laws and standards are intended to keep fire fighter training within a framework of safe practices. Some critics might suggest that the requirements place so many restrictions on the training that it is no longer realistic or relevant. This is simply not true. First and foremost, fire service instructors have a responsibility to keep the training safe and to follow procedures TABLE 9-2. The notion that you have to discard safety requirements to make the training realistic is contrary to the entire idea behind conducting training sessions. The standards, rules, and laws established for safe live fire activities must be adhered to during skills training as well.

Table 9-2	General Training Safety Procedures
Training Location	**Procedures**
Classroom training	Make sure all applicable department SOGs are referenced within your lesson plan. Ensure all participants are aware of facility safety procedures (e.g., fire alarm, smoking, restricted access areas). Identify trip, slip/fall, and other facility hazards.
Practical training	Specify all learning objectives and performances. Use incident action plans as appropriate for high-risk training such as live fire evolutions. State all required PPE for instructors and students. Identify all department procedures for skills being performed (use the same incident command system as required during actual operations). Identify and demonstrate any hazard control activities that have been taken or are in place. Review all applicable safety procedures (e.g., emergency procedures, mayday, backup assignments, communications, rehab, emergency medical services [EMS] standby or aid, accountability, and rapid intervention teams). Check all communications equipment. Follow all applicable standards for the training session (NFPA 1403 for live fire training).
Keys to safe and successful training	Identify companies/members who are in or out of service. Identify and document the lead instructor. Notify supervising officers of all training events. Brief all participants on all training expectations. Monitor the previously mentioned practical training safety factors throughout (act as instructor/safety officer/incident commander). Immediately stop all unsafe acts or conditions; do not try to train through them. Debrief all participants on how training expectations were met and identify opportunities to improve individual and company performance. Return all equipment and personnel to service refueled, recharged, rehabbed, and ready for the next response.

Fire Service Instructor II

Supervising Instructors

One of the responsibilities you have at the Fire Service Instructor II level is supervising other instructors during training that involves a high-risk hazard to either the student or the instructor. NFPA 1403 includes the position of instructor-in-charge, which is the senior instructor who directs and supervises other instructors during a live fire evolution. NFPA 1403 can also be used as an outline for other forms of training.

The process of supervising other instructors during the training process can be influenced by such factors as the material being presented and the qualifications of the instructors, the skill level of the students, and the number of students involved in the training process. A group of senior fire fighters reviewing ropes and knots being taught by one instructor will not need much supervision. In contrast, an instructor teaching advanced SCBA skills to a group of new recruits might need additional help and guidance from other instructors and perhaps someone in charge of the entire training session to make sure groups rotate and the class stays on time.

Many subjective factors can come into consideration when discussing the definition of high-risk training. Most often we associate high-risk training with areas such as high-angle/rope rescue or live fire, but depending on the material to be taught and the experience level of the students and instructors, other areas could be considered high risk or high hazard.

Training BULLETIN

Jones & Bartlett Fire District

Training Division
5 Wall Street, Burlington, MA, 01803
Phone 978-443-5000   Fax 978-443-8000
www.fire.jbpub.com

Instant Applications: Safety During the Learning Process

Drill Assignment

Apply the chapter content to your department's operation, its training division, and your personal experiences to complete the following questions and activities.

Objective

Upon completion of the instant applications, fire service instructor students will exhibit decision making and application of job performance requirements of the fire service instructor using the text, class discussion, and their own personal experiences.

Suggested Drill Applications

1. Review local policies relating to safety in training.

2. Read NFPA 1403, Standard on Live Fire Training Evolutions, and review the sample checklists used for conducting live fire training exercises.

3. Research fire fighter line-of-duty deaths in training accidents and review the recommendations of the investigators regarding ways to prevent similar occurrences.

4. Review the Incident Report in this chapter and be prepared to discuss your analysis of the incident from a training perspective and as an instructor who wishes to use the report as a training tool.

Incident Report

Figure 9-A The Pennsylvania State Fire Academy's non-gas-fired live fire training structure.

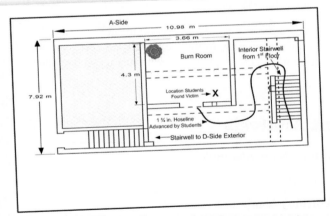

Figure 9-B The floor plan of the basement in the non-gas-fired live fire training structure at the Pennsylvania State Fire Academy.

The Pennsylvania State Fire Academy was conducting a Suppression Instructor Development Program (ZFID), an instructor development program for live fire training. The course covers instructional components including academy policies, NFPA 1403, building fires in the permanent live fire training facility, the instructor's role in student safety, emergency and rescue operations, and more. Instructors taking the course are already Instructor I and Fire Fighter II certified, and have completed the incident safety officer course. Candidates act as students, ignition officers, incident safety officers, incident commanders, and instructors, with an academy instructor evaluating each participant's performance.

The 2½-story residential non-gas-fired structure used for live fire training was built in 1993 and is used for structural firefighting training **(Figure 9-A)**. The building has a basement, ground floor, second floor, and an attic space. Burn rooms are located on the second floor and in the basement, where this incident occurred **(Figure 9-B)**. The burn room in the basement has only one entrance, which can be accessed by either the interior or exterior stairwell. Only wooden pallets and excelsior are used for fuel.

As part of the ZFID program, six scripted (planned) evolutions were organized in the burn building. Certain problems or emergencies were planned, and while the candidates had a basic understanding of the types of events, they did not know which event would occur during each evolution. The candidates were evaluated on their ability to handle the events.

The fire fighter who lost his life in this incident had already served as the evaluator of an instructor candidate in previous evolutions that day. On this last evolution, he was assigned to be the ignition officer for the burn room in the basement. He had more than 30 years in the fire service and had been an adjunct instructor at the facility for 7 years. He was wearing full protective clothing, including coat, pants, helmet, hood, gloves, boots, and SCBA with integrated personal alert safety system (PASS), and he had a portable radio.

The fatal event occurred during the last evolution. The scenario was to simulate a fire that had not been knocked down enough from the previous evolution.

Continued...

Incident Report Continued...

Pennsylvania State Fire Academy—2006

An instructor waiting to monitor the last crew was in the burn building. He reported good visibility with some smoke. While waiting, he heard the basement door open and saw the victim come out of the burn room and lie on the floor, pulling and moving his bunker gear about. The instructor asked the victim twice if he was okay. The victim answered, "It is hot as hell down there!" The instructor asked the victim if he was okay and if he wanted to go outside. Although the victim responded that he was okay, he was told to go outside. The victim declined, got up, and said that he was fine and would see the other instructor in the basement when the crew arrived. The victim went back down the basement steps. (Note: An estimated 1½ minutes elapsed from the time the victim came up the stairwell until he went back down to the basement.)

It is believed that the victim returned to the burn room and was carrying pallets to add to the fire. In NIOSH's report and following National Institute of Standards and Technology's (NIST) experiments, investigators hypothesized that excessive heat in the burn room caused a catastrophic failure of the lens of the victim's SCBA face piece. The face piece failure was a result of the heat conditions within the burn room and not a manufacturer's problem. The victim dropped the pallets and fell forward. The victim then struggled on the ground and crawled toward the exit.

As the next crew approached the burn room while advancing a hose line, they reportedly heard the victim moaning. They found him struggling on the floor in the right corner of the burn room. One of the candidates declared a mayday by radio, and the candidate serving as the incident commander deployed the rapid intervention crew (RIC). There was initial confusion whether this was part of the scenario or an actual emergency. The crew who found the victim immediately removed him from the burn room to outside, and emergency care was initiated. He was transported initially to a local hospital and then transferred by emergency medical services (EMS) helicopter to a regional trauma/burn unit, where he died two days later.

This was the first fatality of an instructor in a permanent live fire training prop, and only the second in such a facility. The fire fighters did not report the PASS device activating, as the victim was still moving. In this case, the PASS device did sound once removed from the victim; due to heat damage it could not be shut off, continuing to sound its alarm two days later.

Post-Incident Analysis

NFPA 1403 (2007 Edition) Noncompliant

No noncompliance issues with NFPA 1403 delineated in NIOSH report

NFPA 1403 (2007 Edition) Compliant

Structure in good condition (6.2.2)

Instructors had experience in the fire service and with live fire training (6.5.1)

RIC crew ready

An ambulance staffed by two EMTs on scene (6.4.11)

Other Contributing Factors

Instructor working alone

Repeated entries by instructor

No monitoring of interior temperatures

No method to rapidly ventilate building

Basement with no "at grade" exit

Wrap-Up

Chief Concepts

- You must always place students' safety ahead of any performance expectations for the training session.
- The 16 fire fighter life-safety initiatives should provide both guidance and motivation to you while you are developing and conducting any training. They highlight the importance of training as a strategy that can help reduce the number of fire fighter line-of-duty deaths.
- A fire service instructor serves as a role model for students. For this reason, he or she should always reinforce safety policies as part of the teaching process.
- Safety during training begins with safety in the classroom.
- In hands-on training, risk management practices must be applied to limit, reduce, or eliminate risk.
- As an instructor, you must always wear the PPE required for the drill being conducted. The appropriate level of PPE depends on both the potential hazards and your department's SOPs.
- Training should mirror as closely as possible the actual fire-ground operation. Medical care and monitoring, hydration, and cooling/warming materials are just a few of the rehabilitation considerations that are required to be available during live fire training events.
- Knowledge of the department's safety policies and any equipment manufacturer's recommendations is the key to keeping students safe during training.
- The example that you set as a fire service instructor sends a far more powerful message to your students than the theories and concepts you present in a lecture or drill.
- The fire service instructor is responsible for inspecting the equipment being used in training to ensure that it meets the required standards prior to conducting the drill. Do not assume that the equipment is in safe and operable condition.
- Live fire training is the pinnacle of all firefighting training, but it is also one of the most hazardous drills.
- NFPA 1403, *Standard on Live Fire Training Evolutions*, is the nationally accepted standard for all live fire burn training.
- Fire service instructors are responsible for both the physical and emotional well-being of students while in training.
- Fire service instructors are held accountable for their planning and adherence to standards and SOPs during the training session.

Hot Terms

Ignition officer An individual who is responsible for igniting and controlling the material being burned at a live fire training.

Instructor-in-charge An individual who is qualified as an instructor and designated by the authority having jurisdiction to be in charge of fire fighter training.

Safety officer An individual who is appointed by the authority having jurisdiction and is qualified to maintain a safe working environment at all training evolutions.

References

National Fallen Firefighters Foundation. (2013). 16 Fire Fighter Life Safety Initiatives. Emmitsburg, MD: National Fallen Firefighters Foundation. http://www.lifesafetyinitiatives.com/

National Fire Protection Association. (2012). *NFPA 1041: Standard for Fire Service Instructor Professional Qualifications*. Quincy, MA: National Fire Protection Association.

National Fire Protection Association. (2012). *NFPA 1403: Standard on Live Fire Training Evolutions*. Quincy, MA: National Fire Protection Association.

National Fire Protection Association. (2007). *NFPA 1403: Standard on Live Fire Training Evolutions*. Quincy, MA: National Fire Protection Association.

National Fire Protection Association. (2013). *NFPA 1451: Standard for a Fire Service Vehicle Operations Training Program*. Quincy, MA: National Fire Protection Association.

National Fire Protection Association. (2013). *NFPA 1500: Standard on Fire Department Occupational Safety and Health Program*. Quincy, MA: National Fire Protection Association.

National Fire Protection Association. (2008). *NFPA 1584: Standard on the Rehabilitation Process for Members During Emergency Operations and Training Exercises*. Quincy, MA: National Fire Protection Association.

You are planning a series of company-based training evolutions on using hand and power tools on various props, scrap materials, and simulators that are readily available to the department. A variety of department members and company types will attend the training sessions, and you want to ensure that a safe and effective training session takes place. Some of the concerns you have are making sure that all proper and required PPE is being used and that the tools and equipment are being carried and operated in a safe manner.

1. What should be referenced in the drill planning to help identify the proper PPE requirements for each planned evolution?
   - **A.** NFPA 1500
   - **B.** Department SOPs
   - **C.** Students' opinions and experiences
   - **D.** Instructors should determine what PPE levels are required based on student experience.

2. Which NFPA standard covers live fire training?
   - **A.** NFPA 1500
   - **B.** NFPA 1403
   - **C.** NFPA 1503
   - **D.** There is no NFPA standard for live fire training.

3. What is the recommended student-to-instructor ratio for high-risk training?
   - **A.** 1:1
   - **B.** 2:1
   - **C.** 5:1
   - **D.** 10:1

4. Which topics should you cover in safety briefing statements prior to beginning the training sessions?
   - **A.** PPE requirements
   - **B.** Proper lifting and carrying techniques
   - **C.** Proper use of tools and equipment
   - **D.** All of the above

5. According to the Life Safety Initiatives, if a fire fighter sees another fire fighter doing something in an unsafe manner or operating in an unsafe area, what should he or she do?
   - **A.** Report it to the supervisor for action.
   - **B.** Discuss it after the incident.
   - **C.** Immediately stop the fire fighter from doing it.
   - **D.** Recommend that a policy be developed to prevent reoccurrences of the action.

Evaluation and Testing

PART

III

Evaluating the Learning Process

Fire Service Instructor I

Knowledge Objectives

After studying this chapter, you will be able to:

- Describe standard testing procedures. (NFPA 4.5.1 , NFPA 4.5.2) (p 232)
- List the types of written examinations. (p 231)
- Describe the methods used by the instructor to grade exams. (NFPA 4.5.3) (p 234)
- Describe how to administer testing. (NFPA 4.5.1 , NFPA 4.5.2) (pp 232–234)
- Describe the legal considerations for testing. (NFPA 4.5.2) (pp 230–231)

Skills Objectives

After completing this chapter, you will be able to perform the following skills:

- Demonstrate how to grade student evaluation instruments. (NFPA 4.5.3) (p 234)
- Demonstrate how evaluations are proctored and results are recorded. (NFPA 4.5.2 , NFPA 4.5.4) (pp 232–234)
- Demonstrate the methods for providing feedback on evaluation performance to students. (NFPA 4.5.5) (pp 234–235)

Fire Service Instructor II

Knowledge Objectives

- Describe standard testing procedures. (p 232)
- Describe how to administer testing. (pp 232–234)
- Describe how to analyze a student evaluation instrument. (NFPA 5.5.2) (pp 237–238)
- Describe the role of testing in the systems approach to the training process. (pp 237–238)
- Describe the types of written examinations. (pp 231, 240, 242–247)
- Describe the legal considerations for testing. (pp 230–231)
- Describe how to develop student evaluation instruments. (NFPA 5.5.1 , NFPA 5.5.2) (pp 238–240, 242–247)

Skills Objectives

After completing this chapter, you will be able to perform the following skills:

- Demonstrate how to grade student evaluation instruments. (p 234)
- Demonstrate how evaluations are proctored and results are recorded. (pp 232–234)
- Demonstrate the methods for providing feedback on evaluation performance to students. (pp 234–235)
- Demonstrate how to prepare an effective exam for student evaluation. (NFPA 5.5.2) (pp 238–240, 242–249)

Fire Service Instructor III

Knowledge Objectives

There are no knowledge objectives for Fire Service Instructor III students.

Skills Objectives

There are no skills objectives for Fire Service Instructor III students.

You Are the Fire Service Instructor

As a fire service instructor, you believe in the four-step method of skill training: prepare, present, apply, and evaluate. Your testing materials, both in the classroom and on the job, reflect a complete and fair assessment of the required knowledge and skills for your course. You have been a fire service instructor for many years and are aware that in each drill you instruct and in many classes you teach, there will be students who do not like to be in a classroom. Other students may enjoy the challenge of learning something new but have difficulty when they have to perform on-the-job skills or take written tests.

You are about to begin a drill session that will conclude with an examination covering the main objectives of the class. You want to make sure that all of the objectives are properly covered and the class members are well prepared for the exam. The students will range from entry-level personnel to seasoned veterans with many years of experience. In addition, these exam results will be compared to other similar classes taught by other instructors.

1. What will you do in each step of the four-step process to ensure you are prepared for your training class?
2. Which evaluation and testing methods should you use for the group learning basic skills?
3. What will you do to help ensure that all members are successful?

Introduction

Testing plays a vital role in training and educating fire service personnel. A sound testing program allows you to know whether students are progressing satisfactorily, whether they have learned and mastered learning objectives, and whether your instruction is effective. Without a sound testing program, there is no way to assess student progress or to determine which learning objectives have been mastered. In addition, it is impossible to determine the effectiveness and quality of training programs without evaluation.

Fire departments across the country continue to encounter problems with testing programs. Common problems in testing include the following issues:

- Lack of standardized test specifications and test format examples
- Confusing procedures and guidelines for test development, review, and approval
- Lack of consistency and standardization in the application of testing technology
- Failure to perform formal test-item and test analysis
- Inadequate fire service instructor training in testing technology

This chapter is designed to provide the necessary information and training to help overcome these and other common weaknesses, thereby strengthening your testing program.

Fire Service Instructor I and II

Legal Considerations for Testing

Testing for completion of training programs, job entry, certification, and licensure is governed by certain legal requirements and professional standards in the United States. These legal requirements and professional standards are readily available and should be important reference documents for anyone involved in the development of test items, construction of tests, administration of tests, analysis and improvement of tests, and maintenance of testing records. Because each state and municipality will have different laws and ordinances, be sure to establish a working knowledge of applicable laws and standards relating to testing and evaluation. The *Legal Issues* chapter covers a much more expanded explanation of legal exposures and laws that govern test administration.

The instructor should be aware of laws and standards that apply to some of the following areas:

- Entry-level testing for new candidates for written and physical ability testing
- Certification testing rules and test candidate requirements
- Promotional testing criteria and the evaluation of promotional test results
- Continuing education testing requirements for emergency medical services (EMS), firefighting and other certifications or licenses
- Record keeping, scoring, and test record storage

The three most important reference documents are *Uniform Guidelines for Employee Selection*, published by the Equal Employment Opportunity Commission; *Standards for*

Educational and Psychological Testing, published by the American Psychological Association; and *Family Educational Rights and Privacy Act*, published by the U.S. Printing Office and other governmental organizations. Each of these references should be available in every Division of Training's library for fire service instructors to study.

Purposes and Types of Tests

Tests should be used for the following three purposes:

1. To measure student attainment of learning objectives
2. To determine weaknesses and gaps in the training program
3. To enhance and improve training programs by positively influencing the revision of training materials and the improvement of instructor performance

The three basic types of tests are written, oral, and performance.

Written Tests

Written tests can be made up of eight types of test items, which are discussed later in this chapter:

- Multiple choice
- Matching
- Arrangement
- Identification
- Completion
- True/false
- Short-answer essay
- Long-answer essay

FIGURE 10-1 Oral tests assess knowledge, skills, and abilities before a performance test.

these aspects are not conveniently measured through other types of tests.

An example of an oral test follows:

Given specific incident types, such as structure fire, motor vehicle accident, and EMS calls, name the types and levels of personal protective equipment necessary for each type of response and the safety-related considerations while wearing them.

Oral tests used in conjunction with performance tests should focus on critical performance elements and key safety factors. The overhead or directed questioning technique seems to work best when a group of students are preparing to take a performance test. Specific oral test questions can be used when a student is taking a performance test. Oral tests used in conjunction with performance tests are usually not graded, but rather are designed to reinforce key safety factors for students and to ensure critical safety factors are being confirmed by the instructor/proctor.

> ### Ethics Tip
>
> The multiple-choice exam is typically the most common form of written test because it is the most objective and thus the most reliable for testing. Because there are many different learning styles, however, some students are better able to demonstrate their competency in a different format than other students. If the primary purpose of the testing process is to determine competence, to what extent should you consider using multiple testing methods to determine competence?

> ### Teaching Tip
>
> It is important to outline the required responses to oral test questions to ensure that proper feedback is given to the student.

Oral Tests

Oral tests, in which the answers are spoken either in response to an essay-type question (oral content) or along with a presentation or demonstration (oral presentation), are not used extensively in the fire service. However, this type of test has a place in technical training. Oral tests are given in a structured and standardized manner to determine students' verbal response to assess knowledge, skills, and abilities important on the job **FIGURE 10-1**. They primarily focus on safety-related skills to be performed. The oral test allows students to clarify answers and you to clarify questions. Oral tests are effective in ascertaining student knowledge and understanding when

Performance Tests

Performance tests, also known as skills evaluations, measure a student's ability to perform a task. This category of tests includes laboratory exercises, scenarios, and job performance requirements (JPRs). Any performance test should be developed in accordance with task analysis information and reviewed by subject-matter experts (SMEs) to ensure technical accuracy in terms of the actual work conditions.

Following is an example of a performance test item:

Perform a daily inspection of self-contained breathing apparatus (SCBA).

Standard Testing Procedures

Within a training department, the test development activities should adhere to a common set of procedures. These procedures should be included in standard operating procedure (SOP) or standard operating guideline (SOG) formats. These valuable documents should offer test development concepts, rules, suggestions, and format examples. The procedures and guidelines specified in the SOP/SOG document should be used by fire service instructors throughout the test-item development, test construction, test administration, and test-item analysis processes.

> **Ethics Tip**
>
> The proper handling of student information, including test grades, may be covered under laws that apply to you and your department. Know your responsibilities and play it safe by maintaining strict confidence of all student information.

Proctoring Tests

You will gain both knowledge and skills in test-item writing and test development when you complete the Fire Service Instructor II requirements. Now, you should explore the other aspects of a sound testing program. Proctoring tests encompasses much more than just being present in the testing environment; it requires specific skills to be performed professionally. Types of tests being proctored require different proctoring skills and abilities.

■ Proctoring Written Tests

The following are some suggested procedures for proctoring written tests:

- Arrive at least 30 to 45 minutes prior to the beginning time for the test.
- Make sure the testing environment is suitable in terms of lighting, temperature control, adequate space, and other related items.
- During the arrival of test takers, double-check those who should be in attendance by checking identification documents and record their presence. (You may develop a log or other means to document arrival time and departure time.)
- Maintain order in the testing facility. Discourage any activities that might potentially be interpreted as cheating, and do not provide any information to any single individual that is not also provided to the entire group.
- Maintain the security of all testing materials at all times.
- Provide specific written and an oral review of all test-taking rules and behavior guidelines during the testing period.
- Remain objective with all test takers. Do not show any form of favoritism.
- Answer questions about the testing process and the test itself.

- Do not answer an individual question about the content of the test unless that information is also shared with the entire group.
- Monitor test takers during the entire period of the test.
- Discipline anyone who becomes disruptive or violates the rules.
- Require all test-item challenges to be made prior to test takers leaving the room. Advise that their challenge will be noted and addressed after all of the tests are scored.
- Collect and double-check answer sheets or booklets to ensure that all information is properly entered and that any supporting materials are returned before allowing test takers to leave the room.
- Never leave the testing environment for any reason.
- Inventory all testing materials and return them to the designated person in the department.

■ Proctoring Oral Tests

Oral tests are generally given in conjunction with performance tests. Even so, there are preferred procedures that need to be followed to maximize the effectiveness of the oral test:

- Determine the oral test items to be used and the type of oral question technique that will be employed (e.g., overhead, direct, rhetorical).
- Make sure the oral test items are pertinent to the task to be performed.
- Focus on the critical safety items and dangerous steps within the performance.
- If the test taker misses the oral test item, redirect the question to another performer. If you are conducting a one-on-one oral test, then provide the test taker with on-the-spot instruction. Safety knowledge and dangerous task performance steps must be clear and mastered before performance begins.
- Use a scoring guide for the oral test and record any difficulties and lack of knowledge on the part of the test taker(s).
- If the group or individual has knowledge gaps regarding the critical safety items or dangerous performance steps, *do not* proceed to the performance test. More training and testing are required.
- Make a training record of oral test results.

■ Proctoring Performance Tests

Performance test proctoring requires considerably different skills than those needed to proctor the written and oral test. Proctors must have keen observational skills, technical competence for the task(s) being performed, ability to record specific test observations (including deficiency and outstanding performances), ability to foresee critical dangers that may lead to injury to the performer or others, and an objective relationship to the test takers **FIGURE 10-2** . Following are some suggested procedures, including pre-test preparations:

- Arrive at the test site at least one hour before the performance test.

Candidate: _____ Date: _____

ID#: _____

Skill Drill 16-25
Fire Fighter I, 5.5.2

No.	Task Steps	First Test		Retest	
Performing an Accordion Hose Load					
Evaluator Instructions: The candidate shall be provided with a fire apparatus, supply hose, and gloves.					
Task: Performing an accordion hose load.					
Performance Outcome: The candidate shall demonstrate the ability to do an accordion hose load.					
Candidate Directive: "Properly perform an accordion hose load."					
		P	F	P	F
1.	Determines whether hose will be used for forward lay or reverse lay.				
2.	For forward lay, places male hose coupling in hose bed first. For reverse lay, places female hose coupling in hose bed first.				
3.	Starts hose lay with coupling at front end of hose compartment.				
4.	Lays first length of hose in hose bed on its edge against side of hose bed.				
5.	Doubles hose back on itself at rear of hose bed. Leaves female end extended so two hose beds can be cross connected.				
6.	Lays hose next to first length and brings it to front of hose bed.				
7.	Folds hose at front of hose bed so bend is even to edge of hose bed. Continues to lay folds of hose across hose bed.				
8.	Alternates lengths of hose folds at each end to allow more room for folded ends.				
9.	When bottom layer is completed, angles hose upward to begin second tier.				
10.	Continues second layer by repeating steps used to complete first layer.				
Retest Approved By:		Retest Evaluation:			

Evaluator Comments: _____   Candidate Comments: _____

_____   _____
_____   _____
_____   _____
_____   _____

_____   _____   _____   _____
Evaluator   Date   Candidate   Date
Retest Evaluator   Date   Retest Candidate   Date

FIGURE 10-2 A skills evaluation sheet can be a useful tool in evaluating a performance.

- Check the test environment to make sure all needed tools, materials, and props are present and in good working order.
- Determine whether the performance test will require an oral test before the actual performance of the task. (Obtain oral test items if required.)
- Check the test takers' identification as they arrive and record their presence.
- Review the test procedures and ask for questions from the test takers.
- Verify that all test takers know what will be expected during performance.
- Begin the performance test. Have the student(s) tell you exactly what they will be doing before they undertake each step in the performance. (This is a critical step—it can help prevent an accident or unsafe performance before it happens.) In certification testing, missing safety-related steps or not knowing critical performances that may cause injury or damage to equipment is cause to terminate testing and refer the test taker for more training. This should also be true for the final performance test in a training program that leads to licensure.
- Remain silent and uninvolved in the task performance. Your job is to verify safe and competent performance.
- Record test results. You may provide technical information and critique the performance if your institution or agency permits proctor interaction after the test.
- Provide test results to the person(s) designated to receive them.

Cheating During an Exam

Student cheating during the examination process is an unfortunate side product of the evaluation process. How cheating is addressed should be based on department policy developed by human resources personnel, the agency's legal advisor or attorney, and the chief of the department. Cheating by an individual reflects on the character of that individual and in many departments can affect employment status, raises, or promotions.

In general, a policy that addresses cheating will contain procedures to follow if a student has been observed cheating. In most cases, if a fire service instructor observes a student cheating, the student should be asked for an explanation of his or her conduct. If the fire service instructor determines that the student was indeed cheating, the student should be asked to leave the test site entirely or be given a verbal warning and be allowed to continue with testing. Again, the actions taken should follow department policy.

The fire service instructor must document what was observed, what the student said, and what the fire service instructor allowed the student to do next. This information should be passed to a supervisor, who will review what happened and then follow up as necessary with additional interviews of the fire service instructor and the student involved. The supervisor will determine what happens from this point. If it is determined that the student did cheat, then department policy will address the consequences. If it is determined

that the student did not cheat, then arrangements should be made to allow the student to take the test again. Regardless of the consequences, the department should have a written policy to address how to handle possible cheating during an examination.

Grading Student Oral, Written, and Performance Tests

The proper grading of oral, written, and performance tests is an essential function of the fire service instructor. Students will desire feedback on their performance; likewise, instructors will want to know how successful their instructional efforts were. Class answer sheets or skills checklists will need to be evaluated against master answer keys and then the answer keys will need to be properly secured for course validity purposes. Graded exams should be stored properly so that completed exams cannot compromise future offerings of the same test and the confidentiality of test scores remains in place. Elimination of bias in the grading process is necessary, especially when dealing with exam question types that do not have an objective format basis in their structure. After the exams are graded and scores are recorded, do not hesitate to secure the completed exam sheets and answer keys.

Reporting Test Results

■ Confidentiality of Test Scores

After the completion of the testing process, student scores must be maintained for various reasons. These test results should be protected using strict security measures. Electronic results should be password protected, and hard copies should be kept safe and secure. Test results should be released only with permission of the student who has completed the testing process.

When you release test scores to a class, you should do so on an individual basis. This sharing of the scores should be done in a private session, on a one-on-one basis, or in a personal letter to each student. The practice of posting scores on a bulletin board with Social Security numbers attached to them as identifiers is unsatisfactory and illegal.

■ Providing Evaluation Feedback to Students

Evaluation of examinations closes with feedback on the testing results with the students. This information must be considered confidential and should be completed in a timely manner using the evaluation data available to the instructor. All feedback should be specific enough for the students to understand the following points:

- The overall score or grade
- Objectives that were missed and those that were met
- Skills that they did not complete to the learning objective standard
- What they may do to modify their behavior and performance in the future
- Any additional study areas or retesting requirements on missed items or failed skill evaluations

Grading scales should be part of the scoring process. When receiving feedback, the student may want access to score tables or percentage breakdowns for letter grades. A written policy should exist that deals with challenges to examination scores or practical skill failures to guide the instructor on which actions should be taken during such a challenge.

Ethics Tip

Determine whether your organization has a policy on how to handle suspicions of cheating and follow it to the letter. Failure to do so is unfair to the student and can get you into trouble.

Fire Service Instructor II

Test Development Standards and Requirements

■ High-Stakes Examinations

According to the Joint Committee on Standards for Educational and Psychological Testing of the AERA, APA, and NCME, high-stakes tests are "used to provide results that have important, direct consequences for examinees, programs, or institutions involved in the testing" (APA, 1999). Some practical examples for fire services and EMS include the following tests:

- **Final training test.** Typically, final training tests are used to support the decision to pass or fail a student at the end of the training program. The importance and consequences of these tests can have an effect on whether the fire fighter or medical first responder can qualify for a certification test or licensure examination. These tests *must* be supported with job-relevant documentation and achieve high levels of test reliability. The consequences of failure from the examinee and training institution's point of view are important and direct.
- **Certification test.** Certification tests often require examinees to meet certain prerequisite conditions, such as completion of training, specified levels of experience in the job domain, or both. For the student to obtain a job in the domain, he or she is required to pass the licensure or certification examination. Failure to pass has important and direct consequences. These examinees typically are removed from job consideration or removed from rank-order lists of eligible applicants. These tests must be supported with job-relevant documentation and achieve high levels of test reliability.
- **Selection tests.** A testing method that determines whether an applicant is accepted or rejected for a particular educational or employment opportunity, a selection test must focus on the "must know" and "need to know" information necessary to perform a job successfully. Job relevance is a major focus of the test. All candidates should have appropriate training or experience prior to taking selection tests. These tests must be supported with job-relevant documentation and must achieve high levels of test reliability. These tests are not intended to be taken by inexperienced or untrained personnel or by the general public.
- **Promotional tests.** As the name implies, promotional tests are used to select the most qualified individuals from a group of candidates to be promoted to a higher position. Promotional tests must also demonstrate high levels of job relevance. In the fire and EMS, it is important that the criteria to be measured are based on either the SOPs/SOGs, the appropriate National Fire Protection Association (NFPA) professional qualification standards, or both. Job relevance must be documented and demonstrated. These tests must achieve high levels of test reliability. Results of these tests are often rank-ordered from the highest score to the lowest. Most organizations then promote from the top of the list to the bottom. It is very important for the list to be updated periodically by using a new series of promotional tests.

Testing in the high-stakes arena requires considerable work, professional guidance, and test-item documentation. It is not wise to undertake the development of high-stakes tests without considerable professional test developer involvement.

■ Low-Stakes Tests

A low-stakes test is one that has only minor or indirect consequences for examinees, programs, or institutions involved in the testing. An example of this type of testing is a test developed by a fire service or medic instructor to use during the training cycle. The trainees often receive immediate feedback and remediation on test items they fail to answer correctly. Such feedback is provided with the intention of raising the competence levels of participants in the training program.

■ Uniform Guidelines for Employee Selection

Uniform Guidelines for Employee Selection (1978, EEOC, Title VII) is the law of the land. It is primarily focused on employee selection/promotion procedures. Training and testing leading to hiring, promotion, demotion, membership in a group such as a union, referral to a job, retention on a job, licensing, and certification are all covered under these guidelines. Written and performance tests should be job related in such training programs. It is up to the user of the testing materials to

JOB PERFORMANCE REQUIREMENTS (JPRS)
in action

For the learning process to be complete and reach its maximum potential, it needs to be evaluated. Evaluation can come in many forms, ranging from written evaluations and hands-on evaluations to ongoing evaluation during the delivery of training. Numerous legal and ethical considerations exist during the evaluation phase of the learning process, and as the fire service instructor, you must be aware of these considerations at all times.

Instructor I	Instructor II	Instructor III
As part of the delivery of training, the Instructor I will be responsible for administrating written evaluation and practical skills sessions.	The Instructor II develops evaluation instruments to measure the student's ability to meet the performance objectives.	There are no JPRs at the Instructor III level for this chapter.
JPRs at Work	**JPRs at Work**	**JPRs at Work**
Administer evaluations of student learning and grade evaluations according to the type of evaluation used. Record evaluation scores and provide feedback to students regarding their performance.	Develop and analyze evaluation instruments to ensure that the student has achieved the learning objectives and that the evaluation is a valid measure of student performance.	The Instructor III would be responsible for the overall evaluation plan for a course. Written and skills evaluations will be used to check knowledge and skill ability related to each course objective.

 Bridging the Gap Among Instructor I, Instructor II, and Instructor III

The Instructor I will administer the evaluations developed by the Instructor II, observing the legal and ethical principles of test administration. The overall course evaluation plan responsibilities lie with the Instructor III, as does the responsibility for evaluation policy development. The Instructor I must learn to identify the desired outcomes of evaluations so as to improve his or her instructional skills. All levels of instructors will have to make sure that the evaluation measures the stated objectives for the lesson plan and that the evaluation correctly measures the learning process.

make sure that the tests are job related. If the tests are acquired from a professional publisher or a distributor of tests, then that source must be able to provide you with the job validity information. For example, if a question is based on the NFPA's professional standards for a particular topic and the question is current, then it would be defined as containing job-validity information.

As a fire service instructor, you must address job-content validity even for the test items that you personally develop and use on a day-to-day basis. Almost all training in the fire service leads to some sort of certification, pay raise, or potential for promotion. Therefore, job-content-related testing should be paramount when using any form of testing.

■ Test-Item Validity

When developing a test, it is crucial to make sure that each test item actually measures what it is intended to measure. All too often, tests contain items that are totally unrelated to the learning objectives of the course. The term *valid* is used to describe how well a test item measures what the test-item developer intended it to measure. Taking steps to ensure validity will prevent your test items from measuring unrelated information.

Four forms of test-item validity are distinguished:

1. Face validity
2. Technical-content validity
3. Job-content validity/criterion-referenced validity
4. Predictive validity

Although this chapter deals primarily with face validity, technical-content validity, and job-content validity, you will encounter other forms during your career as an instructor.

Face Validity

Face validity is the lowest level of validity. It occurs when a test item has been derived from an area of technical information by a subject-matter expert who can attest to its technical accuracy and can provide backup evidence. Each level of validity requires documentation and evidence.

Technical-Content Validity

Technical-content validity occurs when a test item is developed by a subject-matter expert and is documented in current, job-relevant technical resources and training materials.

Job-Content/Criterion-Referenced Validity

Job-content/criterion-referenced validity is obtained through the use of a technical committee of subject-matter experts who certify that the knowledge being measured is required on the job. This level of validity is often called *criterion-related validity*. In the case of criterion-referenced validity, you may use professional standards such as the NFPA professional qualification standards that are based on a job and task analysis. This level of validity should be carefully documented with each test item so that anyone can trace the validity information to the specific part of the NFPA standards and the reference material used to develop the question.

■ Currency of Information

Using the most current information that a student should know and use is as important as the information being relevant to the job.

Test Item and Test Analysis

Test analysis occurs after a test has been administered. Three questions are usually answered in the post-test analysis:

- How difficult were the test items?
- Did the test items discriminate (differentiate) between students with high scores and those with low scores?
- Was the test reliable? (Were the results consistent?)

If a test is reliable and the test items meet acceptable criteria for difficulty and discrimination, it is usually considered to be acceptable. The process used to make this determination is known as quantitative analysis, and it entails the use of statistics to determine the acceptability of a test. If a test is not reliable and the test items do not meet acceptable criteria for difficulty and discrimination, a careful review of the test items should be performed and adjustments made to the test item. This process is known as qualitative analysis, and it entails an in-depth research study performed to categorize data into patterns to help determine which test items are acceptable.

The purpose of test analysis is to determine whether test items are functioning as desired and to eliminate, correct, or modify those test items that are not. Bear in mind that it sometimes takes a cycle or two before real improvements are achieved in a particular set of test items. The best way to achieve acceptable test analysis results is to ensure validity of test items as they are developed and to follow a standard set of test development procedures as the test is constructed and administered.

The Role of Testing in the Systems Approach to Training Process

The systems approach to training (SAT) process was developed by the U.S. military during the early 1970s and 1980s. Many improvements have been made since the early days of this training approach. The effectiveness and efficiency of the SAT have been well documented in training journals, academic studies, and actual practice by leading businesses and industries in the United States and around the world.

Dr. Robert F. Mager, often referred to as "the father of the performance objective/learning objective," played a key role as a leading researcher during the development of performance-based or criterion-referenced instruction. Dr. Mager introduced the idea of learning objectives that have three distinct parts. First, the learning objective is task based by using verbs as part of the learning objective (which implies doing something). Sometimes these verbs are referred to as the action part of the learning objective. The second part of the learning objective deals with the condition(s) under which the action is to be performed. Third, the learning objective should contain a standard or measure of competence. An example of this approach is seen with the following learning objective:

> Given a self-contained breathing apparatus (SCBA), the fire fighter will don the SCBA and place it in operation within one minute.

All forms of testing and training should be based on learning objectives. In technical training, whether the tests are written, oral, or performance based, they must be based on the learning objective.

Within any training program, testing serves three purposes:

1. Measure student achievement of learning objectives.
2. Determine weaknesses and gaps in the training program.
3. Enhance and improve training programs by positively influencing the revision of training materials and improving instructor performance.

In a performance-based training program, the goal is for all students to master all learning objectives. This mastery is evaluated or measured using performance tests and safety-related oral tests. If this is the case, then what purpose does the written test serve in a performance-based training program? Written tests are used to take a "snapshot" of students' knowledge at predetermined points throughout a course of instruction. They may take the form of a precourse exam used to determine a student's level of knowledge upon entering a course, a midcourse test or formative exam to determine a midpoint level of knowledge, and a final exam or summative exam. These snapshots provide information on how well the student is progressing. If a student is not progressing satisfactorily, additional instruction can be provided to bring the student up to the required knowledge level. Testing then allows you to provide needed assistance before the student's lack of knowledge becomes critical. Lack of knowledge mastery is often a primary reason for poor performance or nonperformance of a learning objective and a task on the fire ground.

Test Item Development

The fire service is committed to safety and productivity through improved training programs and courses. Testing, as an important part of any approach to training, comprises two activities:

- The preparation of test questions using uniform specifications
- A quantitative/qualitative analysis to ensure that the test questions function properly as measurement devices for training

The development of objective test items requires the application of specific technical procedures to ensure that the tests are both valid and reliable. Validity means the test items (questions) measure the knowledge that the test items are designed to measure, and reliability means the test items measure that knowledge in a consistent manner.

The fire service is fortunate that it has the NFPA professional qualification standards, which provide an excellent basis for criterion-referenced training and testing. Make sure you carefully consider using the technical references in these standards as part of your own validity evidence and documentation. A test item has content validity when it is developed from a body of relevant technical information by a subject-matter expert who is knowledgeable about the technical requirements of the job. Remember that validity is the ability of a test item to measure what it is intended to measure, and reliability is the characteristic of a test that measures what it is intended to measure on a consistent basis. In other words, a valid and reliable test is one that measures what it is supposed to measure each time it is used.

To develop a test comprehensively, it is important to have a group (bank) of test items. The test-item bank can be organized on a course basis, by standard reference, by job title, by duty, or in any other way that seems sensible and logical. Such a database of test items permits a test developer or fire service instructor to alternate test questions or produce different versions of a test for use between classes or among students within a class. This balancing of test versions is important in today's legal environment and must be applied on a procedural basis to reduce the possibility of any discrimination occurring as a result of the testing program.

To begin building statistically sound tests, test items are written by subject-matter experts, who should be technically competent and actually working in the job for which the test items are being developed. The reason for this level of competence is readily apparent, given that changes are occurring at a more rapid rate than ever in the fire service. This initial activity provides the first level of validity: face validity. From this point, test items are analyzed on the basis of their individual performance in the test using test-item analysis techniques and data collected from responses of students taking the test.

Once data are collected from approximately 30 uses of a test item, the first test-item analysis can be completed. The test-item analysis data provide specific information for establishing a quality test-item bank.

■ Specifications

Test specifications permit fire service instructors and course designers to develop a uniform testing program consisting of valid test items as a basis for constructing reliable tests. Writing and developing test items is difficult for fire service instructors. Indeed, it is difficult to write a test item the first time that is both content valid and reliable. Thus most valid test items and reliable tests are the result of refinement through use. Test-item writers and test developers must follow basic rules or specifications to create a valid and reliable test. This section presents and briefly explains the test specifications, which are intended as a guide for developing test items and tests with initial technical content validity and testing reliability.

Written tests are designed to measure knowledge and acquisition of information. These tests have limitations, however. One major limitation is that testing knowledge and information does not ensure that a student can *perform* a given task or job activity; it simply means that the student has the knowledge *required* for performance. Skill development—that is, the ability to perform tasks or task steps—is obtained through actual performance. It may be achieved during evolutions on the training ground, and performance then documented via a skills evaluation during and at the end of training.

A simple analogy using golf explains the difference between testing for knowledge and information and actual performance of a task. For instance, a person can read numerous books, watch films, view videotapes, attend golf tournaments, and perform other information-gaining activities. By doing these things in the areas of golf, the person could probably make an acceptable score on a written test about golf. However, even with all this information and a perfect score on the written test, the person probably could not actually play a game of golf and score under 100 after 18 holes **FIGURE 10-3** .

This concept can be applied to the many tasks necessary to score well in any course. The basic skills associated with any task must be practiced and kept sharp for the student to be a competent performer. Of course, these same conditions apply to being a public service professional. Whether the job is in firefighting, policing, administration, supervision, or training, success is based on the ability to apply information to achieve levels of professional competence. A balance of must-know and need-to-know information and competent performance is the foundation of performance-based learning. The following specifications, examples, and test development criteria will help ensure that fire service instructors, subject-matter experts, and curriculum developers prepare written test items, construct written tests, and develop performance tests that measure what they should measure and function reliably over a long period of time.

> ## Teaching Tip
>
> There are three justifiable reasons for testing:
> - Determine trainee progress
> - Increase trainee knowledge and skill
> - Improve instructor performance

FIGURE 10-3 Obtaining knowledge and applying knowledge are separate skills.

> ## Ethics Tip
>
> With national standards, some of the areas that are covered may not apply to the students whom you are teaching. For example, part of the fire fighter standards requires a basic understanding of wildland firefighting principles; however, you might be teaching fire fighters in a large, urban department with no wildland interface. Likewise, you might be teaching a very small, rural organization that does not have any standpipes in its jurisdiction. Is it acceptable to omit material from certification courses when the material will not be used by the students?
>
> Professional qualifications are nationally recognized minimum standards for various levels of fire personnel. If you omit material pertinent to these standards, you will have failed as a fire service instructor because you did not teach your students to the standard. Moreover, in doing so, you jeopardize the very theory underlying certifications: ensuring a standardized level of competency. Additionally, while you may feel relatively confident that some areas of the material will not be used by some or all of the students, you will be doing your students a disservice if you do not cover all of the learning objectives for the course. Imagine a student later joining a larger department and not even understanding a standpipe system or a fire fighter attending the National Fire Academy not knowing the term *backfire*.
>
> Professional standards are just what they say. Don't undermine your credibility by not teaching the material.

Written Test

The written test-item type to be developed is determined by two considerations:

- The knowledge requirement to be evaluated
- The content format of the resource material

As you research and identify passages of relevant information in the resource material, ask yourself which type of written test item the material might support. For example, one passage full of terms and definitions might support developing a matching question, while another passage containing a scenario or other complex situation might support multiple-choice questions. You may want to note this observation somewhere in the resource material as you go through the review and research phase. One technique that works well is to use a different-colored highlighter for each type of test item. With this system, when you return to write test items, the analysis for the type of test item is already complete.

■ Complete a Test-Item Development and Documentation Form

Test-item development and documentation forms serve many purposes other than the most obvious—recording the test item. The form is also useful to do the following:

- Connect planning/research with test-item development efforts
- Record pertinent course-related information
- Record the source of test-item technical content
- Provide a record of format and validity approval
- Record the learning objective and test item number for banking

By documenting the required information on the form, you are using it to accomplish all of these purposes.

Collectively, the information you place in the blanks of test-item development and documentation forms will serve to link the identifiers of the program or course and the reference with the test item. Each blank is important to proper identification, so fill in each one carefully with the appropriate information.

A continuation sheet may be used when developing a test item that is too long to fit on a single form.

The reference blank is where the current job-relevant source of the test-item content is identified. In this blank, you should supply enough information to pinpoint the specific location within the resource material where you found the information you are using to develop the test item. Entries might appear as follows:

Starting at the top of page one of the Test-Item Development and Documentation form, complete it as follows:

- **Program or Course Title.** In this blank, put the title of the program or course for which you are writing a test item. An example would be "Fire Officer I."
- **Learning Objective Title.** In this blank, place the learning objective you are working on in preparing test items. An example would be "Fire-Ground Operations."

Components of a Test Item

Test items are made up of components that create a testing tool. The part of the question that asks the information is known as the stem **FIGURE 10-4**. The choices are made up of a correct answer and distracters. An important part of the question is the reference material that documents the standard the question is based on and the source of the information in a textbook.

Selection-Type Objective Test Items

■ Multiple-Choice Test Items

The multiple-choice test item is the most widely used in objective testing. This type of item also contributes to high test validity and reliability estimates. The multiple-choice test item is a favorite among test-item developers because it comes nearer to incorporating the important qualities of a good test than any other type of test item. These qualities include content validity, reliability, objectivity, adequacy, practicality, and utility. Ease of grading of the test items and providing immediate feedback to students make multiple-choice items extremely flexible for the fire service instructor. The most important advantage of the multiple-choice test item is that it can be used to measure higher mental functions, such as reasoning, judgment, evaluation, and attitudes, in addition to knowledge of facts. The multiple-choice test item is so versatile that it can be used to measure almost any cognitive information. The disadvantage is that it may take several well-written test items to cover the learning objective properly.

FIGURE 10-5 provides an example of a properly formatted multiple-choice test item.

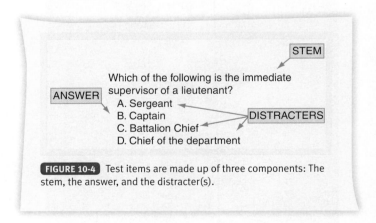

FIGURE 10-4 Test items are made up of three components: The stem, the answer, and the distracter(s).

Fire department synthetic rope, new or used, should be kept

A. warm, dry, and in a sunlit place.
B. cool, moist, and in a sunlit place.
C. neatly coiled and stored out of sunlight.
D. damp and in a dark place.

FIGURE 10-5 An example of a properly formatted multiple-choice test item.

VOICES
OF EXPERIENCE

Complacency can be one of the biggest detriments to a fire fighter's career. When I first got promoted I felt all eyes were on me. I studied hard, enforced all the department rules, and tried to build a good reputation. After a while I realized that being a good officer wasn't just holding myself to a higher standard, but also holding my crew to that same standard. I started to look at the actions and needs of my crew more closely. They seemed to just be going through the motions on many of the repeated or nuisance calls.

One day I decided I was going to start quizzing my crew members about different building features or tactics we would use if this turned into a real incident. They thought I was crazy. Why the change now, they said. I explained to them that if we are to succeed as a team, we'll all need to understand the different types of incidents we go to, as well as our responsibilities within that incident.

After months of pop quizzes, they started to understand my thought process. They soon began questioning me. It was nice to see they weren't taking the nuisance calls for granted anymore. They were learning and preplanning for the future.

> *I explained to them that if we are to succeed as a team, we'll all need to understand the different types of incidents we go to, as well as our responsibilities within that incident.*

One afternoon we were dispatched for a "telephone alarm for a reported building fire." On arrival at a well-known building, we had fire showing from four windows on the second floor. I immediately called for the "250" (meaning length). Yet, I heard from the back step "2 ½" (meaning size). They learned! All the questions and all the explanations had paid off! We took the 2 ½″ hoseline and knocked down the heavy flames. After the visible flames were knocked down, we immediately dropped that line and took a 1 ¾″ inside and finished the extinguishment. That was one of the best stops we ever made. I'll remember that fire for the rest of my career, not just because of the great job we did, but because my crew was ready to make good decisions. I felt like a proud parent.

Testing doesn't always have to be formal. Informal quizzes, whether they are on scene or around the coffee table, give you an idea where your crew's strengths and weaknesses are. I feel the mark of a good officer is one whose crew will succeed even when the officer isn't there.

Dan Sullivan
Captain, Lynn Fire Department
Instructor, Massachusetts Firefighting Academy
Stow, Massachusetts

Teaching Tip

While the multiple-choice question is the most popular type of test item, it is essential to determine the testing method that best ensures competency related to the learning objective rather than to simply fall back on habit.

Directions: Match the terms listed in column A with the definition provided in column B.

Column A	Column B
1. Oxidation 2. Conduction 3. Convection 4. Radiant heat substances	A. Rapid oxidation with heat and light B. Oxygen combining with other C. Heat energy carried by electromagnetic waves D. Heat transfer by circulation through a medium E. Heat transfer by direct contact

FIGURE 10-6 An example of a properly formatted matching test item.

Preparing Multiple-Choice Test Items

Suggested guidelines when preparing multiple-choice test items include the following:

- Select one and only one correct response for each item. The remaining distracters should be plausible, but wrong, to serve as distracters for the correct response.
- Construct the same number of responses for each test item. Four responses are preferable. More than four response choices may be used and sometimes are necessary, but they do not improve the reliability of the test.
- Responses having several words should be placed on one or two lines with space separating them from other responses. If responses consist of numbers or one word, arrange two or more of them on the same line, uniformly placed.
- Provide a line or parentheses on the right or left margin of the page for students to write their responses. This placement permits ease of test taking for students and encourages use of a grading key for more rapid and accurate test grading. Disregard this step if you will be using a machine-scannable answer sheet.
- Change the position of the correct response from test item to test item so that no definite response pattern exists.
- Prepare clear instructions to the student.

Matching Test Items

Matching test items are actually another form of multiple-choice testing. This type of question is limited in use, but functions well for measuring such things as knowledge of technical terms and functions of equipment. Care should be taken during preparation of matching test items because all information must be factual. There should be no more than four items to match with five choices when providing matches.

This type of test item is relatively easy to develop, but requires more space on the test. The matching test item is particularly useful when a question requires multiple responses or when there are no logical distracters. The matching test item is very useful for low-level cognitive information measurement.

FIGURE 10-6 provides an example of a properly formatted matching test item.

Preparing Matching Test Items

Suggestions for preparing matching test items include the following:

- Check to make sure that similar subject matter is used in both columns of the matching test item. Do not mix numbers with words, plurals with singulars, or verbs with nouns.
- Use up to four items in Column A and up to five choices for the match in Column B. This prevents students from earning credit simply by applying the process of elimination.
- Make sure there is only one correct match for each of the items to be matched.
- Arrange statements and responses in random order.
- Check for determiners or subtle clues to the answers.
- Provide a line or parentheses beside each response statement for the student to indicate the answer. Arrange the parentheses or lines in a column for ease of marking and scoring. This step is not required if you are using a machine-scannable answer sheet.
- Prepare directions to the students carefully and observe their following of instructions to ensure that the instructions function as intended.
- Attempt to keep each matching test item on one page.
- When preparing the draft of questions, follow the format guide to expedite typing, reproduction, response by the student, and rapid grading.

Arrangement Test Items

The arrangement test item is an efficient way of measuring the application of procedures for such things as disassembly and assembly of parts, start-up or shutdown, emergency responses, or other situations where knowing a step-by-step procedure is critical. These test items lose strength and efficiency steadily as time lags between the paper-and-pencil solution and the actual performance of the procedure. You should plan to administer an arrangement test soon before an opportunity to perform the actual procedure, disassembly, or assembly.

One disadvantage of arrangement test items is inconsistent grading. A student may miss the order of one step and thus miss the order of all of the steps that follow, even though the remaining steps may be in the correct sequence after the error. Another potential issue is key steps that are included in the SOP or SOG being left out of the test item. The test item should be developed in agreement with actual procedures used on the job if such procedures are required. The test item is not effective if the procedure is much longer than 10 steps. In cases where it is important to test procedures that are longer, it is better to group them at key points in the procedure. Knowledge of emergency procedures, fire drills, triage, or other related procedures where knowing exactly what to do under critical conditions can be measured with high reliability using an arrangement test item.

FIGURE 10-7 provides an example of a properly formatted arrangement test item.

Preparing Arrangement Test Items

Suggestions for preparing arrangement test items include the following:

- Make sure that arrangement test items developed for procedures are based on current procedure. Remember that arrangement test items should relate to emergency performances or to things that must be performed in a certain order. Arrangement test items are useful when you want the performance to occur without references or verbal input.
- Have one or more subject-matter experts review the test items for technical accuracy.
- Make sure the steps of the procedure are disordered and that steps with clues to the correct order are separated.
- State exactly what credit will be given for correct sequencing of all steps.
- Include cautionary statements with the procedure to be rearranged. Example: Order the rapid intervention team to enter a structure. (Caution: Report order to the incident commander.)

Directions: (*Caution—This is a two-part question worth 3 points.*)
The six steps of the basic method that should be used in fire or explosion investigations appear out of order below. **First,** number the steps in the proper order by placing numbers (1–6) in the blanks beside the steps. **Next,** select the answer that matches yours from choices A–D.
_____ Conduct the investigation.
_____ Collect and preserve evidence.
_____ Receive the assignment.
_____ Analyze the incident.
_____ Prepare for the investigation.
_____ Report findings.

A. 1, 4, 5, 2, 3, 6   B. 2, 4, 5, 1, 3, 6
C. 6, 3, 4, 5, 1, 2   D. 3, 4, 1, 5, 2, 6

Answer: D

FIGURE 10-7 A properly formatted arrangement test item.

■ Identification Test Items

The identification test item is essentially a selection-type test using a matching technique. Its major advantage over the matching test item is the ability of the student to relate words to drawings, sketches, pictures, or graphs. These test items do require reasoning and judgment, but basically focus on the ability to recall information. Identification test items can also be considered to be arrangement test items for assembly- and disassembly-type tasks.

Identification test items provide excellent content validity and contribute to high test reliability. Unfortunately, these items are sometimes difficult to develop and may be costly to produce. With today's computer publishing capabilities, digital cameras, and other technologies, however, identification questions can often be rather easily developed using your own fire equipment and apparatus features as the basis for the identification test item. Using this technique makes the test item more job related and makes it much easier to transfer required knowledge from the instructional setting to the job.

FIGURE 10-8 provides an example of a properly formatted identification test item.

Directions: (*Caution - This is a two-part question worth 5 points.*)
   First, label the tools depicted below by placing the number in the blanks provided.
   Then, choose the answer below (A–D) that matches yours.

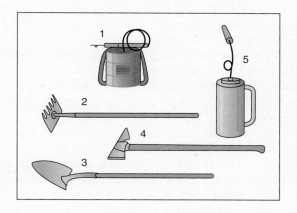

____ McLeod
____ Pulaski
____ Back pack pump
____ Shovel
____ Drip torch

Answers
A. 3, 5, 4, 2, 1
B. 2, 4, 1, 3, 5
C. 5, 3, 1, 2, 4
D. 4, 3, 1, 2, 5

FIGURE 10-8 An identification test item.

Preparing Identification Test Items

Suggestions for preparing identification test items include the following:

- Select graphic, art, or other object for inclusion in the question.
- Place the object into a drawing or illustration software.
- Identify the parts or items you want identified. (Caution: Do not specify more than seven or eight items to be labeled, or even fewer items if the object will be cluttered by arrows or letters. You may choose to make two identification test items using the same graphic and reduce the number of responses from eight to four on each test item.)
- Develop the desired alternatives A–D, with only one answer being the correct identified object.
- The identification test item may give a point value of more than 1 for the correct answer. (A rule of thumb is to award one point for every two correct responses.)
- Have a subject-matter expert review the technical accuracy of the question and alternatives. (Be prepared to provide your validity evidence and documentation.)
- Use the test item several times with the target population to make sure it measures what was intended.

■ Completion Test Items

The completion test item is easy to develop. Completion test items tend to be verbatim from the technical materials, thus they tend to encourage memorization and reinforce rote learning. These test items are appropriate for entry-level positions because they measure low-level cognitive knowledge. The best practice for developing and using completion test items is to follow specific procedures and to develop two test items for every one item that is expected to be included in the final group of test items. Subject-matter experts tend to delete these questions during the validation processes.

Completion test items are of two types: supply and selection. The supply type requires the student being tested to supply the response that completes the test item statement. The selection type requires the student being tested to select from a list of responses for completing the test item. The supply type tests primarily the recall capability of the student, whereas the selection type focuses on recognition and analysis.

FIGURE 10-9 provides an example of a properly formatted selection-completion test item.

Developing Completion Test Items

The following guidelines may help you in developing completion test items:

- Sources of completion test items may include system descriptions, equipment operation manuals, technical specifications, procedures, and safety practices.
- Make the statements as brief and factual as possible. Do not use misleading or contradictory words or terms.

- Avoid using responses that match verb tense for one blank. It is important to have several singular and several plural responses or to use the same verb tense when developing selection-type items.
- Have two or more subject-matter experts review each test item for technical accuracy and quality of response selections.
- Selection-type questions should have two to three more possible responses than there are blanks.
- Be aware that supply-completion test items require manual scoring. They also may require a list of synonyms as possible correct answers. The quality of handwriting can sometimes be a problem when scoring the test as well. It is not unusual for persons taking a test to scribble something in a blank and hope for credit if they do not know the answer or have doubts. The supply-completion test item is also subjective, as you can withhold or provide a point based on whether you like or dislike the student taking the test (the halo effect).

■ True/False Test Items

True/false test items have been controversial for years. Test item developers have experimented with numerous ways to make true/false, right/wrong, and yes/no responses more valid and reliable. The major problem with two-answer selection tests such as the true/false is the guessing factor. A student who marks a response without reading the question has a 50 percent possibility of marking the correct response. Application of two or three rules for taking a true/false test could get the student a higher score just by guessing alone. **FIGURE 10-10** is a typical true/false test item.

A more complex true/false test item can be developed that uses a multiple-choice approach to the test item. **FIGURE 10-11** provides an example of a complex true/false test item.

Foams in use currently are of the mechanical type and must be _____ and _____ before they can be used.

A. proportioned, blended
B. stirred, aerated
C. mixed, proportioned
D. proportioned, aerated

Answer: D

FIGURE 10-9 Selection-completion test item.

Directions: Circle either "T" for true or "F" for false.

As a fuel's surface-to-mass ratio decreases, its ignitability increases.   T   F

Answer: F

FIGURE 10-10 Typical true/false test item.

Teaching Tip

Using multiple types of test questions can increase the efficacy of the test. Use the question formats that best ensure the learning objective is met.

Directions: Read the following statements regarding diversity programs and select your answer from choices A–D.

Statement 1: A diversity program creates an environmental and cultural change in an organization.

Statement 2: Achieving a culturally diverse organization is ethically and managerially a worthwhile accomplishment.

Statement 3: A diversity program is a single effort intended to integrate an organization to meet government mandates.

A. Statement 1 is true; statements 2 and 3 are false.
B. Statements 1 and 2 are true; statement 3 is false.
C. Statement 1 is false; statements 2 and 3 are true.
D. All three statements are true.

Answer: B

FIGURE 10-11 A complex true/false test item.

Developing True/False Test Items

When developing true/false text items, follow these guidelines:

- True/false test items do not cover the subject matter in depth because of the limited application possible in most technical information or courses. Use of true/false test items should be limited to factual information at the basic level of an entry-level course, such as a course for Fire Fighter I candidates. There is little discernible value for the true/false test item in technical instruction when student knowledge and performance measurement are required for safe and efficient operation of equipment and emergency task performance. For this reason, true/false questions should be avoided in operations and maintenance programs.
- The complex true/false test item is much more suited to technical courses and information. These items should be developed from a specific area of knowledge. Statements should be brief and technical in nature. This type of question does require higher-level cognitive skills.
- If used in a multiple-choice format, this test item can be objectively scored.

Essay Test Items

Modern test developers do not consider the essay test to be an objective form of testing. This position is based on research in which the major weaknesses of this testing method were examined. Some of the weaknesses that affect the objectivity of essay tests are listed here:

- Essay tests tend to focus on limited points in a course of study. Because of the length of students' responses and the time needed for the fire service instructor to grade tests, essay tests cannot be comprehensive in coverage of the subject matter.
- The length of the student response tends to cloud the actual knowledge required by the question with extraneous knowledge closely related but not directed to the question. Students can "beat around the bush" without ever really addressing the question.
- The order in which test papers are graded affects the grading. The first papers graded tend to receive lower grades than those graded last. This is especially true when tests are long and are graded one exam at a time. Grading one question at a time for all students tends to improve the consistency of grading on an essay test.
- Handwriting affects grades. The better the handwriting, the higher the grade tends to be, even when the quality of the response varies. Typewritten papers get higher grades than handwritten papers, even when answers are identical.
- Grading varies widely, even among fire service instructors who are recognized subject-matter experts. (Variance identified in research studies has shown ranges as great as three letter grades and 25 points. This was true even when fire service instructors were provided with grading keys and a set of scoring rules for the questions.)
- The "halo effect" influences the grading practices of some fire service instructors. The more they know about a student, either favorable or unfavorable, the more the grade is affected. The personality of the student also tends to have a unique effect on the grade received on an essay test.
- Writing essay questions requires a lot of the fire service instructor's time. Initial essay questions must be refined over several uses before they will function in a reasonably reliable manner.
- Students require much more time to prepare for and write answers to essay questions. Long tests tend to be fatiguing to students and can sometimes encourage lengthy responses containing extraneous information.
- Poorly prepared essay test items have been shown to reinforce negative learning. This means that the answer the fire service instructor receives tends to support the wrong interpretation of facts and conditions, even among students considered to be outstanding.

Research has revealed that a fire service instructor can develop an objective test in approximately the same amount of time that it takes to develop an essay test. However, the objective test will achieve higher content validity, comprehensiveness, discrimination, and reliability than is possible with an essay test.

Although there are many disadvantages to using essay test items, they are used and can be improved to the point that some inherent weaknesses are diminished. The essay test item requires the student to recall facts from acquired knowledge and to present these facts logically and in writing. Writing answers provides the student with an opportunity to express ideas and attitudes, interpret operational situations, and apply knowledge gained to an individual solution. A good technique for answering essay questions can be developed by a fire service instructor and taught to students. Outlining a response to an essay question and marking the key points so that they may be emphasized in the answer is a valuable approach to improve reliability.

Formatting Essay Test Items

It is important to word the essay question properly and to develop a preliminary outline of the key points expected in the response to the question. This exercise helps to determine whether the question will actually elicit the intended responses. The outline also serves as a guide to grading student responses and assigning relative point values to the key points in an answer before the test is given.

A hierarchy of words can be used to develop essay test items, which will provide a range from easy to difficult. For instance, questions—in ascending order of difficulty—may require the student to "describe," "explain," "discuss," "contrast," and "evaluate." That is, asking the student to *describe* an event in plant operations is much easier than asking the student to *evaluate* the same event and *contrast* that event with the implications for plant safety. In the essay test, some of each of these types of questions should be asked so that the relative difficulty of the test is maintained. The easier questions (*describe*, *explain*) should be presented earlier in the test.

Preparing Objective Essay Test Items

Suggested guidelines for preparing objective essay test items include the following:

- Review the learning objectives. Test items should be developed considering the learning objectives and using them as the basis for testing.
- Make a list of major points contained in the body of the course material that is to be covered by the test. This list should focus on major points, rather than on trivial details. Check the list once it is completed to confirm that it is comprehensive in its coverage of the subject matter, keeping in mind that an essay test contains only a few test items and that it requires more time for the student to answer and for you to grade.
- Arrange the list of major points in an order similar to that used when presenting the information or assigning reading material for student outside study.
- Identify the major points that you want the student to describe. These major points should be covered in a less difficult essay question for students to begin with on a test. The use of a relatively easy question for the first item will get students ready for the more difficult questions that follow.
- Next, prepare a question or two requiring the student to describe and explain a key point or major concept in the subject matter. Again, this approach takes the student to the next level of difficulty in an essay exam.

- Prepare the most difficult essay items last and place them toward the end of the exam. These questions usually require the student to contrast, compare, and evaluate a situation or complex set of facts. It is a good testing practice to use the earlier questions in an exam to form the basis for these contrasts, comparisons, and evaluations. Placing difficult questions at the end of an exam has the added value of permitting the well-prepared and knowledgeable student an opportunity to pull the facts together and express personal views in terms of the facts. It is not uncommon for students to expand their knowledge well beyond the level expected by the fire service instructor and to offer plausible solutions that may not have been presented previously. These unexpected results may be rewarded with extra credit.
- Take the test. Answer the questions in terms of the course material presented, assigned reading, and lab and shop activities. Once you have prepared your answers, prepare a detailed outline of each essay test-item response. Determine the amount of credit for each key point in the outline. Determine in advance where extra credit will be awarded for answers that clearly excel and determine the number of points to be deducted for major omissions in test-item answers. The outlines then become the scoring guide and key to help ensure consistent grading from question to question and from one exam to another.
- When essay exam questions are completed, review them once again. While reviewing the exam, it is of extreme importance to keep in mind the exact response behavior desired from the students and the specific abilities and knowledge that students should possess relative to course material covered by the exam.

Grading Essay Test Items

A crucial activity necessary to make an essay exam more objective is the use of a scoring key with associated criteria for consistent grading. Following are a few suggestions to make the grading process more reliable:

- Fold the exam cover sheets back or cover the test takers' names before beginning any grading activity. This permits the application of the grading criteria as anonymously as possible.
- Grade one question at a time on all exam papers before moving to the next question. This tends to keep the focus on the grading key and the response at hand.
- If the exam has an effect on a student's career or promotion, it is good professional practice to have another qualified person give a second grade to the exam. An average of the two grades provides more objectivity.
- Make sure that you provide some flexibility in the grading key or outline for giving additional credit for answers that produce results beyond those expected when the test was developed. Penalties for omission of key points must also be deducted from the test score. Make sure, however, that the omission penalties are uniformly applied to all questions and exams.

■ The *whole method*, which can improve grading of exams, begins with a preliminary reading of the exam papers. During the reading process, a judgment is made of each exam in terms of four groupings: excellent, good, average, below average. Each exam is then sorted into one of these preliminary groups. Once the grouping is completed, grading is accomplished in the same manner as described previously. A shift from group to group of one or more exam papers is not uncommon when the scoring is completed.

Remember—be as objective as possible in the use of the grading criteria. Application of the criteria for grading will improve the objectivity of those applying the grade. An essay test, even under the best of conditions, has low reliability.

Prepare Essay Test Item Grading Key

It is important to develop a preliminary outline of key points expected in the response to the question **FIGURE 10-12**. While helping to determine whether the question will actually elicit the intended response, this exercise also assists the instructor in assigning relative point values to the key points in an answer before the test is given.

In outline format, list the key points that are expected in the answer to the essay test item. For each key point in the outline, assign a weighted point value. These values may vary in weight depending on their degree of importance to the overall answer. Key points of a critical nature or those pertaining to safety-related activities and information should be given higher weighted values. Because of the importance of grading essay test items, the methods for increasing objectivity should be considered.

Theory Into Practice

Putting together a valid testing instrument may be one of the most challenging parts of the entire learning process. This process requires developing good questions where a student who knows the material is able to provide the correct response, whereas a student who does not know the material is not able to provide the correct response. It also should give those students who know the material the best chance to get more correct answers than those who know much less.

To accomplish this goal, take each learning objective and create questions that demonstrate competency for that learning objective. Use various styles of questions to discern the level of comprehension. For example, writing a scenario and having students apply the material concepts is much more difficult than developing questions where a student can recite rote information from the training materials.

Question

State four precautions for handling or working with caustic soda, including the appropriate first-aid actions, in the event contact with caustic occurs.

Grading Key	Point Value
1. Hot water must be used when dissolving caustic soda. Eyes, face, neck, and hands should be protected.	1
2. Wherever caustic soda is stored, unloaded, handled, or used, abundant water should be available for emergency use in dissolving or diluting and flushing away spilled caustic.	1
3. General first aid is of prime importance in case of caustic coming in contact with the eyes or skin. Prolonged application of water to the affected area at the first instant of exposure to caustic is recommended.	2
4. a. If even minute quantities of caustic soda (in either solid or solution form) enter the eyes, the eyes should be irrigated immediately and copiously with water for a minimum of 15 minutes.	2
b. The eyelids should be held apart during the irrigation for contact of water with all the tissues of the surface of the eyes and lids.	1
c. If a physician is not immediately available, the eye irrigation should be continued for a second period of 15 minutes.	2
d. No oils or oily ointments should be applied unless ordered by the physician.	1
Total Point Value	**10**

FIGURE 10-12 Example of an essay test-item grading key.

Performance Testing

Performance testing is the single most important method for determining the competency of actual task performances. For this reason, the major emphasis in performance-based instruction is the validity and reliability of the performance tests. The focus on student performance is the critical difference between traditional approaches to instruction and the use of performance-based instructional procedures. Performance must be evaluated in terms of an outside criterion derived from a job and task analysis. The justification for training program knowledge and skill development activities centers on the concept that the activities are derived from, and will be applied to, training for tasks performed on the job. This focus prevents the inclusion in programs of "nice to know" information and irrelevant skill development activities. The final demonstration of job knowledge and skill application in fire and EMS training programs is the completion of specific on-the-job tasks.

Test Generation Strategies and Tactics

There are many ways to put together tests. You can do it yourself or use a computer or Web-based test-item bank. Doing it yourself is generally referred to as "instructor-made tests."

The primary problem with instructor-made tests, even though they may be valid and reliable, is the tendency to use the same test over and over again. Soon the word gets out about the test content and answers, and then everyone who takes the test will already know the answers. Computer or Web-based testing is rapidly gaining favor in fire and EMS contexts. Many educational publishers and other private organizations, for example, are providing test-item banks with their publications. Although these tools are certainly helpful, you must carefully analyze such test banks to make sure the items in them are properly validated.

Remember that most testing in fire and EMS training environments leads to certification, selection for a job, pay raises, and promotions. If you are testing for any of these reasons, use only test banks that have been rigorously validated and that are known to produce reliable tests. The question then becomes, "How do I know?" A checklist is provided here to help you determine the quality of a test-item bank no matter who is providing the product **FIGURE 10-13**.

Of course, there are many more questions you can add to this checklist. Cost is one of the possible criteria to be added, although it should not be the primary concern. The critical emergency tasks that fire and EMS personnel perform can open them up to costly legal challenges. Training and testing programs are the likely targets for litigation if a suit for liability is filed.

Teaching Tip

All fire service instructors have encountered personnel who have passed certification/promotional examinations but who did not seem truly ready for the job. Conversely, we have seen people who we believed would be very good, but failed the examination. Although some of this reaction may be perceptual bias, often it is accurate. These situations occur because of poor tests. Carefully screen each question for reliability and validity, and the examination as a whole for discrimination, to help prevent these errors.

■ Computer and Web-Based Testing

During recent years, there has been a significant increase in the use of computers for training and testing purposes. Distance learning over the Web is now present in conjunction with most colleges, universities, and two-year institutions. Fire departments and EMS have made great strides embracing this technology, but more needs to be done.

In 2007–2008, approximately 4.3 million undergraduate students, or 20 percent of all undergraduates, took at least one distance education course. According to the U.S.

Questions to Be Answered by Test Bank Provider	Provider 1	Provider 2	Provider 3
Do you claim to have valid and reliable test items in your test banks?	Yes __ No __ Notes:	Yes __ No __ Notes:	Yes __ No __ Notes:
Are there documentation and data available for determining test-bank validity and test reliability?	Yes __ No __ Notes:	Yes __ No __ Notes:	Yes __ No __ Notes:
Do your test banks comprehensively cover NFPA Professional Qualification Standards? If yes, do you have cross-reference tables to document the extent of coverage?	Yes __ No __ Notes:	Yes __ No __ Notes:	Yes __ No __ Notes:
How long have you been providing test banks? Is this your primary business focus?	Yes __ No __ Notes:	Yes __ No __ Notes:	Yes __ No __ Notes:
How many customers do you have, and can you provide user names and telephone numbers?	Yes __ No __ Notes:	Yes __ No __ Notes:	Yes __ No __ Notes:
Do you provide technical support for your test banks and software? If yes, then how?	Yes __ No __ Notes:	Yes __ No __ Notes:	Yes __ No __ Notes:
Is there regularly scheduled training for the test banks and software? If yes, how often and at what locations?	Yes __ No __ Notes:	Yes __ No __ Notes:	Yes __ No __ Notes:
Do you have technically competent staff members in testing technology? If yes, will you provide detailed résumés of their qualifications?	Yes __ No __ Notes:	Yes __ No __ Notes:	Yes __ No __ Notes:
What are your revision and updating strategies for your test bank?	Yes __ No __ Notes:	Yes __ No __ Notes:	Yes __ No __ Notes:
Do you provide qualified expert witnesses supporting your test bank in case of lawsuits? If yes, what are the costs for the service?	Yes __ No __ Notes:	Yes __ No __ Notes:	Yes __ No __ Notes:

FIGURE 10-13 Test bank validation checklist.

Department of Education, nearly 800,000, or 4 percent of all undergraduates, took their entire program through distance education. Many colleges and universities continue to encourage students to take a portion of their program off campus, preferably online. Many of the earlier myths regarding distance learning are being destroyed by educational research. Fire and EMS organizations will likely increase the accessibility of this new technology to their personnel, given that it has the potential to reduce costs for education and training for both the student and the department.

Fire and EMS departments have many options available for putting testing online 24 hours a day, 7 days a week, through computer hardware and professionally developed test banks. Some of these organizations are already conducting online and Web-based certification testing programs. Given the currently high price of fuel, hotels, worker salaries, and other costs, Web-based certification seems to be the wave of the future.

One of the primary benefits of online learning and testing is the convenience for those persons who work full-time jobs. An employer can also extend training and testing time into the available leisure time of willing employees.

As an instructor, you should collect information and look for successful applications of this technology for teaching and testing your students or employees. Find out what is going on with this fascinating technology and make recommendations to your supervisors. Remember—in fire and EMS organizations, you are dealing with situations that require rapid responses and high levels of technical knowledge. Providing your students with the best means for learning and performing is your primary challenge.

The first step is to research the technology available. It is difficult to develop sound applications of technology without conducting some sort of study. Once this first step is completed, you will have a much better idea of the technology you need to meet your specific requirements. There are so many options regarding computers, networking, and online education/training that you can easily become confused, and you may even spend much more money on technology than is beneficial. Collection of data, an objective needs analysis, consideration of your budget, and a thorough assessment of the software and equipment that meet your most immediate needs will ensure that you obtain the desired return on your investment. Be careful when committing to a technology, however: Bookshelves and storage cabinets are filled with equipment and software that were purchased with good intentions but remain unused.

The most significant advantage when using valid and reliable computer-based test banks with large numbers of questions is the reduced risk of test compromise. You can even randomly generate any tests you administer every time you need to give a test. Maintaining large numbers of test items in a test bank can help ensure that no two tests will ever be alike. For instance, if a test bank contains 1000 test items and you randomly generate a new test each time you need one, the chances of getting exactly the same test are negligible.

Safety Tip

As online learning becomes more prevalent, it is important for students to be properly trained in basic ergonomics for keyboarding to prevent repetitive motion and other injuries.

Jones & Bartlett Fire District

Training Division
5 Wall Street, Burlington, MA, 01803
Phone 978-443-5000 Fax 978-443-8000
www.fire.jbpub.com

Instant Applications: Evaluating the Learning Process

Drill Assignment

Apply the chapter content to your department's operation, its training division, and your personal experiences to complete the following questions and activities.

Objective

Upon completion of the instant applications, fire service instructor students will exhibit decision making and application of job performance requirements of the fire service instructor using this text, class discussion, and their own personal experiences.

Suggested Drill Applications

1. Review a recent written examination and compare the student responses to the answer key. Identify any test questions that were too easy (every student answered correctly) and ones that had poor success (high percentage of failures).

2. Review your local policy or practice on the posting of test scores and student completion records.

3. Develop a strategy to improve student performance based on a poor evaluation result.

4. Review the Incident Report in this chapter and be prepared to discuss your analysis of the incident from a training perspective and as an instructor who wishes to use the report as a training tool.

Incident Report

Port Everglades, Florida—2003

Figure 10-A Maritime training prop at Port Everglades, Florida.

Note: This was the first U.S. fatality in a permanent live fire training structure.

A live fire training exercise was being conducted in a prop constructed from shipping containers to represent a seafaring vessel **(Figure 10-A)**. This was the first live fire experience for the students in a class leading to state certification as Fire Fighter II. The students were all new employees of the same fire department.

Three evolutions took place, with no breaks in between the evolutions to allow the all-metal structure to cool. The participants were instructed not to crawl over the metal grating and to avoid holding the handrails, because the metal structure had gotten very hot and there was a possibility of getting burned through their gloves. They were told failure to complete this evolution, or getting injured, would mean termination from the department. There was no preburn briefing or walk-through.

Some members of the third squad received burn injuries during their rotation, and an instructor left during the evolution claiming he had problems with his gloves.

The fire fighter who later died was a member of the fourth squad of students performing the evolution on this particular day. Five students, three instructors, and an observer with a thermal imaging camera entered the enclosed structure on the second level. As in previous evolutions, the students were sent in one by one with the intention being that they would not be able to see or hear one another until arriving at the fire box. According to statements, the instructors did not monitor the students' movement or encourage their progress. Students followed the hose line through a series of three watertight hatches, and then crossed an open-grated catwalk over the gas-fueled fire in the engine room. At the end of the grated catwalk, a combustible materials fire burned in a corner. They then proceeded down a ladder and through the simulated engine room and into the "fire box." The fire box had a raised hearth, similar to a flashover simulator, and the fire was fueled primarily by wood pallets. The students gathered in the room and took turns operating a nozzle in various patterns. They were told to avoid getting water on the fire.

Continued...

Incident Report Continued...

© Greg Henry/ShutterStock, Inc.

Port Everglades, Florida—2003

The fire was knocked down, and an instructor in the burn room directed the crew to remain in the extremely hot area as the fire regained intensity. A "dead man" switch operated an open vent in the room, but it had been disabled so that it remained closed.

The instructor in the fire room was quoted by several trainees as saying the environment was too hot. He instructed the group to hurry up and "get out now." One set of students turned around to exit using the same route through which the group members had entered the room. Now out of sequence, the safety officer led the students up the ladder and over the metal grating in a very high heat environment. Meanwhile, two instructors exited through a side door directly from the fire box to an on-grade side exit, without advising the instructor-in-charge.

There are conflicting reports as to what occurred at the exit, but it was apparent that one student was missing. It is believed that he lost track of the hose, and ended up in a chase where heated gases and smoke were venting. Handprints on the walls indicated that he was trying to find a way out. While a search for him ensued, he collapsed. After seeing the downed student's personal protective equipment (PPE), one officer entered to retrieve him, without protective clothing or SCBA, but had to abandon his attempt due to the heat. Another fire fighter wearing proper protective clothing and SCBA removed the victim to the outside. The victim's personal alert safety system (PASS) device did not sound while he was inside the structure; it is believed he was mobile until just before he was found.

The victim was unresponsive, with no pulse or respirations. After initial treatment by instructors, a medical rescue unit arrived and the victim was transported to a local trauma center, where he was pronounced dead. He reportedly had severe burns on both hands and sloughing of the skin to both knees and hands. He was described as cyanotic from the neck up. Several members of the same crew received second-degree burns to their hands and knees while they were attempting to exit the structure. One student lost consciousness after exiting and was initially cared for by other students. He and three other students were transported to the hospital.

Two separate investigations were conducted by the state fire marshal—one for criminal violations and an administrative investigation of state training codes. No criminal charges were filed, but the matter went to the state attorney. The fire department was cited by the state fire marshal for numerous code violations, and the reports cited 36 specific findings including almost total failure to follow NFPA 1403.

Postincident Analysis: Port Everglades, Florida

NFPA 1403 (2007 Edition) Noncompliant

No written, preapproved plan (5.2.12.3) (6.2.15.3)

No previous live interior firefighting training (5.1.1) (6.1.1)

Safety crews did not have specific assignments (5.4.8) (6.4.8)

No preburn walk-through for the trainees (5.2.13) (6.2.16)

No designated safety officer (5.4.1) (6.4.1)

No communication plan (5.4.9) (6.4.9)

No emergency medical services on site (5.4.11) (6.4.11)

Not all instructional personnel had specific live fire training or instructor certification (5.5.1) (6.5.1)

Postincident Analysis: Port Everglades, Florida

Instructors did not monitor the trainees' movement (5.5.7) (6.5.7)

Noncompliance with NFPA 1402 and 1403 (both required by state code) (5.5.4) (6.5.4)

No temperature monitoring (5.3.5) (6.3.8)

Multiple fires inside structure burning simultaneously, including polypropylene rope and other nonorganic materials (5.3.1) (6.3.1)

Lead instructor determined the environment was excessively hot, but evolution was not terminated; also failed to identify and correct safety hazards (5.3.6) (6.3.10)

NFPA 1403 (2007 Edition) Compliant

Three to four instructors per squad of five trainees (5.5.2) (6.5.2)

Other Contributing Factors

No emergency plans in place

No rapid intervention crew (RIC) assignment

Command structure unknown to students and staff

The fire box vent had been rendered unusable

Wrap-Up

Chief Concepts

- Testing plays a vital role in training and educating fire and emergency medical services personnel.
- Testing for completion of training programs, job entry, certification, and licensure are governed by certain legal requirements and professional standards in the United States.
- Tests should be developed for three purposes:
  - To measure student attainment of learning objectives
  - To determine weaknesses and gaps in the training program
  - To enhance and improve training programs by positively influencing the revision of training materials and the improvement of instructor performance
- Oral tests used in conjunction with performance tests should focus on critical performance elements and key safety factors.
- Any performance test should be developed in accordance with task analysis information and reviewed by subject-matter experts to ensure technical accuracy in terms of the actual work conditions.
- Test development activities should adhere to a common set of procedures.
- Proctoring tests involves much more than just being in the testing environment; it requires specific skills to be performed professionally.
- Performance test proctors must have keen observational skills, technical competence for the task(s) being performed, ability to record specific test observations, ability to foresee critical dangers that may lead to injury to the performer or others, and an objective relationship to the test takers.
- When cheating is suspected, the fire service instructor must document what was observed, what the student said, and what the fire service instructor allowed the student to do next.
- The proper grading of oral, written, and performance tests is an essential function of the fire service instructor. Students will desire feedback on their performance, and instructors will want to know how successful their instructional efforts were.

- Release of test scores to a class should proceed on an individual basis. It should be done in a private session, on a one-on-one basis, or in a personal letter to each student.
- Evaluation of examinations closes with feedback on testing results being delivered to the students. This step must be considered confidential and should be completed in a timely manner using the evaluation data available to the instructor.
- High-stakes examinations include the following types of tests:
  - Final training test
  - Certification test
  - Selection test
  - Promotional test
- A low-stakes test is one that has only minor or indirect consequences for examinees, programs, or institutions involved in the testing.
- Training and testing leading to hiring, promotion, demotion, membership in a group such as a union, referral to a job, retention on a job, licensing, and certification are all covered under the *Uniform Guidelines for Employee Selection*.
- Taking steps to ensure validity will prevent your test items from measuring unrelated information.
- Test analysis occurs after a test has been administered.
- In technical training, whether the tests are written, oral, or performance based, they must center on the learning objective.
- Testing comprises two activities:
  - The preparation of test questions using uniform specifications
  - A quantitative/qualitative analysis to ensure that the test questions function properly as measurement devices for training
- Test specifications permit instructors and course designers to develop a uniform testing program made up of valid test items as a basis for constructing reliable tests.
- The written test-item type to be developed is determined by two factors:
  - The knowledge requirement to be evaluated
  - The content format of the resource material
- Collectively, the information you place in the test-item development and documentation form blanks will serve to link the identifiers of the program or course and the reference with the test item.

fire.jbpub.com/instructor/2e

- The multiple-choice test item is the most widely used in objective testing.
- Matching test items are limited in use but function well for measuring such things as knowledge of technical terms and functions of equipment.
- The arrangement test item is an efficient way to measure the application of procedures for such things as disassembly and assembly of parts, start-up or shutdown, emergency responses, or other situations where knowing a step-by-step procedure is critical.
- The major advantage of the identification test item over the matching test item is the ability of the student to relate words to drawings, sketches, pictures, or graphs.
- The major problem with two-answer selection test items, such as the true/false question, is the guessing factor. A student who marks a response without reading the question has a 50 percent possibility of marking the correct response.
- An objective test will achieve higher comprehensiveness, discrimination, and reliability than is possible with an essay test.
- Performance testing is the single most important method for determining the competency of actual task performance.
- Computer or Web-based testing is rapidly gaining favor in fire and EMS organizations.
- Fire and EMS organizations have many options available for putting testing online 24 hours a day, 7 days a week, through computer hardware and professionally developed test banks.

Hot Terms

Essay test Tests that require students to form a structured argument using materials presented in class or from required reading.

Face validity A type of validity achieved when a test item has been derived from an area of technical information by an experienced subject-matter expert who can attest to its technical accuracy.

Job-content/criterion-referenced validity A type of validity obtained through the use of a technical committee of job incumbents who certify the knowledge being measured is required on the job and referenced to known standards.

Oral tests Tests in which the answers are spoken in essay form in response to direct or open-ended questions. They may accompany a presentation or demonstration.

Performance tests Tests that measure a student's ability to do a task under specified conditions and to a specific level of competence. Also known as a skills evaluation.

Qualitative analysis An in-depth research study performed to categorize data into patterns to help determine which test items are acceptable.

Quantitative analysis Use of statistics to determine the acceptability of a test item.

Reference blank Where the current job-relevant source of the test-item content is identified.

Reliability The characteristic that a test measures what it is intended to measure on a consistent basis.

Skills evaluations See *performance tests*.

Subject-matter expert (SME) A person who is technically competent and who works in the job for which test items are being developed.

Systems approach to training (SAT) process A training process that relies on learning objectives and outcome-based learning.

Technical-content validity A type of validity that occurs when a test item is developed by a subject-matter expert and is documented in current, job-relevant technical resources and training materials.

Validity The documentation and evidence that supports the test item's relationship to a standard of performance in the learning objective and/or performance required on the job.

Written tests Tests made up of several types of test items, such as multiple-choice, true/false, matching, short-answer essay, long-answer essay, arrangement, completion, and identification test items. Answers are provided on the test or a scannable form used for machine scoring.

References and Resources

American Educational Resource Association, American Psychological Association, and National Council on Measurement in Education. (1999). *Standards for Educational and Psychological Testing*. Washington, DC: American Psychological Association.

Equal Employment Opportunity Commission. *Uniform Guidelines for Employee Selection*. Washington, DC: Equal Employment Opportunity Commission. 29 C.F.R. Part 1607.

Joint Committee on Standards for Educational and Psychological Testing of the AERA, APA, and NCME. (1999). *Standards for Educational and Psychological Testing*. Washington, DC: American Educational Research Association.

National Fire Protection Association. (2012). *NFPA 1041: Standard for Fire Service Instructor Professional Qualifications*. Quincy, MA: National Fire Protection Association.

National Fire Protection Association. (2007). *NFPA 1403: Standard on Live Fire Training Evolutions*. Quincy, MA: National Fire Protection Association.

U.S. Department of Education. (2011). *Family Educational Rights and Privacy Act*. Washington, DC: Family Policy Compliance Office.

U.S. Department of Education, National Center for Education Statistics. (2011). *The Condition of Education 2011* (NCES 2011-033), Indicator 43.

You have been tasked with developing the quizzes and final exam for your department's Fire Officer I course, which meets the requirements of NFPA 1021, *Standard for Fire Officer Professional Qualifications*. The department wants to ensure that these tests cover all of the requirements of the standard, but the results will also be incorporated into the upcoming Lieutenant's promotional process scoring.

1. To ensure the test is objective, which of the following formats will you *not* use?
   - **A.** Written
   - **B.** Completion
   - **C.** True/false
   - **D.** Essay

2. Laboratory exercises, scenarios, and job performance requirements (JPRs) that test the student's ability to perform tasks are types of
   - **A.** written examinations.
   - **B.** oral examinations.
   - **C.** performance tests.
   - **D.** multiple-choice tests.

3. What is the most important reason for giving quizzes during the course?
   - **A.** To average out the final exam scores
   - **B.** To make students read their books as they go
   - **C.** To allow the instructor to identify areas that need more attention
   - **D.** To test areas not covered by the final exam

4. If students are able to pass the exam but are not capable of performing well as a company officer, the exam would not be considered
   - **A.** valid.
   - **B.** reliable.
   - **C.** consistent.
   - **D.** unusual.

5. If in one class all students obtain high scores, yet in another class the students all have low scores, the test may not be
   - **A.** valid.
   - **B.** reliable.
   - **C.** consistent.
   - **D.** unusual.

6. What is the major problem with two-answer selection tests such as true/false tests?
   - **A.** They tend to be ambiguous.
   - **B.** They tend to be subjective.
   - **C.** The "guessing" factor can influence results.
   - **D.** They are difficult to construct.

7. If you decide to use a test-item bank:
   - **A.** you can randomly generate the exam and it will be valid.
   - **B.** you must carefully analyze the test bank to make sure the test items are properly validated.
   - **C.** the students will probably not be able to pass the exam.
   - **D.** you will not be able to provide a Web-based version of the exam.

8. You will need to ensure the testing conforms to *Uniform Guidelines for Employee Selection*. Which organization publishes this document?
   - **A.** Equal Employment Opportunity Commission
   - **B.** Department of Labor
   - **C.** International Association of Fire Chiefs
   - **D.** American Psychological Association

Evaluating the Fire Service Instructor

Fire Service Instructor I

Knowledge Objectives

There are no knowledge objectives for Fire Service Instructor I students.

Skills Objectives

There are no skills objectives for Fire Service Instructor I students.

Fire Service Instructor II

Knowledge Objectives

After studying this chapter, you will be able to:

- Describe the purposes of instructor evaluations. (p 260)
- Describe the methods for fire service instructor evaluation (NFPA 5.5.3). (pp 262, 264)
- Describe the evaluation process (NFPA 5.2.6). (pp 264–266)
- Describe the role of providing feedback to the fire service instructor using an evaluation (NFPA 5.2.6). (pp 267, 269)
- Describe how to develop a class evaluation form (NFPA 5.5.3). (pp 269–271)

Skills Objectives

After studying this chapter, you will be able to:

- Perform an evaluation of a fire service instructor (NFPA 5.2.6). (pp 264–266)
- Develop a class evaluation form (NFPA 5.5.3). (pp 269–271)

© Photos.com

Fire Service Instructor III

Knowledge Objectives

There are no knowledge objectives for Fire Service Instructor III students.

Skills Objectives

There are no skills objectives for Fire Service Instructor III students.

You have been tasked with updating the course and instructor evaluation form used in your training division. The current evaluation form is outdated and has not been used for a long time—the information it is meant to gather is not accurate, and neither the students nor the instructors like to use it or fill it out. In addition to the form being out of date, the instructors have a very poor attitude toward the entire evaluative process because they believe that members of the department use the form for "payback." After reviewing the current course and instructor evaluations, you see why both instructors and students dislike the evaluation form. It is several pages long and seems redundant in asking for the same information. The format is cumbersome as well.

1. What would be your first step in addressing the task you have been assigned?
2. Where could you find resources to assist you in updating your evaluation form?
3. What could you do to help your instructors see the value of an evaluation form for use in their classes?

Introduction

At the historic Wingspread Conference in 1966, participants agreed on the need for the fire service to attain professionalism through education. In 1971, the Joint Council of National Fire Service Organizations (JCNFSO) created the National Professional Qualifications Board for the Fire Service (NPQB) to facilitate the development of nationally applicable performance standards for uniformed fire service personnel using the National Fire Protection Association (NFPA) standards. In 1972, the NFPA established a committee charged with developing a standard for the qualifications for a fire service instructor.

The first standard produced by the committee was issued in 1976 as NFPA 1041, *Standard for Fire Service Instructor Qualifications*. The goal of the standard was to establish

... clear and concise job performance requirements that can be used to determine that an individual, when measured to the standard, possesses the skills and knowledge to perform as a fire service instructor.

NFPA 1041 obligates the Instructor II to

Evaluate instructors, given an evaluation form, department policy, and JPRs, so that the evaluation identifies areas of strengths and weaknesses, recommends changes in instructional style and communication methods, and provides opportunity for instructor feedback to the evaluator (NFPA 1041, Section 5.2.6).

An effective and efficient instructor evaluation system is an important component of striving for constant improvement in fire service training. Most training sessions involve some type of evaluation of the student, which can also be viewed as one part of the evaluation of an instructor. If students have poor test results or fail in practical training evaluation, that performance may reflect poorly on the instructor who taught the students the material. Instructors hone their craft and

learn their subject matter well, often becoming subject-matter experts. The skills required to be an instructor go beyond those required to instruct a group of fire fighters. Each instructor will have different comfort levels with different types of instructional methods and audiences.

The instructor should welcome peer review—that is, evaluation by other instructors and members of a training division—as well as student evaluation of the instructor's presentation skills. After all, growth in the ability to present information can occur only when the instructors challenge themselves to do more. When given a subjective evaluation of performance, the instructor should be able to identify strengths and weaknesses in his or her presentation skills.

In reality, the fire service instructor is evaluated at all times during his or her performance, either formally or informally. Informal evaluation occurs during breaks or before or after class, when students compare their current fire service instructor to previous instructors from both inside and outside of the fire department. Formal evaluation is done as part of a fire department's improvement plan and as part of a self-improvement plan for the fire service instructor. In many agencies, these formal evaluations are used as part of the instructor/employee's yearly evaluation process that will impact potential promotions and salary increases. In fact, NFPA 1500, *Standard on Fire Department Occupational Safety and Health Program*, has referenced the evaluation of the department's training program as part of the overall occupational health and safety program.

An important part of any instructor evaluation form is how well the instructor manages fire fighter safety during training. There has been a growing trend of fire fighters being injured during the training process. According to the 2011 NFPA report *U.S. Firefighter Injuries*, in 2011 more than 7500 fire fighters were injured during training activities. Between the years of 2001 and 2010, 108 fire fighters were killed while engaged in training-related activities. Typically, fire fighter deaths during training average more than 10 per year. Some of these deaths occur due to cardiac arrest or some other medical condition; others occur as a result of trauma sustained during a training evolution. Approximately 50 percent of these deaths occur on the drill ground. Given the risks inherent in training, fire service instructors need to be evaluated in each and every setting—not just the traditional classroom, but also during high-risk training on the training ground.

To become an evaluator, job performance requirements (JPRs) must be met. Often, meeting the JPRs includes demonstration of an evaluation of a fellow fire service instructor's presentation and/or reviewing the evaluation with the fire service instructor who was evaluated. The goal is to help the fire service instructor identify strengths and weaknesses and become a better, more effective instructor through recommended improvements.

> ### Teaching Tip
>
> NFPA 1041 requires that the Fire Service Instructor II be capable of identifying another fire service instructor's strong and weak points and providing guidance for improvement.

Fire Service Instructor II

Fire Department Policy and Procedures

NFPA 1041 references departmental policies and procedures as one of the criteria for an evaluation. Many departments have not created their own standard for certifying fire service instructors; instead, they use state or national certification requirements for this purpose.

■ Fire Department Requirements

NFPA 1041 requires that the department or authority having jurisdiction (AHJ) use qualified individuals as fire service instructors. The qualifications include more than just being knowledgeable on the subject: NFPA 1041 also expects that the fire service instructor will have the capability and skill level to demonstrate how to perform the skills included in the training session appropriately. The level of competency of the fire service instructor must be identified in a policy developed and enforced by the AHJ. In addition, this policy must include a method for verifying the qualifications and competency of the fire service instructor.

A department's policy or procedure for fire service instructor evaluation includes many components. For example, the policy should identify who does the evaluation and ensure that the evaluation remains confidential. Confidentiality is important because it allows the evaluator to be critical without embarrassing the fire service instructor being evaluated. A timetable needs to be established for evaluations, so that fire service instructors are evaluated initially and then periodically. The policy should establish an evaluation process for practical instruction as well as classroom presentations.

The fire service instructor evaluation policy should address the fire service instructor's preparations to deliver the material as well as his or her actual presentation of that material. Preparations include the reservation and preparation of the facility and the equipment required. The fire service instructor should ensure that the classroom is appropriately arranged and that any necessary equipment is in place and in working order. He or she should have reviewed the lesson plan and adapted it for the target audience. Also, the fire service instructor needs to ensure that the class starts on time.

Another issue that must be addressed is attire; the fire service instructor must be wearing attire appropriate for the setting. If the coursework will address practical evolutions, the fire service instructor should wear the same level of protection as is required for students **FIGURE 11-1**. The fire service instructor should set a positive example for the students relative to personal protective equipment (PPE) on the drill ground. In the classroom, the fire service instructor should wear attire consistent with the attire expected of the students.

The evaluation policy also should include information on how the presentation is to be delivered. It should require that the fire service instructor follow the lesson outline; keep the class moving; use appropriate audio, video, or display

FIGURE 11-1 The fire service instructor should be in the same level of PPE as the students.

© Keith D. Cullom

equipment; finish material delivery in the allotted time frame; and deliver a summary. The policy must outline positive traits expected of the fire service instructor, including being unbiased, encouraging open discussion, being understandable and clear during the communication process, wearing appropriate attire, remaining flexible, and being honest. Undesirable traits—including the use of verbal fillers or undesirable or distractive mannerisms, inappropriate language, and failure to maintain control of the class—should be identified and discouraged.

The department policy for evaluating a fire service instructor can be as simple as a short paragraph stating when evaluations will be done, by whom, and in which form; conversely, a detailed policy may enumerate the entire evaluation process. In either case, the department's policies and procedures must be incorporated into the evaluation process and then followed as outlined.

Methods for Fire Service Instructor Evaluation

Fire service instructor evaluations may be a requirement for instructors' promotion. For certification at the Fire Officer I level under NFPA 1021, *Standard for Fire Officer Professional Qualifications*, you must obtain Fire Service Instructor I level certification to become a Fire Officer I. Fire service instructor evaluations may also be required to obtain state or national certification, as most states support a certification process for fire service instructors. The National Fire Academy and other institutions require initial and periodic evaluation of instructors. In some cases, fire service personnel may have to go through (and pass) a fire service instructor evaluation process before they are allowed to instruct students. This evaluation may be done by a supervisor, a peer, or a student, or the instructor may do a self-evaluation. Each evaluator puts his or her own knowledge and experience to work to complete the evaluation appropriately.

Teaching Tip

During practice presentations, it is a good idea to videotape the instructor during his or her presentation. After the instructor receives feedback from peers, he or she then goes into another room (with a senior instructor) and begins a critique of the presentation. This practice can be extremely helpful to the instructor.

■ Formative Evaluation

The formative evaluation process is typically conducted for the purpose of improving the fire service instructor's performance by identifying his or her strengths and weaknesses. For a student, this evaluation may take the form of a progress test or unit exam so that learning is measured and feedback can be given on the student's progress to date. For a fire service instructor, a formative evaluation may be done by either a supervisor or another fire service instructor for the purpose of professional development.

The formative evaluation process is intended to assess the instructor's delivery during a class presentation, and the results are reviewed with the instructor to help him or her improve or enhance teaching skills and classroom performance. For this reason, it is often perceived as less threatening than other evaluations. A typical formative evaluation is designed to improve the fire service instructor's abilities by identifying individual skills or deficiencies and allowing for the development of improvement plans.

A formative evaluation incorporates criteria that may allow the evaluator to review and observe the overall performance of the fire service instructor by relating that performance to the expected outcomes of the training session. The instructor's ability to meet these criteria may be measured by the instructional methods used during the training, such as visual aids, transitioning between visual aids and the lesson plan, student participation, and the application of a student testing form to the learning objectives of the class. The evaluator may also participate in observation of student performance and take notes on the fire service instructor's interactions with the students. Much of formative evaluation centers on instructional technique, training content, and the delivery methods used by the fire service instructor. This information is then compared to student success rates.

■ Summative Evaluation

A summative evaluation process measures the students' achievements, through testing or completion of evaluation forms, to determine the fire service instructor's strengths and weaknesses. Many applications of summative evaluations take place at the end of the training process. Such an evaluation is normally conducted for the purpose of making personnel decisions about fire service instructor certification, merit pay, promotion, and reassignment. Student evaluations of fire service instructors are commonly used for assessment as part of the summative evaluation process. Because students may have different reasons for being in the class, precautions are necessary when considering information from this type of evaluation form. Summative evaluations usually are done by an administrator, such as the training director or training officer, rather than by a supervisor or another fire service instructor.

■ Student Evaluation of the Fire Service Instructor

Students are in a position to rate the increased knowledge they have gained and their classroom experience as it relates to a particular fire service instructor. They may provide valuable information about other factors not readily available during an evaluation, such as the fire service instructor's punctuality, use of audiovisual equipment, and typical classroom demeanor.

Some departments include an evaluation form for students to complete at the conclusion of the course. Different delivery

JOB PERFORMANCE REQUIREMENTS (JPRS)
in action

Evaluation of a fire service instructor and the content of a presentation serves as a form of quality assurance. These measurements of the fire service instructor's skills and impressions of how the course material helped students understand and apply the learning objectives are an important part of a fire service instructor's professional development. A student review of the fire service instructor's attributes is one method of performing an evaluation. Another strategy is to have a senior fire service instructor conduct an in-person review of the course delivery.

Instructor I

All instructor levels should develop an appreciation for how course and instructor evaluations benefit the instructional process. At the Instructor I level, you should consider all feedback provided in an evaluation as a measure of professional development and work to improve your performance based on any constructive criticisms provided while building on positive attributes.

Instructor II

Evaluations of instructor performance are a type of personnel evaluation, and their construction and use should reflect the intent of any evaluation—namely, to improve the performance of the person being evaluated. The Instructor II will develop class and instructor evaluations and conduct evaluations of other instructors in an effort to improve both the instructor and the course elements.

Instructor III

Evaluation strategies performed by the Instructor III are covered in the *Training Program Management* chapter and include the development of an ongoing instructor evaluation plan. This may be similar to the annual performance review used for any other level of responsibility, such as fire fighter or company officer, with the criteria in this case focusing on how well the instructor performed his or her duties during that period.

JPRs at Work

NFPA does not identify any JPRs that relate to this chapter. Nevertheless, the Instructor I should understand how to utilize information provided during the evaluation process.

JPRs at Work

Develop evaluation forms and evaluate instructors in an effort to improve instructor skills and the quality of the course materials by providing feedback on class presentation elements.

JPRs at Work

There are no JPRs for the Instructor III.

Bridging the Gap Among Instructor I, Instructor II, and Instructor III

Personnel who meet the Instructor I JPRs will be able to use their experiences in being evaluated when they transition to Instructor II by recalling the elements of the evaluations that they were able to put to use so as to improve their own performance. When conducting evaluations, the Instructor II should be professional and constructive in identifying areas for improvement. Individual strengths and weaknesses should be highlighted, and both instructor levels should use this step in the instructional process in a positive manner. The Instructor III will acquire these evaluations and incorporate their results into an ongoing evaluation plan. Monitoring performance of instructors is a key part of maintaining an effective training program.

schedules and multiple fire service instructors delivering training over a long period of time may require evaluation forms to be distributed earlier, however, so that students can complete the evaluation while specific instructor characteristics are still fresh in their minds. A course evaluation typically includes a section that allows students to evaluate the fire service instructor, classroom setting, instructional material, handouts or audiovisual material, and ability of the material taught to meet their needs. Developing class evaluation forms is discussed in more detail later in this chapter.

Some students may view the evaluation as an opportunity to take a shot at the fire service instructor. Others may just go through the motions of completing the evaluation, without giving it much thought or consideration. Some departments account for these problems by evaluating the students participating in the class—that is, students can be compared in terms of participation, knowledge, and skill level. This information can then be used when reviewing the students' evaluations for the fire service instructor and the course.

Ethics Tip

Fire service instructors naturally tend to become acquaintances because they have a common passion and often attend the same seminars and training sessions. Over time, friendships will develop. In turn, you may be put in the position of having to evaluate a friend. Evaluating a close friend can be difficult. If you find yourself in such a position, you will naturally want to point out the positives and avoid mention of any negatives. Do you have the fortitude to evaluate a friend honestly and fairly? As an evaluator, it is your responsibility to the department, to the students, and to the fire service as a whole to conduct evaluations fairly and honestly. Friendship is a personal relationship that cannot enter into the evaluation process.

Theory Into Practice

Informal feedback from casual conversations by students can be an effective measure of fire service instructor performance, but caution should be used when interpreting those remarks. A single opinion may not necessarily represent the views of all students.

Evaluation Process

■ Preparation

To make an evaluation of a new fire service instructor's performance a positive event, start by selecting a topic from the schedule with which you are familiar and enjoy teaching. Prior to conducting an evaluation, you need to review the evaluation criteria and their relevance to the job description; review the

lesson plans and supporting material; and then schedule the date, time, and location for the evaluation (if deemed appropriate by the AHJ) **FIGURE 11-2**.

If you plan to schedule a visit to a session being conducted by the fire service instructor, state the purpose of your visit and evaluation in positive terms. Identify the areas that you will be reviewing, including the delivery of the provided lesson plan material, the fire service instructor's ability to apply the material to the class, adherence to department and accepted safety practices, and the instructor's overall presentation skills. Encourage the fire service instructor to present normally and not to attempt to involve you in the delivery process. Be sure to arrive early enough to determine whether the fire service instructor is properly prepared. The evaluation standards should be objective, identified prior to the evaluation, focused on the fire service instructor's performance, and part of an improvement program for the fire service instructor and course content, if applicable.

■ Observations During the Evaluation

During the evaluation, look at the fire service instructor's entire skills set. In using all sources of information, you will give a more complete analysis of the fire service instructor's performance. During classroom training evaluations, areas to evaluate include the instructor's dress: Is the instructor's attire professional, compliant with policy, and appropriate for the audience? The demeanor of the fire service instructor should be professional, outgoing, approachable, and appropriate for audience: Does the fire service instructor appear to be engaging the audience? Does he or she inspire confidence and respect?

When evaluating the presentation, note the fire service instructor's ability to relate the topic directly to the students' needs. Note any mannerisms, language usage, and comfort level with technology. Do you feel that you are watching a black-and-white, still-picture show or a full-color, interactive presentation?

One effective method is to observe the classroom presentation while monitoring the lesson plan the instruction is

FIGURE 11-2 Prior to the evaluation, review the lesson plans.

Find a standardized method of performing evaluations if one is not provided by your AHJ. A standardized method ensures a fair and consistent evaluation of all fire service instructors and ensures that learning opportunities are not missed. To perform an evaluation, follow these steps:

1. Identify the appropriate method for the evaluation.
2. Select or develop a form to be used to record the evaluation.
3. Familiarize yourself with the evaluation form.
4. Read the course outline lesson plan.
5. Schedule a time and location for the evaluation.
6. Meet with the fire service instructor prior to the class to explain the intent of the evaluation.
7. Make the atmosphere as friendly as possible. Try to alleviate the fire service instructor's anxiety.
8. Observe the fire service instructor delivering course content.
9. Take notes.
10. Complete the form.
11. Review the fire service instructor's performance with him or her.

If you are evaluating a session that involves PPE, you should wear the same level of PPE as the students and the fire service instructor and you should model how to wear it appropriately.

that the side trip benefits the overall experience and enhances student knowledge. The fire service instructor must be familiar with the lesson plan and its learning objectives for the class—a key point to keep in mind during the evaluation. Effective instructors use the lesson plan as a tool to deliver the desired knowledge. As instructors become comfortable with lesson content, they will appear to use the lesson plan less, but in reality they are so familiar with the lesson content they can teach the desired content without seeming to read the lesson plan. Good and effective instructors deliver the lesson plan but do so with ease and in a comfortable manner. Just because the instructor you are evaluating is not constantly looking at the lesson plan does not mean he or she is not following it.

Closely scrutinizing a fire service instructor's adherence to student safety guidelines should be a paramount concern.

teaching from. The lesson plan components should be reviewed in terms of how they were used and whether the presentation was delivered in a consistent manner. By using a copy of the lesson plan and matching it with the actual presentation, you can also review how well the instructor followed and used the lesson plan.

During practical or drill-ground evaluations, the overall safety of the students is the major focus of the evaluation. Fire service instructors have the twin responsibilities of presenting training in a safe manner and of teaching safe practices. When reviewing a fire service instructor during practical sessions, consider whether he or she was able to anticipate problems in the delivery of instruction such as noise, weather, or equipment availability.

Certain types of training sessions, such as live fire exercises or complex rescue scenarios, may require the assignment of a safety officer and should always follow NFPA guidelines when applicable. Both the fire service instructor and the students must dress appropriately for activity, wear PPE correctly, and clean or inspect the PPE after heavy use. The best review of a practical training session is to determine how closely the fire service instructor's modeling during the training session matched the real fire-ground application of the skill.

■ Lesson Plan

The lesson plan is the fire service instructor's road map. It is occasionally appropriate to stray off the road, providing

■ Forms

Because of the diverse reasons for which evaluations are undertaken, a variety of printed forms are used to document the evaluation process. Fire service instructor evaluation forms vary widely, with some being better than others. When a department adopts a certain form, that document must be used during your evaluation **FIGURE 11-3** . Before adopting an evaluation form, it is prudent to review several different versions of forms to determine which content and criteria best fit the needs of the department. If an agency plans to select an evaluation form that is produced commercially or adopted from another agency, all of the instructors who will be evaluated using the form should be given the opportunity to provide feedback on which form to select. Allowing the instructors the opportunity to share their opinions on the type of form and its content, format, rating system, and results will help in the adoption of the forms. Knowing the purpose and goal of the evaluation process, as well as which information will be gathered, will ensure that instructors are more willing to participate in the evaluation process.

As the old saying goes, "The job isn't over until the paperwork is done." Thorough and accurate records are essential to maintain credibility and help ensure fairness. Procrastination simply makes the job harder. To complete a form for an evaluation, follow these steps:

1. Review the form.
2. Have appropriate writing materials.
3. Take notes and/or fill in the appropriate areas of the form.

4. Review the fire service instructor's performance with him or her.
5. Submit paperwork to the AHJ as directed or required.

Feedback to the Instructor

After completing the evaluation of the instructor, you should meet with the instructor and let him or her know that you

---

Lead Instructor Evaluation Criteria

Please shade the letter that indicates your evaluation of the course/instructor feature using #2 pencil.

Instructor Attributes	Excellent	Very Good	Good	Fair	Poor
Ensure that the daily classroom and practical training evolutions were conducted in a safe and professional manner	(1A)	(1B)	(1C)	(1D)	(1E)
Ensure that all learning objectives were covered with more than adequate resource information	(2A)	(2B)	(2C)	(2D)	(2E)
Coordinate the activities of all students so that the class flow and pace was not interrupted	(3A)	(3B)	(3C)	(3D)	(3E)
The instructor was prepared to present the program	(4A)	(4B)	(4C)	(4D)	(4E)
Provide documentation to your progress in class and in general on a regular basis	(5A)	(5B)	(5C)	(5D)	(5E)
Monitor all class and practical activities and provide a structured learning environment	(6A)	(6B)	(6C)	(6D)	(6E)
Provide a positive role image to you at all times and conduct themselves in a professional manner	(7A)	(7B)	(7C)	(7D)	(7E)
Was accessible and responsive to your individual needs as a student in this program	(8A)	(8B)	(8C)	(8D)	(8E)
Was actively involved in your training and education throughout the program	(9A)	(9B)	(9C)	(9D)	(9E)
Worked to improve program content, policy and procedure	(10A)	(10B)	(10C)	(10D)	(10E)
The instructor is knowledgable about this program and acted accordingly	(11A)	(11B)	(11C)	(11D)	(11E)
The instructor was instrumental in your success in this program	(12A)	(12B)	(12C)	(12D)	(12E)

Instructor(s) Evaluation: Please use the following scale to evaluate the instructor(s).

A = Excellent; clearly presented objectives, demonstrated thorough knowledge of subject, utilized time well, represented self as professional, encouraged questions and opinions, reinforced safety practices and standards

B = Good; presented objectives, knowledgeable in subject matter, presented material efficiently, added to lesson plan information

C = Average; met objectives, lacked enthusiasm for content or subject, lacked experience or background information, did not add to delivery of material

D = Below Average; did not meet student expectations, poor presentation skills, lack of knowledge of content or skills, did not interact well with students, did not represent self as professional

E = Not Applicable; did not instruct in section or course area, do not recall, no opinion

Instructor Name	Excellent	Good	Average	Below Average	Not Applicable
	(13A)	(13B)	(13C)	(13D)	(13E)

FIGURE 11-3   A sample instructor evaluation form.

enjoyed being in the class and that you appreciate the opportunity to participate with the instructor in the evaluation process. Unless you are prepared to meet with the instructor immediately, you should set up a time to discuss your evaluation and to provide feedback. In many situations, the instructor may be tired or in a hurry to prepare for another class, so take the time to make an appointment when both of you can sit down in a relaxed environment and be refreshed. Just as students appreciate quick feedback on test results, so instructors should also have quick access to instructor evaluation tools.

In certain cases, the class may have gone wrong or the instructor may have done a poor job in presenting the lesson plan. In such a situation, you might need to take some time to step back and look for positives in the evaluation before you meet to address negative parts of the evaluation. If the true purpose of the evaluator is to improve the quality and performance of the instructor, then discussing the evaluation after time for consideration works best for all parties involved.

FIGURE 11-4 Review your notes in person with the fire service instructor.

■ Evaluation Review

After the evaluation is complete and you have taken some time to review the evaluation form and any notes you have

---

Theory Into Practice

As an instructor, it is a good practice to retain copies of your evaluations and to review them periodically, especially in cases where you are teaching the same course. You can build upon your strengths and improve your class delivery based on previous comments.

Theory Into Practice

In larger academies and training centers, the training director or administrative staff may summarize student evaluations and provide summary reports to each instructor in the program. Care should be taken to provide each instructor with an individual evaluation that is not shared with all instructors in the program. Rating scales may be provided to show the instructor where he or she stood in comparison with other instructors. If a lead instructor was assigned for the course, he or she may be given access to all instructor ratings to assist in selecting instructors for future offerings.

---

taken, you should be better prepared to provide feedback to the instructor. During your discussion with the instructor, take the time to review your notes in person with the fire service instructor **FIGURE 11-4**. Specifically consider the fire service instructor's strengths and weaknesses. During the review, strive to show professionalism by maintaining a level of formality appropriate for the situation, thereby sending the message that this is a formal process. If agency policy allows, provide the instructor with copies of completed evaluation forms for his or her reference and professional development. At the very least, a summary of evaluation results should be provided to the instructor.

Start the session by thanking the instructor for the opportunity to observe the class. Establishing a dialog with the person you are evaluating will help open the lines of communication, so that when you provide the results of the evaluation there will be a willingness to listen to what you have to share. All feedback should be delivered in a considerate manner, stressing the instructor's positive attributes

first, and then addressing any areas that were observed to be weak. Offer ideas or suggest changes to deal with weaknesses, and ask the instructor for ideas he or she thinks might help in dealing with weak areas. Do not overwhelm the fire service instructor with criticism; instead, provide constructive feedback. Strive for a balance between positive and negative comments, and set goals for fire service instructor improvement when appropriate. Provide examples from your own experiences as a fire service instructor to provide direction. For example, if the instructor struggled with the use of computer equipment, offer time to review the equipment and provide practical teaching tips on that equipment. In this way, you can turn the evaluation session into a learning session. You may also want to include a follow-up review at a later date. As in any evaluation, make sure that any review dates are observed and reevaluations completed to maintain credibility of the evaluation process. Finally, always make yourself available to the new fire service instructor for advice and guidance.

VOICES
OF EXPERIENCE

Evaluations have always been a challenge for me—whether as the one being evaluated or as the evaluator. The form being used and the purpose of the evaluation are always question marks. What the evaluation can actually do is also another question: If the evaluation has no real purpose other than to say we did it, how much credence will it get? If it cannot help you but the negative items can be used to harm you, how will it get viewed? If it impacts whether you are promoted or can keep your job, then what does it mean?

I have been within systems where all of these were true to one degree or another. The greatest challenge to me was to evaluate both full- and part-time people, to do it as fairly as I could, and then to communicate the reasons for the evaluation to the one being evaluated in a way that let them understand my reasons for the scoring of the evaluation. Here is an example to show you what I mean:

> **I asked him what example was being set for his students if he required them to wear proper gear when he did not.**

I visited a fire fighter II training class in which the instructors were teaching ladders. (They were outside doing practical skills.) One instructor was in full turnout gear, as were the students. One instructor—the lead instructor—was not. I had a simple evaluation form that evaluated the conduct of the class, and a portion of it included the use of turnout gear. I talked to the instructor and asked why he was not wearing his gear. He said it was too hot on the roof to stay in turnout gear. I asked him what example was being set for his students if he required them to wear proper gear when he did not. I told him that the evaluation going into his file would indicate that he and I had had the conversation and that the expectation was that he would wear proper turnout gear in the future.

I think that there are some keys to evaluations that I have learned, regardless of the form, the system, or the purpose. The biggest key is communication. Let the person being evaluated know what is on the form, what the form items mean, and what the form can and cannot do. Do not spring the evaluation on the instructor—he or she should know what is coming. If you have been in communication, they should have an idea of what to expect. The other key is to listen. The employee has something to say, so listen. You did not come down from the mountain to convey words of wisdom. So listen to the person; you just might learn something.

Lloyd H. Stanley
Adjunct Instructor
Fire Protection Technology
Guilford Technical Community College
Jamestown, North Carolina

Theory Into Practice

To discuss an evaluation with a fire service instructor, follow these steps:

1. If possible, set up an appointment at a later time to review the evaluation form.
2. At your appointment, meet with the fire service instructor in a private area to discuss his or her performance.
3. Explain the reason for the review and thank the instructor for the opportunity to evaluate him or her.
4. Determine whether the fire service instructor has any comments.
5. Use notes and/or the evaluation form to identify the fire service instructor's strong points.
6. Use notes and/or the evaluation form to identify areas of the fire service instructor's performance that need improvement.
7. Review the overall outcome/final score of the evaluation.
8. Offer the instructor any help or assistance he or she might need and assign follow-up dates for future reviews and reevaluation.
9. Close on a positive note.

Teaching Tip

Remind the fire service instructor that it is his or her obligation to remain professional at all times. The fire service instructor is responsible for what occurs in the classroom.

■ Student Comments

As mentioned previously, a valuable source of feedback is student comments on evaluation forms. When taken objectively, these comments can provide significant information. The challenge for the fire service instructor is to understand that all students are not taking this class to gain better skills; some students need the class or certificate of completion for other means. Read the written comments where provided. If a student took the time to write a remark, evaluate its weight when reviewing the class. Students who provide written feedback in addition to filling in check boxes show a genuine desire to improve the learning environment for the next student who takes that class.

Sometimes students use course evaluation forms to take a shot at the instructor, often due to a personality conflict. As a supervisor, you should weigh these types of comments by comparing them to the complete array of comments received from the entire class. If several students mention similar problems or areas of concern, then as a supervisor this might prompt further investigation into the course itself and the instructor. If there is only one negative comment, however, then it might not be as serious. In all fairness to the student who took the time to make a comment, you as a supervisor need to discuss the comment with the instructor involved. In many cases the instructor will identify the student based on the comment, and in most cases the instructor will be able to provide some additional history or description of the underlying motivation for the comment.

Student comments about the instructor are just one aspect of student feedback that needs to be considered from a course evaluation instrument. Other areas on the form should address course material, course handouts, classroom setting, length of the course, whether the course met the students' needs and expectations, and whether the advertisement for the course was accurate.

Effective evaluation forms are tools used by the instructor and the institution delivering the training as a mechanism to improve how well they perform. If the course was delivered at a training center provided by the department, then questions about advertisement or cost of the course would not have much value, but course content and curriculum would.

Theory Into Practice

You will typically see certain remarks on every student evaluation, such as one indicating that the class could have been shorter. Keep in mind that not all students are in the class to seek knowledge; for whatever reason, some students view themselves as prisoners of the system.

Developing Class Evaluation Forms

One of the skills a Fire Service Instructor II needs to cultivate is the ability to develop a class evaluation form or instrument. An *instrument* is something that is used to measure performance, so this term is more appropriate when considering the process of developing something to measure performance. The most commonly used evaluation instrument is a form. Some departments may provide you with a class evaluation form, whereas others may have you develop your own. Whatever its source, the evaluation form should cover specific topics, such as the training environment and course material **FIGURE 11-5**.

Remember to update the content of instructor and course evaluation forms when newer content or content specific to a particular course is needed. If an existing form does not fit your course, change it to reflect the rating areas on which you want feedback.

Illinois Fire Chiefs Foundation
End of Course Evaluation Form

Course Title: _____ Course Location: _____

Course Evaluation: Please shade the letter that indicates your evaluation of the course/instructor feature using #2 pencil.

Course Feature	Excellent	Very Good	Good	Fair	Poor
1. AV Material / Course materials	(1A)	(1B)	(1C)	(1D)	(1E)
2. Course organization / flow / delivery rate	(2A)	(2B)	(2C)	(2D)	(2E)
3. Observance to safety procedures and practices	(3A)	(3B)	(3C)	(3D)	(3E)
4. Hands on Experience if applicable to course	(4A)	(4B)	(4C)	(4D)	(4E)
5. Time per subject	(5A)	(5B)	(5C)	(5D)	(5E)
6. Evaluation tools	(6A)	(6B)	(6C)	(6D)	(6E)
7. Will this course help you with your professional development?	(7A)	(7B)	(7C)	(7D)	(7E)
8. Did the course meet your expectations?	(8A)	(8B)	(8C)	(8D)	(8E)
9. Would you recommend this course to others?	(9A)	(9B)	(9C)	(9D)	(9E)
10. Overall impression of course	(10A)	(10B)	(10C)	(10D)	(10E)

Add comments on specific areas on reverse

FIGURE 11-5 End-of-course survey.

Evaluating the learning environment is important to student success. Was the room too hot or cold? Was the classroom lit properly? Did the audiovisual equipment work? Were the chairs and tables comfortable? Was the learning environment free from distractions? These are questions that affect students' learning environment, which has a direct impact on students' ability to learn. When used properly, student evaluations of the learning environment can be used to justify additional funds for classroom improvements or upgrades to audiovisual equipment.

Evaluation forms should also include areas for evaluating the course material. Did the material meet the students' needs? Did it meet the learning objectives? Did the training material match NFPA standards, the department's standard operating procedures, and the department's own mission statement? If the course was offered to other agencies or on a regional or national basis, the form might also include questions on the registration process or the advertisement of the course.

The class evaluation form should also include questions regarding the fire service instructor:

- Did the class start and end on time?
- Did the fire service instructor know the course material?
- Did the fire service instructor identify the learning objectives?
- Was the fire service instructor dressed appropriately?
- How well did the fire service instructor meet the needs of the class?
- Did the instructor provide a classroom environment that allowed for open discussion?

In addition to the questions asked on the evaluation form, the format and the manner in which the form is completed represent an important part of the instrument. In general, all evaluation instruments follow a similar format and can be broken down into three areas:

- The heading area, where the students fill in the date, course title, location of course, or other agency-specific information so the evaluation form can be tied to a course and instructor
- The question area, which asks questions related to the following topics:
  - Course content, learning objectives, and course expectations
  - Learning environment, facility, and classroom temperature
  - Whether student expectations were met
  - Instructor performance and conduct
- An open forum where students can write in their thoughts and opinions about the course

A rating scale is a common element on an evaluation form and is used to evaluate the various aspects of the course in the question area. The scale is usually listed 1–5, with 1 indicating the respondent strongly agrees with a statement about the course and 5 indicating the respondent strongly disagrees with the statement. It is very important when administering such a form to make sure the students understand which end of the scale indicates "strongly agrees" or "strongly disagrees." When developing an evaluation form for use in your agency, you need to make sure the questions are asked in a consistent format. Improperly designed evaluation tools will mix questions with the desired responses changing from "agree" to "disagree" for every other question. This becomes confusing to the student and, in frustration or unintentionally, they may mark the wrong response. This, in turn, could have a negative impact on a course or instructor.

Many forms are scanned by a reader or scanner attached to a computer and software that is used to gather information from the form and provide a printout of the results. These types of instruments use a "bubble" form format wherein students fill in or darken a bubble indicating their opinion per the rating scale FIGURE 11-6 . The resulting data can be viewed one evaluation at a time or by assigning a value to each response to provide statistical averages. Optical scanning device readers can computerize these results and make the statistics easier to interpret.

A potential drawback to reliable evaluation occurs when a student evaluation form asks questions that students cannot or do not judge reliably. For example, many times these evaluations may indicate that the class should be shorter when the lesson plan actually calls for a longer presentation.

In addition to the format of the questions, the number of questions asked on an evaluation instrument needs to be considered. The inclusion of too many questions or redundant questions could cause the student to become frustrated, which could then prompt the student to disregard his or her true feelings or opinions and replace them with a hurried attitude and just "bubbling in spots."

The way in which the information gained from an evaluation form is used is just as important as the evaluation form itself.

If a department wants to improve the quality of its programs, then it must act upon the information obtained from the evaluation process. If improvements to the classroom are necessary, then the department can budget for those improvements. If the course material is not meeting students' needs, then revisions to the lesson plan may be necessary. If there are issues with the fire service instructor, they can be addressed as well.

Regardless of how the evaluation form was obtained (purchased, borrowed, or developed in-house), the form needs to be reviewed by the instructors who will be evaluated by it, the supervising instructors to make sure the form addresses the needs of the agency, and the chief because all activities and actions fall back on the leadership of the department. A course or instructor evaluation process is only as important and valuable to an organization as the degree to which the chief supports it. If needs or problems are found during an evaluation process, whether they involve an instructor or a poor classroom environment, progressive and effective chiefs are willing to fix problems and improve their department because ultimately they know they are responsible for the safety and well-being of their fire fighters, officers, and instructors.

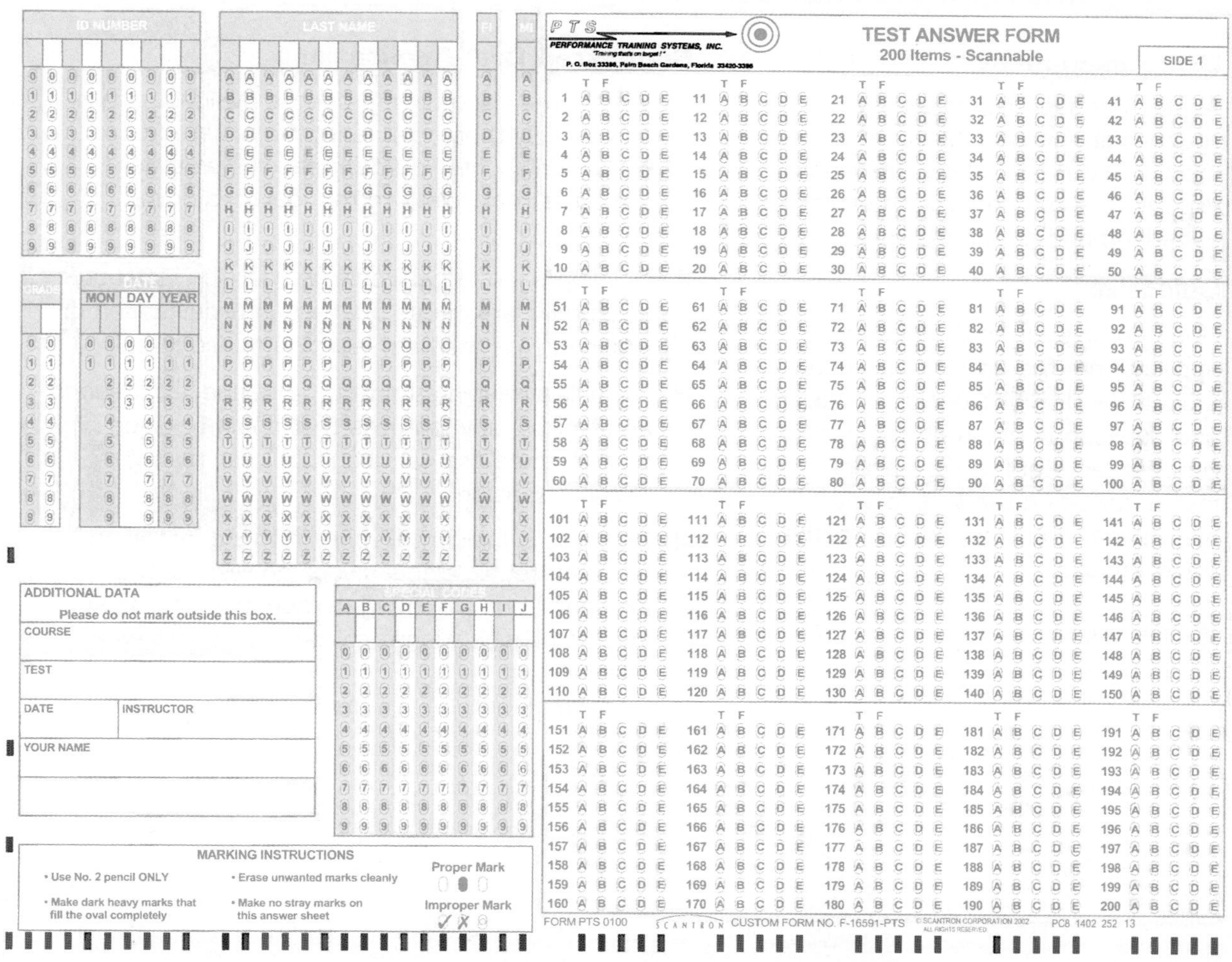

FIGURE 11-6 Sample of a bubble sheet evaluation form.

Jones & Bartlett Fire District

Training Division
5 Wall Street, Burlington, MA, 01803
Phone 978-443-5000 Fax 978-443-8000
www.fire.jbpub.com

Instant Applications: Evaluating the Fire Service Instructor

Drill Assignment

Apply the chapter content to your department's operation, its training division, and your personal experiences to complete the following questions and activities.

Objective

Upon completion of the instant applications, fire service instructor students will exhibit decision making and application of job performance requirements of the fire service instructor using the text, class discussion, and their own personal experiences.

Suggested Drill Applications

1. Review the content of a course evaluation form and identify those elements that relate to fire service instructor qualities as well as those elements that relate to the features of the course.

2. Develop a list of desirable fire service instructor qualities. Rank these qualities in order based on which create the best learning environment.

3. Consult available standard personnel evaluations and identify any features that could be used to evaluate a fire service instructor during a training session.

4. Review the Incident Report in this chapter and be prepared to discuss your analysis of the incident from a training perspective and as an instructor who wishes to use the report as a training tool.

Incident Report

Cambridge, Minnesota—2011

On May 23, 2011, a 35-year-old male volunteer fire fighter died after falling from a rope he was climbing after the conclusion of a ropes skills class.

The department was conducting a ropes and mechanical advantage haul systems training session that consisted of classroom and practical skills intended to provide the fire fighters rope skills. The training was provided by a contract agreement with the local technical college. The drill was a continuation of a two-day, 8-hour rope skills training session that was provided by the technical college. (The first four-hour ropes, knots, and rigging class had been held approximately two weeks prior and most of the students, including the victim, had attended both sessions.)

The instructor for the class was an adjunct instructor for the college and also a fire fighter for a neighboring large city. The college had originally sent two instructors but one of the instructors had to cancel at the last minute due to an emergency. The training consisted of the demonstration of a number of different haul system configurations designed to create a mechanical advantage using ropes and pulleys to lift objects. The fire fighters were rotated through different activities, including viewing the rigging points and pulleys and setting up the systems such as the hauling and anchor points and the safety rope. All of the fire fighters and officers interviewed acknowledged that the instructor kept the students' attention focused on the training activities. There were 19 students participating in the training.

The drill had concluded and the students were in the process of breaking down the drill site and putting the equipment away. The victim and two other fire fighters were standing in front of the 95-ft tower ladder, elevated 20-30 feet, which had served as the anchor point for the rope training, when the victim decided to climb one of two suspended ropes in an attempt to access the other suspended rope. Both the chief of the department and the instructor in charge of the training noticed the fire fighter (victim) climbing the rope and called out to him to stop and descend the rope.

The victim was climbing up a rope that had been used to demonstrate rope haul systems and attempted to grab another rope out of his reach. The fire fighter (victim) likely lost his grip on the rope and fell approximately 6-8 feet to the asphalt pavement, striking his head. The victim was not wearing any PPE when climbing the rope. Emergency medical aid was administered by fellow fire fighters and he was transported to a local hospital where he died from his injuries. The medical examiner reported the cause of death as blunt force head trauma.

Post-Incident Analysis: Cambridge, Minnesota

Contributing Factors

Qualified safety officer (meeting qualifications of NFPA 1521) must be appointed in practical skills training environments

Proper personal protective equipment (PPE) not worn during high-risk activity

Student-to-instructor ratio not maintained during high-risk training drills

Wrap-Up

Chief Concepts

- NFPA 1041 requires that the Fire Service Instructor II be capable of identifying other fire service instructors' strong and weak points, and provide guidance for other instructors' improvement.
- Many departments have not created their own standard for certifying fire service instructors; instead, they use state or national certification requirements for this purpose.
- The level of competency of the fire service instructor must be identified in a policy developed and enforced by the AHJ. In addition, this policy must include a method for verifying the qualifications and competency of the fire service instructor.
- The formative evaluation process is typically conducted for the purpose of improving the fire service instructor's performance by identifying his or her strengths and weaknesses.
- A summative evaluation process measures the students' achievements, through testing or completion of evaluation forms, to determine the fire service instructor's strengths and weaknesses.
- A course evaluation typically includes a section that allows students to evaluate the fire service instructor, classroom setting, instructional material, handouts or audiovisual material, and ability of the material taught to meet their needs.
- Prior to conducting an evaluation, you need to review the evaluation criteria and their relevance to the job descriptions; review the lesson plans and supporting material; and then schedule the date, time, and location for the evaluation (if deemed appropriate by the AHJ).
- During practical or drill-ground evaluations, the overall safety of the students is the major focus of the evaluation. Fire service instructors have the twin responsibilities of presenting training in a safe manner and of teaching safe practices.
- Just because the instructor you are evaluating is not constantly looking at the lesson plan does not mean he or she is not following it.
- Before adopting an evaluation form, it is prudent to review several different versions of forms to determine which content and criteria best fit the needs of the department.
- After completing the evaluation of the instructor, you should meet with the instructor and let him or her know that you enjoyed being in the class and that you appreciate the opportunity to participate with the instructor in the evaluation process.
- During your discussion with the instructor, take the time to review your notes in person with the fire service instructor and consider the instructor's strengths and weaknesses.
- Students who provide written feedback on evaluation forms, in addition to filling in check boxes, show a genuine desire to improve the learning environment for the next student who takes that class.
- In addition to the questions asked on the evaluation form, the format and the manner in which the form is completed represent an important part of the instrument.

Hot Terms

<u>**Formative evaluation**</u> Process conducted to improve the fire service instructor's performance by identifying his or her strengths and weaknesses.

<u>**Summative evaluation**</u> Process that measures the students' achievements to determine the fire service instructor's strengths and weaknesses.

References

National Fire Protection Association. (2009). *NFPA 1021: Standard for Fire Officer Professional Qualifications.* Quincy, MA: National Fire Protection Association.

National Fire Protection Association. (2012). *NFPA 1041: Standard for Fire Service Instructor Professional Qualifications.* Quincy, MA: National Fire Protection Association.

National Fire Protection Association. (2012). *NFPA 1403: Standard on Live Fire Training Evolutions.* Quincy, MA: National Fire Protection Association.

National Fire Protection Association. (2013). *NFPA 1500: Standard on Fire Department Occupational Safety and Health Program.* Quincy, MA: National Fire Protection Association.

National Fire Protection Association. *U.S. Firefighter Deaths Related to Training, 2001–2010.* (2012). http://www.nfpa.org/itemDetail.asp?categoryID=2485&itemID=55950&URL=Research/Statistical%20reports/Fire%20service%20statistics/

FIRE SERVICE INSTRUCTOR *in action*

As the coordinator of a fire fighter recruit training program, you assign a new fire service instructor to assist in the delivery of the training program. Given that this individual is a new fire service instructor, you decide to sit in on an upcoming training session to evaluate the instructor's performance. Your attendance presents a set of possible distractions to the delivery of the material. In the past, some new fire service instructors have directed their instruction to the evaluator instead of the students.

1. Which type of evaluation process is designed to provide improvement recommendations to the fire service instructor?
   A. Formative evaluation process
   B. Personnel evaluation process
   C. Summative evaluation process
   D. Word-of-mouth evaluation process

2. Which type of evaluation process is normally used to determine certifications?
   A. Formative evaluation process
   B. Personnel evaluation process
   C. Summative evaluation process
   D. Word-of-mouth evaluation process

3. When conducting a fire service instructor evaluation, which of the following statements is true?
   A. The evaluation should always be scheduled prior to the class.
   B. The evaluation should never be scheduled prior to the class, but the fire service instructor should be told at the beginning of the class that he or she will be evaluated.
   C. The fire service instructor should not know until after class that he or she was evaluated.
   D. The decision to schedule the evaluation should be determined by department policy.

4. To conduct the evaluation of this fire service instructor, what is the minimum fire instructor certification level you must have achieved, according to NFPA 1041?
   A. Fire Service Instructor I
   B. Fire Service Instructor II
   C. Fire Service Instructor III
   D. Fire Service Instructor IV

5. If you decide to have students provide feedback for fire service instructor improvement, which of the following outcomes is most likely?
   A. The students always tend to go easier on the fire service instructor than they should.
   B. Students always tend to go harder on the fire service instructor than they should.
   C. Some students will not provide honest feedback to the fire service instructor.
   D. Student feedback is very accurate.

6. Which of the following points does departmental policy on evaluation of fire service instructors not need to include?
   A. The order in which the feedback is to be given
   B. Confidentiality requirements for the evaluation
   C. Frequency and timing of evaluations
   D. Who will evaluate the fire service instructor

7. Which of the following is not a reason to conduct an evaluation?
   A. Certification
   B. Promotion
   C. Tradition
   D. Personnel evaluation process

8. During the feedback session, what can you do to help lessen the impact of weaknesses that need to be addressed?
   A. Provide feedback on only the positive aspects of the instructor's performance to reinforce those skills.
   B. Give the worst problem first, so the rest of the weaknesses seem like less of an issue.
   C. Give the least worrisome problem first, so the fire service instructor is desensitized as the problems discussed get worse.
   D. Let the fire service instructor identify problems that he or she is aware of before you mention the problems that the instructor does not cover.

Training Program Management and Curriculum Design

Scheduling and Resource Management

Courtesy of James Gathany/CDC

Fire Service Instructor I

Knowledge Objectives

After studying this chapter, you will be able to:

- Describe how to schedule instructional sessions.
  (NFPA 4.2.4) (pp 280–287)
- Describe the types of training records necessary to document a single instructional session.
  (NFPA 4.2.5) (pp 281–282, 287)
- Describe the process for acquiring training resources.
  (NFPA 4.2.3) (pp 281–282)

Skills Objectives

After studying this chapter, you will be able to:

- Demonstrate scheduling single instructional sessions.
  (NFPA 4.2.4) (pp 281–282)
- Demonstrate completion of training records and report forms. (NFPA 4.2.5) (pp 281, 287)
- Demonstrate requesting training resources.
  (NFPA 4.2.3) (pp 281–282)

Fire Service Instructor II

Knowledge Objectives

After studying this chapter, you will be able to:

- Describe how to schedule instructional sessions.
  (NFPA 5.2.2) (pp 280–291)
- Describe the types of training records necessary to document instructional sessions. (pp 281–282, 288)
- Identify resources needed to implement a lesson plan.
  (NFPA 5.3.3) (pp 281–282)
- Describe the process for acquiring training resources.
  (NFPA 5.2.4) (pp 281–282)
- Describe how to supervise other instructors safely while training. (NFPA 5.4.3) (pp 291, 293)

Skills Objectives

After studying this chapter, you will be able to:

- Demonstrate the ability to schedule training.
  (NFPA 5.2.2) (pp 280–291)
- Demonstrate completion of a training record report form. (pp 281–282, 287)
- Supervise other instructors and students during training.
  (NFPA 5.4.3) (pp 291, 293)

Fire Service Instructor III

Knowledge Objectives

There are no knowledge objectives for Fire Service Instructor III students.

Skills Objectives

There are no skills objectives for Fire Service Instructor III students.

You Are the Fire Service Instructor

Y ou have been appointed to the position of training officer for your department. One of your first tasks is to create a training schedule that identifies the topics that will be covered and the frequency with which instruction on those topics will be delivered throughout the year. You desire to have a balance of hands-on training and classroom training, with evaluations of both the fire fighters in your department and the instructors who will deliver the training. You know there are many topics to cover, including those required by regulatory agencies and other topics that are essential to allow your department to deliver good service safely.

1. Which compliance areas related to training frequency do you know you have to cover in the training schedule?
2. What is the importance of having a balanced program that includes both hands-on training and classroom-based learning?
3. How will you decide what will be included in your training schedule?

Introduction

Training and education are tools used by fire departments to improve efficiency in both their operations and fire fighter safety. Managing a training team in any type of fire department is a staff-level function often fulfilled by a department training officer. The training officer may have a rank position within the department and must possess the managerial skills to accomplish the many job tasks assigned to the training team. The majority of these management- and supervisory-level functions are performed by the Fire Service Instructor II or higher, but an Instructor I is also responsible for handling these duties on occasion. If you wish to be a leader of a training team, you must improve your administrative and leadership skills. Instructional skills are just one part of this job position.

Supervision of other instructors and students during training evolutions will be challenging for a newly assigned fire instructor. Experience in the subject matter helps the instructor prepare lesson plans that clearly identify proper safety behaviors. The lead instructor of a training evolution will assume the main responsibility of supervision during all skills being performed, which will require knowledge of the authority having jurisdiction (AHJ) safety rules, regulations, and practices. All training sessions should adhere to department standard operating procedures (SOPs), and should use your local incident command system (ICS). Each training session, especially those that involve hands-on training, should be developed and delivered in a fashion that most closely resembles the real emergency scene. Examples of SOPs that must be followed during all hands-on training sessions are:

- Proper use of personal protective equipment (PPE)
- Use of the local ICS
- Operational SOPs
- Department and manufacturer guidelines for equipment use, care, and maintenance
- Communications policies
- Supervisory responsibilities for safety

The fire service training program is a critical part of the effort to reduce fire fighter injuries and line-of-duty deaths. Lack of training, failure to train, inadequate training, and improper training are contributing factors to many injuries and fatalities. Put simply, fire fighter injuries and fatalities are prevented through training. Much is at stake and much relies on the quality of the training program. To become an effective manager, you must develop your scheduling, budgetary, decision-making, and supervision skills.

Teaching Tip

Use of the local ICS and reference to all department SOPs are paramount to a training session's success. All applicable SOPs must be referenced in each lesson plan and instructors should carefully monitor their use and practice during training.

Fire Service Instructor I and II

Scheduling of Instruction

As a fire service instructor, you must take into consideration the many aspects of your department's operation when developing a training schedule. A comprehensive schedule includes all job areas and covers elements of both initial and ongoing training. A training schedule must balance the various regulatory requirements with the training sessions needed to help fire fighters deliver safe and effective service. Regulatory requirements that affect the department's training schedule include everything from emergency medical services (EMS) recertification requirements to hazardous materials training. Ongoing skill and knowledge retention and refresher training programs are an important part of the scheduling process. In part-time, paid-on-call, or volunteer organizations, the scheduling and delivery of training can be an extremely difficult task. The amount of time available for delivering training is often compromised by members' occupations and family commitments in such departments.

■ Single Instructional Sessions

An Instructor I may schedule a single instructional session. In some agencies or departments, this type of instruction may be known as a drill or company-level training. It may focus on a subject or topic that is covered only one time based on a needs assessment undertaken by the instructor or an officer. This single instructional session is usually not part of a formal curriculum covering multiple training sessions over a long period of time. Its designated outcome is typically aimed at a very limited and focused set of objectives. In a volunteer or combination department, such a session may be called "drill night" if it occurs once a week or month to refresh and learn skills. It should be considered a formal training session; familiarity with department scheduling procedures, instructional resources, facilities, and timelines for delivery are necessary to accomplish this task.

Another type of single training session may take the form of a quick drill or tailboard drill that covers a single topic, skill, or piece of information passed along among a crew **FIGURE 12-1**. This session might occur during a roll call at shift change or during stand-by time while members are waiting for the return of crews working another emergency (i.e., out on an incident scene).

In other cases, an Instructor I may be assigned to conduct a training session that is part of a regular training schedule. Such instruction may be part of a rotational system including many members of the department who deliver training on a routine basis. At other times, the company may run into a situation or incident where members did not have sufficient knowledge or skill to perform well, indicating that additional training is necessary.

When any of these situations occur, scheduling a single instructional session should include the following considerations:

- Define the goal of the session. Make sure you know the purpose of the training session and the goal to be achieved at the end of the session.
- Determine the time necessary to deliver the session and the department scheduling process. Estimate the time needed to prepare and deliver the training session, and make sure there is sufficient time available in the department training and work schedule for the training session.
- Identify instructional resources that are available for the training session. As an Instructor I, you should locate all of the training materials—including lesson plans,

FIGURE 12-1 A single training session may take the form of a quick drill on a single topic.

skill sheets, training equipment, and other resources—needed to meet the objectives that were prepared as part of the lesson plan. Prepare any requests for resources through the proper channels and ensure that any expendable materials are replaced and accounted for in the agency.
- Gain approval from supervisors as necessary for members to attend the training session. Make sure that company members can attend and any in-service companies are accounted for in terms of service delivery. Determine response procedures and company availability with supervisors and officers.
- Schedule and deliver training sessions according to department procedure. Once a training session is scheduled, make sure that notice is given to any other instructors who will assist and deliver the training session according to the lesson plan objectives. Include all four parts of the four-step learning process in your presentation, and keep your students involved in the learning. Conduct evaluations as necessary.
- Complete training record reports as required by department policy. Create a thorough and accurate training record report according to department policies, including all required data, objectives, attendee signatures, and narratives of the training session. Once complete, forward all reports through the proper channels for recordkeeping purposes.

Surveys focusing on the quality of a training program often reveal that the delivery of training suffers due to poor management of the training schedule. The dynamic nature of the fire service sometimes compounds these problems when incidents occur during training. The best training schedules, calendars, and organizational skills are tested daily when the alarm sounds **FIGURE 12-2**. Few fire departments have the luxury of taking companies out of service for training events and sessions. It can be frustrating when an elaborate training session comes to an abrupt end when a call comes in.

FIGURE 12-2 Fire fighters may have to leave training for an emergency run.

Given that the role of the fire fighter is continually expanding, managing training effectively is a primary function of the Fire Service Instructor II position. All fire fighter duties must be taken into account when developing the training scheduling. NFPA 1500, *Standard on Fire Department Occupational Safety and Health Program*, states the importance of this understanding:

5.1.3. The fire department shall establish training and education programs that provide new members initial training, proficiency opportunities, and a method of skill and knowledge evaluation for duties assigned to the member prior to engaging in emergency operations.

This principle applies to all levels of the fire department—from the chief to the newest fire fighter. Initial training includes training in the job performance requirements (JPRs) that apply to each level of responsibility. As an instructor, it is your duty to ensure that the training takes place at every level, before the fire fighter engages in emergency activities. Today it is no longer acceptable to put an untrained fire fighter into a hostile situation without proper training and evaluation.

Theory Into Practice

Review your department's job descriptions and categorize job functions into similar groups. Your department's training program should have initial and ongoing training content for every job description in the department.

■ Types of Training Schedules

Each department has specific issues when it comes to organizing training schedules. Some cynical fire service instructors have said that a training schedule is not worth the paper

that it is written on, because of all of the things that can go wrong in your attempt to deliver the training. Outside training sessions at the drill tower can be compromised due to weather conditions, for example, or an incident can delay a classroom session.

The two types of training that occur in the fire service are formal training programs and in-service training. At the most basic level, in-service training or on-duty training, often referred to as <u>in-service drills</u>, takes up the majority of a training schedule **FIGURE 12-3**. An in-service drill can comprise a single station drill or involve all on-duty companies. In-service drills can be scheduled to run on specific days or they can take place only when a specific training need is identified. In-service drills are the most common delivery method of training. For the purposes of this chapter, in-service drills are considered to be conducted while fire fighters are available and in service to respond to incidents, whereas formal training courses take place outside the structured workday.

In-Service Drills

In-service drills typically take place at a specified time and location. All types of fire departments use some form of in-service drills. There is very little time to take units out of service to train. Instead, apparatus and personnel must be kept in some form of readiness in the event that an emergency occurs, including during training. On the schedule, some departments identify the topic and level of coverage for the session, whereas others simply state that the company will "train on something" of the fire service instructor's choosing.

In-service training can be broken into different levels:

- Skill/knowledge development occurs when new approaches to firefighting operations are introduced to the department **FIGURE 12-4**. These sessions expand on previously learned information or skill levels to increase the fire fighters' ability to do their job. This level of training is used when a new method, piece of equipment, or procedure is implemented. All new equipment should be used in training by all fire fighters who will use it prior to placing that equipment into service. This approach may take a considerable amount of time to accomplish, but will pay off on the fire ground through better performance and use.
- Skill/knowledge maintenance is one of the goals of a comprehensive training program **FIGURE 12-5**. Both skills and knowledge degrade over time if not used. In a life-safety profession, you cannot afford not to function at your highest level of ability. Skill/knowledge maintenance training is used to develop performance baselines for core duties and functions so as to establish company-level standards and to measure individual skills and weaknesses. All company members should be able to perform basic tasks within a reasonable amount of time. An incident

Des Plaines Fire Department
Monthly Training Brief
January 2013
Wear It: Survive Today & Retire Healthy
PPE ☆ Seat Belts ☆ Safety Vests ☆ Hearing & Eye Protection

January 2013

S	M	T	W	T	F	S
		1	2	3	4	5
6	7	8	9	10	11	12
13	14	15	16	17	18	19
20	21	22	23	24	25	26
27	28	29	30	31		

Weekly Skill Drill Summary

Subject	Date(s)	FH Entry Code	Location	Skill Description
Tie Handcuff Knot on Partner	Dec 30–Jan 5	1WSD1	Stations	Tie a handcuff knot on partner and self
60 Sec. SCBA Donning	Jan 6–12	1WSD2	Stations	Don SCBA from floor in 60 sec or less
Operate Power Saws	Jan 13–19	1WSD3	Stations	Operate and maintain all power saws
Set-up PPV and Smoke Ejector	Jan 20–26	1WSD4	Stations	Assemble RIC Equipment
2 Person Salvage Cover Throws	Jan 27–Feb 2	1WSD5	Stations	Perform salvage cover throws and folds

Company Level Training

Subject	Date(s)	FH Entry Code	Locations	Assignments/Info
Daily Quick Drill	Daily	DQD	Station Roll Call	Assigned to Acting Officer/Driver. List subject covered in notes on FH
Company Readiness Training-Daily	Daily	CRTD	Stations	Daily Company tool, equipment, and apparatus training
Daily Driver Training	Daily	DRIV DAILY	Stations	Daily pre-trip driving inspection, routine and emergency responses experience.
Company Readiness Training-Weekly	Saturdays	CRTW	Stations	Weekly Company tool, equipment, and apparatus training (Saturdays)
Tactical Walk Through Drill	Mondays	TWT	Company Officer Selected	Tactical walk-through for first-due operational information
Daily Physical Fitness Training	Daily	FITS	Stations	Strength, flexibility, cardio and aerobic capacity fitness training
Tool Assignments	Jan 2–4	TOOLS1	Stations	Define standard tool assignments by company type and assignment
MAP Book/CAD Training	Jan 9–11	MAPS	Stations 1 0900 Station 3 1330	Intro to new map book and grids
Fire Behavior	Jan 16–18	BEHAVIOR1	Stations 1 0900 Station 3 1330	Introduction to fire behavior research
RIC – FF Through the Floor	Jan 23–25	1RITOPSFLOOR	TBD	Rescue FF who has fallen through the floor
Ice Rescue	Jan 30–Feb 1	ICE	TBD	Ice Rescue Evolutions

Weekly Tactical Training

Building Name	Date(s)	FH Entry Code	Assignments/Info
Polo Inn	Jan 6–12	2SIMPOLO	Review tactical walk through and complete tactical simulation
Lee Manor Nursing	Jan 13–19	2SIMLEEMNR	
Stonecrest Condos	Jan 20–26	2SIMSTONECRST	

Pump Operator Training

Subject	Date(s)	FH Entry Code	Locations	Assignments/Info
Apparatus Operators Intro	Jan 9–11	2PUMPINTRO	Stations 1 0900 Station 3 1330	Review of apparatus daily, weekly and maintenance procedures. Response policy overview.

HazMat Operations Training

Subject	Date(s)	FH Entry Code	Locations	Assignments/Info
Not Scheduled This Month				

Technical Rescue Awareness Training

Subject	Date(s)	FH Entry Code	Locations	Assignments/Info
Not Scheduled This Month				

Officer Training

Subject	Date(s)	FH Entry Code	Locations	Assignments/Info
Officer Training Session	TBD	TBD		Self-study activities and terminology/SOP review. Assignments via Email groups
Acting Officer Training	TBD	TBD		
Incident Command Training	TBD	TBD		

EMS Training

Subject	Date(s)	FH Entry Code	Locations	Assignments/Info
EMS-Paramedic CE	Jan 8, 15, 22		Stations 1 1330 Station 3 0900	TBA
EMS-EMT Basic CE	Jan 8, 15, 22		Stations 1 0900 Station 3 1330	TBA

Special Teams Training
See Training Calendar on Shared Calendars for Special Teams Drill Information

FIGURE 12-3 Sample training brief.

VOICES
OF EXPERIENCE

As a Fire Department Training Officer, one of the pressing challenges I faced was ensuring training was delivered while making sure units were available for emergency calls. This challenge was more common with EMS incidents that seem to occur all hours of the day. To make things more complicated, our response districts were divided to the point where it became almost impossible to train all personnel from the same shift without taking units out of service.

> **"I realized that I had to become creative to find a way to deliver the training sessions without compromising service levels."**

I realized that I had to become creative to find a way to deliver the training sessions without compromising service levels. Although our fire stations all had kitchens, not all had the proper resources, such as a classroom, to deliver training, and they had very little room outside to conduct driver training.

While trying to look outside the box I came across a complex in our response area where a classroom was needed. Initially I thought I had a solution when I found a business that rents out meeting rooms and office by the hour, but later learned the construction of the building limited our ability to receive and transmit radio communication. I had to keep looking outside the box to find a place to deliver classroom training sessions while keeping units in service.

One day while driving down the road I remembered that I had visited a church that had a classroom that could meet our needs. A short time later I contacted the church and was able to make arrangements to use the classroom at no charge. As a result, our fire department was able to conduct the same training deliveries throughout the fire department's response area.

Although the classroom dilemma was resolved, we still needed a place to conduct driver training. Facing the same challenge, I remembered that we had conducting driver training in a church parking lot during the weekdays in one area of our response area. This was good but it made it difficult to schedule for use by the entire shift.

After thinking about how good use of one church parking lot worked in one response area, I sought out another parking lot in the other. I eventually found a church parking lot that was the right size and received permission to use it for driver training.

As a result of finding the right sites for driver training, I acquired an equal amount of cones and traffic devices to enable our driver training program to be conducted while minimizing interruptions in service. It is helpful to be creative and look outside the box to see what resources are available in your community when it comes to scheduling training for fire fighters.

Bill Guindon
Director
Maine Fire Service Institute
Brunswick, Maine

FIGURE 12-4 Skill/knowledge development occurs when new approaches to firefighting operations are introduced to the department.

FIGURE 12-5 Skill/knowledge maintenance is one of the goals of a comprehensive training program.

commander makes incident assignments knowing how long it takes to stretch a hose line or ventilate a roof, and if the company is untrained or unprepared to meet that standard, other fire-ground assignments may suffer.

- Skill/knowledge improvement is necessary when individual weaknesses become apparent. These drills are used when errors or poor performance have occurred or other undesirable outcomes have been observed. A company may be able to stretch a hose line and place the nozzle in a ready position effectively, but if the pump operator is unable to have the desired gpm and nozzle pressure arrive at the nozzle, the entire company will fail in its assignment. This is a signal for the company officer to train with the pump operator, using fire service instructors, to improve the entire company's skill level.

The in-service drill is the type of training that is most vulnerable to cancellation, delay, or reduction in participation due to emergency response. Consequently, the instructor should have a supplemental training schedule available in the event that some type of mitigating factor requires a change in the original training session **TABLE 12-1**. This should be a published program. The resources needed to accomplish the training should be on stand-by and ready to use when unplanned events occur.

Formal Training Courses

Certain types and levels of training require the use of a set curriculum matched to a directed number of hours or evolutions. Training to NFPA professional qualification standards using JPRs may require that students participating in a course attend a pre-defined-hour program to meet all of the JPRs. This type of training schedule affects the use of the department classroom or training center, and such a program could possibly overlap with the use of the same facilities for in-service training. Formal training courses may also have specific needs related to the facility—for example, audiovisual equipment, props, or appliances—that a department may have to budget for or acquire from other sources.

Theory Into Practice

If your training classroom or training facilities are used by outside groups or to hold department business meetings, make sure that all groups' needs are covered and identified in one specific place so that no conflicts occur. Try to identify a central location where the training room schedule is posted and define who is responsible for keeping the schedule up-to-date.

Table 12-1	A Supplemental Training Schedule		
Supplemental Training Schedule: Available Topics			
Fire Suppression	**Rescue Tools**	**Hazardous Materials**	**EMS**
Nozzles and streams	Air bags	*ERG* book	Vital signs
Extinguishers	Glass removal tools	Terminology	Patient movement
SCBA donning	Hydraulic tools	Decontamination	Documentation
Ladder carries	Hand tools	Spill containment	Initial patient surveys

Other types of training also must be developed when creating a comprehensive training schedule. Special areas of training that need to be considered include the following:

- Technical rescue teams
- Special operations teams
- Hazardous materials teams
- Emergency medical services
- Vehicle operator training

The following areas of department operations may overlap with a training program and should be considered part of in-service training:

- Vehicle and equipment inspection and maintenance
- Hose testing
- Ladder testing
- Pump testing
- Daily/weekly/monthly equipment checks

Other duty areas within a department that must be considered within the framework of training program management include other staff-level functions within the organization. Keeping the concept of training commensurate with duty means that the training needs of staff—personnel such as fire investigators, emergency medical services (EMS) or other staff assignments—must be included in the development of your training plan. Consider the functions of fire-cause determination personnel, fire inspection personnel, public education teams, dispatch and communication personnel, and apparatus mechanics when developing your training plan.

Each of these areas may take up substantial amounts of available time and require the coordination of resources and time management. These areas may also require specialized resources, fire service instructors, or equipment that must be considered when creating the schedule.

Develop a good working relationship between fire service instructors and operational team leaders. These team leaders have backgrounds and expertise that will help in the delivery of instruction. When you work with these team leaders to prepare lesson plans, schedule sessions, and make sure division objectives are incorporated into the course, your training program will be more successful.

■ Scheduling for Success

To meet the needs of the department and its fire fighters, you must develop a schedule that is consistent, easy to understand, and clear **FIGURE 12-6**. To accomplish this goal, you must have

Des Plaines Fire Department—Division of Training & Safety

FIGURE 12-6 Schedules should be posted and visible to all members to ensure consistent delivery of training and proper resource management.

access to, and understand, the department's training policies. A department's training policy or SOP that relates to training may identify who must train, how often, where, and who will provide the training. If this information is already known, the scheduling chore is almost complete.

Subject coverage in a curriculum-based training program should be sequential. When looking at your subjects, consider any prerequisite learning and skill levels. Ask yourself which topic needs to be covered before the next one can be covered. For example, it would be advisable to instruct on ladders before teaching a class on vertical ventilation to a group of new fire fighters. Students should learn and perform skills in a sequential order to maximize their understanding as well as to improve their safety and survival.

Teaching Tip

Review your department's SOPs relating to training participation and scheduling.

Important information in a training policy or SOP that will help you develop a training schedule includes the following elements:

- Who must attend the training session
- Who will instruct the training session
- Which resources are needed for the training session

Information specific to the training session may also be used to help identify the scope and level of the training session. Students appreciate knowing the subject of the training session before they walk into the classroom. Highly motivated students may complete reading assignments or research the topic before the training takes place.

Fire officer development includes the ability to prepare a crew for a training session by practicing skill sets before the drill **FIGURE 12-7**. Advanced polices also detail the responsibilities of those who attend, supervise, and participate in the

FIGURE 12-7 Fire officer development includes the ability to prepare a crew for a training session by practicing skill sets before the drill.

training. Recordkeeping processes are included in this policy as well. Information about the training session itself includes the following points:

- What the training subject(s) is (are)
- Whether it is a practical (hands-on) or classroom type of session
- Whether there are any pretraining assignments
- Where the training will take place

Fire Service Instructor II

Scheduling of Instruction

■ Developing a Training Schedule and Regulatory Compliance

An agency training needs assessment should be performed at the direction of the department administration. The agency training needs assessment helps you, as the Fire Service Instructor II, to identify any regulatory compliance matters that must be included in the training schedule. It may also help identify the topics in which the department wants their

personnel to be trained. The following methods may be used to determine which topics need to be given priority over others:

- Identify skills and knowledge necessary for safety and survival
- Review the skills and knowledge necessary for delivery of the department's mission statement
- Examine the subject areas tied directly to the job description of each job title or function

Every department has specific regulatory issues to address in its schedule, but some issues have a universal

JOB PERFORMANCE REQUIREMENTS (JPRS)
in action

Most fire departments have a training division often led by a training officer or fire officer designated to complete the management-related functions tied to training. Recordkeeping, scheduling, and compliance are some of the functions that must be attended to on a regular basis, regardless of the size of the department. Many resources must be managed, and budgets must be prepared and monitored. Fire service instructors need to be assigned and facilities secured to ensure that instruction and student experiences are of the highest quality. Some of these functions are related to typical fire officer management and organizational principles, but in the area of training program management, many special considerations must be learned and applied to manage the training team.

Instructor I

Teamwork within the training team is as important as it is in the fire company. As the training program is a division within the fire department, proper management requires the Instructor I to act as part of the team and to complete routine administrative tasks in addition to the delivery of training.

Instructor II

Managing the resources, staff, facilities, and budget of a training program requires sound management and leadership skills. Many fire officer JPRs will be put to work while completing these functions.

Instructor III

Working as part of the training team, focus on policy development and good communication flow between instructors on the training team. Supervision and evaluation of instructors and the program are key duty areas.

JPRs at Work

The Instructor I will function as part of the training team by delivering training, scheduling single instructional sessions, managing resources, and ensuring the safety of students during training. Creating and maintaining records, reports, and proper documentation of the learning process will be a large part of the Instructor I's responsibilities.

JPRs at Work

Schedule and acquire the resources necessary to conduct training sessions and assign qualified instructors while administering a budget for these functions. Ensure the safety of the students and instructors by administering department policies and applicable standards.

JPRs at Work

There are no JPRs for the Instructor III for this chapter.

Bridging the Gap Among Instructor I, Instructor II, and Instructor III

The Instructor I can increase his or her ability to work as part of the training team by accepting additional responsibilities, such as scheduling a single instructional session and other management-related functions. The Instructor II can serve in a mentoring position and also reduce his or her workload by utilizing the Instructor I to assist in completing routine tasks. This prepares the Instructor I for increased management responsibilities and helps to fully develop the training team and utilize each member to his or her full potential. An Instructor III will develop processes, policies, and practices that will support the delivery of course goals, agency objectives, and the overall quality of the training program.

impact. Although these areas require interpretation locally, certain types of training requirements are usually specified by many regulatory agencies. Consider the Insurance Service Office (ISO)—it has numerous training requirements that cover initial training for new fire fighters and vehicle operators and ongoing training for fire officers. Although technically not a regulatory agency for fire departments, the ISO may influence the operations of many departments because property classification ratings can be improved by development and implementation of an effective training plan. A good technique is to contact your local ISO representative to have a discussion on the training requirements that are part of the ratings survey.

The Occupational Safety and Health Administration (OSHA) has specific training requirements that should be considered, ranging from respiratory protection training to infectious disease protection training. Some states are considered OSHA states, meaning that enforcement authority for OSHA requirements are with the state. In OSHA states, public entities fall within the authority of the state OSHA. Because OSHA has different levels of enforcement power and authority in each state, you should contact your local OSHA representative to discuss which areas of your program are affected by OSHA training requirements. Fines or other sanctions could be levied against your department if it fails to meet OSHA requirements.

Consider other regulating commissions or agencies that may have training requirements, such as your local insurance carrier or risk management departments, EMS state or licensing agent, NFPA, or the State Fire Marshal or statewide certification entity. These agencies may have significant requirements for both entry-level and ongoing training for your students.

Knowledge of these and other regulatory agency requirements is a key part of the scheduling puzzle. Each agency may require training to be conducted at specific intervals or to last for a certain number of hours. Become familiar with each of these areas and develop ways to cover as many areas as possible by combining resources and sessions to meet the varied requirements.

Part of the Fire Service Instructor II's job description is to keep current with the requirements of the regulatory agencies that affect your department. As new laws and standards are passed, new or different requirements in your department training policy must be included to comply with those new laws and standards. Best practices learned from case studies and other local requirements may require the development of a department training policy. High-risk areas of training, such as live fire training or special operations training, should enforce a student-to-instructor ratio of 5:1, with this requirement being included within the department's written training policy. The department training policy should also specify safety rules during training, the scope of coverage of the training program, and many other areas of administrative responsibility. Many of these concerns may fall within the duties of the Fire Service Instructor III and are covered in the *Training Program Management* chapter.

Theory Into Practice

Research the local, state, provincial, and federal regulatory agencies that impact the management and delivery of your training program. You will find many similar areas of concern, but the frequency and recordkeeping deliveries can vary drastically. Consider the following agencies and their impact to your program:

- OSHA
- ISO
- NFPA
- State Fire Marshal
- EMS system
- State driver's license authority
- State certification or license authority

Create a matrix that identifies the initial training requirements and ongoing training for each agency that impacts your department.

◼ Scheduling Training

A standardized process must be utilized to schedule training sessions. As stated earlier, the schedule must have some flexibility so that you can take advantage of unforeseen opportunities or overcome obstacles in the delivery of training. For instance, you might schedule a block of training on hose lead-outs when a local developer calls you and offers the use of a structure scheduled for demolition. If the property is suitable for use, it may present a great opportunity to enhance your training program through the use of the acquired structure. You might also want to consider covering your hose lead-out drills during the training performed at this structure, along with the multitude of other areas you can train on.

Obstacles can range from weather complications to the scheduled vacation time of students who need training. Additionally, new equipment may be purchased that needs to be trained on before it is put into service, with this consideration taking precedence over other scheduled activities. In addition to the creation of the schedule, a process for the publication, distribution, and delivery of the training schedule must exist so that all who are impacted by it are informed about the training-related activities of the period **FIGURE 12-8** .

◼ Master Training Schedule

A useful method to organize the multiple demands for schedule coverage is a system capturing all the required training areas within a single forum. A simple table form, called a master training schedule, may suffice to begin this process **FIGURE 12-9** . It can serve as the starting point in the development of your in-service training program schedule. When a period of training (typically one month in the year) is planned,

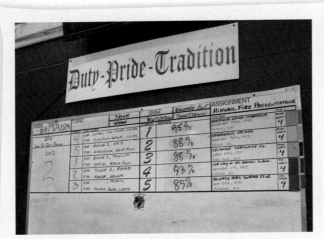

FIGURE 12-8 Schedules should be posted and visible to all members to ensure consistent delivery of training and proper resource management.

you can go to this table to identify the topics that must be included into the schedule.

Here are the steps required to develop a master training schedule for your department:

1. Complete an agency needs assessment of the regulatory agencies that require training on specific topics (OSHA, ISO, NFPA, local agency/authority).
   a. Determine the frequency with which the training must be included.
   b. Determine the type of training that is needed.
      i. Initially upon hire
      ii. Ongoing after hire
      iii. Classroom or hands-on
      iv. Individual, company, or multi-company
      v. Mutual aid training
      vi. Training on special hazards or topics

2. Identify other areas of training that are necessary to train your students on the skills and knowledge necessary to keep them safe at emergencies.
   a. SCBA skills
   b. Fire fighter survival techniques
   c. Minimum company standards or other proficiency-based training

3. Identify ongoing activities that affect the training schedule so that you do not overload the crews with too many activities within a period.
   a. Equipment testing
   b. Fire prevention activities
   c. Apparatus testing
   d. Applicable recertification

4. Create a table that allows you to list the information compiled in steps 1–3 in one column on individual rows and create at least 12 additional columns to represent each month of the year.

5. Insert a checkmark or other indicator of when each type of training needs to be completed into the number of month columns required to meet the standard.
   a. If SCBA needs to be trained on quarterly, insert an X in one month of each quarter.
   b. If protective clothing needs to be donned, doffed, and cleaned semi-annually, insert an X in each of two 6-month periods.

6. Once all elements have been added, balance out your coverage to allow for an even distribution of the content to be covered each month.

7. Develop categories for other types of training by subject matter areas and distribute them throughout the calendar to give your program even coverage.
   a. Examine the subject area coverage in the training materials to establish a well-rounded training plan.
   b. Make sure that all of the JPRs of a given job level have been covered.

8. Review and revise the schedule on a regular basis, preferably annually.

Supervising Other Instructors

Managing fellow fire service instructors requires leadership skills and the ability to recognize training scenarios that require specific supervision relating to safety and standard practice. You must ensure that applicable standards, SOPs, and departmental practices are used in the development and execution of training sessions that involve increased hazards to students. Every fire service instructor has a different style of presenting materials, different background, and different experiences that he or she brings to the training session. As a Fire Service Instructor II, you should be familiar with these characteristics of instructors when supervising training delivery. Allow for individual styles and experiences in the training sessions, but make sure that all objectives, safety-related guidelines, department SOPs, and other performance expectations are covered.

Like any other form of management, management of fire service instructors and students requires leadership and management skills. The quality of instructional delivery and ability of the students to learn or perform a new skill are two of the criteria that may be used to evaluate training success. Too many times in fire service history, instruction has fallen short of this goal and students and fire service instructors have been injured or killed in poorly constructed or poorly supervised training sessions. Likewise, if the quality of the instruction is substandard, then the learning outcomes will be poor—a shortcoming whose aftereffects can persist long

Des Plaines Fire Department
2013 **Master Training Schedule** *2013*

Core Training Sessions	JAN	FEB	MAR	APR	MAY	JUN	JUL	AUG	SEP	OCT	NOV	DEC
Protective Clothing Inspection	X						X					
Officer Training/Acting Officer Training	X	X	X	X	X	X	X	X	X	X	X	X
Minimum Company Standards	*Under Development and Initial Training*											
SCBA Module A (D/D, Care, Maint., Fit Testing)		Don Doff									Fit Tst	
SCBA Module B (Emergency Procedures)			X									
SCBA Module C (Air Management Training)					X							
SCBA Module D (Practical Applications)											X	
Engineer Recertification & Testing		Hydraul. 6–8		Drive 17–19					18–20 25–27			
HazMat Operations Refresher		X								X		
Pump Operator Refresher (Modules)	Maint. & Driving		Mod 1 & 2		Relay	S-Pipe	Foam	Def.			Skids $2\frac{1}{2}$"	Written
Technical Rescue Team Support	Ice			Con Sp		Trench				Vert		
IDOL Compliance Session (BBP/LOTO/RTK)				X								
COMPANY TRAINING												
Company Skill Evolutions (Weekly Skill Drill)	X	X	X	X	X	X	X	X	X	X	X	X
Multi-Company Level Evolutions			X			X					X	
EQUIPMENT TESTING												
Pump							X	X				
Hose				X	X							
Ground/Aerial Ladders									X	X		
SAFETY TRAINING												
Occupational Health & Safety Programs	Mod 1	Mod 2	Mod 3	Mod 4	Mod 5	Mod 6	Mod 7	Mod 8	Mod 9	Mod 10	Mod 11	Mod 12
Trends, Current Events, LODD Analysis	X	X	X	X	X	X	X	X	X	X	X	X
SPECIAL TRAINING PROGRAMS												
EMS Training (Locally Assigned Medic/EMT)	X	X	X	X	X	X	X	X	X	X	X	X
Technical Rescue Team (Local Team)	X	X	X	X	X	X	X	X	X	X	X	X
HazMat Team (Local Team @ Tech Level)	X	X	X	X	X	X	X	X	X	X	X	X
ONGOING TRAINING PROGRAMS												
Min. Company Standards (Individual and Co) Daily Quick Drills Safety Audits	X	X	X	X	X	X	X	X	X	X	X	X
Self Survival and RIT Skill Development	X	X		X	X		X		X	X	X	
Weekly Tactical Walk Through & Exercise	*Every Monday afternoon assigned by Company Officers*											

FIGURE 12-9 Sample master training schedule.

after the student has left the classroom or drill ground. Quality assurance is an important aspect of the management of a training program. Some factors commonly noted with poor training are summarized here:

1. Inconsistency of information between fire service instructors
2. Failure to adhere to or enforce safety practices
3. Unprepared or unknowledgeable fire service instructors
4. Insufficient student involvement in the training session
5. Unclear instructions or learning outcomes
6. Poor presentation skills
7. Repetitive and boring material

Safety Tip

If safety is not adequately addressed, take immediate action. Ensure that students do not leave the training session without a thorough understanding of the safety aspects.

A majority of these factors can be addressed through better oversight of the instruction process. End-of-course surveys and peer review forms **FIGURE 12-10** that are completed by participants can provide valuable insight into the quality assurance part of the instructional process. New instructors will benefit greatly from a peer review of their instructor skills, and this short form will offer valuable insight into their communication skills.

Teaching Tip

Review the *Evaluating the Fire Service Instructor* chapter for a detailed method of gaining insight into the strengths and weaknesses of fire service instructors.

Safety Tip

Always provide adequate fire service instructor staffing levels—that is, levels commensurate with the risks of the evolutions and departmental policy.

Peer Review
Presentation Evaluation

Student Name _____

Department _____

Topic _____

Start Time _____ End Time _____

Communication Skills
1: Poor; 2: Fair; 3: Average; 4: Good; 5: Excellent

Voice Levels:	□1	□2	□3	□4	□5
Eye Contact:	□1	□2	□3	□4	□5
Mannerisms:	□1	□2	□3	□4	□5
Appearance:	□1	□2	□3	□4	□5

Presentation

Introduction:	□1	□2	□3	□4	□5
Body Presentation:	□1	□2	□3	□4	□5
Conclusion:	□1	□2	□3	□4	□5

Comments:

Courtesy of Illinois Fire Chiefs Educational and Research Foundation

FIGURE 12-10   Peer review form used for instant class feedback.

Training
BULLETIN

Jones & Bartlett Fire District

Training Division
5 Wall Street, Burlington, MA, 01803
Phone 978-443-5000 Fax 978-443-8000
www.fire.jbpub.com

Instant Applications: Managing the Training Team

Drill Assignment

Apply the chapter content to your department's operation, its training division, and your personal experiences to complete the following questions and activities.

Objective

Upon completion of the instant applications, fire service instructor students will exhibit decision making and application of job performance requirements of the fire service instructor using the text, class discussion, and their own personal experiences.

Suggested Drill Applications

1. Review your department's training policy for management-related responsibilities that must be completed by the Fire Service Instructor II.

2. Review your department's training schedules. Identify the process for determining training subjects, fire service instructor assignments, and the resources needed as part of the overall training program.

3. Review the Incident Report in this chapter and be prepared to discuss your analysis of the incident from a training perspective and as an instructor who wishes to use the report as a training tool.

Incident Report

© Greg Henry/ShutterStock, Inc.

Kilgore Texas—2009

On January 25, 2009, two career fire fighters, ages 28 and 45, died after falling from an elevated aerial platform during a training exercise in Texas. The fire fighters were participating in the exercise to familiarize fire department personnel with a newly purchased 95-foot mid-mount aerial platform truck. A group of four fire fighters were standing in the aerial platform, which was raised to the roof of an eight-story dormitory building at a local college campus.

In November of 2008, the fire department had received delivery of a newly manufactured 2008 model, 95-foot mid-mount aerial platform apparatus. Each of the three duty shifts received a factory-authorized training program of approximately 8 hours, including both classroom and practical exercises. The hands-on training included raising the aerial platform into the air with fire fighters in the platform. Both victims attended the 8-hour training sessions provided by the fire apparatus manufacturer, as did the fire fighter who was operating the platform controls at the time of the incident.

Three days prior to the incident the department had responded to a working fire where the new aerial device was placed into service and did not perform as desired. It was decided that additional training was needed on the aerial device and a schedule was developed to have on-duty crews participate in this additional department-based training.

During the last evolution the two victims entered the platform along with two other fire fighters and the device was raised. During the training, the operator, who was one of the fire fighters on the platform, proceeded to operate the device near the edge of the dormitory building. While attempting to place the platform on the roof of the building, the operator had to adjust the height of the device. During this process the platform became stuck on the concrete parapet wall at the top of the building. During attempts to free the platform, the top edge of the parapet wall gave way and the aerial ladder sprung back from the top of the building, and then began to whip violently back and forth. Two of the four fire fighters standing in the platform were ejected from the platform by the motion. They fell approximately 83 feet to the ground and died from their injuries.

Post-Incident Analysis: Kilgore, Texas

Contributing Factors

Fire fighters should be fully familiar with new equipment before training under "high-risk" scenarios

Fall protection should be used when fire fighters working on elevated aerial platform

Standard operating procedures for training, including the designation of a safety officer and use of fire apparatus, should be followed

Chief Concepts

- A majority of the management- and supervisory-level functions in an organization's training division are performed by the Fire Service Instructor II or higher, but a Fire Service Instructor I is also responsible for these duties on occasion.
- You must take into consideration many aspects of your department's operation when developing a training schedule.
- A single instructional session is usually not part of a formal curriculum covering multiple training sessions over a long period of time. The designated outcome of such a session is typically aimed at a very limited and focused set of objectives.
- Each department has specific issues that must be taken into account when organizing training schedules.
- At the most basic level, in-service training or on-duty training, often referred to as in-service drills, takes up the majority of a training schedule.
- To meet the needs of the department and its fire fighters, you must develop a schedule that is consistent, easy to understand, and clear. To accomplish this goal, you must have access to, and understand, the department's training policies.
- An agency training needs assessment helps identify any regulatory compliance matters that must be included in the training schedule. It may also help identify the topics in which the department wants fire fighters to be trained.
- A standardized process must be utilized to schedule training sessions, and the schedule must have some flexibility for unforeseen opportunities or obstacles in the delivery of training.
- The multiple demands for schedule coverage can be organized with a master training schedule, which uses a table format to capture all the required training areas within a single forum.
- You must ensure that the applicable standards, SOPs, and departmental practices are used in the development and execution of training sessions that involve increased hazards to students.

Hot Terms

Agency training needs assessment A needs assessment performed at the direction of the department administration, which helps to identify any regulatory compliance matters that must be included in the training schedule.

In-service drill A training session scheduled as part of a regular shift schedule.

Master training schedule Form used to identify and arrange training topics by the number of times they must be trained on or by the type of regulatory authority that requires the training to be completed.

Supplemental training schedule Form used to identify and arrange training topics available in case of a change in the original training schedule.

References

National Fire Protection Association. (2012). *NFPA 1041: Standard for Fire Service Instructor Professional Qualifications*. Quincy, MA: National Fire Protection Association.

National Fire Protection Association. (2013). *NFPA 1500: Standard on Fire Department Occupational Safety and Health Program*. Quincy, MA: National Fire Protection Association.

FIRE SERVICE INSTRUCTOR *in action*

It is the fall of this year and you are charged with developing the department's training program for the upcoming year. You are to identify training needs, specify classroom and drill ground requirements, and schedule the recurring trainings. Once these factors are determined, you must identify recognized standards and procedures on which to base your program and supervise the delivery of training and the adherence to safety practices and procedures by the instructors.

1. Which of the following training activities might require assignment of additional instructors to observe safety-related details?
   a. A video presentation
   b. Any increased hazard training session
   c. A small-company discussion of fire-ground tactics
   d. A single-company drill reviewing water supply options

2. Which type of training do you need to consider when developing a training schedule for the department?
   a. Specialty area training
   b. In-service company training
   c. Outside agencies who use department training facilities
   d. All of the above

3. While scheduling training for the next year, you want to reference professional qualification standards as put forth by NFPA. Which prefix number do these standards begin with?
   a. 10
   b. 14
   c. 15
   d. 19

4. Training that occurs around the station and is often presented by the company officer is called
   a. formal training.
   b. informal training.
   c. special operations training.
   d. in-service training.

Instructional Curriculum Development

Fire Service Instructor I

Knowledge Objectives

There are no knowledge objectives for the Fire Service Instructor I for this chapter.

Skills Objectives

There are no skills objectives for the Fire Service Instructor I for this chapter.

Fire Service Instructor II

Knowledge Objectives

There are no knowledge objectives for the Fire Service Instructor II for this chapter.

Skills Objectives

There are no skills objectives for the Fire Service Instructor II for this chapter.

Fire Service Instructor III

Knowledge Objectives

After reading this chapter, you will be able to:

- Discuss the methods of training needs assessment.
  (NFPA 6.3.2) (pp 301, 303)
- List the steps in the curriculum development model.
  (NFPA 6.3.3 , NFPA 6.3.7) (pp 303–305)
- Discuss the importance of properly stated program and course goals. (NFPA 6.3.3 , NFPA 6.3.5) (pp 306–307)
- Describe the content of properly constructed learning objectives. (NFPA 6.3.3 , NFPA 6.3.6) (pp 310–312)
- Explain the purpose of modifying existing curricula.
  (NFPA 6.3.4) (pp 307, 309–310)

Skills Objectives

After reading this chapter, you will be able to:

- Perform an agency needs assessment based on job observation. (NFPA 6.3.2) (pp 301, 303)
- Perform an agency needs assessment based on regulatory compliance. (NFPA 6.3.2) (pp 301, 303)
- Perform an agency needs assessment based on skill and knowledge development. (NFPA 6.3.2) (pp 301, 303)
- Write a clear, concise, and measurable program or course goal to address training needs assessment gaps identified by the developer. (NFPA 6.3.3 , NFPA 6.3.5) (pp 306–307)
- Write properly constructed learning objectives to support the course goal and intended outcome. (NFPA 6.3.3 , NFPA 6.3.5) (pp 310–312)
- Review and modify an existing curriculum to meet agency training needs assessment. (NFPA 6.3.4) (pp 307, 309–310)
- Convert a job performance requirement (JPR) into a valid learning objective. (NFPA 6.3.5) (p 312)

Your department has expanded its mission by forming a special operations team to respond to structural collapse incidents. Previously, your department has had no capability in this area and had to wait for a county team to respond to these emergencies. You have received grant funding for a response unit, equipment, and training of your department members for this mission.

1. Which training needs exist for your new mission?
2. Which resources should you review as you begin to plan the training for this mission?
3. How will you develop a training plan and select instructional staff for the mission?

Introduction

Planning and developing an instructional program begins with determining which training is necessary and who needs the training. Each fire service instructor must develop a system to plan instruction, just like an elementary or secondary school teacher does. The duration of the training is determined by the goal of the program, the number of objectives, and the number of participants. You may find yourself planning instruction as short as a few minutes (e.g., a quick drill or tabletop discussion), a semester-long course run at a local college, and every combination in between. Ultimately, instruction needs to be centered on specific outcomes that benefit the user—that is, the student.

Few curricula are developed to benefit the instructor or the agency hosting the event. Training and learning are user centered, a principle that must always be observed. NFPA 1041, *Standard for Fire Service Instructor Professional Qualifications*, charges the Fire Service Instructor III with planning, developing, and implementing comprehensive programs and curricula. To plan instruction properly, first a training need must be identified, and then a clear, student-centered goal must be established. The course development and program management model identifies the training needs assessment as the center of the development and management process **FIGURE 13-1**.

Policies

Program Evaluation

1. Preparation
- Course Goal
- Course JPR's & Objectives
- Course Content
- Evaluate & Modify Existing Curriculum

2. Presentation
- Lesson Plan
- Media
- Student Resources
- Instructional Methods
- Instructional Staff

Training Needs Assessment
- Job Observation
- Compliance and Regulation
- Skill/Knowledge Improvement

4. Evaluation
- Course Evaluation Plan
- Instructional Staff Evaluation
- Quality Assurance

3. Application
- Class Activities
- Progress Checks
- Practice Environment

Financial Considerations & Processes

Records and Reports

FIGURE 13-1 Training needs assessment diagram.

Fire Service Instructor III

Agency Training Needs Assessment

An agency training needs assessment can be performed for many reasons. Department administration may instruct you to perform an assessment to identify any regulatory compliance matters that must be included in the department training program. The chapter *Scheduling and Resource Management* refers to this process for the purpose of developing the department training schedule. It is important to your organization to make sure that all regulatory compliance topics are covered at the correct frequency during the year, as determined by the agency. This assessment will also help prioritize topics and coverage based on safety and survival and the relationship that the subject has to the department mission.

Breaking down this process even further, you will discover three typical methods of conducting a training needs assessment. These methods encompass both methods of conducting the assessment and outcomes of the assessment, and they are used to help plan and establish the need for instruction:

- Job observation
- Compliance and regulation
- Skill/knowledge development and improvement

With all three methods, the Instructor III may be the person actually conducting the assessment in part or in whole, or he or she may receive feedback from a chief or a fire officer in the department relating to the need for skill and knowledge improvement.

These analyses or assessments are designed to identify gaps or deficiencies in skill and knowledge. Any such gaps should be viewed as opportunities to improve individuals' skills and abilities, rather than being subject to correction that is punitive in nature.

Teaching Tip

By seeking input from line officers regarding training needs, the instructor can focus on training that is relevant and needed by the department and mission.

■ Job Observation

The most common method of determining training needs is the job observation method. Watching how members operate at an incident scene reveals many opportunities for the instructor to plan instruction. Opportunities to improve both individual performance and crew performance occur at almost every incident. Major operational errors or failure to apply department standard operating procedures (SOPs) are signals that training is needed. A skilled and objective eye is needed to provide the correct feedback and determination of training needs.

Sometimes individuals make mistakes during an operation; that in itself may not signal a department-wide training need, only an individual training need. A simple view of this method might play out like this:

Who needs what training delivered by which instructor using which methods because _____ occurred at an incident?

Ethics Tip

When evaluating an incident and determining possible training needs, remember to look at the big picture and not allow personal biases to influence your assessment.

■ Compliance and Regulation

The list of agencies that specify and recommend training can be almost endless. The instructor must have a well-rounded and in-depth knowledge of the local and state requirements and the subjects and content areas to be covered. Agencies that may specify training content, and even offer resources to conduct training, include the following organizations:

- State fire service certification agencies
- Insurance Service Office (ISO)
- National Fire Protection Association (NFPA)
- Occupational Safety and Health Administration (OSHA)
- Local insurance carriers
- State and regional emergency medical services (EMS) providers
- Equipment manufacturers

The Instructor III and each department must do their homework in this area and identify the hour requirements and subject area coverage needed for compliance. Recordkeeping for each of these agencies may impact the administrative side of division management as well. Another simple way to view this method may be to say:

_____ says who needs which training to meet _____ standard or regulation.

Ethics Tip

When assessing your agency's training needs based on external requirements, it is important to find out exactly what is required to ensure your agency is in compliance with those requirements.

VOICES
OF EXPERIENCE

The Insurance Service Office (ISO) has very specific grading criteria for fire departments, and specifically for training programs. As a new training officer, I didn't know what to expect when our department was being graded for a possible reduction in protection class. I knew we had weekly drills, officer meetings, periodic preplan training sessions, and a pretty good file cabinet full of certifications and class certificates. When the first audit occurred, I was surprised to see the emphasis placed on, and many questions directed toward, the curricula that were used and referenced as part of the training program delivery.

When the first audit occurred, I was surprised to see the emphasis on the curricula that were used and referenced as part of the training program delivery.

Questions about which references were used for the training, which objectives were covered, and whether the training was practical or classroom or a combination of both, were not easily answered by the available recordkeeping system. The number of participants and the number of training hours they accumulated in initial and ongoing training, according to the job level they filled within the department, also were not very evident. Many things that I thought we would gain credit for—such as EMS training—were not counted toward the hours accumulated in a month, nor were some of the officer meetings where no fire operations were being discussed.

I learned that the ISO rating schedule has many categories of training criteria, including resources, use of resources, the types and frequency of drills that are held annually, and preplanning of structures within the department. In subsequent reviews, we prepared well in advance for the inspection and used a master training schedule that marched out the various intervals of each type of drill and kept close tallies on each member's certifications and courses attended. Each training sheet now has a reference to an NFPA JPR or state fire marshal learning objective tied to a certification course and well-defined start, end, and total training time indicators evident. Each visit is a new learning experience, and all have resulted in helping the department improve its rating evaluation.

Training officers and instructors should become well versed in the impact and criteria of the many compliance and regulatory agencies that review training program management and delivery. Every level of instructor has a role to play in the management and delivery of training, and proper delivery, recordkeeping, and use of available resources will go a long way toward improving safety, survival, and good service, as well as help your department during these audit processes.

Forest Reeder
Division Chief of Training
Des Plaines Fire Department
Des Plaines, Illinois

■ Skill and Knowledge Development and Improvement

Every instructor desires to improve the skill and knowledge levels of his or her students. To do so, it is necessary first to determine the existing levels and then to work from there. This assessment method is a two-phase process: (1) determine the students' baseline knowledge and skills and (2) establish goals for improvement.

For example, suppose you have recently discovered a helpful standard and training resource, NFPA 1410, *Standard on Training for Initial Emergency Scene Operations*, and identified a recommended baseline of individual skills to be the ability to don a self-contained breathing apparatus (SCBA). You recommend a time limit of 60 seconds and use a series of timed evolutions following a prescribed donning sequence (skills checklist) to identify and set your department's baseline for this skill level. After the timed evolutions are completed, you review the data from the evolutions and establish that the department average (baseline) is 55 seconds. This means that some of the members probably did not meet the recommended time of 60 seconds or less. A skills improvement need has been identified. Because all of the members already had developed the skill to don the SCBA, the next phase—that is, the skill improvement phase—will allow the department to work toward the betterment of the entire organization. A new standard of 45 seconds or less can be established, and through repetitive and ongoing practice, each member can make individual improvements moving toward the new goal.

The determination (or baseline) component of this analysis is an intensive and even exhaustive process. Looking at your department members to determine where their skill levels are is not something you can do from your desk. Other forms of knowledge and skill assessment may also be utilized at this level, including surveys, question-and-answer sessions, review of job performance evaluations, and detailed analysis of data and training records/reports. This method fills in the blanks in this example:

A review of recent skill tests and written exams has identified that <u>who</u> need additional training in <u>which</u> subject area to improve <u>which</u> skill or knowledge so as to perform <u>which</u> service and ensure member safety.

Curriculum Design Model

To introduce the curriculum design and program management model (CDM) for this text, consider a process that revolves around a central core (see Figure 13-1). The central core in this case is the training needs assessment, which drives each of the curriculum design steps. Surrounding the curriculum design model, and visualized as a frame that encircles it, are the administrative duty areas that allow the curriculum to be delivered and the overall training program to be managed. From the central core, arrows show the forward direction of the CDM, starting from "Step 1: Preparation" and moving clockwise. Arrows to and from the training needs assessment show that each of the four steps relates back to (and references) the training need that has been identified.

Many other models for curriculum development exist, of course, including the following:

- The ADDIE model (analyze, design, develop, implement, and evaluate)
- The National Fire Academy's (NFA) five-step model of determining training needs, design, development, administration, and evaluation

You may find it useful to review these methods as well and adapt them to create your own strategy for developing curriculum and course content. After all, you are designing a course based on your particular agency needs, and every department has different patterns of delivery and resources available to it.

■ Overview of the Four-Step Teaching Method

This CDM uses a well-established <u>four-step method of instruction</u> as its core for curriculum development steps **TABLE 13-1** . The four steps in this teaching method are preparation, presentation, application, and evaluation, and each step has a separate and distinct function in the learning process. Each instructor uses these steps to present material effectively and to guide their students to a successful outcome. The four-step method of instruction is the method of instruction most commonly used in the fire service.

Step 1: Preparation

The <u>preparation step</u> is the first step in the four-step method of instruction; it refers to preparing or motivating the student

Table 13-1	The Four-Step Method of Instruction
Step in the Instructional Process	**Instructor Action**
1. Preparation	The instructor prepares the students to learn by identifying the importance of the topic, stating the intended outcomes, and noting the relevance of the topic to the student.
2. Presentation	The presentation content is usually organized in an outline form that supports an understanding of the learning objective.
3. Application	The instructor applies the presentation material as it relates to students' understanding. Often the fire service instructor will ask questions of the students or ask students to practice the skill being taught.
4. Evaluation	The students' understanding is evaluated through written exams or in a practical skill session.

to learn—hence it is also sometimes called the motivation step. When beginning instruction, the instructor should provide the students with information explaining how they will benefit from the class. This consideration is especially important for adult learners, as very few adults have time to waste sitting through a presentation that will not directly benefit them. The benefit of a class can be explained in many ways:

- The class may count toward required hours of training.
- The class may provide a desired certification.
- The class may increase the student's knowledge of a subject.

Whatever the benefit may be, the instructor should explain it thoroughly during the preparation step. In a lesson plan, the preparation section usually contains a paragraph or a bulleted list describing the reasons for the class. During this step, the instructor needs to grab students' attention and prepare the students to learn. Suggested preparation points include safety and survival-related information, suggestions for local examples, and explanations of how the material will help improve the students' ability to do their job.

Step 2: Presentation

The presentation step is the second step in the four-step method of instruction. This is the most well-known step because it is the actual presentation of the lesson plan. During this step, the instructor presents the material through lecture, leads discussions, uses audiovisual aids, answers student questions, and uses other techniques to present the lesson plan. The *Methods of Instruction* chapter in this text discusses instruction methods used during this step. In a lesson plan, the presentation section usually contains an outline of the information to be presented. It may also contain notes indicating when to use teaching aids, when to take breaks, where to obtain more information, and which references and other sources provide additional information.

Step 3: Application

The application step is the third step in the four-step method of instruction. This step is often considered the most important step because it is when the student applies the knowledge presented in class. The learning process becomes practical during this step as the student practices skills and applies the new knowledge. The student may make mistakes and may need to retry skills. The instructor should provide direction and support as the student performs this step. A common tactic used during the application step is "I do" (instructor), "we do" (student/instructor), and "you do" (student) as a student learns a new skill. During the application step, the instructor must ensure that all safety rules are followed.

In a lesson plan, the application section usually lists the activities or assignments that the student will perform. In the fire service, it is very common for the application section to require the use of skill sheets for evaluation. The experienced fire service instructor uses the application step to make sure

that the student is progressing along with the lesson plan. This step also allows students to participate actively and to remain involved in the learning process.

Step 4: Evaluation

The evaluation step is the final step. This step is intended to ensure that the student correctly acquired the knowledge and skills presented in the lesson plan. The evaluation may be in the form of a written test or may be a skill performance test. No matter which method of evaluation is used, the student must demonstrate competency without assistance. In a lesson plan, the evaluation section indicates the type of evaluation method and the procedures for performing the evaluation. The Instructor III will develop systems to evaluate student success/failure, instructor performance, and overall program success.

■ Curriculum Development with the Four-Step Model

As stated earlier, the CDM uses the recognizable four-step method of instruction to guide the development of training identified in the needs assessment. Each step in the teaching method will be expanded by the developer to create or modify the curriculum and address the gap identified by the needs assessment. The steps within the process assist in the development of a comprehensive curriculum as specified in the job performance requirements (JPRs) for this level.

Step 1: Preparation Step (Assembling Course Materials)

In the first step in the CDM process, you must step back to the training needs assessment to identify the focus of your curriculum. Questions that have been answered within the needs assessment include the following:

- Who needs training (audience)?
- Which gap or need has emerged, creating the need for the program?
- Which results are desired at the completion of the program?

The preparation step in the CDM includes several tasks that help collect needed information and resources to begin developing your curriculum:

1. Determine the course goal (end result).
2. Reference or construct applicable course JPRs based on a task analysis.
3. Outline course content.
4. Evaluate and modify the existing curriculum or create new curriculum.

Consider this step to be the Instructor III's "ready time" to research and refine the end result of the development process. Research methods will come in handy during this phase; experienced developers know the value of identifying resources from existing curricula and NFPA standards that can be used as reference points in their development. Revisit your Instructor II skills in modifying existing curricula as well. Professionally developed textbooks and instructor toolkits are available and ready for modification to your training needs assessment and will save hours of preparation time for the developer **FIGURE 13-2**.

Step 2: Presentation Step (Materials Used to Deliver the Course)

In the second step of the CDM, the developer assembles the materials needed to deliver the course. Many of the tasks at this level can be completed with skills at the Instructor II level:

1. Development/modification of lesson plans
2. Development/modification of media
3. Creation and identification of student resources
4. Determination of appropriate instructional methods
5. Selection of instructional staff

As you can see from these tasks, the developer relies heavily on Instructor II skill sets and expands each of those sets into larger segments until the entire course content requirements have been met. Again, research and reference will be key to limiting the development of materials from scratch. Best practices in the development area suggest that use of established references and standards should always be part of your development process.

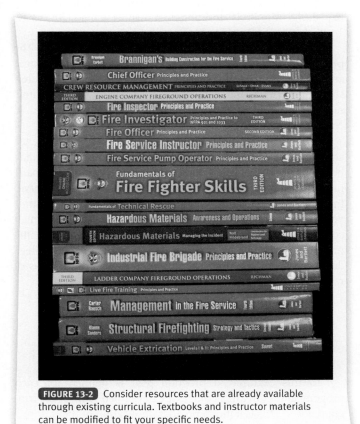

FIGURE 13-2 Consider resources that are already available through existing curricula. Textbooks and instructor materials can be modified to fit your specific needs.

Step 3: Application Step (Activities Used to Enhance Learning and Engage the Learner)

Step 3, which focuses on application, may be the most important step in the CDM process. Adult learners are more frequently engaged and participate in learning under very specific conditions. Adults enjoy participating in their learning and have experiences and previous learning that can and should be used to improve the process. During this phase, the developer will complete the following tasks:

1. Development of class activities
2. Development of progress and comprehension check methods
3. Construction of a practice environment

Experienced instructors know the true value of engaging their learners and advocating the "less instructor and more instruction" philosophy. Even the shortest training session can involve the student interacting with the instructor and immediate feedback being provided through interesting class activities. Open-ended questions, case examples, and student participation activities allow this to happen. (See the *Methods of Instruction* and *Communication Skills* chapters for important skills necessary for instructor success.)

Step 4: Evaluation Step (Materials Used to Evaluate the Learning Process)

The final step in the model is the evaluation step, in which the developer completes a form of quality assurance by creating instruments that evaluate the course as a whole and the delivery of the course. A well-rounded evaluation plan allows the developer to see whether the outcomes specified by the needs assessment were met and to what degree. Both the course content and delivery methods are evaluated in this quality assurance measure. Tasks completed in this step include the following:

1. Development of a course evaluation and student feedback plan
2. Creation of an instructional staff evaluation
3. Development of an overall quality assurance measure of the course to provide feedback for future course use

Course Development Sequence

An order model can be used to help the developer proceed through the course development process in a step-by-step manner. In some cases, steps may be omitted if information or material exists and is ready to use to meet course intent. For example, you may have an existing program on pump operations that can readily be used as the basis for updating the skills of new pump operators. Many of the design requirements may already have been completed, and the developer will find that a majority of his or her work involves only modification and updating of the previous content. The step-by-step model takes the developer from start to finish and can be used as a checklist **FIGURE 13-3**.

❏   Conduct a training needs assessment.
    ❏   Identify a training need.
    ❏   Define the audience.
❏   Present findings for approval or authorization.
❏   Design the program.
    ❏   Modify an existing curriculum.
        ❏   Review available curricula and references.
    ❏   Develop course goals.
    ❏   Write course objectives.
    ❏   Construct a content outline.
    ❏   Develop delivery materials.
        ❏   Lesson plans
        ❏   Media
        ❏   Student resources
        ❏   Class activities
        ❏   Practice environments
        ❏   Evaluation instruments
        ❏   Delivery schedule
        ❏   Documentation process and record keeping
❏   Select instructional staff.
    ❏   Determine qualifications.
    ❏   Schedule instructors and support.
❏   Deliver the program.
    ❏   Monitor a pilot course.
    ❏   Modify the program as indicated.
❏   Evaluate the program.
    ❏   Student evaluation
    ❏   Program evaluation

FIGURE 13-3 Course development checklist.

Developing Instructional Materials

Preparing instructional materials to be used for a training program or course is probably the most difficult and most time-consuming CDM step to complete. In this step, developers must write, rewrite, and retool many thoughts and ideas about how they think their program will meet the needs assessment gap.

Having a defined program and course goal sets the stage for a well-designed program. A program goal can be viewed as the final outcome of a series of courses, classes, drills, or training sessions conducted over a period of time. A course goal may be a smaller picture of the desired outcome of the training. For example, suppose a program goal is to establish a rapid intervention team. The outcome specified for this program goal might be to have a trained and capable rapid intervention team ready to respond and prevent fire fighter injuries and death on the fire ground. To accomplish that program goal, several courses may have to be conducted, each with its own course goals that allow the participants to understand their duties and responsibilities. (These may also be thought of as modules or units.) Examples of course goals in this scenario may be:

- Improve skills in downed fire fighter rescue.
- Perform size-up for rapid intervention teams.

Each of these modules will contain its own behavioral objectives, lesson plans, student activities, and evaluation measures. Behavioral objectives are written to meet the course goals for each course within a larger program. These objectives follow a standard format and specify the level of performance and the desired outcome. They must also be measurable. When all of the courses are completed, the student will have met the program goal.

You may also be familiar with how a college or university degree is structured. You register to begin your formal education for a degree in fire science management (program goal = obtain degree) and have to take several courses of study (course goal = each course meets degree requirements) to meet the degree requirements. At the completion of the entire program, you are rewarded with a diploma. This depiction of course development may seem to complicate a basic process, but it is important for the developer always to be working toward a defined goal that meets the agency's goals and mission.

■ Writing a Program Goal

A program goal may also be implied by the courses within it. Consider taking a cue from the NFPA JPR statements and using the duty areas to guide your program goal titles. Additionally, help may come from the traditional task analysis process, in which occupations are used to determine job functions. Programs could be titled by task analysis (e.g., fire fighter, driver/operator, fire officer, fire instructor) to begin this development step, and then course goals would follow. They may come from NFPA training documents such as 13E, 1404, 1405, 1407, 1410, 1451, 1452.

Consider the steps to write a fire officer program. The program goal may be "to prepare fire fighters for the challenges and responsibilities of company-level supervision and leadership and to direct and lead assigned companies of fire fighters safely." Once the program goal has been established, individual courses can be written to meet the program goal.

Program goals are not always necessary if the audience and delivery are contained within one organization. It may be understood that any course delivered is part of a department training program and the program itself is self-evident. In contrast, certification entities, college or university course developers, and fire academies may have to identify which courses fall under particular programs. In these cases, specific program goals will be necessary. A fire academy or college delivery system may break its program delivery into areas of responsibility such as the following:

- Firefighting programs
- Hazardous materials programs
- Officer training programs
- Special operations team programs
- Public education and code enforcement programs

In this example, a program goal is defined for each program and individual courses are developed to meet the program goals. This is a much broader application of the terminal objective and enabling objective development process, defined in NFPA 1041.

◼ Writing a Course Goal

To write a course goal, you will need to refer to what brought you to this step in the CDM—the training needs assessment. Through some form of assessment, such as job observation, you have identified that a program needs to be developed or modified to improve the knowledge or skills of individuals within your organization. Whether you plan to develop a course from top to bottom or to modify an existing curriculum, a clear and concise goal needs to be established.

Each Instructor III's teaching and delivery vision will influence how he or she develops a course goal. Some developers prefer to run through a list of questions relating to the course proposal, whereas others simply define a specific outcome for the intended audience. In either case, the developer should consider three specific groups and be able to state in specific terms how the course goal will affect the following groups' relationships with it:

- The students in the course
- The instructor(s)
- The agency or organization

The students in the course should be the first consideration in the identification and statement of a course goal. After all, it is the knowledge, skills, and abilities of those students that the course is meant to improve. A program/course goal may be stated in a paragraph form or in a short sentence or two. Another tie-in to Instructor I and II JPRs that may help you develop your course goal would be use of the ABCD method of objective writing to guide your goal statement development. Ask the following questions as you write your program goal:

- Who is the intended **audience**?
  - Fire fighters
  - Instructors
  - Chief officers
  - Special teams members
- Which **behaviors** (skills, ability, knowledge) do you intend to modify with the program?
  - Improve ability to ventilate a peaked roof
  - Understand motivation and leadership responsibilities and principles
- Under which **conditions** will the program take place?
  - In a classroom setting
  - In live fire training
- At which **degree** will the program be delivered?
  - To a timed standard or number of hours
  - Being able to apply the skills or knowledge in case-driven scenarios

A completed course goal may look like the following example using the ABCD method of writing:

The fire academy course goal for the Incident Commander Bootcamp is to provide aspiring and existing officers responsible for incident command duties with skills and knowledge in the assumption of command, declaration of strategy, assignment of tactical objectives, and other command duties. Participants will complete 24 hours of intensive and interactive classroom and simulator training and complete a day-long hands-on command simulation during a live fire training exercise.

Some agencies may prefer to see a course goal written into specific objective forms like this:

The fire academy course goal for the Incident Command Bootcamp is:

- To prepare incident commanders for their role in emergency scene management
- To provide incident commanders with the opportunity to simulate duties and responsibilities in a controlled environment
- To instruct incident commanders in the proper methods of command functions during emergency incidents

In either case, the reader (student, instructor, agency) can understand what the course requires of the student and what the outcomes will be. In some college settings, this information may be found in a course syllabus.

Evaluating and Modifying an Existing Curriculum

Developers should always strive to find ways to reduce the time devoted to developing programs from scratch. In many cases, existing curriculum or training materials may already be in place within the organization, and may simply need updating to current standards and practices and to current students' abilities and needs. You might also find that a neighboring organization or local fire academy has a program or course similar to what you are trying to develop. Such programs and courses might include some specific procedures that do not match those adopted by your own organization, so some modification may be necessary to make the program compatible with your agency. Some curricula will need to be updated to new standards or qualifications because NFPA professional qualification standards are typically reviewed and updated every five years. When a standard or professional qualification is updated, new JPRs are introduced and existing ones are modified to meet the current practices and responsibilities of the person performing the task. Textbooks and curriculum packages are updated by major publishers corresponding to updated standards, usually in a timely fashion.

Both of these scenarios—updating an existing program and "borrowing" one from another source—require the developer to evaluate the currency of materials being used while planning instruction. Although differences exist in writing styles and organizational structures of textbooks, most of these references continue to focus on a specific version of a standard. The developer, in turn, should review multiple sources when doing research for program development. Endorsements of major fire service organizations such as NFPA, International Association of Fire Chiefs (IAFC), and other nationally recognized agencies

JOB PERFORMANCE REQUIREMENTS (JPRS)
in action

Planning, developing, and implementing comprehensive programs and curricula is the major instructional development duty area for the Fire Service Instructor III. Although other instructors will be responsible for delivery and use of the materials, the instructor who develops any curriculum used by the department must complete an agency needs analysis and conduct research while designing the program. Knowledge of which agency goals are addressed by the course is essential, and modification of existing curricula allows for program and course goals to be met. When writing course objectives, keep in mind that you are creating "big picture" outcomes, rather than individual learning objectives. These high-level skill applications require a large amount of time and research to create usable and effective programs.

Instructor I	Instructor II	Instructor III
The Instructor I functions at the delivery level of course materials. Make sure you are familiar with all course elements and are well prepared to deliver the training. Review prepared application exercises and make sure you are ready for questions and possible performance problems.	As an Instructor II, one of your first responsibilities is to modify materials from existing curricula to accommodate student characteristics and even time schedules. Review class rosters and available time to break down larger courses into manageable class segments. Make sure you are able to ensure consistency and continuity of material flowing from class to class.	The Instructor III develops the entire curriculum package, including course goals, course objectives, lesson plans, media, and evaluation materials. As the course is developed, ensure that you are selecting the right instructors to deliver the material and that they have adequate time to review and prepare. Conduct a pilot course with a small audience of content experts, if possible, to work out any problems that may occur before finalizing the product.
JPRs at Work	**JPRs at Work**	**JPRs at Work**
There are no JPRs at the Instructor I level; however, all of the curricula produced by the Instructor III may eventually be delivered by the Instructor I.	There are no JPRs at the Instructor II level; however, all of the curricula produced by the Instructor III may eventually be modified by the Instructor II.	Discuss the methods of training needs assessment. List the steps in the curriculum development model. Discuss the importance of properly stated program and course goals. Describe the content of properly constructed learning objectives. Explain the purpose of modifying existing curricula. Write a clear, concise, and measurable program or course goal to address training needs assessment gaps identified by the developer. Write properly constructed learning objectives to support the course goal and intended outcome.

Bridging the Gap Among Instructor I, Instructor II, and Instructor III

All levels of the instructional delivery team play varying roles in the curriculum process. While the Instructor III develops the material, the Instructor I and II will make the lesson plans achieve the course goals. The outcome-based system of the CDM revolves around input and constant evaluation of program material to strive toward making all programs more effective and efficient. Communications skills within the instructional delivery team will be vital in ensuring the success of the course goal. Those persons who adapt and modify materials must communicate the reasons why such changes are necessary. The designer should seek input and actively participate in the delivery process.

provide content validity for textbooks because they indicate that panels of content experts have reviewed the material.

■ Evaluating Existing Curriculum

Instructors who are responsible for evaluating resources should develop or follow an objective-based decision-making process when comparing resources. As part of such an assessment, you must evaluate the quality, applicability, and compatibility of the resource for the department. Many high-quality resources have been developed in recent years, whereas some time-proven resources still hold sway in certain market areas. When evaluating resources for use within your curriculum, consider the following questions:

1. **How will this product/resource be used in your department?** Consider this question to be a form of audience analysis. Consider both the audience and the fire service instructor(s) who will use it. Will this material require training for the fire service instructors who will use it? Will it require new learning skills of the students?

2. **With which standard(s) is the product/resource developed in conjunction or compliance?** This question may be very important when considering adoption or reference of a textbook or training package. Certain state training agencies use specific references for certification training.

3. **Is this a "turn-key" product (i.e., ready to use as is), or must the fire service instructor adapt the product to local conditions?** Almost all material purchased as part of a curriculum must be adapted to local conditions and student needs. Always check to see if the product allows you to edit all components.

4. **Is this product/resource compatible with your agency procedures and methods?** Consider the content-specific issues of the product/resource. Does it include the methods you use to operate, or are there significant differences in the equipment or methods used or advocated by the product/resource?

5. **Does the product/resource fit within your budgetary restrictions?** Every agency has some sort of budgetary oversight, and the instructor should be prepared to complete a budget justification for the material.

6. **Are there advantages to purchasing larger quantities of the product/resource?** Many vendors will discount certain training materials when they are purchased in larger quantities. Allowances for shipping or enhanced vendor support are also opportunities for comparison of materials.

7. **Has your research regarding the product/resource that you are considering included both references from vendors and your own investigation within your local area or statewide instruction networks?** You should seek out users who are not necessarily on the list provided by vendors. This may take some extra work, but instructor associations and chiefs groups may have users who have varied opinions on the material being evaluated.

8. **Will the product/resource be available for your use within the prescribed timeline within which you are working?** Some publications and resources may become backordered due to shortages and demand. Make sure that the vendor selected has a sufficient quantity of the material ready to ship to meet your timetable.

9. **Are there any local alternatives, such as your own design of a product/resource or identification of other local agencies that may be willing to share the use of the product/resource?** Some fire academies have lending libraries that may be available for your use if your budget demands require this approach.

10. **Are manufacturer demonstrations, product experts, and other types of developer assistance related to the product/resource available to the instructor?** Consider the availability of on-site assistance, phone support, Web site information, product literature, and suggested user guidelines or instructions.

Also consider the level to which the resource was written. You may want to compare the level of writing, the level of objectives, and any correlations that may have been done on the curriculum TABLE 13-2 . Networking with other developers may provide insight into the benefits and shortcomings of any curriculum being considered for adoption.

Teaching Tip

When evaluating curricula from vendors or other agencies, it is important to have two or three instructors assist in the evaluation process. By undertaking this process and coming to a consensus, bias is diminished and objective opinion will help select the best material for your agency.

■ Modifying an Existing Curriculum

The Instructor III is also able to modify a curriculum, making basic or fundamental changes to it. Such changes may include changing the performance outcomes, rewriting learning objectives, or even modifying the content of the lesson. An existing curriculum can include any previously established lesson plan or content of a lesson plan. Reasons a developer may modify an existing curriculum include the following:

- Update of information
- Changes in students' ability or knowledge in the current program
- Outdated references to equipment, technology, or techniques
- Application of an individual department or organization's policy and procedures
- Safety improvements
- Changes in applicable standards
- Audience analysis factors
- Time constraints
- Available resources

Table 13-2	Resource Evaluation Form		
Criteria	Vendor/Product 1 Rating (1 = poor; 5 = excellent)	Vendor/Product 2 Rating (1 = poor; 5 = excellent)	Vendor/Product 3 Rating (1 = poor; 5 = excellent)
Ease of use within our department Instructor Student			
Amount of modification needed for our department			
Standard compliance NFPA OSHA ANSI Other			
Cost of product/resource			
Similar products already in use by department			
Quantity discount available			
References			
Availability			
Vendor support			
Local alternatives			
Other			

Many publishers release instructor support materials, such as the *Instructor's ToolKit for Fundamentals of Fire Fighter Skills*, in a format that allows the user to modify and adapt those materials to his or her specific needs. This approach represents a tremendous time and money saver for the user, and significantly reduces the amount of time necessary to create and prepare material. Modification is always necessary and recommended to ensure compliance with local practice and procedures.

Ethics Tip

When modifying existing curricula and/or adapting curricula for your agency's use, make sure all changes and modifications are documented.

Writing Course Objectives

In the CDM, writing course objectives uses and references skills developed in Fire Service Instructor II training. The enhancement at the Instructor III level is to ensure that the objectives that are written are part of a course within a training program instead of a specific class of instruction or training session. Most of your experience in this area has probably been in developing objectives and lesson plan materials for smaller *classes* of instruction instead of a *course* of instruction. The expansion of duties within the Instructor III position requires a better understanding of how JPRs are referenced and broken down and how terminal and enabling objectives are used in course development.

After the program and course goals have been written, instructional planning and development continue with the identification of the desired outcomes, or objectives. What do you want the students to know or be able to do by the end of class? Learning objectives and behavioral objectives are defined as goals that are achieved through the attainment of knowledge, a skill, or both. The goal in each case can be observed or measured. Sometimes these learning objectives are referred to as performance outcomes or behavioral outcomes. If students are able to achieve the learning objectives of a lesson, they will achieve the desired outcome of the class. Consider the learning objectives of a lesson to be the steps necessary to get to the top of the staircase, which is the course goal.

■ Components of Learning Objectives

Many different methods for writing learning objectives have been proposed. One method commonly used in the fire service is the ABCD method. ABCD stands for **a**udience (Who?), **b**ehavior (What?), **c**ondition (How?), and **d**egree (How much?). (Learning objectives do not always need to be written in that order, however.) This method was introduced in the book *Instructional Media and the New Technologies of Education*, written by Robert Heinich, Michael Molenda, and James D. Russell (Macmillan, 1996). (The *Lesson Plans* chapter discusses the ABCD method in terms of writing lesson plans.)

The audience aspect of the learning objective describes who the students are. If the lesson plan is being developed specifically for a certain audience, the learning objectives should be written to indicate that fact. In contrast, if the audience is not specifically known or includes individuals with varied backgrounds, the audience part of the learning objective could be written generically, such as by referencing "the fire fighter" or even "the student."

Once the audience has been identified, the behavior is listed. The behavior must be an observable and measurable action. A common error in learning objective writing is using words such as *know* or *understand* for the behavior. Is there really a method for determining whether someone understands something? It is better to use the words *state, describe*, or *identify*—these are actions that you can see and measure. It is much easier to evaluate the ability of a student who is identifying the parts of a portable fire extinguisher than to evaluate how well the student understands the parts of a portable fire extinguisher.

The condition describes the situation in which the student will perform the behavior. Items that are often listed for the conditions include specific equipment or resources given to the student, personal protective clothing or safety items that must be used, and the physical location or circumstances for performing the behavior. For example, consider these fragments of objectives:

> "… in full protective equipment, including self-contained breathing apparatus."
>
> "… using the water from a static source, such as a pond or pool."

The last part of the learning objective indicates the degree to which the student is expected to perform the behavior in the listed conditions. With what percentage of completion is the student expected to perform the behavior? Total mastery of a skill would require 100 percent completion—this means perfection, without missing any steps. Knowledge-based learning objectives are often expected to be learned to the degree stated in the passing rate for written exams, such as 70 percent or 80 percent. Another degree that is frequently used is a time limit. This can be included on both knowledge and skill learning objectives.

ABCD learning objectives do not need to contain all of the parts in the ABCD order. Consider this example:

> In full protective equipment including SCBA, two fire fighter trainees will carry a 24-foot extension ladder 100 feet and then perform a flat raise to a second-floor window in less than one minute and thirty seconds.

In this example, the audience is "the fire fighter trainees." The behaviors are "carry a 24-foot extension ladder" and "perform a flat raise." Both "carry" and "raise" are observable and measurable actions. The conditions are "full protective equipment including SCBA," "100 feet," and "to a second-floor window." These describe the circumstances for carrying and raising the ladder. The degree is "less than one minute and thirty seconds." The fire fighter trainees must demonstrate the ability to perform the behaviors to the proper degree to meet the learning objective successfully.

Strictly speaking, well-written learning objectives should contain all four elements of the ABCD method. Shortening learning objectives should be done only when the assumptions for the missing components are clearly stated elsewhere in the lesson plan. It is not very likely that a learning objective would omit the behavior component because this component is the backbone of the learning objective.

Teaching Tip

Using the ABCD method in the development process will add consistency to your development process and will increase the quality of the end product.

■ Level of Achievement

Another important consideration is the level to which a student will achieve the learning objective. This is most often determined by applying Bloom's Taxonomy (*Taxonomy of Educational Objectives*, 1965), a method used to identify levels of learning within the cognitive domain. For the fire service, the three lowest levels in Bloom's Taxonomy are commonly used when developing learning objectives. In order from simplest to most difficult, these levels are knowledge, comprehension, and application. Knowledge is simply remembering facts, definitions, numbers, and other data. Comprehension is displayed when students clarify or summarize the important points. Application is the ability to solve problems or apply the information learned in situations.

To use this method, a Fire Service Instructor III who is developing objectives must determine which level within the cognitive domain is the appropriate level for the student to achieve the course objectives. For example, if an Instructor III is developing objectives for a portable extinguisher class, the following objectives could be written for each level:

- Knowledge: "The fire fighter trainee will identify the four steps of the PASS method of portable extinguisher application." For this objective, the student simply needs to memorize and repeat back the four steps of pull, aim, squeeze, and sweep. This is a very simple knowledge-based objective and easily evaluated with a multiple-choice or fill-in-the-blank question.
- Comprehension: "The fire fighter trainee will explain the advantages and disadvantages of using a dry-chemical extinguisher for a Class A fire." This objective requires the student first to identify the advantages and disadvantages of a dry-chemical extinguisher, and then to select and summarize those that apply to use on a Class A fire. This is a higher-level objective. Such an objective may be evaluated by a multiple-choice question but is better evaluated with a short answer-type question.
- Application: "The fire fighter trainee, given a portable fire extinguisher scenario, shall identify the correct type of extinguisher and demonstrate the method for using it to extinguish the fire." This is considered the highest level of objective, because it requires the student to recall several pieces of information and apply them correctly based on the situation. Such an objective is often evaluated with scenario-based questions that may be answered with multiple-choice items or short answers.

There is no one correct method used to determine which level or how many learning objectives should be written for a lesson plan. Typically, a lesson plan will contain knowledge-based learning objectives to ensure that students learn all of the facts and definitions within the class. Comprehension objectives are then used to ensure that students can summarize or clarify the knowledge-oriented material. Finally, application objectives are used to ensure students' ability to use the information that was learned in the lesson. As the course goals and course objectives are created, the learning objectives used in the lesson plan will satisfy the course goal.

> ### Safety Tip
>
> When developing practical skill objectives, ensure that all safety issues and concerns are addressed for both the student and the instructor.

■ Converting JPRs into Learning Objectives

Often, a Fire Service Instructor III needs to develop learning objectives to meet JPRs listed in an NFPA professional qualifications standard. This occurs when a course is being developed to meet the professional qualifications for a position such as Fire Officer, Fire Instructor, or Fire Fighter. The JPRs listed in the NFPA standards of professional qualifications are not learning objectives, but learning objectives can be created based on the JPRs. Each NFPA professional qualification standard has an annex section that explains the process of converting a JPR into an instructional objective and provides examples of doing so **FIGURE 13-4**. By following this format, the Instructor III can develop learning objectives for a lesson plan to meet the professional qualifications for NFPA standards.

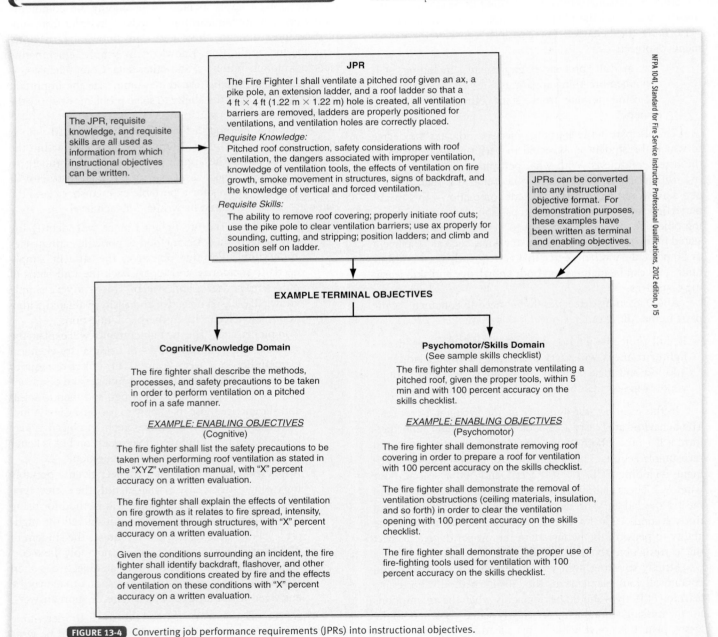

FIGURE 13-4 Converting job performance requirements (JPRs) into instructional objectives.

NFPA 1041, Standard for Fire Service Instructor Professional Qualifications, 2012 edition, p 15

Determining Course Content

At this stage in the CDM, the Instructor III has written a program goal, determined the course goals for the courses of instruction required to meet the program goal, and written learning objectives to meet the course goals. The developer now has to determine the course content elements for each course of instruction. This step is a mapping of what will be necessary to complete the course and helps the developer specify the length of the course, identify the resources necessary, and begin the process of determining the methods of instruction. It is as much about developing a sequence or building the steps necessary to accomplish the course goal as anything else. The developer should consider the following factors in completing this process:

- Time available and required to deliver the material
- Experience and knowledge of the audience
- Chronological order of delivery (Which material has to be taught before other material is introduced)
- Relationships between objectives and units
- Research of resources used in the course
- Availability of instructors and instructional resources
- Environmental considerations
- Deadlines on completion of course
- Methods of instruction used

■ Organization of Course Content

Developers will have to modify the decision-making process used to determine course content every time they develop or modify a course. Many factors play into the decision making, the most prevalent being time constraints. The sequence of delivery is also very important to this process. A student cannot expect to move to step 4 if he or she has not learned steps 1, 2, and 3 adequately. For example, a student should learn about building construction, fire behavior, ground ladders, and tool use *before* the student is instructed in peaked-roof ventilation. The sequence provides the chronological order of delivery that

enhances understanding. We will return to the ABCD model for help in determining content TABLE 13-3 .

Your content's organization must be adaptable and flexible, especially with new courses or content. A pilot course should be used, during which students should be made aware that they are participating in a pilot or first-run program. This approach allows the developer to adapt to course flow determinations. A breakdown of organizational order might look like this:

- Determine available and required time to deliver the course
- Identify the audience and audience delivery factors
- Assemble the course into units (days or hours) to deliver
- List the objectives for each unit
- Determine the logical sequence of related tasks
- Determine and locate the resources needed for each unit
- Identify qualified instructors needed to deliver each unit
- Develop the lesson plan, instructional media, and activities

A course content outline can now be written based on this breakdown of information. Such an outline can help the developer schedule the blocks of time in which material is to be presented, as well as determine the specific order in which it is to be presented. During the pilot course, the course content outline will be adjusted according to the rate and flow of the information to the students. Functional groupings of information can be implemented to organize the information being presented and can also be used in the creation of lesson plans, typically at the level of the larger heading areas. As a course developer, you should estimate the time necessary to deliver the material and determine which activities, such as student applications, are necessary to make the learning effective for the learner. Once this breakout is created, you have developed a delivery schedule for the course; this matching of time estimates and activities will also help keep you on task to complete material on time.

Table 13-3	The ABCD Method in Organization of Course Content
Audience factors	Number of students who will attend each session Learning level of the students • If this is new material, it may take longer to achieve the desired level of understanding or a refresher may allow for shorter delivery.
Behavioral factors	The outcomes for the program • Cognitive or psychomotor. • Determines the number of instructors needed and the amount of skill practice that can be completed in a session. Remember the 5:1 ratio of students to instructors for hands-on programs. • Realistic view of how much material can be absorbed in a session.
Conditions of delivery	Resources necessary for delivery • Include the classroom, drill ground, and any props. Environmental issues in learning environment Availability of other resources (e.g., texts, handouts, computers)
Degree	Your performance expectations • Students must pass all practical exams as individuals or in crews/companies. • Classroom measurements with evaluation measures.

Conducting Research and Identifying Resources

Conducting research into current issues and trends and obtaining up-to-date statistics are part of the developer's responsibility and ensure that all objectives reflect current practices and are relevant to the topic. The Internet brings the firefighting world to each developer's desktop and allows for even personnel in the busiest of departments to learn from one another. Adult learners prefer to be able to relate learning to where and how they will use it. Case studies and current events allow for that to happen if they are used in the right way. Instructors must make sure they blend these types of discussions into the lecture or demonstration, but must also be cautious that they do not result in "Monday morning quarterbacking" of incidents and operations. It is easy for a class to become sidetracked on these types of discussions unless the instructor keeps the group on task.

The *Legal Issues* chapter covers the legal applications and copyright information that must be adhered to by the developer when citing and referencing materials researched or used in the development of class material. Photos, lesson plans, PowerPoint® presentations, articles, books, and other studies and reports, while beneficial in enhancing the quality of the class, need to be used and referenced in the correct manner.

The peer support network established by many developers is another great resource. Instructors are often willing to share their work and will help you when you are stuck for resources or ideas. The National Fire Academy (NFA) has a free, subscription-based resource network titled TRADE (Training Resource and Data Exchange). Developed by Chief Robert Bennett

(retired), this network allows developers and others to post "help needed" messages into a large e-mail listing that goes to several thousand users each week **FIGURE 13-5**. When a developer needs resources, it is remarkable to see the outpouring of support that the TRADE list supplies after just one e-mail listing.

State and local training bulletins and manuals may also represent time-saving resources for the developer. Many states have baseline curricula available for use, with most of those materials referencing NFPA standards and state-specific requirements for training. It is a best practice to reference and incorporate these curricula into your instructional planning. The Illinois Office of the State Fire Marshal uses an Ad Hoc committee process to develop referenced baseline curricula for fire departments to reference and use in the development and delivery of training programs. Objectives, lesson plans, basic PowerPoint® presentations, and sample evaluation tools are available to departments that have the appropriate instructors to develop the program.

Please reply to this message if you are experiencing technical difficulty. For all other inquiries, please <u>Contact USFA</u>.

You are subscribed to Training, Resources and Data Exchange Network (TRADENET) for U.S. Fire Administration. This information has recently been updated.

Welcome to TRADENET: TRADE's Information Exchange Network

A National Fire Academy Sponsored Activity

In This Week's TRADENET

- <u>Member Requests for Information</u>

1. SOP/SOG request
2. (**EFOP**) Manning survey request (**EFOP**)
3. (**EFOP**) Command Officer Performance Appraisal survey request (**EFOP**)
4. Insurance reimbursement form letter request
5. (**EFOP**) Dispatch automatic aid survey request (**EFOP**)

- Emergency Medical Services Page
- Featured This Week

NIC Requests Resource Identification
Free Webinar: Policy is Not Enough...
Career Opportunity: Cogswell College
Lakeland Community College Offers Free ERG Slide Sets
"Bridging the Gap: Guide to Implementing Residential a Sprinkler Requirement" 2nd Edition Arrives
State/Local Partner Sponsored 2-Day Off Campus Courses

- <u>National Events Calendar</u>
- <u>TRADE Regional Pages</u>

Region I: MFSI Newsletter

- <u>TRADE Regional NFA Course Calendar</u>
- <u>Important Links</u>

Member Requests for Information

If you have a question or comment for inclusion in the weekly TRADENET newsletter, please send it and your contact information to <u>FEMA-trade@fema.dhs.gov</u>. Please be sure and include your Department or Organization name, your e-mail address, and your preferred method of receiving feedback. USFA reserves the right to edit content or reject any material submitted.

The United States Fire Administration recommends everyone should have a comprehensive fire protection plan that includes smoke alarms, residential sprinklers, and practicing a home fire escape plan.

Follow USFA updates on <u>Twitter</u>

Update your subscriptions, modify your password or e-mail address, or stop subscriptions at any time on your <u>Subscriber Preferences Page</u>. You will need to use your e-mail address to log in. If you have questions or problems with the subscription service, please contact <u>support@govdelivery.com</u>.

This service is provided to you at no charge by the <u>U.S. Fire Administration</u>.

<u>Privacy Policy</u> | GovDelivery is providing this information on behalf of U.S. Department of Homeland Security, and may not use the information for any other purposes.

This email was sent using GovDelivery, on behalf of: U.S. Fire Administration · U.S. Department of Homeland Security · Emmitsburg, MD 21727 · (301) 447–1000

FIGURE 13-5 The NFA's resource network, TRADE, allows developers and others to post "help needed" messages into a large e-mail listing that goes to several thousand users each week. Here is a sample.

Training BULLETIN

Jones & Bartlett Fire District

Training Division
5 Wall Street, Burlington, MA, 01803
Phone 978-443-5000 Fax 978-443-8000
www.fire.jbpub.com

Instant Applications: Instructional Curriculum Development

Drill Assignment

Apply the chapter content to your department's operation, the training division, and your personal experiences to complete the following questions and activities.

Objective

Upon completion of the instant applications, fire service instructor students will exhibit decision making and application of job performance requirements of the fire service instructor using the text, class discussion, and their own personal experiences.

Suggested Drill Applications

1. Using the job observation method, make a list of recent operational errors or inconsistencies that may warrant the development of a training curriculum aimed at improving member performance.

2. List the current curricula your department uses for basic entry-level certification training, and identify the edition of the material and the latest NFPA standards that exist at that level. Are they up-to-date? Are other products available that are more current?

3. Build an organization chart of topics within a subject area for which a sequential subject coverage arrangement would be required to teach the student all of the material. Which subjects need to be taught and mastered before you can move onto the next? (Example: Peaked roof ventilation requires knowledge of which topics before you can instruct on the actual skills and knowledge of how to ventilate a peaked roof?)

4. Review the Incident Report in this chapter and be prepared to discuss your analysis of the incident from a training perspective and as an instructor who wishes to use the report as a training tool.

Incident Report

Indianapolis, Indiana—2002

On June 14, 2002, an Advanced Open Water SCUBA Diver course was being completed, which consisted of conducting a final dive at a privately owned lake. The dive class included two certified National Association of Underwater Instructors (NAUI), six dive masters, and a total of 22 students. An instructor briefed the students on the scenario, which would test their skills in manipulating equipment to lift an object from the bottom of the lake in zero visibility and cold water.

The victim was a 37-year old male who had been a member of the fire department for seven years and had completed several training courses during his career. He was working to complete the divers training to qualify to be part of the department's Dive and Rescue Team. Prior to this course he had no previous diver training.

At approximately 1243 hours, the victim and his partner were attempting to locate and retrieve an object in approximately 50 feet of water. The pair had located the object and were using a lifting device, in zero visibility, to assist in bringing the object to the surface. The victim's partner was in a kneeling position on the bottom of the lake attaching his line to the object when the victim knocked him over on his side and attempted to remove his mask. The victim's weight belt landed on top of his partner. The victim may have released his weight belt in an attempt to surface.

The partner surfaced and informed the personnel in the safety boat that the victim was in trouble. The safety diver immediately entered the water and began searching for the victim. Additional divers were shuttled to the site by the safety boat to continue searching for the victim. The victim was recovered over 2 hours later using a sonar device. Advanced life support was initiated en route to a local hospital where he was pronounced dead.

During the investigation, the equipment worn by the victim was tested. The second-stage regulator failed an exhaust flow test due to the exhaust diaphragm sticking shut. A protective shield ring on the mask, which was designed to keep debris out of the second-stage area, was missing. Sand was found inside the diaphragm area. The exhaust valve was disassembled and cleaned for testing; it worked as designed. It is not known if the protective shield ring was in place during the dive or if the problems with the diaphragm existed during the dive or resulted from losing the protective ring during recovery efforts, allowing sand to enter the device.

Post-Incident Analysis: Indianapolis, Indiana

Contributing Factors

Equipment checks not performed thoroughly before dive

Defective equipment not repaired or replaced before dive

Victim had not practiced the specific evolution in a controlled environment such as a swimming pool before attempting in an open water or limited visibility environment

Wrap-Up

Chief Concepts

- Planning and developing an instructional program begins with determining which training is necessary and who needs the training.
- Department administration may instruct you to perform an assessment to identify any regulatory compliance matters that must be included in the department training program.
- Three methods may be used to help plan and establish the need for instruction:
  - Job observation
  - Compliance and regulation
  - Skill/knowledge development and improvement
- Remember that you are designing a course based on your particular agency's needs and that every department has different patterns of delivery and resources available to it.
- The four-step method of instruction includes the following elements:
  - Preparation
  - Presentation
  - Application
  - Evaluation

- An order model can be used to help the developer proceed through the course development process in a step-by-step manner.
- A program goal can be viewed as the final outcome of a series of courses, classes, drills, or training sessions conducted over a period of time. A course goal may be a smaller picture of the desired outcome of the training.
- Programs could be titled based on task analysis (e.g., fire fighter, driver/operator, fire officer, fire instructor) to begin the development of instructional materials, and then course goals would follow.
- Whether you plan to develop a course from top to bottom or to modify an existing curriculum, a clear and concise goal for the course needs to be established.
- In many cases, existing curriculum or training materials may already be in place within the organization, and may simply need updating to current standards and practices and to current students' abilities and needs.
- Instructors who are responsible for evaluating resources should develop or follow an objective-based decision-making process when comparing resources. You must evaluate the quality, applicability, and compatibility of each resource in light of the department's unique needs and characteristics.
- At the Instructor III level, the instructor is expected to ensure that the objectives are written as part of a course within a training program instead of being

written to a specific class of instruction or training session.

- Knowledge is simply remembering facts, definitions, numbers, and other data. Comprehension is displayed when students clarify or summarize the important points. Application is the ability to solve problems or apply the information learned in situations.

- Determining course content entails a mapping of what will be necessary to complete the course and helps the developer specify the length of the course, identify the resources necessary, and begin the process of determining the methods of instruction.

- Many factors play into the decision-making process for course development, with the most prevalent being time constraints. The sequence of delivery is also very important to this process.

- Conducting research into current issues and trends and obtaining up-to-date statistics are part of the developer's responsibility to make sure that all objectives reflect current practices and are relevant to the topic.

Hot Terms

ABCD method Process for writing lesson plan objectives that includes four components: audience, behavior, condition, and degree.

Application step The third step of the four-step method of instruction, in which the student applies the information learned during the presentation step.

Behavioral objectives Goals that are achieved through the attainment of a skill, knowledge, or both, and that can be measured or observed; also called learning objectives.

Class of instruction An individual lesson or drill that is a smaller part of the entire course. In an academy-style course, you will receive many classes of instruction that prepare you for completion of the course or certification.

Course goal The end result and desired outcome of multiple classes of instruction and lessons.

Enabling objective An objective that allows the student to meet the terminal objective identified in a course design document. Several enabling objectives may be required to understand or perform the terminal objective.

Evaluation step The fourth step of the four-step method of instruction, in which the student is evaluated by the instructor.

Four-step method of instruction The most commonly used method of instruction in the fire service. The four steps are preparation, presentation, application, and evaluation.

Learning objectives Goals that are achieved through the attainment of a skill, knowledge, or both, and that can be measured or observed.

Pilot course The first offering of a new course, which is designed to allow the developers to test models, applications, evaluations, and course content. Many pilot courses do not fully represent the final course product, as much feedback is exchanged between students and instructors in the pilot program to improve the final product.

Preparation step The first step of the four-step method of instruction, in which the instructor prepares to deliver the class and provides motivation for the students.

Presentation step The second step of the four-step method of instruction, in which the instructor delivers the class to the students.

Terminal objective One way of framing a JPR or outcome of a set of enabling objectives or learning steps. If several enabling objectives are mastered, then the terminal or end point knowledge or skill is achieved.

References

Bloom B. S. (1965). *Taxonomy of Educational Objectives, Handbook I: The Cognitive Domain*. New York: David McKay Company.

Heinich, Robert, Michael Molenda, and James D. Russell. (1998). *Instructional Media and the New Technologies of Education*. 6th ed. Upper Saddle River, NJ: Prentice Hall College Division.

National Fire Protection Association. (2012). *NFPA 1041: Standard for Fire Service Instructor Professional Qualifications*. Quincy, MA: National Fire Protection Association.

National Fire Protection Association. (2010). *NFPA 1410: Standard on Training for Initial Emergency Scene Operations*. Quincy, MA: National Fire Protection Association.

FIRE SERVICE INSTRUCTOR *in action*

You are the Fire Service Instructor III who has identified a specific training need, and you are going to begin the process of developing a refresher training course on fire-ground size-up. Recent fires have featured inconsistent and poorly performed size-ups, which could have contributed to close calls and poor tactics being employed.

1. Considering this scenario, which type of training needs assessment have you performed?
   A. Behavioral based
   B. Job observation
   C. Compliance or regulatory
   D. Certification based

2. What audience level factors need to be considered when developing your curriculum content?
   A. Age and experience
   B. Willingness to learn new material
   C. Previous experience in the course content
   D. All of the above

3. You are developing a simulation-based exercise for use in a class, which will give the students the chance to practice proper size-up techniques. Which step in the four-step method of instruction is this?
   A. Preparation
   B. Presentation
   C. Application
   D. Evaluation

4. What is another way of expressing the course goal for this training course?
   A. End result
   B. Certification program
   C. Behavioral objective
   D. Course application

5. A final examination for the course, instructor evaluations, and student satisfaction surveys may all be included in which part of your course?
   A. Course goal
   B. Evaluation system
   C. Lesson plan
   D. Course outline

6. You are going to write the course goal for this training program. Which factors should be considered in preparing a proper goal statement?
   A. The department's need for the training
   B. The students' need for the training
   C. The outcome that is expected for both the student and the organization
   D. All of the above should be considered as part of the goal

7. "Perform an initial incident scene size-up, identifying the size, height, occupancy type, problem description, and initial actions of the first arriving company" is an example of which part of the CDM process?
   A. A course objective
   B. A student application
   C. An evaluation criterion
   D. All of the above

Managing the Evaluation System

Fire Service Instructor I

Knowledge Objectives

There are no knowledge objectives for Fire Service Instructor I students.

Skills Objectives

There are no skills objectives for Fire Service Instructor I students.

Fire Service Instructor II

Knowledge Objectives

There are no knowledge objectives for Fire Service Instructor II students.

Skills Objectives

There are no skills objectives for Fire Service Instructor II students.

Knowledge Objectives

After studying this chapter, you will be able to:

- Describe record keeping systems needed for test results. (NFPA 6.5.2) (pp 324–325)
- Discuss applicable laws and methods of providing feedback to students about test results. (NFPA 6.5.2) (pp 324–325)
- Discuss creation and implementation of course and program evaluation plans. (NFPA 6.5.3) (pp 325, 327–332)
- Explain test validity, reliability, and item analysis. (NFPA 6.5.5) (pp 332, 334–339)
- Discuss the use of performance evaluations. (pp 339–342)
- Discuss the interpretation of evaluation results. (p 340)

Skills Objectives

After studying this chapter, you will be able to:

- Analyze student evaluation tools. (NFPA 6.5.2 NFPA 6.5.5) (pp 332, 334–339)
- Create a program evaluation plan. (NFPA 6.5.3 NFPA 6.5.4) (pp 325, 327–328)
- Create and implement performance evaluations. (pp 339–342)
- Interpret evaluation results. (p 340)

You Are the Fire Service Instructor

You have just completed a fire academy training course, serving for the first time as the program coordinator. As part of the course, you used a new curriculum for the first time. Now you would like to review the test results for your 10 fire academy recruits and complete a full review of the evaluation and testing process used during the academy. The operations manual for the academy does not contain any direction to help you start on this effort, but you know that the success of future academies can be improved by undertaking such a review of the evaluation system used. Test scores are available for many subject quizzes, unit exams, and final examinations for you to work from. Once the evaluation is completed, you will have to develop a method for the storage and maintenance of the tests as well.

1. What is the value of evaluating test scores for this fire academy?
2. Which changes would you make if you discover that certain exam questions had very low success rates or that others had very high success rates?
3. During the course of the academy, how should you release test scores to students and staff instructors?

Introduction

Evaluation systems, commonly known as tests, come in a variety of forms, as described in the chapter *Evaluating the Learning Process*. Much is at stake for the student as well as the instructor when the evaluation system is used to demonstrate students' competence and knowledge or for certifications. Evaluation systems must be developed to comply with many laws and standards, and the Instructor III will be responsible for understanding these requirements.

When designing an evaluation system, many facets of evaluation and testing will require careful attention to both student results and instructor performance. A system to collect acquired evaluation results needs to be established that will allow the course developer to review success and failure rates on particular test questions and to track trends over time, comparing various classes on the same criteria. In many agencies, these data can be provided by the examining authority, such as a state fire marshal's office, training academy, or other certifying entity. In other cases, the local department will be responsible for data analysis of the results. Many records will be produced as part of testing and evaluation, and a formal process of result disclosure and storage should be available to guide the instructor on how to handle this information properly. As the *Legal Issues* chapter discusses, test records should be made available to the student but should not be readily accessible and visible to anyone else. A course and program evaluation plan will help the agency make sure that the goals of a program were met and the course components, instructors, and evaluations met the expectations of the agency.

Fire Service Instructor III

Acquiring Evaluation Results

Evaluation results should flow from the students who have completed the test, through the instructor who administered the exam, and then to the individual or agency responsible for grading the results. Once test results are graded and data compiled on the success or failure rate on a particular question, skill, or objective, the instructor can analyze those results. A simple flowchart may direct those who come in contact with this information on the next step in the process. Some systems will use an electronic data summary.

Storing Evaluation Results

Examination results and student assignment results should be treated as confidential every step of the way. This confidentiality should be part of your organization's policies and made clear to the student. The physical score sheets, practical skill sheets, and other summary reports and performance documentation may need to be retained for defined periods of time depending on local laws or ordinances. Other considerations related to the storage of evaluation results include the following:

- Evaluation records may need to be made available for inspection by examining or auditing authorities.
- Access to evaluation results should be available in case a challenge to student performance occurs.
- Data should be available for recurring tests of the same nature for the purpose of establishing and identifying trends in the tests' results.

Each fire department should have a policy that states how long evaluation results must be stored and the manner in which they must be kept. It may be acceptable to scan results into a data storage system or to compile the results onto spreadsheets and then destroy the original evaluation forms or checklists. It is a good practice to establish some type of backup system in the event one system fails. The *Legal Issues* and *Training Program Management* chapters discuss record retention in more detail.

Theory Into Practice

Check with your local legal counsel or records management agency, such as the Secretary of State's office, for policy on test record retention and proper disposal.

Dissemination of Evaluation Results

To reemphasize a point made earlier, evaluation results should be kept confidential and policy should dictate which staff members and administrative staff have access to these records. A best practice in this area may be to have any request for release of evaluation records approved by the fire chief and the student(s) involved. Summary tables that do not include individual names or other identifiable data of the students might be acceptable for the purpose of assisting instructors and staff in reviewing the course participants' progress FIGURE 14-1. It is best not to single out any individual student during a mass release of data to instructors; only key administrators of a program should have full access to individual records.

Students should be granted full access to their evaluation results, but not necessarily to the evaluation tools themselves. Some agencies prepare summary reports of tests administered that show which objectives were missed. Students can then review these reports as part of their personal exam analysis, without having the actual exam questions. The *Legal Issues* chapter discusses several legal areas of concern for posting and releasing evaluation records and should be reviewed by both the local department and legal counsel when developing a policy.

Course and Program Evaluation Plans

Evaluating a training course should be part of the ongoing and constant improvement goal of any training division. Program evaluation may consider the following aspects of the course:

- Performance
- Reliability or consistency
- Facilities and equipment
- Reputation of the program
- Conformance with standards
- Response to needs of the students and the community at large

The department training officer or a course developer should establish a system for reviewing key course and program components on a regular basis as soon as possible after a course is completed. Among the items that should be reviewed as part of a program evaluation plan are the following:

- Administrative details
  - Costs
  - Paperwork processing
- Instructor ratings and reviews by students
- Instructor reviews of students
- Peer review or supervising instructor review of instructional staff
- Course components and features
  - Media
  - Textbook
  - Tests, quizzes, and evaluations
  - Time allotments
  - Homework and study assignments
- Facilities and equipment used in course
- Written and verbal comments by students for course improvements

Program evaluation tools are designed to assess the performance of the course. They identify strengths and weaknesses of a course and assess student attitudes. Program evaluations are also designed to look internally at the institute's objectives (e.g., How did the program measure up? Were the goals met?) and externally at students' opinions and community-wide impressions (e.g., Did the program meet your expectations? What did you like or dislike about the program?) to determine the overall degree of excellence within the program. Program evaluations are a continual process; a key aspect of a program assessment is to provide ongoing evaluation throughout the entirety of the course, not just at the end of the course. In addition, the training division should consider developing a matrix for each student, explaining the student's areas of strength and areas that need improvement and offering a pathway for mastery of skill levels.

Course Result Summary					
Student ID	Building Construction	Ladders	Ventilation	Forcible Entry	Water Supply
1871	88	94	94	80	97
6112	94	96	92	95	100
3809	82	96	88	83	94
4018	97	84	92	93	91
1796	95	90	90	88	96

FIGURE 14-1 Evaluation result summary tables are a resource for instructor and course evaluation that maintains student confidentiality.

JOB PERFORMANCE REQUIREMENTS (JPRS)
in action

Developing an evaluation plan should not be confused with developing evaluation items such as test questions or practical skill sheets. The former process is a management function that usually takes place after a course or program is completed; as part of this process, the instructor—who may be the developer of the program—evaluates the effectiveness and efficiency of the tools used for the course.

Instructor I

The Instructor I who administers tests and course evaluations must do so in a professional manner. Test item security and confidentiality of test results must be maintained.

Instructor II

The Instructor II may develop test items and course evaluation forms that will be used to evaluate student knowledge and skill levels.

Instructor III

The Instructor III is responsible for evaluating the learning process using tests, instructor evaluations, and other performance measures.

JPRs at Work

There are no JPRs for the Instructor I for this chapter.

JPRs at Work

There are no JPRs for the Instructor II for this chapter.

JPRs at Work

Develop an evaluation plan that reviews test data, course components, facilities, and instructors and manages the results of evaluation measures.

Bridging the Gap Among Instructor I, Instructor II, and Instructor III

Instructors at all three performance levels will develop, administer, and analyze the evaluation systems created for each course. The Instructor I should adhere to the agency policies and procedures for test administration, while the Instructor II develops test items based on course objectives. The review of all evaluation data becomes the responsibility of the Instructor III, and a comprehensive course review mechanism is used to identify program and content strengths and weaknesses.

Defining Program Goals

A program evaluation looks at the level of proficiency that the students, instructors, and administrators have attained. Specific program goals and benchmarks are set for specific educational programs. These goals should be assessed both during the program and at its conclusion. The goals that are established as part of this process must be both measurable and attainable.

The following are examples of program goals:

- Upon completion of the Fire Fighter I certification program, the course's final written examination will be administered. The minimum passing score for the examination is 75 percent, with an ideal class average of 80 percent.
- A 50% balance of classroom and practical training will occur.
- Instructors will provide a 10-minute break for every 1 to 1½ hours of classroom instruction.
- The course coordinator will be present for at least 85 percent of the class sessions.

Goals that are set too high or too low for the type of course and students being taught are not helpful; the same is true of goals that cannot be evaluated. The preceding goals can easily be measured by using a statistical review of the examination results, student surveys, or an administrative review of time sheets.

Program goals should be written well before a program starts, ideally by a group of instructors and course administrators. They may encompass the institute's goals for a particular time period as well as long-range program goals to help guide overall student and instructor performance.

Course Objectives

Course objectives should be singled out for specific review at the completion of a course whenever possible. Because course objectives are stated as outcomes, this part of the review and analysis process may be one of the best measures of identifying whether the course did what it was supposed to do. Data analysis techniques can be used to summarize test results and organize them by objectives **FIGURE 14-2**. A review of practical skill sheets can identify any skill area for which retakes or second

Office of the State Fire Marshal
Division of Personnel Standards and Education
Exam Results by Employing Fire Department/School

January 17, 2013
03:26:07 pm

Fire Dept/School: CBA Fire Department   FDID: QQ001
Exam: Basic Operations Firefighter Mod A - BFA2
Course Approval: Basic Operations Firefighter Mod A (08/01/2010) FD: Walter Davis Community C
Exam Site: Character Village
Exam Date/Time: 01/20/2012 3:00 pm

Dog Bounty Hunter c/o CBA Fire Department

Total Question: 98
Passing Score: 69

Firefighter Name		Score	Battery
Bird, Terri	IKNKLSH	No Show *	
Brown, Cody	ZYXWVUTSRQ	80.00	
Duck, Denny	ZYXWVU	75.00	
Frog, Kenny D	ZYXWV	84.00	
Fudd, Emmett	ZYXWVUT	No Show *	
Kent, Clark	ZYXWVUTSR	84.00	
Land, Candy	ZYXWVUTS	83.00	
Log, Bernie	Z	89.00	
Menace, Donald D	MNBGFCJCV	91.00	
Mouse, Mandy	ZYX	89.00	
Mouse, Michael	ZYXW	68.00 *	
Mouse, Mitchell R	ZY	79.00	
Parker, Pete R	VGHCYDG	84.00	
Pirate, Jacob D	XYZWSTED	88.00	
Sponge, Bubba	ZXYWSTRU	77.00	

Total Firefighters: 15   **Total Firefighters Tested:** 13

The following objectives were missed on this exam:

Missed	Objective	Objective Description	
10	2-1.1	Identify the organization of the fire department (5.1.1)	
		Reference: J&B, Delmar, IFSTA	
1	2-1.2	Identify the basic firefighter's role as a member of the fire service (5.1.1)	
		Reference: J&B, Delmar, IFSTA	
13	2-1.5	Identify the function of Standard Operating Procedures / Guidelines (5.1.1)	
		Reference: J&B, Delmar, IFSTA	
1	2-2.1	Definitions dealing with Fire Behavior	
		Reference: J&B, Delmar, IFSTA	**Multiple Questions**
1	2-2.1	Definitions dealing with Fire Behavior	
		Reference: J&B, Delmar, IFSTA	**Multiple Questions**
8	2-2.3	Identify the relationship of the concentration of oxygen to combustibility and life safety (5.3.11A)	
		Reference: Delmar	

Office of the State Fire Marshal
Division of Personnel Standards and Education
Exam Results by Employing Fire Department/School

January 17, 2013
03:26:07 pm

Fire Dept/School: CBA Fire Department   FDID: QQ001
Exam: Basic Operations Firefighter Mod A - BFA2

2	2-2.4	Identify the products of combustion commonly found in structure fires that create or indicate a hazard (5.3.5A, 5.3.11A)	
		Reference: J&B, Delmar, IFSTA	
2	2-2.7	Identify the methods of heat transfer (5.3.12A)	
		Reference: J&B, Delmar, IFSTA	
3	2-2.9	Identify common fire conditions (5.3.11A)	
		Reference: J&B, Delmar, IFSTA	
3	2-2.11	Identify the process of thermal layering as it relates to a structure fire (5.3.12A)	
		Reference: J&B, Delmar, IFSTA	
1	2-2.12	Identify the development and prevention of a backdraft (5.3.11A)	
		Reference: J&B, Delmar, IFSTA	**Multiple Questions**
1	2-2.12	Identify the development and prevention of a backdraft (5.3.11A)	
		Reference: J&B, Delmar, IFSTA	**Multiple Questions**
2	2-3.1	Identify common structural components of buildings (5.3.10A, 5.3.12A)	
		Reference: J&B, IFSTA	**Multiple Questions**
6	2-3.1	Identify common structural components of buildings (5.3.10A, 5.3.12A)	
		Reference: J&B, IFSTA	**Multiple Questions**
1	2-3.2	Identify basic structural characteristics of the following types of building construction A. Fire Resistive (Type I) B. Non-Combustible (Type II) C. Ordinary (Type III) D. Heavy Timber (Type IV) E. Wood Frame (Type V)	
		Reference: J&B	**Multiple Questions**
2	2-3.2	Identify basic structural characteristics of the following types of building construction A. Fire Resistive (Type I) B. Non-Combustible (Type II) C. Ordinary (Type III) D. Heavy Timber (Type IV) E. Wood Frame (Type V)	
		Reference: J&B	**Multiple Questions**
2	2-3.3	Identify the methods of framing used in Type V construction	
		Reference: J&B, IFSTA	**Multiple Questions**
1	2-3.3	Identify the methods of framing used in Type V construction	
		Reference: J&B, IFSTA	**Multiple Questions**
5	2-3.4	Identify construction features of roofs (5.3.12A)	
		Reference: J&B	
1	2-3.5	Identify the components of a truss	
		Reference: J&B	
1	2-3.7	Identify dangerous conditions created by fire and fire suppression activities (5.3.10A, 5.3.12A)	
		Reference: Outline	
2	2-3.8	Identify indicators of building collapse (5.3.12A, 6.3.2A)	
		Reference: Outline	
2	2-3.9	Identify the effects of the fire on the building materials (5.3.10A, 5.3.12A)	
		Reference: Outline	

FIGURE 14-2 Sample test report for an academy test exam.

attempts were needed for successful completion. Additionally, a review of referenced National Fire Protection Association (NFPA) job performance requirements (JPRs) should be conducted to ensure that course objectives are up-to-date and that all requisite knowledge and requisite skills are covered adequately. A periodic audit of NFPA standards and the course objectives should be undertaken to confirm that the reference being used is the most current edition. Objectives should also be reviewed for proper format and correct conditions if equipment or tasks have been updated or changed.

■ Plan of Action

The plan of action, or strategy, should be developed at the same time as the program's goals. This document outlines how the program goals will be assessed and identifies the evaluation tools that will be used to obtain this information. A variety of methods may be applied to program evaluation. Some institutions use the final examination results as the benchmark for the assessment of a program's effectiveness; others use student opinion surveys; in still others, a course coordinator monitors the actual instruction and rates the instructor's performance. No one evaluation tool can capture the entire picture of an educational program's overall performance. Taken together, however, program evaluations can determine the overall program quality, assign accountability for poor student performance, and outline ways to improve the program.

Commonly used evaluation tools include statistical evaluations, student surveys, and forecasting surveys (which are sent to the community to identify future educational needs and to assess the reputation of the institute). Each tool provides different types of information. Only after reviewing all the information can any conclusions be reached regarding the final evaluation.

■ Agency Policies

Policies relating to course administration and delivery should be reviewed on a regular basis to ensure they adequately do the job for which they were written and to determine whether any updates are needed. Some agency policies relating to course delivery and evaluation may include the following **FIGURE 14-3** :

- Administrative titles and duties
- Disciplinary procedures
- Records and reporting procedures
- Flowchart or area of responsibility diagrams
- Teaching curriculum
- Quality improvement measures
- Grading policies
- Code of conduct for students and instructors

In most cases, a well-written policy manual will be a decision-making guide for all members of the training program delivery team. Students should also be briefed, both verbally and in written form, on agency policies that relate to their performance and conduct in large course environments, and perhaps be required to sign acknowledgments of receipt and understanding of these policies.

■ Program Evaluation Plans

A program evaluation plan will evaluate many facets of course delivery. The course evaluation forms filled out by the students who were the direct users of the materials are often the first component of the program evaluation plan. Instructors, agency policies, and procedures relating to the course; course components; and facilities are all often covered by such student-provided evaluations. Test results can be part of the overall evaluation as well and should be included in the program evaluation plan. The overall program evaluation plan should encompass the following items:

- Course evaluations completed by the students in the class. Course features should be evaluated for ease of use, amount of work, time to complete work, and application to the content related to the job.
- Instructor evaluations identifying the strengths and weaknesses of each instructor in the program.
- Test results that have been properly analyzed based on the following criteria:
  - Reliability
  - Discrimination
  - Percentage scores against the pass/fail rate established
- Agency policies and procedures:
  - Attendance requirements
  - Testing and grading policies
  - Student and instructor codes of conduct
- Facilities and equipment:
  - Up-to-date
  - Comfortable and accessible to all students
  - Equipment used was safe, serviceable, and contemporary
- Safety analysis
  - Injuries and accident data
  - Adherence to safety policies

Theory Into Practice

Make sure evaluation plans include analysis of all accidents, injuries, and damage to equipment, as well as documentation of any near misses that occurred during training. Consider reviewing your department's safety plans, notices, and communicating tools with the entire class. Ensure that each student is aware of the overall safety obligations and the risks associated with not taking safety into consideration at all times.

Course Evaluation

In addition to statistical evaluation, course evaluations are a helpful tool for assessing whether a program has achieved its established goals. Responses to objective questions posed to students and faculty can identify how the institute is meeting its goals **FIGURE 14-4** . Unlike direct observation of students,

Testing Policy

Written Tests	Oral Tests	Performance Tests
• Arrive at least 30 to 45 minutes prior to the beginning time for the test. • Make sure the testing environment is suitable in terms of lighting, temperature control, adequate space, and other related items. • During the arrival of test takers, double-check those who should be in attendance by checking identification documents and record their presence. • Maintain order in the testing facility. Discourage any activities that might be interpreted as cheating and do not provide any information to anyone that is not also provided to the group. • Maintain the security of all testing materials at all times. • Provide specific written and oral reviews of all test-taking rules and behavior guidelines during the testing period. • Remain objective with all test takers. Do not show any form of favoritism. • Answer questions about the testing process and the test itself. • Do not answer an individual question about the content of the test unless the information is shared with the entire group. • Monitor test takers for the entire period of the test. • Discipline anyone who becomes disruptive or violates the rules. • Require all test-item challenges to be made prior to test takers leaving the room. Advise that their challenge will be noted and addressed after all of the tests are scored. • Collect and double-check answer sheets or booklets to ensure that all information is properly entered and that any supporting materials are returned before allowing test takers to leave the room. • Never leave the testing environment for any reason. • Inventory all testing materials and return them to the designated person in the department.	• Determine the oral test items to be used and the type of oral question techniques that will be employed (e.g., overhead, direct, rhetorical). • Make sure the oral test items are pertinent to the task to be performed. • Focus on the critical safety items and dangerous steps within the performance. • If the test taker misses the oral test item, redirect the question to another performer. If you are conducting a one-on-one oral test, then provide the test taker with on-the-spot instruction. Safety knowledge and dangerous task performance steps must be clear and mastered before performance begins. • Use a scoring guide for the oral test, and record any difficulties and lack of knowledge on the part of each test taker. • If the group or individual has knowledge gaps regarding the critical safety items or dangerous performance steps, *do not* proceed to the performance test. More training and testing are required. • Make a training record of oral test results.	• Arrive at the test site at least 1 hour before the performance test. • Check the test environment to make sure all needed tools, materials, and props are present and in good working order. • Determine whether the performance test will require an oral test before actual performance. (Obtain oral test items if required.) • Check the test takers' identification as they arrive and record their presence. • Review the test procedures and ask for questions from the test takers. • Verify that all test takers know what will be expected during performance. • Begin the performance test. Have the students tell you exactly what they will be doing before they do each step in the performance. • Remain silent and uninvolved in the task performance. Your job is to verify safe and competent performance. • Record test results. You may provide technical information and critique the performance after the test. • Provide test results to the person(s) designated to receive them.

FIGURE 14-3 Sample testing policy.

instructors, and administration, course evaluations are not affected by evaluator subjectivity.

Course evaluations should contain the following key elements:

- Student opinions
- Instructor viewpoints
- Course coordinator/administration views
- Institute/facilities

The survey could be a comprehensive form that asks questions about each of these four categories, or it can be specifically geared to just one category. This second approach is more time consuming, but it provides better information.

FIGURE 14-4 An important component of course evaluations is student feedback.

Student Grades and Surveys

An instructor needs to provide a fair grading system. When an instructor assigns a score, he or she can use one of three scoring categories—either a letter grade, pass/fail, or a numerical score. The letter grade (A, B, C, D, F) is probably more familiar to most adult learners, while the numerical score (out of 4.0 or 100 percent) is closely associated with institutes of higher learning. The instructor needs to clearly identify which grading system will be used at the beginning of the class and define a passing score. An instructor should try to limit the use of the term "failure" when describing inadequate performance, instead preferring a term such as "unsatisfactory," which is interpreted as not meeting the accepted criteria at that specific time. Whether a practical or written examination is being administered, the instructor must maintain a consistent scoring system. Some students will use scores as a competitive gauge to better their own personal performance. The majority of students, however, will use their scores as a gauge of their individual position in the class.

The survey is an aspect of the program evaluation strategy. With the development of the evaluation strategy comes the development of specific surveys—for example, geared toward students or toward instructors—that will be used to determine the effectiveness of the program. Because this is an overall program evaluation, the questions are directly associated with an established goal or standard for the educational institute. Explain to students that their individual evaluations are critical for continuous improvement of the program. Their input will help ensure future success.

Rating Scales

Several rating scales may be used in surveys. The most commonly used scales are the numerical rating scale and the graphic rating scale. The numerical scale is the simplest type. The following is an example:

5 = Outstanding
4 = Above average
3 = Average
2 = Fair
1 = Poor

For example:

The primary instructor's preparation of course material was:
1   2   3   4   5
The quality of the audiovisual materials was:
1   2   3   4   5

A program administrator can quickly tally this kind of data and identify the common trends and opinions expressed in the survey **FIGURE 14-5**. For example, if two surveys were administered, one designed for students and the other designed for instructors, it would be appropriate to assess the same objectives for both surveys. The responses to the same objective from the two different perspectives can then be compared. The numerical rating scale would make this type of comparison very easy to conduct because specific values are listed on the rating scale.

The graphic rating scale is not as neatly defined as the numerical version. It uses a line graph to plot the respondents' answers to a specific question. With this rating scale, it is possible to obtain values that are between a given value (e.g., 3.3, 4.5, 2.3). Instead of locking a respondent into a specific number, a specific value assessment for a question can be attained. Although this rating scale is much more difficult to use for tallying purposes, the information it provides better reflects the respondents' actual opinions for the question.

A variation on the two previously mentioned rating scales is the multiple-choice rating scale. It is written using the same format as a multiple-choice examination; that is, specific answers are preidentified. The respondent selects the answer that best matches his or her opinions. On a typical form, four answers are preidentified and a fifth answer is left blank. When respondents believe that the preidentified answers do not accurately reflect their opinion, they can write in a unique response. The following is an example of a multiple-choice survey item:

The audiovisual materials that were used during the program were:

Poor. Most were old and outdated. They did not enhance the course material.
Fair. A few of the audiovisual materials were outdated, but overall they complemented the lesson material.
Good. The materials were current and enhanced the lesson materials.
Excellent. The materials greatly enhanced the course materials.

(Write in the response on the attached form.)

The responses to a multiple-choice survey are very easy to tally. If necessary, the questions can be reworded to allow a comparison between two different categories—for example,

Illinois Fire Chiefs Educational and Research Foundation

Fire Officer Training Program
QUALITY ASSURANCE REPORT SUMMARY
COURSE NAME: Fire Service Instructor I

Course Evaluation

Course Feature	Number of Surveys	Excellent (5)	Very Good (4)	Good (3)	Fair (2)	DNA (0)
1. AV Material / Course materials / Textbook	20	18	2			
2. Course organization / flow / delivery rate	20	16	2	2		
3. Observance to safety procedures and practices	20	14	3			3
4. Hands on Experience *If applicable to course*	20	19	1			
5. Time per subject *Add comments on specific areas on reverse*	20	12	4	4		
6. Evaluation tools	20	16	2	2		
7. Will this course help you with your professional development?	20	19	1			
8. Did the course meet your expecations?	20	18	2			
9. Would you recommend this course to others?	20	18	2			
10. Overall impression of course	20	17	2	1		

Instructor(s) Evaluation

A = Excellent 5.0; clearly presented objectives, demonstrated thorough knowledge of subject, utilized time well, represented self as professional, encouraged questions and opinions, reinforced safety practices and standards.

B = Good (4.0); presented objectives, knowledgeable in subject matter, presented material efficiently, added to lesson plan information

C = Average (3.0); met objectives, lacked enthusiasm for content or subject, lacked experience or background information, did not add to delivery of material

D = Below Average (2.0); did not meet student expectations, poor presentation skills, lack of knowledge of content or skills, did not interact well with students, did not represent self as professional

E = Not Applicable; did not instruct in section or course area, do not recall, no opinion

Instructor Name	Number of Surveys	Excellent (5)	Good (4)	Average (3)	Below Average (2)	Not Applicable	Evaluation Score
Chief Smith	20	20					5.0
Capt. Brown	20	15	5				4.5
Bill Engine	20	5	12	3			3.8

LIST STUDENT COMMENTS IN THIS SPACE

* Good class, lots of practice speeches helped my confidence
* Too much PowerPoint® on day 1
* Chief Smith's insight was very valuable and his evaluations of the speeches were very helpful
* Thanks to all instructors who helped out
* Should get reading assignments before class
* Doing workbook assignments and preparing practice speeches as homework was a lot when I was on shift
* Looking forward to Instructor II, opened my eyes to instructors job

FIGURE 14-5 Survey results can be tallied to identify trends in responses.

students and instructors. The preselected responses allow a respondent who is not familiar with educational concepts to respond to the question. The downside to using preselected responses is that the survey author points a respondent toward a conclusion that may not totally reflect the individual's opinion. The respondent often selects an answer simply because he or she does not want to write in a response. For this reason, care must be used when analyzing the results of a multiple-choice survey.

■ Survey Questions

The key to a successful survey is the way the questions are written on the survey form and interpreted by the person completing the survey. Program evaluation questions are based on accepted educational concepts or on an accepted curriculum. In other words, the goals for a program become the evaluation mechanism for the program.

Each question covers a specific goal. The question is to be written in simple English. Just like written examination questions, a program evaluation question should be a complete sentence whenever possible. Remembering these concepts can increase the validity of the survey.

Survey Format

The following basic guidelines are used by most surveys. First, the overall survey should appear simple, not overwhelming. Many surveys are too complex in their appearance. The respondent should be able to look at the survey and readily identify the information being requested. In addition, the questions must be written to the point. There is no need for fancy terms or lengthy sentences. The questions for a survey should be based on accepted educational objectives or accepted standards.

Second, only one rating scale should be used. This consistency will lessen the respondent's confusion as to which rating scale is being used for a particular section. The evaluator should write the rating scale at the top section of each page of the survey. If the rating scale has the scale built into each question—for example, in graphic type—then the respondents have all the information they need right in front of them.

Third, because subjectivity can influence a survey, the respondent should remain anonymous. Thus a survey should not request the name of the respondent.

A sound survey reflects the clarity of the survey's directions. Students or faculty members often complete surveys with no monitor present to clarify a particular question. The directions to a survey must contain enough details so that anyone who completes the survey can do so without difficulty. Poor directions result in poor information because the respondent, if unsure or confused, may provide inaccurate information. This outcome invalidates some of the information gathered from the survey.

The use of these guidelines not only helps in writing a survey, but can also lead to improved validity for the survey. A survey is constructed to assess the opinions of a group of individuals. From this survey, key programming decisions will be made. If a survey has inaccurately assessed the individuals, then the decisions based on its results may be made in error. Care must be used when developing, using, and evaluating

with a survey. The information attained from a survey should be helpful, not harmful.

A survey can only collect opinions from the respondents. Often, this information is not enough to base program changes on. Another evaluation tool, a performance evaluation, can provide more substantial information on which to base program revisions.

Theory Into Practice

Consider having a different instructor than the one who taught the course administer the survey. It is sometimes difficult for a student to offer feedback on a course when the lead instructor is present. Take the fear away up front.

Analyzing Evaluation Tools

A great deal of attention has been given to the procedure for developing valid and reliable test items. It is equally important, however, to follow up each administration of the test items with a post-test item analysis. This analysis will allow you to look at each item on the test in terms of its difficulty, or P+ value, and its ability to discriminate; this analysis also allows you to ascertain the reliability of the overall test. You can then use these data to identify potentially weak or faulty items, scrutinize each item to identify the possible cause of a weakness or fault, and make necessary improvements.

Theory Into Practice

The P+ value is computed for each question using statistical calculations and allows the administrator to generally evaluate whether the test item is reliable.

Analyzing Student Evaluation Instruments

One of the areas that educators most commonly examine when assessing an institute's or program's quality is the written examination scores. A written examination is designed to objectively assess the degree of learning that has occurred. Because it is based on the course curriculum or textbook, the results on this examination are one of the first indicators reviewed. Of course, test results alone are not a sufficient measure of overall program quality. Performance evaluations of the instructors, opinion surveys of the students, and course coordinator reports regarding student grievances should also be considered.

■ Item Analysis

An item analysis is a listing of each student's answer to a particular question. This evaluation tool is useful because examinations may be flawed. Performing an item analysis of the examination can determine whether it measured what it was supposed to, had poorly worded questions, did not contain the correct answer, or had other problems. An item analysis should be done for every examination.

VOICES
OF EXPERIENCE

After every class I have ever taught I ask myself, "Did I do a good job? Did learning take place? Was the material appropriate, informative, and presented in an understandable format?" How can one determine the answers to these questions? To answer these questions you must evaluate the results presented, quiz scores, results of practical examinations, and the cumulative results of the final examination, along with student and peer evaluations. The results of this evaluation will provide a clear picture of your success or failure at delivering information to your fire fighters. These results should also show the strengths and weaknesses of your instruction and that of the materials presented. Armed with this information you may then go forward and re-enforce your strengths and work to correct your weaknesses; doing so should lead to progressively better instruction on your part and consistent improvement in the materials used. This improvement is vitally important in the fire service. We are preparing our students to go into harm's way, or to lead fire fighters into harm's way. Their safety may depend upon the mastery of the knowledge that has been presented to them. This is why I truly believe we as instructors owe our fire fighters no less than our very best.

> **We must constantly evaluate the materials and instruction methods used to present the best practices available at any given time.**

My career has progressed to the point that I must not only evaluate myself and what I teach, but also manage and evaluate a cadre of instructors and an array of classes. I must understand the need to take great care and to study constantly the materials presented and the results obtained to ensure the consistent quality of courses and instruction. We must always be prepared for state audit and ensure the privacy of our students. We must constantly work and strive to improve the courses we present, updating materials and the technology used in presentation of these materials.

Firefighting is not a static endeavor; it is fluid and ever changing. This fact dictates that we must constantly evaluate the materials and instruction methods used to present the best practices available at any given time. We as teachers and coaches must constantly work to improve our mastery of the skills of our chosen trade so that we may impart this information to those who will go into combat armed with the knowledge they have received from us. I believe it should always be our goal as instructors to send out into the public the best trained and most professional firefighting force possible given the resources available. After all, the safety of our brothers and sisters and the public they serve depend on it.

The importance of the evaluation and management of the information we produce cannot be overstated. Yesterday's near-perfect instruction and materials may be missing vital information in the classroom of tomorrow. As Chief Van Dorpe always says, "we must constantly work to master our craft."

Robert D. Shaw Jr.
Chief, Du Quoin Fire Department
Director, Rend Lake College Fire Science Department
Ina, Illinois

For example, if fewer than 50 percent of the students answered a particular question correctly, the question should be reviewed for problematic sentence structure, misspelling, or lack of a clear answer. If a particular question was answered correctly by almost all students, it should also be reviewed to determine if it is too easy or its answer was given away by another question in the examination.

Item analysis can indicate how well the students understood the course material. The wealth of information provided by the item analysis makes it worth the time it takes to create this kind of analysis.

Theory Into Practice

If item analysis reveals an incorrect question or answer, all the scores for the examination should be adjusted (e.g., by dropping the question from the examination and recalculating students' scores).

Item Difficulty

Exam questions should be reviewed for level of difficulty. Building on the information provided by an item analysis, each question should be examined by looking at the number of correct answers. A high item-difficulty score means that it is an easy question. Conversely, a low item-difficulty score indicates a difficult question. Identification of the item difficulty can enable the instructor to improve an examination for future use by linking the item difficulty to the test blueprint. For each question, an item-difficulty score can be identified. When it comes time to modify an examination, the instructor can look at this score and insert a new question with a similar item-difficulty score. This allows the examination blueprint to remain nearly unchanged.

The ideal examination will demonstrate balanced item-difficulty scores. As a general guideline, any item that has an item difficulty score of less than 0.55 should be reviewed for question validity. At 0.55, almost half the group answered the item incorrectly. An examination that consists of questions with item-difficulty scores ranging between 0.60 and 0.95 has a normal distribution of scores and is considered to be a balanced examination.

Balancing the question difficulty using an item-difficulty analysis offers a clear-cut way to determine the actual item performance for each question. When an examination is used repeatedly, a sense of how a class should perform on the examination develops. This predictability enables the instructor to look at the performance of one class versus another class.

The comparison between the two classes allows the instructor to gauge each class's performance for a particular point in its education. Changes can be based on this comparison, if deficiencies are noted between the two classes.

Item difficulty is a powerful tool with which to evaluate each question on an examination. Instructors should respect its usefulness and routinely analyze examination questions in this way. The objective information provided is invaluable in determining the validity of the examination.

A test-item analysis helps determine whether the test items discriminate between respondents with high or low scores and whether the results were consistent. Such an analysis can use data from one administration of a test and formulas for computing item difficulty (P+ value) and discrimination. The analytical objective will be met when the P+ value and discrimination index have been computed for at least 10 items on the test, and each item has been identified as acceptable or needing review.

A post-test item analysis should be approached methodically, using the following process:

1. Gather data.
2. Compute the P+ value.
3. Compute the discrimination index.
4. Compute the reliability index.
5. Identify whether items are acceptable or need review.
6. Revise test items based on post-test analysis.

■ Gather Data

Immediately following the administration and grading of a test, it is important to collect information on the responses to each item. For each item, you need to know how many times each alternative was selected—that is, whether the alternative was the keyed (correct) response. On a tally sheet, tally each participant's response to each alternative for each test item.

To see how this process works, we will use a multiple-choice test item from a 10-item test that was administered to a group of 12 students. This particular test item had four alternatives— A through D—and B was the keyed response (i.e., correct answer). After tallying each participant's response to each test item, you are able to determine how many correct responses were marked and how many incorrect responses were marked for each distracter. Now, as you begin to analyze one test item at a time, isolate the data TABLE 14-1.

Other types of test items will require different tally-sheet formats or provisions for more alternatives than the four (A–D) shown in Table 14-1. The example in TABLE 14-2 provides space for 10 alternatives. Matching, arrangement, and identification test items require five alternatives; therefore, you will need to provide adequate space on your tally sheet.

Table 14-1	Isolation of Data					
Item Number	Alternatives				Responses	
	A	B	C	D	Right	Wrong
1	1	*8	2	1	8	4

* Bold indicates keyed response.

Table 14-2	Test-Item Analysis Tally Sheet												
Item Number	**Alternatives (Where * indicates the answer)**											**Responses**	
	A	B	C	D	E	F	G	H	I	J		Right	Wrong

Total:

Score Distribution

Score distributions may be depicted in many ways. One popular method utilizes a listing of high to low scores in one column. Then, beside each score, a second column lists the number of students who attained that score on the examination. An example of such a distribution is shown in **TABLE 14-3**. This list of numbers is not easily visualized when it is collected as a random set of numbers. If the numbers are placed onto a chart, however, the resulting graphic provides a visual depiction of the students' scores. Recording the scores on a chart is a common method for showing how well students performed.

Educators like to see a perfect distribution of scores—referred to as a normal distribution—from an examination. In a normal distribution, the majority of the scores cluster around the average score for the examination. The mean, or average score, is positioned in the center of the distribution. In ideal circumstances, 95 percent of the class will have scores within the major area covered by the distribution, which leaves only 2.5 percent on each side of the curve for extremely high and low scores. **FIGURE 14-6** shows the normal curve.

In practical applications, the normal distribution is not the rule, but rather the exception. Most examination distributions do not conform to the normal distribution. Often, the scores may vary so dramatically that no clustering of the scores is evident. Alternatively, they may be off in one direction, either to the high side or to the low side. In such situations, the distribution

Table 14-3	Sample Score Distribution
Score	**Number of Students**
10	1
9	0
8	**2**
7	**4**
6	**3**
5	**2**
4	1
3	0
2	2
1	0
0	0

Note: The main cluster of students falls between 8 and 5, as shown in **bold**.

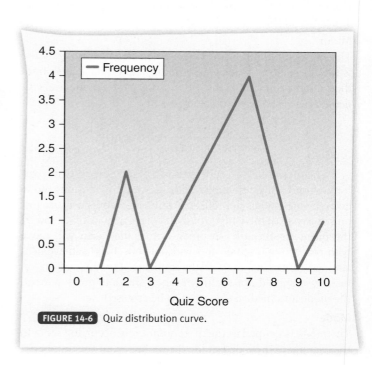

FIGURE 14-6 Quiz distribution curve.

is said to be skewed. Two specific types of skews are important to an instructor: a positive skew and a negative skew.

The discovery of a skewed distribution immediately alerts an instructor to a potential problem with an examination. With a positive skew, very few scores appear toward the top of the distribution, and the majority of the scores are below the mean. Interpreting the score distribution chart reveals potential causes of this distribution. Perhaps the examination was too difficult for the majority of the class. To determine whether this perception is accurate, look at the individual student scores. In particular, examine the scores of those students who were performing exceedingly well before the exam and those who were performing poorly before the test. Compare their performance on the examination. If the individuals who were expected to do well on the examination actually did poorly, it could be an indicator that the examination was too difficult.

A negative skew has the opposite effect of a positive skew. In a negative distribution, scores for the majority of students are above the mean. An immediate interpretation is that the examination was too easy. Only a few individuals scored poorly on the examination. A negative distribution needs to be subjected to the same kind of investigative review as a positive skew; that is, the instructor needs to review the scores and to validate the examination results. If an examination has overevaluated or underevaluated students' knowledge, then it has not fairly represented the amount of learning achieved by that group of students. For this reason, instructors should routinely review the distribution results.

Average Scores

For an instructor to assess the true average score for an examination, the mean, mode, and median scores should be calculated. In a normal distribution, all three should be found on the center line that divides the distribution. In a skewed distribution, each of these values may have a different line location within the distribution. An instructor should use the measurement that best represents the distribution of the students' score as a gauge of the test's effectiveness.

Mean

The most commonly used measurement for the average score is the mean. It is calculated by adding up all the scores from the examination and then dividing by the total number of students who took the examination:

$$\frac{\text{Total scores}}{\text{Number of students}} = \text{Mean}$$

The mean offers a crude estimate of how the students performed on the examination. With skewed distributions, however, the mean does not reflect all students' scores because just one or two extremely high (or low) scores may raise (or lower) the mean significantly. In other words, the mean will move with the extreme scores and will not reflect the achievement attained by the majority of the students. For this reason, an instructor should not use the mean as the only measurement; the mode and median also need to be assessed.

Mode

The mode is defined as the most commonly occurring score in the examination. To determine the mode, count the number of

student scores for each examination score. The largest number of single examination scores is the mode. If the majority of student scores are clustered in a distribution, the likelihood of the mode appearing in that cluster is high. If the distribution is skewed, the mode may appear anywhere. In a very large distribution, multiple modes may exist.

FIGURE 14-7 shows the mean and the mode for a distribution of exam scores. The mean is the more reflective average score for this particular distribution. More than 11 students attained at least the mean score, whereas only 4 students were at a score of 18. The mode does not reflect all the student scores for this distribution.

Median

The median score is a more involved measurement than either the mean or the mode. The median score is the score located exactly in the middle of the distribution. It is based on a percentage of the total number of students taking the examination. To calculate the median score, the first step is to create a column called a "cumulative frequency" (cf) by adding up the number of student scores at each level. The top value is equal to the actual number of students who took the examination. Then, each cumulative score is divided by the total number of students taking the examination. The score that is found at the 50 percent indicator is called the median score.

Many educators refer to the C% column as a percentile. Percentiles show student scores based on a 100 percent scale **TABLE 14-4** . Notice that the median score is between the scores of 15 and 16 in Table 14-4. A statistical method is used for determining the exact score. For the purpose of this example, it is important to note that the mean and the median are nearly identical. Thus, for this distribution, either the mean or the median score would be equally reflective of the actual average score for the distribution.

Identifying the true average for the students' scores assists the instructor in locating the correct average for the

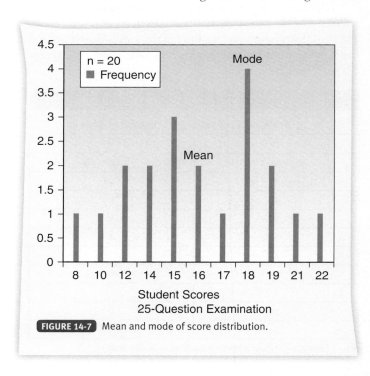

FIGURE 14-7 Mean and mode of score distribution.

distribution. As noted earlier, the mean may be significantly affected by just one or two extraordinarily high or low scores. To find the true average score requires all three measures of the distribution—mean, mode, and median—to be assessed and compared. At that point, the instructor can select the measurement that best reflects the average score for the distribution.

■ Compute the P+ Value

The P+ value of a test item indicates how difficult that item was for the class taking the test. Your calculation of the P+ value will be useful later on as you revise a test item based on the results of your analysis. Here is the formula to calculate the P+ value of a test item:

$$P + value = \frac{Number\ right\ (R)}{Number\ taking\ the\ test\ (N)}$$

$$P + = \frac{R}{N}$$

Now, let's apply that formula to our example test item. Eight of the 12 respondents answered the question correctly.

$$P + = \frac{8}{12}$$
$$= 0.666$$
$$= 0.67$$

Thus the P+ value of that item is 0.67.

The P+ value is merely a statistical indicator and should not be used as the sole judgment of the relative merit of a test item. Instead, it should be used in conjunction with the discrimination and reliability indexes.

■ Compute the Discrimination Index

The discrimination index refers to a test item's ability to differentiate (or distinguish) between test takers with high scores on the total test and those with low scores. Several formulas may be used to compute the discrimination index. Our example formula divides the group tested into three subgroups, as equal in number as possible, based on their grades on the overall test. The example test was administered to a group of 12 participants. We arrange their scores in descending order and divide them into three groups TABLE 14-5 .

Once the groups have been determined, you can then calculate each test item's discrimination index. Another tally sheet, such as the one shown in TABLE 14-6 , is useful for this purpose.

In this example, the four best grades constitute the high group and the lowest four grades constitute the low group. We now count the number of respondents in the high and low groups who answered the item correctly. In the high group, three respondents answered correctly; in the low group, two answered correctly. The number of respondents in the low group who answered correctly is subtracted from the number in the high group who also answered correctly. The difference (1) is then divided by one-third of the total number of respondents (4). The formula for computing the discrimination index follows:

$$D_S = \frac{H - L}{N/3} = \frac{3 - 2}{12/3} = \frac{1}{4} = 0.25$$

where

H = number of respondents from the high group who answered the item correctly = 3
L = number of respondents from the low group who answered the item correctly = 2
N = number of respondents taking the test = 12
   Number of groups = 3

Table 14-4	Average Scores		
Score	Score Frequency	Cumulative Frequency (cf)	C% (Percentile)
22	1	20	100%
21	1	19	95%
19	2	18	90%
18	4 (mode)	16	80%
17	1	12	60%
16	2 (mean)	11	55% (median 15–16)
15	3	9	45%
14	2	6	30%
12	2	4	20%
10	1	2	10%
8	1	1	5%

Table 14-5	Group Determination	
Score	Number of Respondents	Group (12/3 = 4)
100	2	High
90	2	High
80	4	Middle
70	4	Low

Notes: When working with a group size that is not evenly divisible by 3, place equal numbers in the high and low groups and place the remaining numbers in the middle group.
  It is important to have equal numbers in the high and low groups.

Table 14-6	Test-Item Discrimination Index Tally Sheet										
Groups by Test Score	Test-Item Alternatives (Indicate number of students choosing each response choice)										Total Number of Respondents Answering Correctly
	A	B	C	D	E	F	G	H	I	J	
High third	1	3	0	0							3
Low third	0	2	1	1							2

The procedure follows these steps:

1. Sort test papers or answer sheets into high, middle, and low groups.
2. Tally all high-group and low-group responses to the item on the tally sheet in the space under "Test-Item Alternatives."
3. Carry the number of correct responses for each group to the right column.
4. Insert the numbers representing the "respondents answering correctly" for the high and low groups into the formula to replace H and L, respectively.

$$\frac{H - L}{N/3} = \frac{3 - 2}{N/3}$$

5. Insert the total number of test respondents to replace N.

$$\frac{3 - 2}{N/3} = \frac{3 - 2}{12/3}$$

6. Complete the calculation to derive the discrimination index for that test item.

$$\frac{3 - 2}{12/3} = \frac{1}{4} = 0.25$$

The index of 0.25 indicates the degree to which the item discriminates between those respondents who obtained high total-test scores and those who obtained low total-test scores. Theoretically, the range of the discrimination index is from −1 to +1.

- An index of +1 results when all respondents in the high group and no respondents in the low group answer correctly.
- An index of −1 results when all respondents in the low group and no respondents in the high group answer correctly.
- If the same number of respondents from both groups answer either correctly or incorrectly, the index is 0 (the exact center of the range).

Here are some assumptions that can be made after studying discrimination-index trends:

- When more respondents from the high group than from the low group answer correctly, the item has positive discrimination.
- When more respondents from the low group than from the high group answer correctly, the item has negative discrimination.
- When the same number from both groups answer the item correctly or incorrectly, there is neutral (zero) discrimination.

Test items that have very low positive discrimination, zero discrimination, or negative discrimination need to receive a quality review to determine their validity. If neutral or negative discrimination occurs with a test item of medium difficulty, you should determine the reason and correct that item before it is used again.

As mentioned earlier, the discrimination index has more meaning when it is used in conjunction with the P+ value (i.e., the measure of difficulty). Our example test item has a P+ value of 0.67 and a discrimination index of 0.25. Thus, although the P+ value is high, the difficulty level and the discrimination index are moderate to low. The answer to the question, "Is this good or bad?" depends on the purpose of the test item. Generally speaking, if both the difficulty level and the discrimination index are in the same range (low–low, medium–medium, or high–high), the test item is doing its job. An intentionally difficult yet valid test item should have a high discrimination index; an intentionally easy test item should have a low discrimination index.

Theory Into Practice

A test's reliability depends on three factors: the number of test items, the standard deviation, and the mean (average) of scores.

■ Compute the Reliability Index

The first two indexes are used to rate the difficulty and discrimination characteristics of a single test item. The reliability index, by comparison, rates the consistency of results for an overall test. Statisticians use a variety of measurements for determining whether an examination is reliable. They apply these measurements to evaluate whether the examination has consistency among the questions. Through the use of a mathematical formula, a determination of the examination's reliability is made.

One of the formulas most frequently used for this purpose is the Kuder–Richardson reliability formula, which determines the internal consistency of the examination. The range for scores computed from this formula is from 0 to 1. An examination is said to be reliable with a score of 0.60 or higher.

Kuder–Richardson Formula 21

$$\text{Reliability} = \frac{ns^2 - M(n - M)}{ns^2}$$

where

n = Number of test items
s = Standard deviation
M = mean

Another formula to calculate reliability is the Spearman–Brown formula, also known as the split-half method.

Spearman–Brown Formula (Split-Half Method)

$$r_{11} = 2\left(\frac{1S_a^2 + S_b^2}{S_t^2}\right)$$

where

r_{11} = Reliability of total test
S_a = Standard deviation of the first half of the test (odd-numbered items)

S_b = Standard deviation of the second half of the test (even-numbered items)

S_t = Standard deviation of the whole test

Given that both of the preceding formulas use the standard deviation, this value must be calculated before proceeding with the reliability index. The formula for computing standard deviation is

$$\sqrt{\frac{\Sigma(x - x_2)^2}{n}}$$

where

Σ = Sum of

x = Each score

x_2 = Mean

n = Number of scores

The reliability index computation, using any formula, is a complex and time-consuming process that is best accomplished using a computer and is beyond the scope of this text. Your computer technician should have no trouble implementing the formula.

Ideally, the reliability index should be as close to 1.00 as possible. In reality, anything greater than 0.80 is generally acceptable. If the reliability index for the test does not reach 0.80, you need to look for possible causes, such as the following problems:

- Nonstandard instructions
- Scoring errors
- Non-uniform test conditions
- Respondent guessing
- Other chance fluctuations

Educators also measure an examination's reliability through repeated administrations of an examination. Theoretically, if an examination is given to a group of students one week, and the same examination is given to the same students a week later, the scores should be nearly identical. Although fire service instructors would not perform this type of educational test in their classes, an instructor should keep track of an examination's performance each time it is administered. Although different students are taking the examination, if the material presented in the course is the same, the scores on the examination should be similar. This can be a factor to consider when evaluating a program.

Instructors should check the reliability of their examinations. Assessing the examination reliability is part of ensuring a valid measurement of a student's learning experience.

■ Identify Items as Acceptable or Needing Review

To identify test items as acceptable or in need of review, the first step is to perform a quantitative analysis of the test item. This analysis uses difficulty (P+), discrimination, and reliability data to identify potentially weak or faulty test items. To accomplish the quantitative analysis, compare the test item's statistical data against the following screening criteria:

- Difficulty level (P+ value) between 0.25 and 0.75
- Discrimination index between +0.25 and +1.0
- Reliability of test 0.80 or greater

When a test item fails to meet these criteria, identify the item so that you can later conduct a qualitative analysis of it. During the qualitative analysis, you take a close look at the test item in terms of its technical content and adherence to the format guidelines for its type. Keep in mind that some test items identified as potentially weak or faulty during the *quantitative* analysis may not be found to be weak or faulty when reviewed more closely during the *qualitative* analysis. Some items on each test cover the "core" body of knowledge that everyone must possess. The instructor appropriately emphasizes this information during training, and the test item may then be intentionally written at a low enough difficulty level to ensure that the majority of the trainees "get it right." Therefore, when conducting the quantitative analysis, try to remain objective about the uses of statistical data.

■ Revise Test Items Based on Post-Test Analysis

When you have determined the probable causes of your test-item faults, make changes accordingly. Even after second and third revisions, some items may remain problematic and require additional revision. The analytical item-analysis method does not guarantee immediate success in all cases, but it does promise gradual success over time.

Performance Evaluations

A performance evaluation is a tool that can provide more in-depth information than is possible to collect via a survey alone. The categories used in such evaluations are the same as those for surveys. Performance evaluations can be conducted among students, instructors, course administration, or facilities. The difference between a performance evaluation and a survey is that a trained evaluator physically monitors each group. The monitor uses prepared objectives-based checklists. As with the survey approach, these objectives reflect the broad goals of the educational institute.

The approach used to evaluate students' ability to perform skills is the same as that used in the program evaluation. The same pitfalls also apply, however (e.g., the halo effect, subjectivity, and poor evaluators). As with any educational evaluation, the information that is obtained through this method needs to be closely examined. The performance evaluation is a tool that, when used in conjunction with a survey, can provide answers to objectives not addressed by the survey.

The use of a performance evaluation versus the use of a survey differs somewhat, as can be seen in the following example. Suppose students provide their impressions of the instructor's performance on a survey. They may not be able to fairly or objectively assess the abilities of an instructor, however, nor do they know what should and should not be taught to a class. To obtain this information, an educationally oriented evaluator needs to monitor various class sessions. The instructor can be evaluated by this impartial evaluator, and the information collected in this manner can then be compared with the students' impressions.

Performance Assessment Checklists

Checklists may be written either based on the main goals for a program or to reflect accepted standards of performance. Both of these elements are important when developing instructor, administrative, and facility checklists. The instructor, administrative, and facility checklists are written in nearly the same format; the significant differences between them reflect the objectives being assessed.

The performance evaluation for the instructor is nothing more than a modified survey **FIGURE 14-8**. With this type of tool, a trained evaluator measures the performance of an instructor against specific educational objectives. Most checklists consist of the following components:

- Basic demographic data
- Date(s) of the evaluation
- Evaluator(s) name
- Rating scale
- Specific objectives
- General comments
- Strengths and weaknesses
- Tally column
- Didactic presentation
- Skill presentation

For administration checklists, objectives oriented to assess customer satisfaction are often used. Either the administration is seen as responsive to a customer's needs (e.g., students, faculty, or the community) or support provided to the customer is perceived as lacking. These business concepts are appropriate objectives to utilize when assessing the performance of the administration because any educational program inevitably has bureaucratic aspects. For example, how the paperwork is handled often influences the quality of instruction. How can paperwork affect a program? Perhaps handouts should be made but are forgotten. Or perhaps audiovisual equipment

that was supposed to be in the classroom is missing. Maybe the classroom is too small for the size of the class. The list goes on. All of these details are administrative in nature, yet all influence the quality of the program. Because of their impact, administrative staff members should be assessed in the same manner as other program components are.

The administrative assessment being referred to here is an overall assessment—one that is not targeted at a specific individual's performance. Such an assessment should focus on those administrative tasks that directly affect the classroom programs.

Interpreting Results

In ideal circumstances, a committee consisting of faculty, administration, and members from the community at large should review the program material. This committee should look at the areas of concern that have been identified in any of the assessments. For example, a "No" response to a question about the adequacy of the audiovisual materials used in the classroom is a potential area of concern. Cross-checking the various surveys—for example, student opinion and instructor surveys—may provide additional information about the area of concern. In addition to negative responses, positive achievements should be reviewed. For example, if a program goal was "Students will achieve a score of 80 percent or higher on the state certification examination," and the class average was 82.4 percent, then this outcome is a positive achievement. The people involved in the program should be given praise for meeting this goal.

Once the committee has reviewed both the positive and negative aspects of the program, then a final report needs to be written. In this report, findings related to each assessment area are summarized. A discussion of the positive and negative aspects of the program is then initiated.

The report concludes with a plan of action, or program recommendations. This plan of action specifies which parts of a program need to be changed to correct a problem or to improve the educational experience. As stated in the first section of this chapter, to have a quality program, assessment and reassessment of the program are required. This means rewriting objectives, policies, and procedures; fixing broken equipment and buying new equipment; and redefining the assessment process so that these improvements can be assessed in the next program.

National Fire Academy Course Evaluation Form

Part 2 of 4 - Course Instructors / Overall Training

Considering the instructors for this course, to what extent do you agree that...
1=Strongly disagree, 2=Disagree, 3=Unsure, 4=Agree, 5=Strongly agree, NA=Not Applicable
Instructor: John Smith

	1	2	3	4	5	NA
knew the material well.	○	○	○	○	○	◉
regularly clarified course and assignment expectations.	○	○	○	○	○	◉
encouraged independent thinking from students.	○	○	○	○	○	◉
fostered a collaborative "team-based" learning experience.	○	○	○	○	○	◉
supplemented course with helpful experience.	○	○	○	○	○	◉
answered students' questions clearly.	○	○	○	○	○	◉
presented engaging lectures.	○	○	○	○	○	◉
led the learning process without dominating it.	○	○	○	○	○	◉
exhibited a positive attitude toward students.	○	○	○	○	○	◉
conducted the class in a professional manner.	○	○	○	○	○	◉
worked well with co-instructor.	○	○	○	○	○	◉
is worth recommending to others.	○	○	○	○	○	◉

Considering the instructors for this course, to what extent do you agree that...
1=Strongly disagree, 2=Disagree, 3=Unsure, 4=Agree, 5=Strongly agree, NA=Not Applicable
Instructor: Richard Karlsson

	1	2	3	4	5	NA
knew the material well.	○	○	○	○	○	◉
regularly clarified course and assignment expectations.	○	○	○	○	○	◉
encouraged independent thinking from students.	○	○	○	○	○	◉
fostered a collaborative "team-based" learning experience.	○	○	○	○	○	◉
supplemented course with helpful experience.	○	○	○	○	○	◉
answered students' questions clearly.	○	○	○	○	○	◉
presented engaging lectures.	○	○	○	○	○	◉
led the learning process without dominating it.	○	○	○	○	○	◉
exhibited a positive attitude toward students.	○	○	○	○	○	◉
conducted the class in a professional manner.	○	○	○	○	○	◉
worked well with co-instructor.	○	○	○	○	○	◉
is worth recommending to others.	○	○	○	○	○	◉

Is there particular feedback you have regarding the instructors for this course?

(Continues)

Courtesy of FEMA

Considering the training overall, to what extent do you agree that this training experience...
1=Strongly disagree, 2=Disagree, 3=Unsure, 4=Agree, 5=Strongly agree, NA=Not Applicable

	1	2	3	4	5	NA
will help me do my current job better.	○	○	○	○	○	◉
was consistent with my department's training expectations.	○	○	○	○	○	◉
will be useful for a department the size of mine.	○	○	○	○	○	◉
will be applicable to my future work.	○	○	○	○	○	◉
included material helpful to my department's prevention efforts.	○	○	○	○	○	◉
will help reduce the fire-related risks in my community.	○	○	○	○	○	◉
provided sufficient opportunities for networking.	○	○	○	○	○	◉
provided information my department can use when responding to an all-hazards and/or terrorist event.	○	○	○	○	○	◉
is worth recommending to others.	○	○	○	○	○	◉

How satisfied are you with NFA's...

classroom and learning facilities.	○ Satisfactory	○ Unsatisfactory	○ N/A
Learning Resource Center (LRC).	○ Satisfactory	○ Unsatisfactory	○ N/A
dormitory rooms.	○ Satisfactory	○ Unsatisfactory	○ N/A
student center facilities.	○ Satisfactory	○ Unsatisfactory	○ N/A
dining facilities.	○ Satisfactory	○ Unsatisfactory	○ N/A
course registration and administrative details.	○ Satisfactory	○ Unsatisfactory	○ N/A
workout/weight room facilities.	○ Satisfactory	○ Unsatisfactory	○ N/A

| Save For Later | Back | Continue |

The Save for Later button will allow you to save answers already provided and come back to the form at a later time (provided the evaluation period hasn't expired) to complete it.

FEMA Form 064-0-5

Paperwork Reduction Act Notice | OMB No. 1212-0065

Last Reviewed: December 28, 2006

U.S. Fire Administration, 16825 S. Seton Ave., Emmitsburg, MD 21727
(301) 447-1000 Fax: (301) 447-1346 Admissions Fax: (301) 447-1441

FIGURE 14-8 Sample course instructor and overall training experience rating survey which is completed on-line at the end of a course.

Jones & Bartlett Fire District

Training Division
5 Wall Street, Burlington, MA, 01803
Phone 978-443-5000 Fax 978-443-8000
www.fire.jbpub.com

Instant Applications: Managing the Evaluation System

Drill Assignment

Apply the chapter content to your department's operation, its training division, and your personal experiences to complete the following questions and activities.

Objective

Upon completion of the instant applications, fire service instructor students will exhibit decision making and application of job performance requirements of the fire service instructor using the text, class discussion, and their own personal experiences.

Suggested Drill Applications

1. Conduct a test-item analysis using a recent test that you or another instructor administered to a student group. Identify any questions that need review or revision based on the evaluation results.

2. Review a set of class evaluation forms and identify any course features that need review and improvement to better meet the needs of the students.

3. Create a test administration policy for proctoring oral, written, and performance-based tests that defines student and evaluator expectations.

4. Review the Incident Report in this chapter and be prepared to discuss your analysis of the incident from a training perspective and as an instructor who wishes to use the report as a training tool.

Incident Report

Loretto, Tennessee—2003

On May 18, 2003, a 28-year-old male volunteer training/safety officer was seriously injured when he fell from a moving pick-up truck. He was participating in a three-day live fire training course when the incident occurred. The victim served as a sergeant with the sheriff's department and had been a volunteer fire fighter for three years. He had served for the past two years as Training and Safety Officer for the fire department.

The three-day training course began on Friday, May 16, and ended on Sunday, May 18, 2003. The 22-hour course consisted of interactive practical training activities and live-fire simulations. The course was conducted at an official fire service training facility with most participants residing on campus for the duration of the training. The program included ladder drills, maze trailer/self-contained breathing apparatus (SCBA) exercises, search and rescue sessions, sprinkler system review, and hose drills. Live fire included training exercises on pressure gas flange fires, flammable liquid spill fires and vehicle fires, and interior evolutions. The final event was a large fire scenario on the last day of the training.

Witnesses report that at one point during the training exercises on Saturday, the victim complained of feeling ill and considered dropping out of the class and going home. However, it was reported that after discussing the content of remaining training evolutions with fellow fire fighters, he believed he had completed most of the strenuous exercises and decided to stay. On Sunday morning the victim told several fire fighters that he had chosen not to eat breakfast prior to the live-burn training evolution because he feared he would again become ill.

On May 18, the victim completed the last evolution, exited the burn building, and walked to the parking area where he got into the bed of the pick-up truck. He sat facing backward, on the passenger side, with his legs hanging over the edge of the lowered tailgate. The driver proceeded out of the parking area and after coming to a stop at the end of the entrance road he made a left turn and proceeded south toward the training center building. The pick-up truck had traveled approximately 140 feet from the stop sign when the victim fell off the tailgate and landed on the roadway pavement. It is reported that he appeared to slide off the tailgate feet first, landing in a face-up position, with the back of his head striking the pavement last. According to witnesses, neither speed nor the motion of the vehicle appeared to be a factor. No official estimate of speed was recorded on the police report.

Fire fighters at the scene immediately treated the victim for severe head trauma. An emergency medical services unit that was stationed on-site responded within minutes and provided advanced life support. A nearby medical helicopter was on-scene within five minutes and transported the victim to a local trauma center where he died from his injuries six days later on May 24, 2003.

Post-Incident Analysis: Loretto, Tennessee

Contributing Factors

Personnel being transported when on duty were not securely seated and restrained in an approved vehicle passenger compartment. (*Although it is unclear whether a medical or physical condition contributed to this fatal incident, fire departments should consider implementing these safety and health recommendations based on the physical demands and medical requirements of firefighting.*)

Pre-placement and annual medical evaluations, consistent with NFPA 1582, should be mandatory for all fire fighters to determine medical fitness for duty and training exercises.
Periodic physical capabilities testing should be performed to ensure that fire department personnel meet the physical requirements for duty and training exercises.

Chief Concepts

- Much is at stake for both the student and the instructor when the evaluation system is used to demonstrate competence and knowledge or for certifications.
- Evaluation results should flow from the students who have completed the test, through the instructor who administered the exam, and then to the individual or agency responsible for grading the results.
- Examination results and student assignment results should be treated as confidential every step of the way—from the time they are completed and submitted to the instructor by the student to the time they are permanently stored or disposed of.
- It is best not to single out any individual student during a mass release of data to instructors; only key administrators of a program should have full access to individual records.
- Evaluating a training course should be part of the ongoing and constant improvement goal of any training division.
- Goals that are set too high or too low for the type of course and students being taught are not helpful; the same is true of goals that cannot be evaluated.
- Because course objectives are stated as outcomes, this part of the review and analysis process may be one of the best measures for identifying whether the course did what it was supposed to do.
- No single evaluation tool can capture the entire picture of an educational program's overall performance.
- Policies relating to course administration and delivery should be reviewed on a regular basis to ensure they adequately do the job for which they were written and to determine whether any updates are needed.
- Items that should be included in the overall program evaluation plan include course evaluations completed by students, instructor evaluations, test results, agency policies and procedures, facilities and equipment, and safety analysis.
- Course evaluations should contain the following key elements:
  - Student opinions
  - Instructor viewpoints
  - Course coordinator/administration views
  - Institute/facilities
- The instructor needs to clearly identify which grading system will be used at the beginning of the class and define passing scores.
- The most commonly used rating scales are the numerical rating scale and the graphic rating scale.
- The key to a successful survey is the way the questions are written on the survey form and interpreted by the person completing the survey.
- A post-test item analysis allows the instructor to look at each item on the test in terms of its difficulty, or P+ value,

and its ability to discriminate; this analysis also reveals the reliability of the overall test.
- A written examination is designed to objectively assess the degree of learning that has occurred.
- Performing an item analysis of the examination can determine whether it measured what it was supposed to, had poorly worded questions, did not contain the correct answer, or had other problems.
- Immediately following the administration and grading of a test, it is important to collect information on the responses to each item. For each item, you need to know how many times each alternative was selected—that is, whether the alternative was the keyed (correct) response.
- The P+ value of a test item indicates how difficult that item was for the class taking the test. Your calculation of the P+ value will be useful later on as you revise a test item based on the results of your analysis.
- The discrimination index refers to a test item's ability to differentiate between test takers with high scores on the total test and those with low scores.
- Statisticians use a variety of measurements for determining whether an examination is reliable—that is, whether the examination has consistency among the questions.
- To identify test items as acceptable or in need of review, the first step is to perform a quantitative analysis of the test item.
- The analytical item-analysis method does not guarantee immediate success in all cases, but it does promise gradual success over time.
- The performance evaluation is a tool that, when used in conjunction with a survey, can provide answers to objectives not addressed by the survey.
- Checklists may be written either based on the main goals for a program or to reflect accepted standards of performance. Both of these elements are important when developing instructor, administrative, and facility checklists.
- In ideal circumstances, a committee consisting of faculty, administration, and members from the community at large should review the program material.

Hot Terms

Discrimination index The value given to an assessment that differentiates between high and low scorers.

Item analysis A listing of each student's answer to a particular question used for evaluation.

Mean A value calculated by adding up all the scores from an examination and dividing by the total number of students who took the examination.

Median The score in the middle of the score distribution for an examination.

Mode The most commonly occurring value in a set of values (e.g., scores on an examination).

P+ value Number of correct responses to the test item. Example: P = 67 means that 67 percent of test takers answered correctly.

Reliability index Value that refers to the reliability of a test as a whole in terms of consistently measuring the intended material.

Standard deviation The value to which data should be expected to vary from the average.

References

National Fire Protection Association. (2012). *NFPA 1041: Standard for Fire Service Instructor Professional Qualifications*. Quincy, MA: National Fire Protection Association.

A series of courses have been completed under your direction as training officer. Students attending the course have provided feedback about the course and course features using an end-of-course evaluation form and have also challenged end-of-course exams and state certification tests. A lot of data and information are available to you, and your responsibility is to ensure that the users of the course have been successful in meeting the course objectives and that the agency's goals for the course have been met. Future offerings of this course need to be scheduled, and you plan to implement any changes before the next offering of the course is posted.

1. Students should be provided with a process to provide their impressions on course features and instructors to the agency sponsoring the program. What should a form for this process contain?

   A. Course feature evaluation

   B. Instructor ratings

   C. Space for written narrative on selected course areas

   D. All of the above

2. In the context of analyzing examination results, what does the term *discrimination* imply?

   A. The test blocks certain groups from being successful in passing an item

   B. The test item separates percentages of the top-performing students from lower-performing students in the class

   C. The test item is not able to be completed by anyone other than the top student

   D. The instructor finds it easy to score the test item

3. When you review examination results, you identify a test question that has a very low success rate. What should you do with that test item?

   A. Review it for proper grammar

   B. Review course content and reading assignments to make sure it was covered

   C. Discuss the item content with the instructors to make sure they covered it adequately

   D. All of the above

4. Which type of agency policy should be established to help you determine the actions to be taken with exam and class data when you complete your review of it?

   A. Test administration policy

   B. Records storage and disposal policy

   C. ADA compliance policy

   D. Test record dissemination policy

5. If you wish to find out the average score on the end-of-course examination or on a state certification exam, which statistical data point would you look at?

   A. Mode

   B. Median

   C. Discrimination index

   D. Frequency of scores

Training Program Management

Fire Service Instructor I

Knowledge Objectives

There are no knowledge objectives for Fire Service Instructor I students.

Skills Objectives

There are no skills objectives for Fire Service Instructor I students.

Fire Service Instructor II

Knowledge Objectives

After studying this chapter you will be able to:

- Describe the budget process, the creation of a bid, and budget management. (**NFPA 5.2.3**) (pp 350–353)
- Describe the procedures for creating a training budget. (**NFPA 5.2.3**) (pp 351–352)
- Describe the process for acquiring training resources. (**NFPA 5.2.4**) (pp 353–358)

Skills Objectives

After studying this chapter you will be able to:

- Demonstrate the procedures for creating a training budget. (**NFPA 5.2.3**) (pp 351–353)
- Demonstrate acquisition of training resources for the use in delivering training. (**NFPA 5.2.4**) (pp 353–358)

Fire Service Instructor III

Knowledge Objectives

After studying this chapter, you will be able to:

- Describe methods used to evaluate equipment and resources used for training delivery. (NFPA 6.2.6) (pp 353–358)
- Define the policies and procedures for the management of instructional resources, staff, facilities, records, and reports. (NFPA 6.2.1)(NFPA 6.2.2) (pp 359–369)
- Recommend needed policies to support the training program. (NFPA 6.2.3) (pp 365–367, 369)
- List staff selection and instructional responsibilities. (NFPA 6.2.4) (pp 369, 371–373)
- Discuss methods of constructing instructor evaluation plans. (NFPA 6.2.5) (pp 365–367, 373)

Skills Objectives

After studying this chapter, you will be able to:

- Write purchasing specifications for training resources. (NFPA 6.2.6) (p 355)
- Administer a training record system. (NFPA 6.2.2) (pp 359–369)
- Select instructors for delivery and management of the training program. (NFPA 6.2.4) (pp 369, 371–373)
- Construct performance-based instructor evaluation plans. (NFPA 6.2.5) (pp 365–367, 373)
- Present training program evaluation findings to various audiences. (NFPA 6.2.7) (p 369)

You Are the Fire Service Instructor

An audit of the training division's training record keeping system has been scheduled by the state fire marshal's office. The purpose of this routine visit is to verify hours of attendance and subject-area coverage for members of your department who are applying for various certifications. As a relatively new training officer, you are concerned that the recordkeeping practices of the department are outdated and not consistent among members. You have heard stories from other nearby departments about their audits and know that penalties assessed for incomplete or missing records could delay certifications being issued to the members.

1. Which types of training records should be kept by the department to document regular and special training events?
2. Which laws pertain to record storage and access that would apply to a fire department training division's recordkeeping system?
3. Which standards are available to guide you in developing and maintaining a better records management system for the training division?

Introduction

Managing the training division is a crucial responsibility with much at stake if it is not done well. Many regulatory compliance matters must be addressed, including filing and recordkeeping of the many documents related to the delivery of training and multitudes of policies that must be developed and enforced. Training records and reports are also becomwing more important for use as evidence in liability law suits brought against an organization's fire-ground activities. The fire service instructor who is assigned the responsibility for managing the training division may be referred to as a training officer, director of training, training program manager, chief of training, or similar title designating a specific level of job performance above that of an instructor who delivers a lesson plan. For the purposes of this chapter, the title of Training Officer (TO) will be used to identify the person responsible for these functions.

Typical management cycles and functions are implemented on a regular basis to ensure that effective and efficient training takes place. All too often, a failure to deliver competent service to the community during an incident may trace its roots back to how the personnel were trained. If an injury or line-of-duty death (LODD) occurs, significant investigation into the contributing factors and the members' history in training are among the critical pieces of evidence that help identify the cause of that incident.

The job description of the TO covers all four of the main duty areas identified by NFPA 1041, *Standard for Fire Service Instructor Professional Qualifications*:

- Program management
- Instructional development
- Instructional delivery
- Evaluation and testing

This chapter covers the program management component.

Fire Service Instructor II and III

Budget Development and Administration

The budget process of a fire department can be very confusing, if not overwhelming. Understanding the budget development process and your role in purchasing, specifying resources and materials, and justifying the funding requirements of a training project or program is crucial to successful training program management.

■ Introduction to Budgeting

A budget is an itemized summary of estimated or intended revenues and expenditures. Revenues are the income of a government from all sources appropriated for the payment of public expenses and are stated as estimates. Expenditures are the money spent for goods or services and are considered appropriations that allow the expenditure authority. Every fire department has some type of budget that defines the funds that are available to operate the organization for a particular period of time, generally one year. The budget process is a cycle:

1. Identification of needs and required resources
2. Preparation of a budget request
3. Local government and public review of requested budget
4. Adoption of an approved budget

5. Administration of approved budget, with quarterly review and revision
6. Close out of budget year

Budget preparation is both a technical and a political process. The funds that are allocated to the fire department define which services the department is able to provide for that year. The technical part relates to calculating the funds that are required to achieve different objectives, whereas the political part is related to elected officials making the decisions on which programs should be funded among numerous alternatives.

■ Budget Preparation

Successful budgeting requires justifications of the amount of money being requested. A budget justification is useful in defending your position when requested money is allocated for a specific line item or category. A training division may be an example of a line item in a budget document. When developing the line item for the training division, the justification can be based on previous expenses in this area, the agency training needs assessment, and the forecasting of the training needs for the upcoming period **FIGURE 15-1** . Note that the *Modifications and Justification* column in this table includes just a synopsis of the justification, and it is likely a more in-depth justification will be needed.

■ The Budget Cycle

Every department has a budget cycle, which is typically 12 months in length. The budget document describes where the revenue comes from (input) and where it goes (output) in terms of personnel, operating, and capital expenditures. Annual budgets usually apply to a fiscal year, such as starting on July 1 and ending on June 30 of the following year. Thus the budget for fiscal year 2013, referred to as FY13, would start on July 1, 2013, and end on June 30, 2014. Some agencies use a true calendar-year budget process beginning on January 1 and ending on December 31.

Theory Into Practice

Developing budgets is a great way to demonstrate what the training division does and how it adds value not only to the department, but to the community as a whole. For this reason, you should approach this process by viewing it as an opportunity rather than a necessary evil. Know the goals, and be able to outline the direct relationship between the attainment of those goals and your need for the requested resources. Be articulate in your justification by being to the point but also thorough. Don't fluff the numbers to give "negotiation" room. Also, don't sell yourself short. Budget requests are a business decision. Demonstrate your professionalism by helping the department live within its means and by making good decisions based on the budget process.

Category/Budget Code	Justification Explain the goal or target for each line item category	2012–13 Requested	2012–13 Actual	2013–14 Requested	2013–14 Modifications and Justification
Contingency 7100-01	Unforeseen expenses related to training program	1000.00	1000.00	1000.00	Status quo, no changes
Outside Seminars 7100-02	FF attendance at certification programs and seminars	17,500.00	15,000.00	20,000.00	Increase fire officer training course work for officers
Firefighter I Academy 7100-03	1 new full-time member to attend academy	3000.00	3000.00	6000.00	2 hires on schedule for retirements in 2013
Hosted Courses 7100-04	Expenses for in-house hosted classes; coffee, pop, snacks	1000.00	1000.00	1000.00	Status quo, no changes
Text/Publications/Video 7100-05	Purchase of updated manuals, magazine subscriptions	8500.00	7000.00	9000.00	Library updates are needed for stations 2 and 3
Drill Site Maintenance 7100-06	Upkeep of training center props and buildings	30,500.00	25,000.00	30,500.00	Upkeep and replacement of burn tower linings and prop construction
Supplies/Maintenance 7100-07	Smoke fluid, office supplies, repairs of equipment	8,000.00	5375.00	5000.00	LP fuel, smoke fluid and repair to training equipment
Part-Time Instructors 7100-08	Part-time instructional staff wages	35,000.00	25,000.00	35,000.00	Increase post-academy training program hours
Instructor Seminars 7100-09	Attendance at local, state and national seminars, conf.	15,500.00	7000.00	10,000.00	Conference price increases and travel expenses.
TOTALS	**Support of Division of Training Mission and Assignments**	**120,000**	**89,375**	**$117,500**	

FIGURE 15-1 Training budget and justification columns.

In many cases, the process for developing the FY13 budget would start in 2012, a full year before the beginning of the fiscal year **TABLE 15-1**. The timeline in Table 15-1 shows that there is a long period from the time you make a budget request to the time the money is available to spend. As a consequence, all training needs for the forthcoming fiscal year may not be apparent to you during the budget development process. Effective budget management requires you to anticipate as much as possible when preparing the budget and to be able to develop contingency plans for unforeseen events. Some actions and purchases may have to be cancelled or delayed when others take priority.

Mandated continuing training, such as hazardous materials and CPR recertification classes, is also factored into the operating costs portion of the budget. That figure includes the cost of the training per fire fighter per class. The cost to pay another fire fighter overtime to cover the position during the mandated training could also be included. As new mandates

Table 15-1	Fiscal Year 2013 Timeline
August–September 2012	Fire station commanders, section leaders, and program managers submit their FY13 requests to the fire chief. Their concentration is on proposed new programs, expensive new or replacement capital equipment, and physical plant repairs. Depending on how the fire department is organized, a station commander might request funds for a new storage shed or a replacement dishwasher for the fire station at this point. Documentation to support replacing a fire truck or purchasing new equipment, such as an infrared camera or a hydraulic rescue tool, would be submitted, and a proposal for a special training program would be prepared. In most fire departments, these requests are prepared by fire officers throughout the organization and submitted to an individual who is responsible for assembling the budget proposals.
September 2012	The fire chief reviews the budget requests from fire stations and program proposals from staff to develop a prioritized budget "wish list" for FY13. This is the time when larger department-wide initiatives or new program proposals are developed. For example, purchasing a new radio system or establishing a decontamination/weapons of mass destruction unit would be part of the fire chief's wish list.
October 2012	The city's budget office distributes FY13 budget preparation packages to all agency or division heads. Included in each package are the application forms for personnel, operating, and capital budget requests. The budget director includes specific instructions on any changes to the budget submission procedure from previous years. The senior local administrative official (mayor, city manager, county executive) also provides specific submission guidelines. For example, if the economic indicators predict that tax revenues will not increase during FY13, the guidelines could restrict increases in the operating budget to 0.5 percent or freeze the number of full-time equivalent positions.
November 2012	Deadline for agency heads to submit their FY13 budget requests to the budget director.
December 2012	The budget director assembles the proposals from each agency or division head within the local government structure. Each submission is checked to ensure that it complies with the general directives provided by the senior local administrative official. Any variances from the guidelines must be the result of a legally binding agreement, settlement, or requirement or have the support of elected officials.
January 2013	Local elected officials receive the preliminary proposed budget from the budget director. For some budget items or programs, two or three alternative proposals may be included that require a decision from the local elected officials. Once the elected officials make those decisions, the initial budget proposal is completed. This is the "Proposed Fiscal Year 2013 Budget."
February or March 2013	The proposed budget is made available to the public for comment. Many cities make the proposed budget available on an Internet site, inviting public comment to the elected officials. In smaller communities, the proposed budget may show up in the local newspaper.
March or April 2013	Local elected officials conduct a public hearing or town meeting to receive input from the public on the FY13 proposed budget. Based on the hearings and on additional information from staff and local government employees, the elected officials debate, revise, and amend the budget. During this process, the fire chief may be called on to make a presentation to explain the department's budget requests, particularly if large expenditures have been proposed. The department may be required to provide additional information or submit alternatives as a result of the public budget review process.
May 2013	Local leaders approve the amended budget. This becomes the "Approved" or "Adopted" FY 2013 budget for the municipality.
July 1, 2013	FY13 begins. Some fire departments immediately begin the process of ordering expensive and durable capital equipment, particularly items that have long lead times for delivery. Many requests for proposals are issued in the early months of the fiscal year.
October 2013	Informal first-quarter budget review. This review examines trends in expenditures to identify any problems. For example, if the cost of diesel fuel has increased and 60 percent of the motor fuel budget has been spent by the end of September, there will be no funds remaining to buy fuel in January. This is the time to identify the problem and plan adjustments to the budget. Amounts that have been approved for one purpose may have to be diverted to a higher-priority account.
January 2014	Formal midyear budget review. The approved budget may be revised to cover unplanned expenses or shortages. In some cases, these changes represent a response to a decrease in revenue, such as an unanticipated decrease in sales tax revenue. If the money is not coming in, expenditures for the remainder of the year might have to be reduced. Sometimes, revenues exceed expectations and funds become available for an expenditure that was not approved at the beginning of the year.
April 2014	Informal third- and fourth-quarter review. Activity within the third quarter can be reviewed and projections for the remaining quarter can be made. Final adjustments in the budget are considered after 9 months of experience. Year-end figures can be projected with a high degree of accuracy. Some projects and activities may have to stop if they have exceeded their budget, unless unexpended funds from another account can be reallocated.
Mid-June 2014	The finance office begins closing out the budget year and reconciling the accounts. No additional purchases are allowed from the FY13 budget after June 30th.

are established, finding the money to cover these expenses can prove very difficult. New mandated training is one of the most difficult challenges for you to budget and justify.

■ Capital Expenditures and Training

Capital expenditures refer to the purchase of durable items that cost more than a threshold amount and last for more than one budget year. Local jurisdictions differ on the amount and the time period that these items are supposed to last. Items such as hydraulic rescue tools, apparatus, and self-contained breathing apparatus (SCBA) are examples of equipment purchased as capital items. Each of these items requires training when delivery takes place, and additional costs may be incurred to prepare both fire service instructors and fire fighters to train on the new equipment. A Fire Officer III may be responsible for assigning values and depreciation rates to fixed assets and for developing policies that define these items.

The training budget should not be approached any differently than any other part of the budget document and process in a fire department. Budgets must be prepared, justified, and managed throughout the life of the budget cycle. You must be able to understand the budget process used by your department and be able to perform an administrative review of your budget area. When budgets need to be cut or reduced, administrations may seek out areas that can be trimmed by reducing activities or resource purchases. In the training area, you might lose funding for new textbooks or curriculum packages, for class fees and seminar registrations, or for the purchase of a much-needed training prop. Capital budget money for training towers or training sites is often an early target for elimination in a cash-strapped department.

■ Training-Related Expenses

Expense areas that may be included in the training line item include the following:

- Fire service instructors' salaries, benefits, and expenses
- Student expenses for class attendance
- Tuition, overtime if applicable, transportation, lodging, and meals
- Textbooks, publications, and DVDs
- Office supplies
- Equipment and training aid maintenance
- Course and seminar fees
- Facility maintenance and improvements
- Capital expenditures
- Subscriptions and memberships
- Contracts
- Educational assistance programs
- Tuition reimbursement
- Grant programs

Ongoing budget maintenance activities should take place to monitor the training division's expenditures to date as part of your fiscal responsibilities. Your performance may be measured in part on your ability to operate within the financial constraints of your program. Good budget management starts with an understanding of the department's budget process and ends with you following the purchasing procedures and tracking of expenses related to the program on a regular basis.

Acquiring and Evaluating Training Resources

Purchasing of resources and products used in the instructional area is a management skill. With shrinking budgets, trial and error does not suffice as a means of selecting training resources. Understanding of the purchasing responsibility and your authority in this process must be evident and documented.

Resource management—whether of equipment, materials, facilities, or personnel—requires a full understanding of the training goals of the department. Purchasing new training equipment every year may not be an option for every department. Often, front-line equipment that has been replaced by newer equipment may be designated for training use. This practice helps with the resource allocation process but can require additional resource management skills on behalf of the training division.

■ Training Resources

Training resources can be defined as any equipment, materials, or other resources that are used to assist in the delivery of a training session. They can be either consumable or reusable. Consumable items can be used only once, such as a student workbook that will be written in. Reusable items can be refilled or recharged, such as smoke fluid used for a smoke-generating machine used in SCBA training. Prop structures may have certain durable components, while other parts are disposable (such as a ventilation prop) **FIGURE 15-2** .

FIGURE 15-2 An instructor should carefully explain all steps of an evolution using props and equipment before it is performed in a controlled environment in which it is easy to see and hear.

Books and curriculum packages may have reusable sections in the form of PowerPoint® slides and reproducible handout materials. As noted throughout this text, you as the instructor must identify the types of resources needed to conduct a training session each time you develop or prepare to instruct from a lesson plan **FIGURE 15-3**. The equipment and materials that are usually identified in a lesson plan that are considered resources include the following items:

- Materials that the students need to complete learning objectives
- Materials that the fire service instructor needs to present, apply, and evaluate the learning objectives
- Props, aids, equipment, or other resources necessary to complete the session

Preparing for an effective training session requires you to review the availability, usability, and reliability of each resource. All equipment must be checked for proper operation and safety and be ready to use for the duration of the training session. Enough fuel, batteries, blades, paper, flipchart paper, or whiteboard markers need to be available for the entire training session. Have a backup plan in case the resources you identified are not available or do not work properly.

■ Hand-Me-Downs

When a power saw is taken off duty as front-line equipment and given to the training division to use in training, the fire service instructor is able to teach a forcible entry session without worrying about creating wear and tear on in-service

Ethics Tip

The fire service has lived off its "make do" attitude in getting the mission accomplished with inadequate resources. As a fire service instructor, you may have to use materials or equipment to serve as proxies for reality. This necessity often reinforces the need for creative thinking. But what do you do when you must use training methods that may lead to training automation on the fire ground? Training automation occurs when students drill until actions become second nature. The problem with this outcome is that some actions exist only for training purposes. For example, if the gate is opened only halfway during hose stream drills, fire fighters will not get the feel of a hose at normal pressure. As a fire service instructor, you have an obligation to know when substandard resources create risks.

equipment. The downside of this practice is the reason why the power saw was taken off front-line service in the first place. Was it difficult to start? Did it have broken parts? Did it leak fuel or oil? Was it obsolete? If the answer to any of these questions is "yes," then you must prepare for the impact of that issue.

A good practice may be to have in-service equipment ready to use at the training site so that if the hand-me-down

Instructor Guide
Lesson Plan

Lesson Title: Use of Fire Extinguishers   ←   **Lesson Title**

Level of Instruction: Firefighter I   ←   **Level of Instruction**

Method of Instruction: Demonstration

Learning Objective: The student shall demonstrate the ability to extinguish a Class A fire with   ←   **Learning Objective**
a stored -pressure water-type fire extinguisher. (NFPA 1001, 5.3.16)

References: Fundamentals of Firefighter Skills , 3rd Edition , Chapter 8   ←   **References**

Time: 50 Minutes

Materials Needed: Portable water extinguishers, Class A combustible burn materials, Skills ←   **Instructional Materials**
checklist, suitable area for hands-on demonstration, assigned PPE for skill   **Needed**

Slides: 73–78*

Step#1 Lesson Preparation:

- Fire extinguishers are first line of defense on incipient fires
- Civilians use for containment until FD arrives
- Must match extinguisher class with fire class
- FD personnel can use in certain situations, may limit water damage
- Review of fire behavior and fuel classifications
- Discuss types of extinguishers on apparatus
- Demonstrate methods for operation

FIGURE 15-3 Lesson plan example showing resources needed for training evolution.

does not work, the flow of the training session is not interrupted. For example, that old rotary cut-off saw that has been given to the training division should be backed up by the new saw that was purchased to replace it. It is also important to make sure that any operational differences between the two models are highlighted. Notify the proper company officers or shift commanders before removing in-service equipment from apparatus for use in training. It may be hard to justify a delayed response to an incident because the company had to repack hose before they were able to respond on a fire run.

Reserve apparatus may assist in delivery of your training session, but you must also be sure to identify any differences between what students would normally use in the performance of their duties and the equipment being used for the training. Make sure that significant differences in safety equipment, SOPs, and other dynamic factors are spelled out and understood during the training session **FIGURE 15-4**. It would be great to allocate duplicate resources to the training division that are the same as the front-line equipment, but few departments are able to afford such luxuries.

Purchasing Training Resources

All resources used in the delivery of training should be identified during the construction of a lesson plan. Books, projection equipment, training aids, computer resources, and even straw or hay used to generate smoke in a burn tower—that is, all of the resources identified in the lesson plan—in turn need to be considered within the budget process.

When given the budgetary approval to purchase training resources, you must follow all departmental guidelines. Solicitation of bids for resources costing more than a stated amount of money is often a requirement that you have to observe, for example. Many departments are required to solicit bids from competitive vendors in an effort to get the best price for a product. Textbooks, smoke fluid for smoke generators, and training foams are all examples of materials that could have different price points from vendors. Other materials may be identified as <u>single-source</u> products that are manufactured or distributed by only one vendor. In that case, you may not have to go through the bid process.

Safety Tip

When using reserve apparatus, ensure that all safety features, such as pressure relief valves or pressure governors, are working properly.

Theory Into Practice

When purchasing training resources, be sure to understand your department's purchasing policy in relation to bid requirements.

Courtesy of Carey King

FIGURE 15-4 Reserve apparatus and in-service company.

■ Evaluating Resources

Instructors who are responsible for evaluating resources for purchase must observe and follow all purchase policy requirements. Before the submission of a request to purchase a resource is made, you must evaluate the quality, applicability, and compatibility of the resource for the department. Many high-quality resources have been developed in recent years, whereas some time-proven resources still dominate in certain market areas. When considering and evaluating resources for purchase, consider the following questions:

1. How will this product/resource be used in your department?
   a. This is a form of audience analysis that was discussed in the *Lesson Plans* chapter.
   b. Consider both the audience aspect and the fire service instructor(s) who will use the resource. Will it require training for the fire service instructors who will use it? Will it require new learning skills of the students?
2. Which standard(s) is the product/resource developed in conjunction or compliance with?
   a. This question may be very important when considering adoption or reference of a textbook or training package. Certain state training agencies use specific references for certification training.
3. Is this a "turnkey" product—meaning it is ready to use as is—or does the fire service instructor have to adapt the product to local conditions?
4. Is this product/resource compatible with your agency procedures and methods?
   a. Consider the content-specific issues of the product/resource. Are these the methods you use to operate, or are there significant differences in the equipment or methods of the product/resource?

5. Does the product/resource fit within budgetary restrictions?
6. Are there advantages to purchasing larger quantities of the product/resource?
7. Can you get references from vendors on the product/resource that you are considering, or do you have to do your own investigation within your local area or statewide instruction networks?
8. Can the product/resource be available for your use within the prescribed timeline within which you are working?
9. Are there any local alternatives, such as your own design of a product/resource or identification of other local agencies that might be willing to share the use of the product/resource?
10. Is manufacturer demonstration, product expert, or other developer assistance available to the instructor about the product/resource?
    a. On-site assistance
    b. Phone support
    c. Website information
    d. Product literature
    e. Suggested user guidelines or instructions

Experienced fire service instructors may develop a resource evaluation form to help justify this process and validate the selection process TABLE 15-2. A simple checklist of these questions may assist in your decision making. Effective and efficient training program management requires the development and use of some standardized practice to evaluate products and resources. In today's budget-conscious departments, doing your homework before submitting budget requests can pay off by getting your resources/products purchased quickly FIGURE 15-5.

Table 15-2   Resource Evaluation Form

Criteria	Vendor/Product 1 Rating (1 = poor to 5 = excellent)	Vendor/Product 2 Rating (1 = poor to 5 = excellent)	Vendor/Product 3 Rating (1 = poor to 5 = excellent)
Ease of use within our department • Instructor • Student			
Amount of modifications needed for our department			
Standard compliance • NFPA • OSHA • ANSI • Other			
Cost of product/resource			
Other similar products already in use by dept.			
Quantity discount available			
References			
Availability			
Vendor support			
Local alternatives			
Other			

CITY OF DES PLAINES

1420 MINER STREET
DES PLAINES, ILLINOIS 60016
TELEPHONE: (847) 391-5300

PURCHASE ORDER

ORG. NO.	ACCOUNT NO.	PROJECT CODE	DEPT / DIVISION	VENDOR NO.	PURCHASE ORDER NO.

			SHIP TO	
VENDOR NAME				
ATTN				
ADDRESS				
CITY, STATE, ZIP				

QUANTITY	ITEM/DESCRIPTION	UNIT COST	EXTENDED
			0.00
			0.00
			0.00
			0.00
			0.00
			0.00
			0.00
			0.00
			0.00
			0.00
			0.00
			0.00
			0.00
		SUB-TOTAL	0.00
		SHIPPING/OTHER	
		TOTAL	$0.00

DATE	REQUESTED BY	DEPARTMENT HEAD	FINANCE DIRECTOR	CITY MANAGER

(Continues)

CITY OF DES PLAINES

1420 MINER STREET
DES PLAINES, ILLINOIS 60016
TELEPHONE: (847) 391-5300

1.	PRICE		CONTACT PERSON	
	VENDOR NAME		PHONE	
	ADDRESS		FAX	
	CITY, STATE, ZIP		EMAIL	

2.	PRICE		CONTACT PERSON	
	VENDOR NAME		PHONE	
	ADDRESS		FAX	
	CITY, STATE, ZIP		EMAIL	

3.	PRICE		CONTACT PERSON	
	VENDOR NAME		PHONE	
	ADDRESS		FAX	
	CITY, STATE, ZIP		EMAIL	

VENDOR RECOMMENDED		TOTAL AMOUNT BUDGETED	
OTHER PROJECT COSTS		ONGOING OPERATING COSTS	

THIS PURCHASE AND OTHER PROJECT COSTS ARE EXPECTED TO REMAIN WITHIN THE BUDGET: (CHECK BOX)	☐ YES	☐ NO

PLEASE MARK ALL BOXES BELOW THAT APPLY TO THIS PURCHASE:

☐	APPROVAL OF LOWEST RESPONSIBLE BIDDER	☐	PROFESSIONAL SERVICES CONSULTING
☐	EMERGENCY PURCHASE	☐	SOLE SOURCE SUPPLIER
☐	EQUIPMENT STANDARDIZATION	☐	TECHNICAL NATURE OF ITEMS MAKES COMPETITION IMPOSSIBLE
☐	JOINT GOVERNMENT PURCHASING PROGRAM	☐	OTHER (PLEASE EXPLAIN BELOW)

EXPLANATION:

FOR PURCHASES OVER $10,000

A FORMAL REQUEST FOR A RESOLUTION HAS BEEN SUBMITTED TO LEGAL: (CHECK BOX)	☐ YES	☐ NO
THIS ITEM HAS BEEN PLACED ON CONSENT AGENDA: (CHECK BOX)	☐ YES	☐ NO

DATE	REQUESTED BY	DEPARTMENT HEAD	FINANCE DIRECTOR	CITY MANAGER

FIGURE 15-5 Purchase request form.

Fire Service Instructor III

Training Record Systems

As mentioned throughout this text, training record systems are one of the most important management functions assigned to the TO. A well-known saying—"If it isn't documented, it didn't happen"—sums up almost every aspect of a fire department training record system. A record system must be developed by the Instructor III, who may function as the TO, to document all aspects of the training program. Most instructors are familiar with the training record report, which is the most common type of training record, but they may not be aware of the many types of training reports created to summarize and detail other training activities. NFPA 1401, *Recommended Practice for Fire Service Training Reports and Records*, details best practices for these training record systems and provides structure for the management functions of this duty area **FIGURE 15-6**. An excerpt of NFPA 1401 is included in Appendix D.

■ Using Training Records

Training records are used for many purposes. In some organizations they are analyzed to identify personnel assignments or

1401

NFPA® 1401
Recommended Practice
for Fire Service Training
Reports and Records
2012 Edition

NFPA®,1 Batterymarch Park, PO Box 9101, Quincy, MA 02269-9101, USA
An International Codes and Standards Organization

FIGURE 15-6 NFPA 1401, *Recommended Practice for Fire Service Training Reports and Records*, details best practices for training record systems and provides structure for related management functions.

to make pay-scale adjustments, whereas in other organizations the number of training sessions attended are tallied to allow for continued assignment to emergency response duties. A formal department policy—created by the training division and approved by the fire chief—should exist to determine the uses, preparation requirements, release of records, and storage of all types of reports.

Compliance

A major use of training records is to ensure regulatory compliance. Many local, county, state, provincial, and national organizations require very specific types of training, and the documentation of those training sessions is often reviewed in audits. The TO should become very familiar with the highly dynamic world of compliance mandates and should ensure that every effort is made to follow the guidelines or procedures necessary to meet the intent of the agency. In some cases, emergency service agencies have been denied federal and state grants due to the lack of training level documentation. In other cases, fire departments have seen their Insurance Service Office (ISO) rating either increase or decrease as a result of careful examination of training record reports that document the numbers of hours of individual fire fighters' training completed.

TABLE 15-3 identifies several regulatory agencies and some examples of the types of training records they may require the emergency service agency to create and make available during audits or investigation and for compliance purposes.

■ Types of Training Records

Many types of training record reports exist throughout the emergency service community, the most common of which is the training record report used to document the ongoing training of the members. Although each agency should create records and reports that meet its own specific needs and criteria, NFPA 1401 does provide some direction on the content and use of training records and reports.

Electronic or Paper Records and Reports

Efforts to reduce paper use and storage are a basic environmental and space-saving management decision. Training officers should be aware, however, that some agencies still require the permanent storage of certain types of training records, especially ones used to document initial training on occupational health and safety. It is acceptable to maintain a combination of electronic and paper-based record systems and to provide a clear identification of which types should be kept in which form. This arrangement should be identified in the training policy.

Many software programs specific to emergency service training documentation storage exist and make instant recall and custom reporting easy for the department. Paper-and-pen systems may be easy to use, but they require a lot of file storage

Table 15-3	Training Record Examples per Regulatory Agency		
Insurance Service Office (ISO)	**State Certification Boards**	**Local EMS Providers**	**OSHA**
• Training record reports • Summary of training hours • Recruit, pump operator, and officer training reports • Mutual aid training record reports • Preplan review training • Equipment and procedure training	• Certifications earned by members • Hours completed toward certifications • Testing record reports	• Certifications and licenses earned • Continuing education hours • In-service training records • CPR training • AED training • Policy training	• Mandated training area documentation • Respiratory protection training • Hazardous materials training • Infection control • Confined space and special hazard training • Injury and accident reports

AED = automated external defibrillator; CPR = cardiopulmonary resuscitation.

space and further management decisions about the length of time for which records must be kept. Some documentation aspects that may need to be decided by local legal counsel include the following:

- The types of records and reports that must be kept in original form
- The length of time that records and reports must be kept before disposal
- Access and permissions to electronic and paper records and reports

The TO should work with the local agencies responsible for reviewing the content of a training record, as they may provide ruling to determine these requirements. For example, the Illinois Office of the State Fire Marshal (OSFM) has made a specific ruling that allows for an electronic daily training record (paperless training record) to be used as long as a hard-copy monthly summary report is created for each member, then reviewed and signed by the member and a qualified instructor **FIGURE 15-7**. This summary report must be kept in the member's training record jacket and made available for audit and inspection by the OSFM.

Performance Tests, Exams, and Personnel Evaluations

Performance tests, examination score reports, and personnel evaluations related to training may be some of the more confidential types of records and reports the TO will be responsible for. These documents may be covered by the Privacy Act of 1979 or other state and local acts that deal with provisions of confidentiality and disclosure. A person's training file may be classified as a personnel file and should be treated as confidential information by the training division. As discussed in the chapter *Managing the Evaluation System*, performance tests and examination records will include a detailed breakdown of success and failure of the member. Some learning management systems (LMS) include features that will electronically score and store examination and quiz score records.

Storage and maintenance of personnel evaluations may not fall directly under the responsibilities of the TO. Nevertheless, a TO may be able to provide valuable training-related information to other officers for personnel evaluations by supplying records of training hours, certifications attained, and

any performance deficiencies related to training. Local policy will dictate this area of recordkeeping responsibility.

Progress Reports

Much in the same way that school districts track the academic progress of elementary and secondary school students, the TO will track the progress of fire fighters as they pursue certifications and licenses. Many state certification agencies have specific hour- and content-related requirements that must be documented. A progress report may help track these pursuits by logging hours completed, test scores, and practical skill testing toward the certification or license. In some cases, a certification may require completion of several courses, and a progress record will assist the member and TO in tracking which additional courses are needed.

During a fire academy, weekly progress reports may be issued back to the employing fire department because the recruit may be attending the academy at some distance from the department **FIGURE 15-8**. In such a case, it would be important to keep the department up-to-date on the progress of its members in an effort to identify any performance issues early into the training process.

State Certification Records

Each state training entity will have specific requirements for obtaining a certification, including the records that relate to these criteria. It may be a handy and useful practice to keep this portion of the training file separate from other areas for quick access, as it may be more frequently utilized than other training files **FIGURE 15-9**.

■ A Typical Training File

At a minimum, the following information should be kept in a state certification record file, and in all types of training files in general:

- Master individual training accomplishments
- Dates, hours, locations, and instructors of all special courses
- Monthly summaries of all departmental training

Required Signatures

Almost every type of training record must identify both the instructor of the course and the person who was instructed.

WILLOW SPRINGS FIRE-RESCUE

Monthly Training Record

DECEMBER 2012

Fire Chief Larry Moran

Date	Time	Hours	Class Description
12/04/2012	09:00	03:00	MABAS 10 Haz Mat Training
12/04/2012	13:00	02:00	Christ Hospital Continuing Education
12/05/2012	19:00	03:00	Incident Command System
12/07/2012	08:00	05:00	Communications and Information Management
12/11/2012	09:00	03:00	IMAT Training
12/13/2012	11:00	02:00	MABAS 10 Training Meeting
12/17/2012	09:00	03:00	Fireground Communications and Mayday Procedures
12/19/2012	10:00	02:00	Mabas 10 Chief Meeting
12/27/2012	10:00	01:00	District Familiarization
		24:00	**Total Hours**

_____   _____   _____

Member's Signature   Training Officer's Signature   Chief's Signature

FIGURE 15-7 Sample monthly summary training record report.

In most circumstances, initials are not acceptable for this purpose on permanent training records and state certification training records. It may be acceptable to use a preprinted roster listing a department roster, shift, company, or unit, but a column should be available for each participating member to sign in as a record of attendance and participation. Certificates of completion for coursework should include the trainee's name, the title of the training session, the dates of attendance, and the signature of a lead instructor or program coordinator who is validating all of the information on the certificate.

Elements of Training Record Information

A standard training record report must be developed and used anytime training takes place in the department **FIGURE 15-10**. Some information may be preprinted, whereas other portions of the record are a fill-in-the-blank format. NFPA 1401 recommends the areas in **TABLE 15-4** as the minimum areas that should be included on the record. These requirements pertain to both electronic and paper-based training records.

A summary of the objectives and content of the training session should also be included, as should a narrative provided by the instructor in charge.

Additional Information on Training Record

Training record reports should be easy to complete. Nevertheless, the more thorough the information, the more useful the document will be. Additional information that might prove beneficial could include source information used as the basis for training, such as textbook title and edition and any available lesson plan information. If a local policy name and version is used as a reference, then be sure to include the reference number in the record. Some departments use a coding system related to state certification objectives and insert the objective numbers into the record, as the objective number is a known factor in the state program. As distance-learning programs become more common as part of overall training curricula, remember that distance-learning sources should be cited and referenced just like textbooks or other curricula.

NIPSTA FIRE ACADEMY
Weekly Evaluation Report
January 25, 2013

Candidate Name: Firefighter Newguy   Company: 5

Candidate Department: J & B Fire Rescue

Week 2	Monday	Tuesday	Wednesday	Thursday	Friday
Subject	Safety	Fire Behavior	SCBA 2	Ladders	Ladders
Attendance	Present	Present	Late/Excused	Present	Present
Quiz	Orientation	Terms	SCBA 1	Ropes	Tools
Score	100%	97%	100%	90%	88%

Overall Weekly Evaluation		Comments
Discipline	Meets Expectation	
Effort	Needs Work	Step up and complete tasks on time
Tools	Meets Expectation	
Ability	Improving	Showing progress from Week 1
Initiative	Needs Work	Ask for additional assignments
Ladders	Meets Expectation	
SCBA	Meets Expectation	Donning time is currently :48 sec.

Additional Information		Comments
Prepared for each day:	Yes	
Obeys rules and regs:	Yes	
Appearance and hygiene:	Very Good	
Overall attitude:	Good	Respectful to instructors and others
Level of participation:	Needs Work	Waits to be told what to do
Quality of work:	Good	
Ability to work w/others:	Very Good	Is involved in group work
Injuries Reported:	None	
Areas of Concern:	Study habits should improve	
Corrective Action Plan:	Increase use of workbook and on-line resources	

Additional Comments:
Continue to work on understanding of terminology and tool names/uses. Increase your participation in house duties, classroom and drill ground set-up and group team building exercises. Physical fitness level is improving, keep it up. Document additional on-line study and workbook completion before unit 2 exam.

Upcoming:	Monday	Tuesday	Wednesday	Thursday	Friday
Week 3	Hose 1	Hose 2	Live Fire 1	Ventilation	Live Fire 2

Academy Director: Deputy Chief Drew Smith
Academy Coordinators: Scott Exo   Mike Fox   Ken Koerber

FIGURE 15-8 Sample fire academy weekly progress report.

Postincident analysis is a very valuable training tool and should be completed at the conclusion of both significant and routine events. Opportunities for improvement can be inserted into the training program using postincident analysis, and these opportunities should be noted in the narrative. As a best practice, a specific code can be created in an electronic recordkeeping system to identify the use of postincident analysis in the training program.

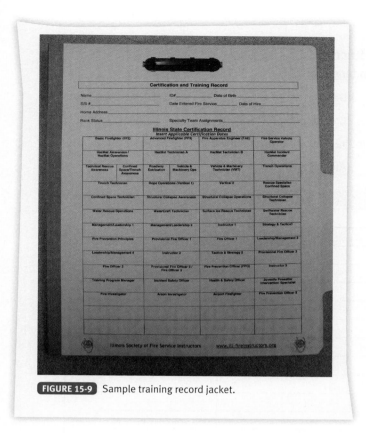

FIGURE 15-9 Sample training record jacket.

The delivery method used for training should be identified as part of the training record. Some program evaluations comprise an assessment of the percentages of each method of instruction used. A good balance among the various methods used should be apparent. Common methods of instruction that should be noted in the training record report include the following:

- Lecture
- Demonstration
- Skills training
- Discussion
- Role-play
- Labs/simulations
- Case studies
- Out-of-classroom assignments
- Self-study
- Online/video/computer based

If a combination of methods is used in the training session, a breakdown of how much time is spent doing lecture and then skills training, for example, should be noted within the report. Another overlooked part of additional information is any evaluation of training objectives that takes place. Written tests or quizzes, skills examinations, or other tools used to assess the student's understanding or ability to meet the learning objectives should be noted. Maintain score confidentiality by keeping scores separate from the training record; all that is needed on the training record is a notation of the evaluation's completion. This notation could take the form of a check box indicating whether an evaluation was completed or as a note in the narrative section of the report.

Types of Training Records

Many types of training records can be created, each with a different purpose. Consider the following NFPA 1401 recommended types of training records (NFPA 1401, Section A.5.3):

- *Departmental training record.* A report that serves as a permanent record showing all the training fire personnel receive. These reports usually are completed on a yearly basis. Company officers usually make entries on this form; however, training officers in small departments might take on this responsibility.
- *Individual special course record.* Special schools or courses made available to fire fighters. Their attendance and certification should be recorded.
- *Individual training record.* A record containing a chronological history of an individual's progress from the time of entry into the organization until separation.
- *Progress Chart.* A record form that provides an accurate and complete picture of all class activities and work accomplished by both the instructor and the students. The chart also shows, at a glance, how the class or program is progressing with respect to calendar or time schedules.
- *Certification training record (fire fighter).* A worksheet that maintains a record of each fire fighter's progress during the pursuit of certification.
- *Educational Courses.* A documentation of courses taken outside the department, such as at institutions for higher education.
- *Vocational Courses.* A documentation of courses, most of which are provided by state or regional programs in the form of workshops or demonstrations outside the department.
- *Seminars and other Training.* A documentation of all seminars, short courses, and other individual or group development meetings attended. A certificate of achievement or certificate of completion obtained in this area of training should be made a part of the individual's training file.
- *Periodic company summary.* A report showing all other training conducted or attended by the company. The number of hours spent by each individual on each subject should be recorded, and this report should be submitted through proper channels to the officer in charge of training. Such reports should be submitted monthly.
- *Chief officers' periodic training summary.* A report showing all training conducted by fire companies within a division, battalion, or district. This report serves as confirmation for chief officers that company officers are properly conducting company training, and it can be submitted to the officer in charge of training for inclusion in monthly and annual summaries.
- *Group training records and evaluation.* Because most fireground operations are accomplished by more than one fire fighter, group performance of basic evolutions should be an important part of any training system. A means or method of measuring the effectiveness of the organization's evolutions and how well they prepare fire fighters to make an attack on a fire should be established. The group or company performance standard evaluation report form should be designed to allow for

Training Record
Training Attendance Form

Date _____   Station(s) _____   Description _____

Start Time _____   End Time _____   Credit Hours (Total Time) _____

Method of Training: ☐ Classroom   ☐ Practical   ☐ Self-Directed   Certification Credit: ☐ Yes   ☐ No

Lead Instructor _____   Additional Instructor(s) _____

Print Name	Dept ID #	Signature	Hours Attended

Objectives: _____

Description of Training (Notes) _____

_____

_____

_____   _____

Instructor Signature   Training Officer Approval

Type of Training (ISO Category)				
Company	Multi-Comp.	Officer	Mutual Aid	Night
Tower Burn	Classroom	Practical	Combo.	Driver

Equipment Used in Training Session	Feet of 1 1/2" Hose Used	Feet of 2 1/2" Hose Used	Supply Hose Used/Ft.	Feet of Ladders	Number of Engines	Number of Trucks	Gallons of Water	Number of SCBA	Total Number Of Firefighters

FIGURE 15-10 Sample training record report.

quick and accurate determination of the group's ability to meet minimum basic requirements. The report should also enable the individual group or company to check on the progress of its abilities in carrying out standard operating procedures, and it should recommend areas in which additional training is needed.

Reports for the Training Program

Periodically, the training division must compile and submit reports on the activities, inventory, statistics, and successes of the division. These reports can be used for budget justification, equipment purchasing, and many other useful management

Table 15-4	Minimum Areas for a Training Record Report
Who	Who was the instructor(s)? Who was included in the training (individuals, company, multiple companies, or organization)?
What	What was the subject covered? Which equipment was utilized? What were the stated objectives?
When	When did the training take place? • Date and start/end time
Where	Where did the training take place? • Station, classroom, drill ground, off-site, academy, school
Why	Why did the training take place? • In-service training, certification based, compliance training, professional development, performance improvement

functions. Generally speaking, training reports—like other reports submitted to a supervisor or administrative body—are considered technical reports, and specific guidelines for their creation may exist. Reports should be clear and concise and follow specific organizational steps.

■ Format for Training Reports

A logical sequence of the information contained should be followed throughout each report, and the same sequence should be used whenever the report is used. With some frequently submitted reports, such as monthly training activity reports, the data fields may simply be updated on a monthly basis once the format is approved by the receiver of the report **FIGURE 15-11**.

The organization of a report can follow a five-step process that can be helpful in identifying, investigating, evaluating, and solving a problem. These five steps should be completed before the report is written:

1. Determine the purpose and scope of the report.
2. Outline the method or procedure.
3. Collect the essential facts.
4. Analyze and categorize the facts.
5. Arrive at the correct conclusions and make the proper recommendations.

Other Useful Reports

The training officer may be required to submit any number of training reports to a superior officer or regulating body. The frequency and content of these reports are determined by the local authorities and should be included in the overall training policy. Some reports, including those profiling assets assigned to the training division and detailed plans to improve the training division, can be used for budget request purposes. Another useful document is a periodic report on the progress of any probationary fire fighters; this type of a progress report can help identify strengths and weaknesses and document hours of training toward a certification **FIGURE 15-12**. If a probationary fire fighter must be released from service due to poor performance, this report will help justify these difficult but necessary management decisions.

A monthly summary of all activities of the training division is one of the most frequently used reports completed by the TO. This report can then be used to complete an annual report of all activities of the training division.

■ Legal Aspects of Training Records

The *Legal Issues* chapter discusses many laws and standards that apply to recordkeeping in the training division. The TO and department should make it a regular management practice to consult with legal representation about the quality, type, and quantity of information contained within the training record report. An example of how NFPA standards are used to specify training records in different ways can be found by examining NFPA 1403, *Standard on Live Fire Training Evolutions*. NFPA 1403 requires specific records and reports to be completed by the instructor in charge and also requires participants' training reports to be submitted to verify prerequisite training of participants. Failure to follow these guidelines can result in civil law suits and criminal charges in the event that a participant or instructor is injured or killed during the training event.

Privacy of records must be maintained and observed over the life of the organization. Training records are considered to be personnel records and, therefore, can be disclosed only with written permission from the student, unless required by law, statute, or court order. The use of Social Security numbers in such documents should be strictly prohibited, as should inclusion of other confidential individual information. Older training record jackets may display and contain Social Security number information, and those files should be retroactively stripped of such information to avoid potential legal issues. No confidential medical information should be kept within the training record. In fact, medical records should be kept separate from all other types of records. Personally identifiable or proprietary information is restricted by several privacy laws.

Records Storage and Retention

A records retention schedule that details the life expectancy of the various records and reports maintained and stored by the training division is required to be created; in fact, creation of this document may be required by state regulatory offices, such as the Secretary of State **FIGURE 15-13**. Do not dispose of any form of training record or report without first checking the records retention schedule. Computerized forms can be kept forever and backed up on multiple hard drives and other electronic media. Keep in mind that some records must be kept in their original form, including some completion certificates and licenses.

Training Program Policies

A formal training program policy should be created to help manage the training division activities. This policy can take the form of a standard operating procedure (SOP), given that few deviations or exceptions to the execution of the policy will occur. When developing training and program policies, the Instructor III may find technical writing coursework beneficial.

Des Plaines Fire Department
Division of Training & Safety

Monthly Training Report Summary January 2013

	Month	Year-To-Date
	2012/2013	2012/2013
Total In-Service Training Hours	3127/3377	3127/3377
Total Classes in Period	309/430	309/430
Total Attendees in Period	2061/2791	2061/2791
Acting Officer Training Hours in Period (Lt)	0/48	0/48

In-Service Training / Activities Scheduled

1. Company Tool Assignments
2. Grid Map Book Training
3. Fire Behavior 1
4. Apparatus and Equipment Maintenance
5. Ice Rescue
6. EMT/Paramedic Refresher

Out of Department Training or Schools Attended

Leadership I (40 hours)	FF/PM Sam Foster, Eng Terry Dillon, FF/PM Collin Moon
Ice Rescue (16 hours)	FF/PM Sam Foster, FF/PM Eric Zack, FF/PM Ryan Petty
Peer Fitness Trainer (40 hours)	FF/PM Mike Shaughnessy, Eng Matt Matzl

Total Out of Department Training: 248 hours

Division Related and Attended Meetings & Events

MABAS Division 1 Training Officers Meeting 1-15-13
IFCA Educational Research Foundation 1-17-13
Promotional Assessment Orientation 1-23 & 24, 2013
NWC Mass Casualty Training 1-17-13
Monthly Staff and Operations Mtg 1-28-13

Certifications Received

Ed Rogers	Provisional Fire Officer I

Division Chief Forest Reeder

FIGURE 15-11 Sample monthly training activity report.

The resulting policy should be reviewed on a regular basis, and all members of the organization should be familiar with its contents. Each member of the organization will have differing levels of involvement with training, and these roles should be clearly specified. A sample training policy includes some of the following areas (see also the appendix A: *Resources for Fire Service Instructors*):

1. **Purpose and scope of the policy.** The purpose for which the policy is created must be stated, as should the scope of whom it covers.
2. **Personnel responsibilities.** Each member involved in the development, delivery, participation, and administration of the training program should be identified, along with that person's level of responsibility **TABLE 15-5**.

Courtesy of Forest Reeder, Des Plaines Fire Department

Des Plaines Fire Department
DAILY INSTRUCTOR EVALUATION OF CANDIDATE
Operational Readiness Period – Probationary FF/PM

Date	Evaluator	Company Officer	Company Assigned	Eval. Start	Eval. End

List training modules completed:

Scale	Not Applicable N/A	Unacceptable 1	Needs Work 2	MEETS EXPECTATION 3	Above Average 4	Excellent 5

Candidate Name

Work Ethic	N/A 1 2 3 4 5	Injuries Received/Reported:
Judgment	N/A 1 2 3 4 5	Areas for Concern:
Teamwork	N/A 1 2 3 4 5	
Time Management	N/A 1 2 3 4 5	
Skill Level	N/A 1 2 3 4 5	
Physical Ability	N/A 1 2 3 4 5	Provide examples of positive and negative performance observed throughout evaluation period.
Communication	N/A 1 2 3 4 5	
Initiative / Motivation	N/A 1 2 3 4 5	
Accepts Guidance	N/A 1 2 3 4 5	
Adaptability / Stress Mgt.	N/A 1 2 3 4 5	
Community Awareness	N/A 1 2 3 4 5	
Empathy	N/A 1 2 3 4 5	
Appearance	N/A 1 2 3 4 5	

Evaluator Signature/Date

Candidate Signature

FIGURE 15-12 Periodic reports on the progress of probationary fire fighters can be helpful.

3. **Training assignments.** The types of training assignments in which members need to participate should be identified as mandatory, regular, company-level, or department-level training sessions.

4. **Documentation of training.** The requirements for documentation of training should be clearly spelled out in the policy. Signatures, record retention, display of progress and completion records, and access to information

**APPLICATION FOR AUTHORITY TO
DISPOSE OF LOCAL RECORDS**

(CONTINUATION SHEET)

APPLICATION NO. 84:156C

PAGE 39 OF 67 PAGES.

Courtesy of Forest Reeder

ITEM NO.	DESCRIPTION OF ITEMS OR RECORD SERIES	ACTION TAKEN
	Some records contained on this application are subject to state and/or federal audits. If that is the case the retention period will be followed by this symbol *. The audit clause which applies in this case is as follows.	
	Provided audit completed according to Illinois Revised Statutes 1983, Chapter 24, Article 8, Division 8.	
	Some records contained on this application are to be retained permanently. However, the original records may be disposed of if microfilmed in accordance with the regulations and standards of the Local Records Act, and providing the microfilm is retained permanently. If that is the case the retention period will be listed as "Retain permanently." followed by this symbol + . If microfilming is not an option, the retention period will be listed as "Retain permanently."	
195.	SAFETY TRAINING PROGRAM FILE – STATE CERTIFICATION RECORDS Dates: 1974 – Volume: 6 cu. Ft. Annual Accumulation: 1 cu. Ft. Arrangement: Chronological by year Recommendation: Retain permanently.+	
196.	ENGINEER TRAINING RECORDS Dates: 1980 – Volume: 2 cu. ft. Annual Accumulation: 1/2 cu. ft. Arrangement: Chronological by year Recommendation: Retain permanently.+	
197.	PARAMEDIC TRAINING FILE, DEPARTMENT TRAINING RECORDS (State Certifications) Dates: 1976 – Volume: 2 cu. ft. Annual Accumulation: Negligible Arrangement: Chronological by year Recommendation: Retain permanently.+	
198.	DAILY DRILL RECORDS Dates: 1973 – Volume: 6 cu. ft. Annual Accumulation: 1 cu. ft. Arrangement: Chronological by year Recommendation: Retain 5 years and dispose of.	
199.	PERFORMANCE EVALUTIONS Dates: 1981 – Volume: 2 cu. ft. Annual Accumulation: 1/2 cu. ft. Arrangement: Chronological by year Recommendation: Retain 5 years after termination of employment and dispose of.	

FIGURE 15-13 Sample records retention schedule.

Table 15-5	Personnel Responsibilities
Personnel	**General Responsibilities**
Training officer	Responsible for the development, scheduling, resource management, and delivery of the training program
Instructor in charge	Designated as the lead instructor in training sessions requiring multiple instructors, and identified as the final decision maker and responsible person
Instructor	Assigned to an instructor in charge or may instruct a training topic on his or her own following a lesson plan and set of objectives
Shift commander	Responsible for the administration and monitoring of the department training plan for the assigned shift; may coordinate company rotations and participation in training and provide feedback to the TO
Other officers	Line and staff officers, assigned to companies, who may assist in the safety evaluation of the evolution, evaluation of the members of their crew, and coordination of shift activities (including the training schedule). They may also participate in training.
Fire fighters	Crew members who participate in training sessions; their individual responsibility in training participation and the chain of command should be outlined

contained in the policy should be specified. Evaluation of training activities and the ways in which results are compiled and stored should be outlined.

5. **Other policy areas.** Requests to attend training sessions and school selection policies are often called into question and need clarification by members and officers. These policies should be very specific, including any forms that are used to request participation in training. Performance evaluations and periodic progress reports are another area that should be detailed. If individual performance deviations are identified, personal improvement plans and their follow-through should be included.

Selecting Instructors

A management function for the TO in large and small departments alike is the assignment and selection of instructors to deliver training. In some larger organizations, a formal process and assignment to the training division may exist. In most cases, specialization in a particular topic and a combination of field experience and educational background are often used to select instructors for a training session.

In some states, a certification program may require a specific level or type of instructor. In most cases, certification as an Instructor I is required to participate as an assisting instructor and certification as an Instructor II is required to serve as lead instructor. Lead instructors in a course or certification program should have their specific duties spelled out and, in best practice, reviewed to by the TO **FIGURE 15-14**. Highlighted areas of responsibility cover the safety, evaluation of conditions, control of evolutions, and record and time keeping.

An instructor's ability to effectively communicate and utilize instructional media and equipment may also determine which instructor is selected to present training. An instructor who may be highly skilled in classroom teaching techniques may not be as skilled during hands-on training, and vice versa.

Making Program Evaluation Presentations

A program evaluation presentation can take many forms. In some organizations, it may be a matter of a quick review of an annual training report highlighting significant activities, milestones, and trends over a selected period of time. In other organizations, you may be responsible for producing and delivering a very formal summary of goals, objectives, data, and milestones accomplished within a defined period of time. Take the time to learn which presentations or reports are necessary so you can compile and track required data throughout the year instead of trying to capture these data all at once. Some departments utilize a training committee for guidance and direction and, at the very least, a review of your annual training program should take place with them.

Your audience will determine the delivery method and often the medium for the program evaluation presentation. Be careful not to speak in technical terms or in training terminology that may be unfamiliar to those members of your audience who are not exposed to it. Even simple terms used in drills and evolutions may be foreign to some members of the audience. Describe how each term relates to job performance and effective and safe delivery of services.

Using multimedia may also enhance the presentation and clarify your key points. When a training program is evaluated and the results presented to an authority or other professional members, try to remove your personal involvement in the process and discuss the findings specific to the audience. Prepare yourself for questions from the audience by putting yourself in their shoes and considering which points of the presentation need clarification or will generate the most discussion.

JOB PERFORMANCE REQUIREMENTS (JPRS)
in action

The management of a training program requires the creation of policies and procedures that direct all members, including the training staff, on the management of instructional resources, staff, facilities, records, and reports that will help determine staff selection and better define instructional responsibilities. Additional requirements of this position include methods of constructing instructor evaluation plans and the way in which a training program manager should evaluate equipment and resources used for training delivery.

Instructor I

As part of the instructional team, manage resources, time, and materials efficiently to conserve available funds while providing quality training.

Instructor II

When developing training materials and completing training reports and records, ensure that they are all completed according to department guidelines and practices.

Instructor III

Create policies and manage the training division so that effective training that meets the agency goals is delivered and all applicable compliance issues are met. Instructor selection and performance reviews should be conducted on a regular basis to ensure quality delivery of training.

JPRs at Work

There are no JPRs at this level within this chapter.

JPRs at Work

The Instructor II will plan and budget for training resources. This will include deciding when replaced equipment can be used instead of newer in-service equipment for a course or training session.

JPRs at Work

The Instructor III will create the policies and procedures for the management of instructional resources, staff, facilities, records, and reports for the training program. Methods of constructing instructor evaluation plans will be a critical function at this level. Knowledge of the methods used to evaluate equipment and resources used for training delivery will allow the Instructor III to present training program evaluation findings to various audiences.

Bridging the Gap Among Instructor I, Instructor II, and Instructor III

All levels of instructors should be considered members of the training division team. Program management functions will include many recordkeeping and quality assurance reviews and measures. All instructors should complete records and reports thoroughly and correctly.

Illinois Fire Chief's Educational and Research Foundation
Lead Instructor Agreement & Course Planning Document

Lead Instructor
Information & Agreement

Name _____   **Course** _____

You have been selected to be a lead instructor for the Illinois Fire Chiefs Educational and Research Foundation. This selection was based on your professional reputation, past achievements within our training program and our trust and confidence in your ability to fulfill the responsibilities of this position.

As lead instructor, you represent the IFCA and the Foundation and are often the first and lasting contact that our students will have with the Foundation. We rely on your trust and judgment to do the right thing, do the safe thing and to ensure a quality program that meets student and sponsoring department expectations. You assume a role of management in this position and are responsible for the management of your program as well as the delivery of the program. This means you have fiscal responsibility to hold course costs in line with their budget, you have personnel responsibility to manage your instructional staff to ensure that they are prepared to deliver their course according to the schedule and objectives for the day(s) they are assigned.

The key responsibilities of this position are as follows:
1. Ensure that all training sessions are conducted with student safety as the primary concern.

2. Ensure that all learning objectives are covered according to OSFM, IFCF or other regulatory agency requirements. You are responsible for validating that all material is covered and student comprehension exceeds minimum standards. We do not accept minimal performance levels in our program.

3. Coordinate all activities with your instructional staff and document all activities using the **Training/Project Planning Form**. Any changes to instructor assignments, location or other logistics must be approved by the Fire Officer Program coordinator or designee.

4. Coordinate the pick-up, distribution, completion and return of all course administration materials in a timely fashion. This may include:
   • Instructor pay sheet for course
   • Student information rosters
   • Student address mailing envelopes
   • Class payments
   • Student sign-in sheets
   • Waiver/Release forms
   • OSFM forms
   • Course evaluation forms
   • Other individual course forms

5. Monitor class/course activities and complete any quality assurance forms or reports as assigned by the Fire Officer Program Coordinator.

6. Review course progress by contacting instructors each day and review end of course evaluations for quality assurance purposes.

7. Be responsible. Period.

A stipend of $xxx.xx will be provided for you at the end of your course. This is paid in addition to hourly rates incurred in classroom hours that you instruct. This stipend is intended to compensate you for these additional responsibilities but is paid on a performance basis as determined by the Fire Officer Program Coordinator. We thank you for your continued dedication to fire service excellence and desire to lead our instructional staff.

Lead Instructor Agreement
I _____ have read and acknowledge the position responsibilities of the lead instructor for the Illinois Fire Chiefs Foundation. I agree to assume these duties for the course(s) listed below and will complete these duties to the best of my ability.

_____   _____
Print Name   Signature   (Date)

Lead Instructor for Course: _____

Scheduled date(s): _____

(Continues)

Courtesy of Illinois Fire Chiefs Foundation

Illinois Fire Chief's Educational and Research Foundation
Lead Instructor Agreement & Course Planning Document

Course Planning & Project Management

Course/Training Name _____   **Date(s)** _____

Class Location _____   Enrollment ____ as of _____   Min # to Run _____
Lead Instructor _____   Phone # _____   Email _____
Lead Instructor Cellular _____   Home Phone # _____   Shift _____

Additional Instructors Assigned:

	Name	Date(s) Assigned to Instruct	Contact #
☐			
☐			
☐			
☐			
☐			

Training Delivery Support Materials: (Describe in detail in Tracking Notes) Indicate dates needed & when complete

☐ Copies ☐ Manuals ☐ Binders ☐ Textbooks ☐ Coffee ☐ Donuts ☐ Books
☐ Off-Site ☐ Host Info ☐ _____

Purchases Needed

Item	Vendor	PO Request #	Complete	Cost

Tracking Notes: (Discussions, Assignments, Commitments, Problems, Concerns, Decisions)   Date

_____
_____
_____
_____
_____
_____
_____
_____
_____
_____

(Continues)

Illinois Fire Chief's Educational and Research Foundation
Lead Instructor Agreement & Course Planning Document

Courtesy of Illinois Fire Chiefs Foundation

Contact Listing

Name	Affiliation	Phone	Address	Fax	Email

Administrative Requirements

☐ Class Roster   ☐ Instructor Pay Sheet   ☐ Training Record Reports   ☐ OSFM Roster

☐ Course Costing   ☐ Course Evaluations   ☐ Mailing Labels   ☐ _____

FIGURE 15-14 Sample instructor contract and project planning forms.

Developing Instructor Evaluation Plans

A critical element of all training programs is having an instructor management plan for evaluating the instructors within a program or division. Program evaluation includes evaluation of the curriculum, facilities, resources, and instructors who teach a course or program. Evaluating these various components of a program improves the course management and delivery. The training officer is responsible both for providing on-the-spot instructor performance reviews during training he or she witnesses and for supporting and encouraging instructors over the long term. This is similar to what company officers would do for members they are responsible for.

A key area that needs to be addressed is the evaluation of the instructors who teach within a program. Instructor evaluation can take place at different times during a course, such as when a new instructor is giving his or her first presentation or when a seasoned instructor finishes the end of a course. Regardless of when the evaluation takes place, the key to effective program management is having a justification of why the evaluation is being administered. The evaluation should be based on set criteria established by policy and known by both the evaluator and the instructor. The policy should also include how the results are to be used and who should see the results. The instructor evaluation plan is critical to successful management of a training program. A cycle of when instructor evaluations are completed should be established and followed for consistency.

To develop criteria for an instructor evaluation plan, one must first review the content of the end-of-course evaluations (see the *Evaluating the Fire Service Instructor* chapter).

Class evaluation forms can be very helpful in providing data and commentary from students who have participated in the program.

An instructor evaluation form assists in identifying how well program managers performed during course delivery. These criteria are specific to management-level functions above those of typical course delivery evaluations and may prove helpful in overall program analysis. Agencies that provide a specific job description for instructors can add those duty areas to the analysis plans. A sample method of constructing an instructor evaluation plan might appear as follows:

1. Review the job description duties of the instructor.
2. List the desired criteria to be evaluated.
3. Assemble a form or spreadsheet to collect data.
4. Review the evaluation plan with a supervising officer and gain approval for it.
5. Review the evaluation plan with instructors and inform them of its purpose.
6. Conduct a review of various courses and provide feedback with recommendations for improvement and acknowledgment of positive performance indicators.

For company-level instruction, it is not a common practice to have on-duty or on-call personnel complete written evaluations of instructor performance. As a program manager or training officer simply participating in the drill or in-service program, you can make personal notes on the instructor's performance and provide constructive feedback in appropriate areas. This ongoing evaluation of in-service training instructors should not be overlooked and should be included in the overall evaluation of the training program.

VOICES OF EXPERIENCE

One of the many responsibilities assigned to an instructor or training officer is the task of managing a training program. The fire service has evolved drastically over the years and has now taken on many new responsibilities. It is extremely important to prepare your students with as much pertinent information as possible without creating mass confusion. Today's fire fighters are trained with an all-hazards approach, encompassing everything from structural firefighting to technical rescue.

One of the biggest mistakes an instructor can make is believing they must know the answer to every question that is asked of them. You must understand that in order to create a thorough and balanced program you will need to reach out to other instructors. Many instructors have devoted countless hours in their field of study and are invaluable resources. You can either know a lot about a little or a little about a lot; it is up to you to decide which path you will take. Attempting to provide excellent instruction for your students in every discipline is nearly impossible. You will have to develop a support network of instructors to help you provide the best training programs for your students.

> **At all costs, avoid the temptation of producing "cookie cutter" training programs.**

Ronald Reagan once said, "Surround yourself with the best people you can find, delegate authority, and don't interfere as long as the policy you've decided upon is being carried out." Reagan's approach in effectively managing his staff provides a blueprint for instructors to manage their training programs. When I became a Training Officer I remember the difficulties of trying to provide our department with a solid, well rounded training curriculum. I realized very quickly that we had an extremely good group of instructors who could provide training in each of their focused disciplines. Your students will reap the benefits of receiving instruction from other instructors who have devoted themselves to a particular field of study.

At all costs, avoid the temptation of producing "cookie cutter" training programs. Your training programs should be tailored to meet the needs of your target audience. The fire service is always changing and your training programs will require constant evaluation to remain relevant and accurate. Having an instructor resource network available will enable you to bypass many of the hurdles instructors face while developing training programs.

Matt Hinkle
Training Officer, Lafayette County Fire Department
Adjunct Instructor, Mississippi State Fire Academy
Oxford, Mississippi

Training BULLETIN

Jones & Bartlett Fire District

Training Division
5 Wall Street, Burlington, MA, 01803
Phone 978-443-5000 Fax 978-443-8000
www.fire.jbpub.com

Instant Applications: Training Program Management

Drill Assignment

Apply the chapter content to your department's operation, its training division, and your personal experiences to complete the following questions and activities.

Objective

Upon completion of the instant applications, fire service instructor students will exhibit decision making and application of job performance requirements of the fire service instructor using the text, class discussion, and their own personal experiences.

Suggested Drill Applications

1. Review your department's record storage policy and determine the length of time that training records and reports must be kept before destruction.

2. Compile a list of the agencies that may audit training records for your department and the criteria that will be used during the records review.

3. List the titles or job areas in your department and their responsibilities in participating in and delivering training.

4. Review the Incident Report in this chapter and be prepared to discuss your analysis of the incident from a training perspective and as an instructor who wishes to use the report as a training tool.

Incident Report

© Greg Henry/ShutterStock, Inc.

Manteca City, California—1999

On June 16, 1999, a 38-year-old male fire fighter/Captain died after falling approximately 20 feet from the top of a ladder which had been previously raised to the second-story window of a fire building at a fire training center. The victim had been involved in the fire service for 18 years, 14 years with his current agency. At the time of the incident the victim was wearing full turnout gear, including SCBA, but the face piece was not worn.

On the morning of the incident, several fire departments were involved in a multi-jurisdictional, multi-company training exercise. The exercise was conducted by three divisions performing separate evolutions simultaneously. Division A demonstrated proper tactics and procedures during live fire-attack operations and proper search-and-rescue techniques within a simulated single-family residential occupancy. Division B demonstrated proper search-and-rescue techniques and ladder-rescue operations from a second-story elevated platform and/or window, and Division C demonstrated proper tactics and procedures for advancing a fire-attack hoseline to gain access to a third-floor fire by entering a second-floor window via a ladder and extending the hoseline up a stairwell to the fire.

The victim and fire fighters from Division C were assembled on the second story when the air horn sounded to evacuate the building as previously planned. Fire fighters on the second floor began to exit the building by way of the stairway, but the victim and other fire fighters discussed using a "ladder bail," which was new to the department. The new procedure involved a head-first advance over the top of the ladder, hooking an arm through a ladder rung and grasping a side rail, swinging the legs around to the side of the ladder, and sliding down the ladder to the ground.

One fire fighter approached the ladder and noted that it was not footed at the bottom and a call was made over the radio to "foot" the ladder. Before the ladder was properly "footed" the victim, for unknown reasons, approached the ladder and attempted to exit the second floor using the ladder bail technique. The victim, who was about 3 feet away from the top of the ladder, took one step and leaped over the top of the ladder. The victim was unable to adequately hook the ladder rungs or grasp a ladder side rail and fell about 20 feet headfirst to the concrete landing. Although the victim received immediate attention from fire fighters and medics in the area, the victim was transported to the local hospital where he was pronounced dead, about 40 minutes after the incident.

Post-Incident Analysis: Manteca City, California

Contributing Factors

All new training programs should undergo a comprehensive review prior to the implementation of the program.

Before a program is implemented, instructors should collaborate with other organizations regarding the feasibility of all new training procedures.

All aspects of safety should be adhered to per established standards and recommendations while training is being conducted.

A safety officer should be designated at all training exercises to observe operations and to ensure that safety rules and policies are followed.

Chief Concepts

- Understanding the budget development process and your role in purchasing, specifying resources and materials, and justifying requirements of a training project or program is crucial to successful training program management.
- Budget preparation is both a technical and a political process.
- A budget justification is useful in defending your position when requested money is allocated for a specific line item or category.
- The budget document describes where the revenue comes from (input) and where it goes (output) in terms of personnel, operating, and capital expenditures.
- The training budget should not be approached any differently than any other part of the budget document and process in a fire department. Budgets must be prepared, justified, and managed throughout the life of the budget cycle.
- Ongoing budget maintenance activities should take place to monitor the training division's expenditures to date as part of the training manager's fiscal responsibilities.
- Purchasing of resources and products used in the instructional area is a management skill.
- Training resources can be defined as any equipment, materials, or resources that are used to assist in the delivery of a training session. These resources can be either consumable or reusable.
- Reserve apparatus may assist in delivery of a training session, but instructors must be sure to identify any differences between what students would normally use in the performance of their duties and the equipment being used for the training.
- Resources used in the delivery of training should be identified during the construction of a lesson plan.
- Instructors who are responsible for evaluating resources for purchase must observe and follow all purchase policy requirements. Before submitting a request to purchase a resource, they must evaluate the quality, applicability, and compatibility of the resource for the department.
- Many regulatory compliance matters must be addressed in managing the training division, including filing and recordkeeping of the many documents related to the delivery of training and multitudes of policies that must be developed and enforced.

- A record system must be developed by the Instructor III, who may function as the TO, to document all aspects of the training program.
- In some organizations training records are analyzed to identify personnel assignments or to make pay-scale adjustments, whereas in other organizations the number of training sessions attended are tallied to allow for continued assignment to emergency response duties.
- Many types of training record reports exist throughout the emergency service community, the most common of which is the training record report used to document the ongoing training of the members.
- At a minimum, the following information should be kept in a state certification record file, and in all types of training files in general:
  - Master individual training accomplishments
  - Dates, hours, locations, and instructors of all special courses
  - Monthly summaries of all departmental training
- Many types of training records can be created, each with a different purpose.
- Periodically, the training division must compile and submit reports on the activities, inventory, statistics, and successes of the division. These reports can be used for budget justification, equipment purchasing, and many other useful management functions.
- The organization of a report can follow a five-step process that can be helpful in identifying, investigating, evaluating, and solving a problem:
  1. Determine the purpose and scope of the report.
  2. Outline the method or procedure.
  3. Collect the essential facts.
  4. Analyze and categorize the facts.
  5. Arrive at the correct conclusions and make the proper recommendations.
- The TO and fire department should make it a regular management practice to consult with legal representation about the quality, type, and quantity of information contained within the training record report.
- A records retention schedule that details the life expectancy of the various records and reports maintained and stored by the training division should be created; in fact, this document may be required by state regulatory offices, such as the Secretary of State.

Wrap-Up, continued

- A formal training program policy should be created to help manage the training division activities. This policy can take the form of an SOP given that there are rarely any deviations or exceptions to the execution of the policy.
- Specialization in a particular topic and a combination of field experience and educational background are often used to select instructors for a training session.
- Take time to learn which presentations or reports are necessary in your department so you can compile and track required data throughout the year instead of trying to capture these data all at once.

Hot Terms

Budget An itemized summary of estimated or intended revenues and expenditures.

Expenditures Moneys spent for goods or services that are considered appropriations.

Revenues The income of a government from all sources, which is appropriated for the payment of public expenses and is stated as estimates.

Single source Materials that are manufactured or distributed by only one vendor.

References

National Fire Protection Association. (2012). *NFPA 1041: Standard for Fire Service Instructor Professional Qualifications*. Quincy, MA: National Fire Protection Association.

National Fire Protection Association. (2012). *NFPA 1401:Recommended Practice for Fire Service Training Reports and Records*. Quincy, MA: National Fire Protection Association.

National Fire Protection Association. (2012). *NFPA 1403: Standard on Live Fire Training Evolutions*. Quincy, MA: National Fire Protection Association.

An accident occurred during a training event that you were supervising, and a member was injured. The department has decided that an investigation must take place to learn the cause of, and any possible contributing factors to, the accident and injury. During the investigation, the agency training records are to be reviewed as part of the process. The intent of the training record review will be to identify whether adequate and up-to-date training was provided to the member who was injured. You should cooperate completely during the investigation and provide the training records needed for the investigation.

1. Which NFPA standard applies to training records and reports?
   - A. 1401
   - B. 1041
   - C. 1500
   - D. All of the above

2. During the review of the training records, the investigators request a training report summary of the hours completed in training on a set of specific subject areas. How might this report be used in the investigation?
   - A. To determine whether the member was certified in the job being performed
   - B. To identify whether the member had previous training in the tasks that were being performed
   - C. To help provide direction for new policies on recordkeeping practices for the department
   - D. All of the above

3. If the incident occurred during live fire training, specific training record reports are required for members participating. Where can the instructor find direction on reports to be kept of live fire training?
   - A. OSHA Live Burn Regulations
   - B. NFPA 1041
   - C. NFPA 1403
   - D. ISO Guidebook

4. At the completion of the investigation, it is found that the member did not have sufficient training in the skill area being performed. Which steps should be taken to ensure that this type of accident does not occur again?
   - A. Establish prerequisites for each level of skill that must be completed before the member moves on.
   - B. Provide a skills check on the material being covered prior to engaging in the evolution.
   - C. Conduct a postincident review and review the findings of the investigation as a learning tool for all members.
   - D. All of the above may help reduce the possibility of a reoccurrence of the accident.

The Learning Process Never Stops

Fire Service Instructor I

Knowledge Objectives

After studying this chapter, you will be able to:

- Discuss the importance of continuing learning for the fire service instructor. (pp 382–386)
- Identify professional organizations that will help in the professional development of the fire service instructor. (pp 386, 388)
- Identify and discuss the value and importance of coaching and mentoring the next generation of fire service instructors. (pp 388, 390–391)

Skills Objectives

There are no skills objectives for Fire Service Instructor I students.

Fire Service Instructor II

Knowledge Objectives

After studying this chapter, you will be able to:

- Discuss the importance of continuing learning for the fire service instructor. (pp 382–386)
- Identify professional organizations that will help in the professional development of the fire service instructor. (pp 386, 388)
- Identify and discuss the value and importance of coaching and mentoring the next generation of fire service instructors. (pp 388, 390–391)

Skills Objectives

There are no skills objectives for Fire Service Instructor II students.

Fire Service Instructor III

Knowledge Objectives

After studying this chapter, you will be able to:

- Discuss the importance of continuing learning for the fire service instructor. (pp 382–386)
- Identify professional organizations that will help in the professional development of the fire service instructor. (pp 386, 388)
- Identify and discuss the value and importance of coaching and mentoring the next generation of fire service instructors. (pp 388, 390–391)

Skills Objectives

There are no skills objectives for Fire Service Instructor III students.

You made it! You have just been appointed as the senior instructor for your fire department. After passing through the training division as a captain and then back out to the floor and into several other positions, you have returned to the training division and feel that you are "back home." After the initial shock of your promotion wears off, you start to review the current status of the division and begin to formulate some initial goals and plans for your staff. One concern you had several years ago when you first worked in the training division as a captain was your desire to be a better instructor and the means by which you could improve your skills as an instructor. Other instructors shared your concern then—and now you as the senior instructor have the opportunity to address these feelings and concerns.

1. Which assessment tools could you use to assess instructors' skills?
2. Based on this assessment, which professional development opportunities exist to improve an instructor's skills?
3. How will improving the skills of the training division staff affect the department as a whole?

Introduction

The purpose of this book is to prepare you to be an effective fire service instructor. So far, you have learned the information and skills necessary to prepare a lesson plan, teach from a lesson plan, arrange for resources, ensure the safety of both students and instructors, and evaluate both students and other fire service instructors. You have learned about budgeting techniques, supervisory skills, communication strategies, and ways to use today's technology in presenting material. So what comes next? Have you reached the height of your career, or is there more to accomplish? At all instructor levels you should be actively engaged, have a desire to learn new teaching techniques, understand the latest instructional technology, and continue to develop professionally.

For some fire service personnel, teaching at a recruit school or being assigned to the training division is a momentary and brief assignment in their upward progression within the fire department. Indeed, some of those assigned to teach may simply tolerate that duty. For others, however, the training division is exactly where they want to be—filling the role of a fire service instructor who is in a prime position to teach and influence current and future fire fighters and to shape the future of their department. The role of a fire service instructor can be a formal designation or an informal one. Regardless of the designation, you are in a position to influence those around you through your actions, your words, and—most importantly—your example.

Once you are in the fire service instructor position and have received the training and certifications to prepare you for the job, you need to continue with your own growth and development. Reaching the position is just the beginning; it is the ongoing learning process that will make you a truly excellent fire service instructor. Learning is a lifelong process. As part of that process, many fire service instructors nationwide go on to publish reports of their work in trade journals or books. When their ideas and lesson plans become publicly known through the literature, other fire service instructors have a starting point from which to build their own ideas and lesson plans. Publishing also adds to the creditability of the fire service instructor and helps in ensuring that the fire service is recognized as a profession.

Fire Service Instructor I, II, and III

Lifelong Learning and Professional Development

Recall the fire service instructors who shaped and influenced your development (both good and bad). What is the common thread among these individuals who had a positive impact on you? Most likely, they always had a magazine or an article in their hand or their nose in a book **FIGURE 16-1**. The individuals who actively sought opportunities to learn were often the ones called upon to teach at a moment's notice because they had a wealth of knowledge that others wanted to hear. The same individuals attended conferences to learn and focused on networking with their peers. Eventually, they became presenters after obtaining a thorough knowledge base

FIGURE 16-1 Reading is a great way to continue personal development.

Teaching Tip

When teaching a class, title an easel board "Resources." Over the course of the class, add resource information to this list and encourage students to do the same. Typically, this information will include Web site addresses, books, and names of other fire service instructors.

Safety Tip

As a fire service instructor, you may feel that you have finally "arrived" and are at the top of your game. The fire service is constantly evolving, however. If you are not continuously learning, you are falling out of touch, and your knowledge base is growing outdated. Become an avid lifelong learner to avoid this complacency.

and developing their own ideas and theories to share with the fire service.

Another way to look at this idea is through an analogy. The speaker who is often heard at various fire service conferences did not start out as the keynote speaker. At one time he or she was just a fire fighter sitting in the audience, taking notes, and listening to every word. This person took what was learned at the conference, returned to the firehouse, and looked for ways to put that new knowledge into practice. As he or she learned more and advanced into higher positions within the department, the fire service instructor had opportunities to teach new concepts and ideas to the next generation. The ideas, concepts, or theories that this individual shared stood the test of time and eventually became the basis for new policies and procedures followed within the department. Soon, word of mouth spread about these new ideas. The fire fighter who once sat in the audience taking notes is on the stage leading the discussion and is respected by his or her peers as a leader in the profession. All of this came about because of the individual's desire to learn and to develop—because the fire fighter was a true lifelong learner.

An important part of becoming a great fire service instructor is having the basic desire to always improve and to become a better fire service professional. Who benefits when you develop into a respected fire service instructor? Both you and the fire department, because a good fire service instructor supports the department's goals and, in turn, the department recognizes the instructor's critical value to the fire service.

As grand and glorious as being a fire service instructor sounds, it also carries a great deal of responsibility. A department is often judged by the performance of its fire fighters. When a fire fighter is injured or killed in the line of duty, some of the first items reviewed and scrutinized are the training records and the qualifications of the instructors who taught that fire fighter during recruit school and during his or her time in the department. Keeping up-to-date on all changes and improvements in the fire service is a difficult, but critical task.

How do you stay informed, do your job effectively, and still maintain your personal life?

■ Time Management

As a fire service instructor, one of the first skills you need to learn and develop is time management. Several methods or techniques are available to help you make the best use of your time. You should select those techniques that suit your personality and are easy and convenient to use. For example, setting a goal to read one magazine article per shift, or per week, is a good start. Attending one fire service conference a year and earning a new certification are goals that will help in your development.

One helpful method is to set time aside each day to create a "to do" list. Prioritize the items on this list, and then work your way through them, checking off what you have accomplished. Included on this list might be reading an article or reviewing a department standard operating procedure (SOP). Another method is to set goals for the year and break the goals down into months and weeks. Use whatever method works for you: The key is to develop a system that fits your needs. When creating your personal "to do" list, be mindful of any recertification requirements for your certifications or job position. As a fire fighter and instructor, you might have required hours of continuing education or "re-cert" hours that must be earned in a set period of time. As you are setting your goals and priorities for the year, make sure that your own credentials do not expire! Budget the time and resources needed to obtain new or renewed certifications, and do not wait until the last moment to complete these requirements.

Regardless of which system you use or develop, perhaps the most important part of successful time management is learning to prioritize your "to do" list. Too often we spend a great deal of time on low-priority items that are easy to handle but distract us from the more important items on our list. Learning how to prioritize your list takes self-discipline and practice. Another challenge to accomplishing the items on your "to do" list is balancing your list against that of your organization. When developing this list, make sure every item supports the mission and goals of your agency. Finding this balance will help you develop both as an individual and as a professional in your organization.

■ Staying Current

Subscribe to trade magazines to stay up-to-date on trends and new developments in the fire service. Read other professional magazines in related fields, such as homeland security or business, as well. The benefit from reading trade journals is that this practice helps you stay current in the fire service. The benefit from reading material outside the fire service profession is that it helps you develop into a well-rounded individual and professional.

Another tool for you as a fire service instructor is the Internet. The Web has opened the door to resources that were difficult to obtain in earlier years, but now are readily available to anyone who has a computer and Internet access. The numerous Web sites hosted by colleges, professional journals, professional organizations, the National Fire Academy (NFA) and state fire academies, and government agencies offer volumes of information at minimal or no cost. Another online resource is the National Fire Protection Association (NFPA), which provides a variety of reports, online training, conferences, and workshops. This organization is also the source for the Fire Service Professional Qualifications Standards. Some material on the NFPA Web site is free, but other access requires a subscription.

One drawback to the information found on the Internet is the sometimes questionable accuracy of the content presented. If you use material from the Internet, you need to verify the authenticity of the source and the information presented. More than one fire service instructor has taken information from the Internet and used it in a lesson plan, only to find that it was inaccurate or completely wrong. This, in turn, hurts the reputation of the instructor and may put students in harm's way. Another challenge when using the Internet is the fact that a great deal of information is available at the click of a mouse, and you can spend too much time sifting through all you find. The Internet is a tool and, just like with any other tool, you need to know how to use it properly and when to use it.

Safety Tip

Information is only as good as its source. Be sure to verify your source of information before you include that information in your lesson plans, particularly if it is potentially harmful to students. Be aware, however, that even a reputable source may occasionally present outdated or otherwise flawed information.

■ Higher Education

Another method you can use for personal and professional development is to pursue a degree from an institution of higher education. Most departments require fire fighters to have a high school diploma or a GED as a condition of employment. In addition, many community or state colleges now offer degrees in fire science, emergency management, homeland security, or communication. Although this coursework is within your discipline, colleges also require you to take general education courses such as history and English literature as part of the degree program so that you will become a well-rounded thinker.

Higher education has become more accessible today than ever before, with degrees being offered by traditional colleges as well as online programs offered on the Internet. Both of these delivery methods (traditional or online) provide the opportunity to earn an advanced degree and leave you the choice of which methodology you wish to pursue. One big advantage of pursuing an online degree is the convenience and flexibility it offers the student. You can "attend" class on your own schedule, while at the station or at home. Some higher education programs also offer a blended learning format, which combines the use of the Internet for the majority of the classroom work but requires students to meet face-to-face with the instructor during certain points of the course. (Online courses and blended courses are discussed further in the *Technology in Training* chapter.) Check the institute's accreditation before obtaining an online degree. Some programs are not accredited, so your degree will not be recognized by other schools or your employer.

An increasing number of departments have begun requiring an advanced degree for promotion, such as an associate's degree for the captain's position and a bachelor's degree for the battalion chief's position. Other departments offer financial incentives for engaging in continuing education, paying for college or even giving fire fighters a raise for each college-level course they complete.

A major incentive for obtaining a bachelor's degree is the requirement in place at the National Fire Academy that requires an advanced degree to be eligible to enter the Executive Fire Officer Program (EFOP). Beginning October 1, 2009, a bachelor's degree was required for entrance into the

Theory Into Practice

Attending college can be a time-consuming and expensive proposition. You may be willing to make the time commitment, but your finances may not be flexible enough to handle the entire cost. If so, take some time to investigate the many possible funding sources for which you may qualify. For example, your fire department or city may have a tuition reimbursement program. The federal government provides grants that are awarded to thousands of students every year. If you served in the armed forces, you may qualify for funding as a veteran. Many local charitable organizations also provide scholarships to students, including nontraditional students. Do not automatically assume that you cannot attend college because of money concerns. If you are willing to make the commitment and expend the effort, funding sources are available to support your continuing education.

EFOP. The EFOP is considered by many to be a key to promotion as a chief officer and is a desirable qualification to add to your toolbox.

■ Conferences

Attendance at conferences, workshops, or seminars also offers you a chance to increase your skills and knowledge as a fire service instructor FIGURE 16-2 . The many professional conferences put on every year provide opportunities for growth and development at all levels. Some conferences provide hands-on training opportunities, whereas others are limited to classroom lectures or presentations. One of the largest annual conferences that all instructors should attend if given the opportunity is the Fire Department Instructors Conference (FDIC), which is held each spring in Indianapolis, Indiana. FDIC offers a wide range of courses for everyone in the fire service, from chief officer to the newest fire fighter. Many courses are directed at the fire service instructor to improve your skills as an instructor—including your classroom presentation skills, research techniques, new learning tools for classroom instruction, and a variety of other skills.

Other instructor conferences are held at the local and regional levels and provide opportunities to learn new skills, enhance your existing teaching skills, and learn from others. Regardless of which type of conference you choose to attend, take advantage of each and every opportunity you have to learn from your peers. Be willing to look outside the fire service and look for instructional learning opportunities in clubs, community organizations, and community college courses.

Take advantage of the opportunities at these conferences and workshops to network with other instructors from both within and outside the fire service community. Such conferences are great ways to learn new presentation skills that you can add to your own toolbox. Networking is a fundamental strategy used by great fire service instructors to learn from others and to bounce an idea off a peer to determine whether it is valid. Networking is a powerful addition to your toolbox that can open the door to sharing or collaborating on an idea and developing a new methodology or course.

A major factor affecting today's fire service is the limited budget that every type of department faces. When an instructor asks the chief permission to attend FDIC or to pursue an advanced degree, the chief is thinking, "How does the department benefit from this investment?" The real benefit comes from instructors who return from the conferences or workshops and improve the way they teach or share new information with their fellow fire fighters and instructors. The relationship between the chief and the instructor is one of respect and trust, and one of investment and return on investment.

FIGURE 16-2 Conferences are an excellent self-improvement activity.

Theory Into Practice

Although networking is often associated with politics and "schmoozing," what you are really doing when you network is creating a web of friends. The bigger the web, the more likely you are to have access to the precise resource you need when the time comes. Networking is designed as a give-and-take system.

The best way to create a large web is to be a giver. When someone needs something that you can provide, offer it. If someone has a question, answer it. Helping others be successful creates a powerful network that can work to your benefit. By giving, you become the person who is sought out: The more sought out you are, the more contacts you will have. Develop your network not by schmoozing, but by being genuinely interested in helping others succeed.

■ Lifelong Learning

Lifelong learning is an endless path marked by endless opportunities. For the one person who says he or she has seen it all, there are a hundred other individuals who ask for more. A major career killer is adopting the attitude that you cannot teach students anything new. Your greatest challenge is to create the desire to learn in all of your students.

In your pursuit of becoming a lifelong learner, you need to remember the basics. Your teaching skills got you where you are, so continually brush up on your existing presentation skills and learn new ones. Remember the teacher in high school whom no one liked because he taught as if he was still locked in the past? Remember the high school teacher who stayed current and fresh to students? Those teachers who worked at truly "teaching" a subject and encouraging students to learn were the same teachers whose classes were the first to fill during registration. Learning can be enjoyable. Instructors who are willing to learn new methodologies to present information become valuable assets to the organization because of their willingness to learn and develop themselves into strong instructors.

Keeping your teaching skills sharp and current is just as important as reading magazines, networking, or going to college. If you are the senior fire service instructor in the training division, make a point of teaching one subject once a month so that you keep your teaching skills up-to-date. As your knowledge base about the fire service has grown, have you stayed current with the types of students entering the fire service today? Are you still an effective fire service instructor for the new generation of fire fighters? Test yourself by teaching a class. Be honest with yourself about how effective you were. Great fire service instructors are lifelong teachers who love to teach.

An important skill to add to your toolbox is the acknowledgment and acceptance of the changes we are seeing in the fire service. The new generation of fire fighters is more diverse and has a strong understanding of technology. Effective instructors should recognize that today's students are different and learn how to teach them by using methods that will allow these students to learn the important skills that will save their lives and the lives of others. Perhaps instructional methodology needs to change to assist the new and diverse learners, but what we teach has not changed. Firefighting is a dangerous profession, and fire will kill either a young or old fire fighter without regard to how they learn. Our job as instructors is to learn whatever methods will enable us to teach life-saving skills to those students whom we are responsible for teaching and training.

Teaching Tip

Part of being an effective fire service instructor is learning to scan the horizon for industry trends and to anticipate which will pan out. Part of this scanning process involves looking for new ways to enhance your skills and abilities. Attending a one-day skills development program and reading a new book are two ways to stay on the cutting edge of your profession.

Professional Organizations

There are many professional organizations that you can join to help in your professional development. Some organizations operate on the national level, but most are found at the state, local, regional, provincial, or county level. (Among these are state training organizations.) Regardless of the level, these organizations provide many opportunities to network, share resources, and develop contacts. Some are very formal and highly structured, whereas others are more informal and relaxed.

The International Society of Fire Service Instructors (ISFSI) is an organization that is committed to the development of fire service instructors. ISFSI actively supports the professional development of the fire service through its push for qualified fire service instructors teaching quality material. In 2011, the ISFSI introduced a Professional Development Credential for the Fire Instructor. This credentialing program is designed to provide instructors with a direct pathway to develop and advance as professional educators. Criteria for this program are available at the ISFSI Web site: www.isfsi.org.

The National Fire Academy, which is part of the National Emergency Training Center in Emmitsburg, Maryland, is the nation's "fire university." At the National Fire Academy, you can attend 6-day or 2-week courses that teach you how to develop lesson plans and build courses. In addition, other courses are available to help in all aspects of your professional and educational development. The National Fire Academy also supports two other programs that will help in your professional development: the Fire and Emergency Services Higher Education (FESHE) initiative and the Executive Fire Officer Program (EFOP).

JOB PERFORMANCE REQUIREMENTS (JPRS)
in action

This chapter is the culmination of this entire book and sets the stage for the ongoing process of instructor professional development. As a professional fire service instructor, you have a responsibility to keep up-to-date on the educational trends, legalities, and professional organizations and resources available to you. Just as we hold fire fighters, driver/operators, and officers responsible for their professional development, fire service instructors are also required to continue to learn and practice their craft.

Given the rapid rate at which technological changes are occurring and the increased workloads being placed on instructors, you should identify a peer work group and organizations that will be able to assist in your continuing education mission. In addition, part of your professional development includes the selection and mentoring of future instructors. This is an opportunity to pay back the fire service and improve it by sharing your experiences and guidance.

Instructor I

The Instructor I must keep current on the latest instructional delivery techniques, media, and resources. Membership in state and national instructional organizations, if possible, is key to your professional development.

Instructor II

The Instructor II will continue his or her learning by keeping up-to-date on recent legal decisions and methods of preparing instructional materials. Networking through trade shows, attending conferences, and taking courses designed to bring new information back to your department will help keep you motivated.

Instructor III

The Instructor III will now take a broader view of programs and curricula that could be used within the department, the budgetary impact of programs, the concept of "return on investment," and professional development of instructional staff of the department.

JPRs at Work

NFPA does not identify any JPRs that relate to this chapter. Nevertheless, the concept of continuing educational and professional development in every job level is embraced by virtually every major fire service organization. All levels of instructors must continually refine their skills and knowledge throughout their entire career.

JPRs at Work

NFPA does not identify any JPRs that relate to this chapter. Nevertheless, the concept of continuing educational and professional development in every job level is embraced by virtually every major fire service organization. All levels of instructors must continually refine their skills and knowledge throughout their entire career.

JPRs at Work

NFPA does not identify any JPRs that relate to this chapter. Nevertheless, the concept of continuing educational and professional development in every job level is embraced by virtually every major fire service organization. All levels of instructors must continually refine their skills and knowledge throughout their entire career.

Bridging the Gap Among Instructor I, Instructor II, and Instructor III

Each level of instructor should now be prepared for the next level of instructional certification. The Instructor I has learned important skills in the area of course delivery, and the Instructor II has developed a series of skills in writing lesson plans, conducting evaluations, and developing media for many areas of coursework. The Instructor I should now begin taking the coursework necessary for certification at the Instructor II level and, with the help of a mentor or experienced instructor, be able to cross the gap to Instructor II successfully. The Instructor II is now ready to begin certification training for Instructor III responsibilities in curriculum design, budget and finance, and administrative areas of managing a training program. In all cases, each level of instructor has a responsibility to the fire service and his or her organization to mentor the next generation of instructors. You can learn from them just as much as you will help them learn.

© Greg Henry/ShutterStock, Inc.

A key part of what the National Fire Academy is doing to professionalize the fire service can be seen in the FESHE initiative. Across the United States, many different community and state colleges and universities offer degrees in different aspects of fire service management or engineering, but there are no standardized model degree programs. The FESHE initiative is intended to bring together representatives from colleges and universities to develop a standardized degree format and model curricula. If college degree programs follow the same set of rules, students will be able to transfer coursework from one school to another more readily. In addition, courses would present a standard curriculum, which increases the ability to transfer college credits and standardizes the information taught. Another purpose of the FESHE initiative is to develop a career path for those seeking to be chief officers.

Part of this development process has resulted in the National Professional Development Model. This model summarizes the combination of education, training, certification, and experience that leads to the development of a well-rounded and fully developed fire officer who has the skills and ability to lead the fire service into the future.

The EFOP is another program developed by the National Fire Academy that has helped to professionalize the fire service. Its purpose is to provide senior officers with the tools they need to become leaders of the future. Acceptance into the EFOP is by application process only; that criteria can be found at www.usfa.fema.gov/nfa/efop/selection.shtm. One of the key criteria for application in the EFOP is the individual's "potential for future impact on the fire service." Of the many positions in the fire service, fire service instructors have one of the most crucial roles in influencing the future of their organizations and the fire service in their communities.

The EFOP requires a major time commitment from both the fire officer and the fire department. The program takes four years to complete, with attendance at the National Fire Academy for four 2-week courses. At the end of each course, the student is required to submit an applied research project (ARP) within 6 months of the course's ending date. This ARP is evaluated by peer reviewers against a set of criteria established by the National Fire Academy staff. Once the ARP is approved, the student is allowed to register for the next course in the series.

Each spring, the National Fire Academy hosts the annual EFOP Graduate Symposium on its campus in Emmitsburg, Maryland. Each year the symposium features a topic of interest to the fire service, and EFOP graduates give presentations on the symposium topic. This symposium serves as an opportunity for EFOP graduates to renew friendships, network, and "charge their batteries" as leaders in their organizations.

The Next Generation

A true mark of a great leader is his or her willingness to plan to be replaced. Succession planning is the process of identifying others within the department who could be mentored and coached to replace you and to carry out the functions you perform.

The concept of training your replacement can be very difficult and challenging to handle on an emotional level. It may be easier if you understand that the process of training another person to take your place usually means that you are also moving up and onward to something new. Another benefit of training your replacement is that you can select someone who shares your own ideals and goals, thereby ensuring that your training program will continue after you leave. Through this legacy, your influence could be felt for many decades in your department **FIGURE 16-3** .

How do you train your replacement and help him or her be successful? Four basic concepts will aid you in teaching your replacement:

1. Identifying
2. Mentoring
3. Coaching
4. Sharing

Theory Into Practice

Become certified by an accredited organization to give credibility to your competencies.

Theory Into Practice

A method that might help an individual to "self-identify" as a potential fire instructor is to publish certification requirements (Fire Instructor I) or course requirements (completion of an Instructional Methodology course from NFA) for the job. This information will allow those who want to teach to indicate to their supervisor they want to prepare for future promotions.

FIGURE 16-3 Succession planning ensures that the goals and objectives for the department or organization continue.

VOICES
OF EXPERIENCE

A challenge I encountered as a new training officer in my combination fire department was teaching the "senior" members of my department. There were several members who had been in the department for years and had a great deal of experience as fire fighters, and this was very intimidating to me as a new training officer.

When given the assignment by the fire chief to train all department members and prepare them for the state Fire Fighter I certification exam, I was apprehensive, to say the least. Considering my lack of experience when compared to many members of the department, I felt inadequate to carry out my assignment.

I developed a training schedule with the guidance and approval of the chief and posted it on the training board. As I prepared for the first class, one of the "senior" members of the department offered to assist me if needed, and I jumped at the opportunity to use his assistance. He offered me some great advice that has served me throughout my career; it seems so simple looking back now, but it wasn't so obvious to me at the time. He suggested that I involve the senior fire fighters in my training schedule, let them help in the breakout sessions, and let them use their experience in teaching the younger fire fighters. We developed a list of senior fire fighters and approached each of them to ask for their assistance. Some accepted the offer to assist and others turned me down, but the net result was that once the training started, everyone participated because they were either assisting or had been given the opportunity to assist.

> *By the end of the 6-month training schedule, our department had prepared for the state certification exam but had also become a closer and stronger department.*

By the end of the 6-month training schedule, our department had prepared for the state certification exam but had also become a closer and stronger department. A new level of respect for the senior fire fighters developed among the younger members, but in turn the senior members realized the next generation of fire fighters wasn't so bad either. Everyone passed the practical and written certification exams. The department as a whole moved up a level in professionalism, but as a department we became more of a family.

Looking back to this time, I learned that embracing others' strengths and differences made me a better instructor and improved my ability to teach. Part of the lifelong learning process is learning to learn from others, and learning comes to us in many different shapes and forms.

Alan E. Joos
Fire and Emergency Training Institute
Louisiana State University
Baton Rouge, Louisiana

Identifying

Identifying is the process of selecting those whom you would like to mentor, coach, and develop. The identification process almost requires a "crystal ball" approach of looking at individuals—that is, you need to see the person today and forecast who the person could be tomorrow. Look for these qualities:

- The quick learner. Look for those who truly understand how, why, and when (and when not).
- Individuals who expect 110 percent and inspire others by leading by example.
- Individuals who are competent and knowledgeable as fire fighters. This type of fire fighter stands out and is respected by his or her peers.
- Individuals who possess organizational and preparation skills.
- Individuals who understand and exercise time management skills.
- Fire fighters with a passion for the fire service. Passion is contagious and can push a crew or a training division to move to the next level.

There are many important questions to ask the potential instructor. Oral interview questions may include the following:

1. Assume that you are hired to be the training officer and have been approached by members of a fire company who complain that their officer treats training as an unimportant issue and spends as little time as possible on drills. The members respect the officer as a good fireground commander, but claim that he has no appreciation for training and does little to support the training program. Which action would you take in this situation?
2. Why are you interested in becoming the training officer? What have you done to prepare for this position?
3. Which personal attributes that you possess will be of the greatest benefit to you in fulfilling the duties of this position?
4. Do you have any plans for personal development that will benefit you in this position?
5. Throughout the course of the year, members of the department may need motivation, direction, and guidance regarding professional training and education needs. Do you see this as part of your role as training officer? Why or why not? If so, briefly describe your thoughts on how you can best provide this need.
6. What is your top priority or priorities that would have to be accomplished in the first 3 months?
7. Describe specifically how you will gain the support, backing, and assistance of those company officers who do not consider training to be a high priority.
8. One of the responsibilities of this position will be to function as the on-scene safety officer at incidents. Have you had prior experience in this type of position? Also, explain your philosophy and approach to the position of scene safety officer.
9. How would you structure, document, and provide follow-up to ensure that all personnel complete required training?
10. How would you structure training sessions to accomplish training and maintain district coverage?
11. Assuming this department is a combination department of full-time and part-time personnel, which additional challenges do you anticipate in developing and conducting training for this type of structure?
12. What is your philosophy regarding the role of the company officer in training?

Written/practical evaluation questions may include the following items:

1. Prepare a written plan (not to exceed three pages—single-sided) describing how you will structure, conduct, and document the self-contained breathing apparatus (SCBA) training program.
2. Full-time personnel hired by the department are required to have prior certification as a Fire Fighter II. These individuals, once hired, are placed directly on shift. A current training issue is to develop a 40-hour program to provide an orientation to the department and complete a basic assessment of skill levels. Prepare a written plan (not to exceed three pages—single-sided) of how you would structure, conduct, and document a new fire fighter orientation/assessment program.
3. An important aspect of department training programs will be the addition of certification courses that are in compliance with the state's certification or training requirements. Prepare a written plan (not to exceed three pages—single-sided) detailing the steps that will be needed to implement an in-house fire apparatus engineer program.

Ethics Tip

Is it ethical to choose who gets your personal attention and support for success and who does not? This can be a painful question to answer. Although you want everyone to have equal opportunities and chances of success, you also have limited resources at your disposal. The reality is that you have an obligation to the fire service. Develop a process of screening to ensure that those persons who are most capable and willing to carry the torch to future generations are given additional support and mentoring.

Mentoring

Mentoring is a tool you can use to nurture and develop a fire fighter who has the basic skills and abilities to become a great fire service instructor. As a mentor, you select an individual to begin the process as either a formal or an informal protégé. Your role as a mentor is to demonstrate how to perform functions or skills and to explain and answer questions. Most importantly, you act as a sounding board if your protégé has a problem. The following ideas form the foundation for a solid mentoring program:

- The mentor's job is to promote intentional learning and to provide opportunities for the protégé to learn and grow. This does not mean that you throw your protégé to the wolves and walk away; rather, it means that you give your protégé opportunities to teach or demonstrate a skill while you watch.

- As a mentor, you need to share your experiences, both good and bad. Learning comes from both positive and negative experiences.
- The mentoring process takes time, and patience needs to be shown on the part of both the mentor and the protégé.
- Mentoring is a joint venture and requires both parties to be actively involved in the process.

Teaching Tip

Sometimes the identification and selection process does not work as anticipated. For example, you might select a person whom you feel is ready for mentoring, but he or she does not respond to the process or fails to perform as desired due to lack of skill or ability. In this scenario, there is still an opportunity to learn and grow from the experience by understanding why the process failed and then learning from it. This allows everyone to part on good terms and leaves the door open to future contact.

■ Coaching

Coaching is a process usually associated with sports. The coach is a person who has experience and assists players in developing and improving their skills. Coaching is most often tied to manipulative skills development and physical training. The difference between mentoring and coaching is subtle. Coaching is the process of helping an individual see his or her abilities and come up with solutions independently. A coach encourages through words but allows the player to develop his or her own skills. A good coach helps a player see his or her potential and then gives the player the freedom and support to achieve it.

As a fire service instructor, you must have expertise in numerous skills, such as firefighting skills for all levels you could be expected to teach. In addition to your technical skills, you have another set of skills at your disposal—namely, the skills used in the presentation of lessons, lesson and curriculum development, and use and development of evaluation tools. Part of the coaching process consists of allowing the fire fighter you are mentoring the opportunity to develop these skills as well.

Coaching also requires that you evaluate those persons whom you are grooming to replace you. These evaluations may take the form of either formal or informal evaluations or critiques. Just as when teaching a child to ride a bike, you must help your protégé up to a point, and then allow him or her to go on alone. You might run alongside the bike for a while, but as the child gets the hang of it, you back off. The same process applies when coaching a new fire service instructor: You teach your protégé how to do a presentation during training, and then the fire fighter gets to do it for real in a recruit class while you stand in the back of the room and watch. Afterward, you give some advice, highlighting good points and offering suggestions for rough areas. This feedback helps your protégé grow and improve. Coaching involves taking a "big brother" or "big sister" approach toward helping a person use the skills that he or she has and then coaching/teaching the person to do even better.

■ Sharing

The last skill in succession planning focuses on the basic concept of sharing. Sharing is a key part of mentoring and coaching. To be successful in succession planning, you must mentor and coach with a giving attitude. Not only do you need to take your protégé under your wing, but you also need to do so with a sharing attitude. You must be willing to share everything that has made you such a successful fire service instructor. This arsenal of experiences includes your skills and knowledge, as well as all of the little nuances that have made your job easier and fun. All too often, those persons charged with the task of training their replacements hold back the little tricks of the trade. Some have the attitude, "I had to learn it myself; so can you!" This attitude hurts both the department and your reputation. Share your knowledge and insight with your protégé as he or she takes on the responsibilities of training others. Give your protégé the extra step up rather than making him or her repeat all of your mistakes and waste all of that extra time and effort. When you share what you know with others and help others excel, you benefit both personally and professionally.

When you approach succession planning with this positive attitude and mentor the next generation of fire service instructors by teaching and coaching with a caring and sharing attitude, both you and your department move forward. Leave a legacy of professionalism that will outlive your time on the job.

Training BULLETIN

Jones & Bartlett Fire District

Training Division
5 Wall Street, Burlington, MA, 01803
Phone 978-443-5000 Fax 978-443-8000
www.fire.jbpub.com

Instant Applications: The Learning Never Stops

Drill Assignment

Apply the chapter content to your department's operation, training division, and your personal experiences to complete the following questions and activities.

Objective

Upon completion of the instant applications, fire service instructor students will exhibit decision making and application of job performance requirements of the fire service instructor using the text, class discussion, and their own personal experiences.

Suggested Drill Applications

1. Identify at least three statewide organizations and three national organizations that are available to assist all levels of instructors in their professional development.

2. Review your local or state certification process to learn which affiliations or correlations exist in accordance with the Fire and Emergency Services Higher Education (FESHE) program, discussed in this chapter; "The 16 Firefighter Life Safety Initiatives" developed by the National Fallen Firefighters Foundation; and NFPA standards relating to training and education.

3. Review the JPRs specific to your next level of instructor certification, and identify opportunities to experience learning in those areas.

4. Select an instructor or instructor candidate who has specific talents or background that can be used to improve a particular course. Urge that person to become involved in the training program.

5. Review the Incident Report in this chapter and be prepared to discuss your analysis of the incident from a training perspective and as an instructor who wishes to use the report as a training tool.

Incident Report

© Greg Henry/ShutterStock, Inc.

Parsippany, New Jersey—1992

At the end of a Fire Fighter I program, search and rescue drills were being conducted in a converted school bus by members of the Greystone Park Fire Department for members of several area departments.

The windows of the converted school bus were covered with welded steel plates, and the interior seats were removed. The fire department that owned the bus used it for hot smoke training, normally using kerosene-soaked wood chips in drums. This practice was noncompliant with the legal requirements of the county where the training site was located.

Shredded paper was added to a couch in the bus and ignited with a road flare. The fire was allowed to burn for about 10 minutes to produce heat and smoke in the bus. The students were told the purpose of the drill and were instructed to enter the bus through the rear emergency exit carrying a forcible entry tool, pass the burning couch, conduct a "primary search," and then exit through the front bifold bus door.

The instructor followed three students into the bus, and two more students followed behind them, leaving the back door open. After roughly 2 minutes, a flashover occurred in the bus. The instructor and the last two students escaped through the back door without injury. Students trying to exit through the front door were thwarted when the door became jammed, and students on the outside pried the door open with forcible entry tools. Two students escaped, but interior conditions were so bad that their protective clothing was smoking.

At this point, the sound of the third student's low air alarm echoed through the bus. Students pulled a preconnected 1½″ (38.1-mm) hose off the engine, but the hose lines did not have nozzles. Another student used the booster line and extinguished the fire from the back door. Without donning his protective gloves or SCBA, the ignition officer entered through the front door, found the trapped student against the front partition to the stairs, burned his hands when trying to remove the student, and had to exit. The lead instructor and a student then entered through the front door and tried to drag the student out through the front door, but were unable to do so due to his size. Another student, in full PPE, entered through the rear door and was able to drag him out that way.

The injured student went into respiratory arrest and EMS was summoned. He was revived and transported, after having received second- and third-degree burns to roughly 30 percent of his body, as well as respiratory burns. The other two fire fighters who were initially trapped both received second- and third-degree burns, one to 15 percent of his body and the other to 20 percent of his body.

The fire was investigated by the county prosecutor's office, and eventually taken to a grand jury, although a criminal indictment was not issued. The lead investigator for the prosecutor's office said, "This training facility violated all applicable standards, such as those governing emergency ventilation, emergency lighting, fuel sources, communications, and emergency evacuation."

Continued...

Incident Report Continued...

Parsippany, New Jersey—1992

Postincident Analysis: Parsippany, New Jersey

NFPA 1403 (2007 Edition) Noncompliant

Student-to-instructor ratio was 17:1 (7.5.2)

No written emergency plan (7.2.24.3)

Students not informed of emergency plans/procedures (7.2.24)

Interior personnel without hose line (7.4.8)

No established water supply (7.2.22)

No incident safety officer (7.4.1)

Flashover and fire spread unexpected (7.3.5)

No incident commander or instructor-in-charge (7.5.4)

Absence of ventilation to prevent flashover (7.3.6)

Ingress and egress routes not monitored (7.2.23.5)

No walk-through by students for familiarization, escape routes, etc. (7.2.25)

No EMS standby (7.4.11)

No inspection of props to ensure safety devices were operable (front vehicle passenger door) (7.2.8)

© Greg Henry/ShutterStock, Inc.

Wrap-Up

Chief Concepts

- Learning is a lifelong process. An important part of becoming a great fire service instructor is having the basic desire to always improve and to become a better fire service professional.
- Professional development can take the form of training, certifications, and formal education.
- Professional development can also be achieved by participating in local, state, and national organizations and by attending training conferences.
- Networking is a powerful tool in your toolbox that can open the door to sharing or collaborating on an idea and developing a new methodology or course.
- For all fire service instructors, it is essential to keep their teaching skills sharp and current. Also, as you grow in your knowledge, stay current with the types of students entering the fire service. Test yourself by teaching a class and be honest with yourself about how effective you were. Great fire service instructors are lifelong teachers who love to teach.
- Succession planning is the process of developing individuals who can be mentored and coached to take the lead in the future of the fire department.
- One benefit of training your replacement is that you can select someone who shares your own ideals and goals, thereby ensuring that your training program will continue after you leave. Through this legacy, your influence could be felt for many decades in the department.

Hot Terms

<u>Coaching</u> The process of helping individuals develop skills and talents.

<u>Identifying</u> The process of selecting those persons whom the fire service instructor would like to mentor, coach, and develop.

<u>Mentoring</u> A relationship of trust between an experienced person and a person of less experience for the purpose of growth and career development.

<u>Networking</u> An activity or process of like-minded individuals meeting and developing relationships for professional and personal growth through sharing of ideas and beliefs.

<u>Sharing</u> The basic concept of giving to others with nothing expected in return.

<u>Succession planning</u> The act of ensuring the continuity of the organization by preparing its future leaders.

References

National Fire Protection Association. (2012) *NFPA 1041: Standard for Fire Service Instructor Professional Qualifications.* Quincy, MA: National Fire Protection Association.

National Fire Protection Association. (2007) *NFPA 1403: Standard on Live Fire Training Evolutions.* Quincy, MA: National Fire Protection Association.

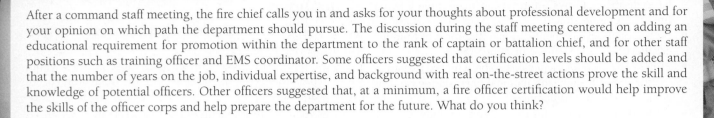

After a command staff meeting, the fire chief calls you in and asks for your thoughts about professional development and for your opinion on which path the department should pursue. The discussion during the staff meeting centered on adding an educational requirement for promotion within the department to the rank of captain or battalion chief, and for other staff positions such as training officer and EMS coordinator. Some officers suggested that certification levels should be added and that the number of years on the job, individual expertise, and background with real on-the-street actions prove the skill and knowledge of potential officers. Other officers suggested that, at a minimum, a fire officer certification would help improve the skills of the officer corps and help prepare the department for the future. What do you think?

1. What is the process of training your replacement called?

   A. Mentoring

   B. Coaching

   C. Succession planning

   D. Promotion

2. What is the process of helping a fire fighter to learn and develop a new skill by helping the fire fighter to set his or her own goals called?

   A. Mentoring

   B. Coaching

   C. Promotion

   D. Evaluation

3. What is the goal of a mentor?

   A. To provide intentional opportunities for his or her protégé to learn and grow

   B. To evaluate his or her protégé to determine if that individual is ready to take the helm

   C. To identify individuals who may be able to lead the department

   D. To develop every individual in the department

4. Which of the following is a vast new source of information that is available but is not always reliable?

   A. Fire Emergency Training Network

   B. *Firehouse* magazine

   C. YouTube

   D. U.S. Fire Administration Web site

5. Which of the following is not a method of professional development?

   A. Reading *The Wall Street Journal*

   B. Attending fire service conferences and seminars

   C. Reading fire-related material on the Internet

   D. All are methods of professional development

6. Which of the following is *not* a trait to look for when selecting a replacement?

   A. Someone who is willing to learn

   B. Someone who works for someone else

   C. Someone who is friendly

   D. Someone who has humility

Resources for Fire Service Instructors

■ Contents

- Training Record Attendance Report
- Training and Education Report
- Sample Training Policy
- Instructor Training and Experience Validation Form
- Fire Fighter Skill Performance Ratings Template
- Examples of Verbs Often Used to Write Learning Competencies
- Personal Improvement Agreement
- Outside Training Request Form
- Firefighter Performance Expectations
- JPR Skill Sheet Template
- Resources for Training Officers

Training Record
Attendance Report

Date_____   Station(s)_____   Description_____

Start Time_____   End Time_____   Credit Hours (Total Time)_____

Method of Training: ☐ Classroom   ☐ Practical   ☐ Self-Directed   Certification Credit: ☐ Yes ☐ No

Lead Instructor_____   Additional Instructor(s)_____

Print Name	Dept ID #	Signature	Hours Attended

Objectives:_____

Description of Training (Notes):_____

_____

_____

_____   _____
Instructor Signature   Training Officer Approval

Type of Training				
Company	Multi-Comp.	Officer	Mutual Aid	Night
Tower Burn	Classroom	Practical	Combo.	Driver

Equipment Used in Training Session	Feet of 1 ¾″ Hose Used	Feet of 2 ½″ Hose Used	Supply Hose Used / Ft.	Feet of Ladders	Number of Engines	Number of Trucks	Gallons of Water	Number of SCBA	Total Number of Firefighters

FIGURE AA-1

Training and Education Report

Name_____   ID#_____ Date of Birth_____

S/S#_____   Date Entered Fire Service_____ Date of Hire_____

Drivers License# _____   Expiration Date_____ D/L Classification_____

Certification Record
Indicate date of Certification in Boxes Below

Basic Fire fighter	Advanced Fire fighter	Apparatus Operator (FAE)	Hazmat Awareness
Hazmat Operations	Hazmat Technician	Hazmat Specialist	Hazmat Incident Command
Technical Rescue Awareness	NFPA 1006–Trench Ops	NFPA 1006–Trench Tech	NFPA 1006–Rope
NFPA 1006–Rope 2	NFPA 1006–Confined Space	Structural Collapse Operations	Structural Collapse Technician
Instructor 1	Instructor 2	Instructor 3	Training Program Manager
Provisional Fire Officer 1	Fire Officer 1	Provisional Fire Officer 2	Fire Officer 2
Provisional Fire Officer 3	Fire Officer 3	Fire Prevention Officer	Vehicle Machinery Operations
Vehicle Machinery Technician	Water Rescue Operations	Watercraft Technician	IS 100 (IS100)
IS 200 (IS200)	IS300 (IS300)	IS400 (IS400)	IS 700 (IS700)
IS 800 (IS800)	Emergency Medical Technician (EMT)	Advanced Emergency Medical Technician (AEMT)	Paramedic

Formal Education Record

High School Attended_____ Year Graduated_____ GED_____

College Attended_____ Credits Earned_____ Degree_____

Advanced Degree_____ Credits Earned_____ Degree_____

Advanced Degree_____ Credits Earned_____ Degree_____

FIGURE AA-2

Sample Training Policy

PURPOSE AND SCOPE:

This directive provides definition as to the training responsibilities of each rank within the department. This directive also establishes the basic format of the department's training program, including training requirements and documentation.

RESPONSIBILITIES:

TRAINING OFFICER
The Training Officer is responsible for the overall administration and management of the department's Training Division. Through working with the staff and line officers, the Training Officer is to develop and implement a comprehensive yearly training plan. The Training Officer has overall responsibility and accountability to ensure that department training activities are current and consistent with applicable standards and practices.

INSTRUCTOR-IN-CHARGE
The Instructor-in-Charge is the designated person who is responsible for the overall delivery of a specific lesson plan or training objective. All resources assigned to the training session are under the responsibility of the Instructor-in-Charge, including personnel, apparatus, and facilities and equipment. The use of all required and best-practice safety procedures throughout the training session will be the responsibility of the Instructor-in-Charge. All elements of documentation will also be overseen by the Instructor-in-Charge.

INSTRUCTOR(S)
The instructor is responsible for assisting the Instructor-in-Charge at any high-risk training event or routine training session where it has been determined that additional instructors are needed. This may be for the purpose of safety, accountability, or reduction in the student-to-instructor ratio. The instructor is responsible for delivering lesson plans, evaluating performance, and providing feedback and documentation of the evolution or objectives as necessary. The instructor will be responsible for the safety of the members that are assigned to them during the evolution.

SHIFT COMMANDER(S)
The Shift Commander is responsible for administering and monitoring the department training plan within his or her assigned shift. This includes assisting in the coordination, presentation, and evaluation of specific department level training sessions. The Shift Commander should make every attempt to attend high-risk training events, training events based on procedural operations where his or her presence is needed to simulate operations, and any other training event where he or she can assist in the evaluation of member performance.

Through periodic evaluation of companies and/or individuals during training and emergency operations, the shift Deputy Commander is responsible for identifying training deficiencies and providing recommendations to the Training Officer regarding the specific training needs of his or her shift.

FIGURE AA-3

CAPTAIN(S)/OTHER MID-LEVEL SHIFT SUPERVISOR(S)

The Captain is responsible for implementing and monitoring the department's training plan within his or her assigned area of authority (example—shift 2, west side). This includes reviewing monthly training reports to ensure company and individual compliance with training assignments. He or she also may be assigned by the Training Officer to assist in the coordination, presentation, and evaluation of specific department level training sessions. The Captain is responsible for completing the training responsibilities of a Company Officer.

COMPANY OFFICER(S)/ACTING COMPANY OFFICER(S)

The Company Officer is the key to the department's training program. He or she is the individual most responsible for the training and readiness of personnel. The Company Officer is required to complete monthly training assignments and submit all necessary documentation, including training reports and skills checklists. The Company Officer is to coordinate the various daily company activities so that training assignments are completed. He or she is responsible for coordinating company level training so that all members receive the training regardless of time off, vacations, Kelly days, etc.

The monthly training assignments represent the minimum of what must be done. The Company Officer is not limited to this, as each individual has strengths and weaknesses which must be addressed by the Company Officer. It is the responsibility of the Company Officer to improve the performance of the personnel assigned to him or her and to foster an environment that encourages the company toward continuous improvement.

The monthly Training Calendar and Company Officers' Training Packet lists the assigned training activities for the month. Some activities will have specific dates and/or time periods designated. For company level training assignments, the Company Officer has the authority to vary from the published schedule, if necessary for valid reasons. The Company Officer shall be responsible for scheduling and completing the training. The objective is for all training assignments to be completed by the end of each month.

The Company Officer is responsible for the safety of his or her personnel while training.

The Company Officer is responsible for maintaining the licensure required of the position. This includes EMT-B or Paramedic and appropriate drivers license classification.

FIRE FIGHTERS

Department fire fighters are expected to maintain a high level of preparedness through regular training and individual study. This includes keeping current on both fire service and departmental changes and notifying the Company Officer of training needs.

Fire fighters are responsible for participating in an aggressive, safe, and positive manner in all classroom and practical training.

Fire fighters are responsible for maintaining the licensure required of the position. This includes EMT-B or Paramedic and appropriate drivers license classification.

FIGURE AA-3

MONTHLY TRAINING ASSIGNMENTS

The Training Division is responsible for providing monthly training assignments through the Training Bulletin, Training Calendar, and the Company Officers' Training Packet issued at the beginning of each month. This shall specify the assigned and make-up dates for training (if applicable), who is required to attend, and other necessary information.

Monthly training will be classified as follows:

MANDATORY

Training that must be accomplished by all members. This may also include 40-hour personnel. Mandatory training will include Federal- and State-mandated courses and courses deemed as mandatory by the Anytown Fire Department. These training sessions and make-up sessions will be scheduled by the Training Division and coordinated through the Shift Commanders.

REGULAR

Company level—Training that is to be completed by each company during the month. Company training designated as part of the Essential Skills program shall be completed by each company member. Other types of company level training will specify whether make-up sessions are required for personnel who miss the training. Company Officers are responsible for monitoring this, and for scheduling and conducting any necessary make-up training.

Department level—Training that is primarily scheduled and conducted through the Training Division. Department level training that is designated as mandatory will have make-up sessions scheduled by the Training Division.

DOCUMENTATION OF TRAINING

The Training Officer shall be responsible for providing to the Shift Commanders and Captains a report as to the completion of training assignments from the previous month. The report shall be provided within the first 10 days of the next month. The Shift Commanders and Captains are responsible for following up with their Training Officers on the training that has not been completed.

The instructor of a specific training activity is responsible for completing and submitting the training report to the Training Division.

The Company Officer is responsible for ensuring that all skills checklists that may be required as part of a training assignment are completed and submitted to the Training Division.

For individual-type training activities (i.e. independent study, reviewing fire service publications, physical fitness, etc.) the individual is responsible for completing the report and submitting it to his or her Company Officer for signature.

Individuals attending classes outside of the fire department (i.e. fire officer classes, tactics seminars, etc.) are responsible for completing a training report upon returning from the course.

BY ORDER OF: _____

DATE: _____

FIGURE AA-3

Instructor Training and Experience Validation

Name: _____ Department: _____

Years of Fire Service Experience: _____ Rank: _____

Certification Levels

☐ Instructor 1 ☐ Instructor 2 ☐ Instructor 3 ☐ Instructor 4 ☐ TPM ☐ Advanced Degree

☐ Fire Officer 1 ☐ Fire Officer 2 ☐ Fire Officer 3 ☐ EFO ☐ Other

> Copies of certifications must be available upon request

Previous Live Fire Training Educational Experience

☐ NFPA 1403 course/class at state training facility, local academy, or agency

☐ Attended FDIC Live Fire Training programs

☐ Attended other national conference live fire training courses

> Copies of course/class completion certificates are available upon request

Previous Live Fire Training Practical Experience

Approximate # of Live Fire Training Exercises conducted at:

☐ **Training Tower (Class A)**   ☐ **Training Tower (Gas Fired)**

☐ **Acquired Structures**   ☐ **Container/Simulator/Portable Unit**

☐ **Exterior Burn Prop**   ☐ **Class B Fuel Burn**

☐ **Approximate # of Years of Live Fire Training Experience**

Validation / Attestation Statement

I have read and understand all components of NFPA 1403, Standard on Live Fire Training Evolutions, and agree to abide by all requirements specified for the positions/roles that I am assigned for this training event. I understand and accept responsibility for the position requirements for the activities I am assigned to complete and do so knowing the hazards and dangers associated with these assignments.

_____
Signature   Date

FIGURE AA-4

Fire Fighter Skill Performance Ratings

- <u>Unskilled (0)</u>
    - ○ Member failed evolution or skill (Mandatory re-evaluation will take place)
        - ▪ Exceeded time limit
        - ▪ Missed step in procedure
        - ▪ Created safety hazard to self or other member
        - ▪ Unable to perform task
        - ▪ Repeated failure of task attempt
    - ○ Requires Personal Improvement Agreement and formal documentation on standard evaluation form
    - ○ No credit is given for purpose of progress reporting or evaluation towards applicable certification

- <u>Moderately Skilled (1)</u>
    - ○ Performance meets the minimum requirement for the task and is performed on first attempt. With additional practice, improved performance levels can be attained
    - ○ Member's performance of skill may require supervision on fireground
    - ○ Time performance near minimum requirement
    - ○ All appropriate safety precautions are taken
    - ○ Instructor/evaluator determines need for additional training or repeat of skill

- <u>Skilled (2)</u>
    - ○ Performance is above the minimum level because:
        - ▪ The time was above average
        - ▪ Skill meets all performance levels and could be performed on fireground without supervision
        - ▪ No serious/critical errors were committed
        - ▪ All appropriate safety practices were observed and performed

- <u>Highly Skilled (3)</u>
    - ○ Performance is at a high level of competence because:
        - ▪ It was error free
        - ▪ It was the fastest time
        - ▪ Member could supervise others doing same task and identify errors or suggest improvements
        - ▪ Member knows the role and importance of this task in relation to other fireground operations

Example

Subject	Skill	Subset	Equipment	Performance Rating
Forcible Entry	Force an Inward-Swinging Door	Hand Tools	Halligan Bar/Flat Head Axe	3

FIGURE AA-5

Data from Miami-Dade College

Sample Objective Verbs by Domain	
Domain Level	**Verbs**
Affective Domain	
Receiving	Ask, choose, describe, follow, identify, select
Responding	Answer, comply, conform, discuss, read, write
Valuing	Complete, explain, initiate, justify, propose, share
Organization	Adhere, defend, integrate, modify, organize, synthesize
Value Complex	Act, discriminate, influence, qualify, question, serve
Psychomotor Domain	
Observation	Observe, view, watch
Imitation	Adjust, assemble, change, clean, combine, connect, display, operate
Manipulation	Arrange, control, design, fasten, fix, follow, maintain, organize, support, work
Precision	Administer, derive, identify, locate, manipulate, modify
Articulation	Conduct, document, encircle, regulate, set, transfer, vocalize, work
Cognitive Domain	
Knowledge (Remember)	Define, describe, identify, label, list, match, name, outline, provide, select, state, reproduce
Comprehension (Understand)	Acquire, distinguish, estimate, examine, explain, extend, give, infer, paraphrase, summarize
Application (Apply)	Apply, change, compute, create, develop, modify, predict, prepare, relate, solve, use
Analysis (Analyze)	Break down, correct, diagram, differentiate, discriminate, discuss, illustrate, outline, relate, separate, study, subdivide
Synthesis (Create)	Categorize, combine, compose, devise, generate, organize, plan, reconstruct, reorganize, rewrite, sequence, summarize
Evaluation (Evaluate)	Apprise, compare, conclude, contrast, criticize, diagnose, enhance, interpret, justify, research, support
Fire Service-Specific Verbs	
Activate, advance, calculate, charge, climb, connect, construct, cut, describe, discharge, disconnect, doff, don, draft, fill, fold, force, hoist, load, lower, lubricate, maintain, maneuver, name, perform, prepare, pull, pump, raise, recharge, refill, reset, search, tie, transmit, transport, unload, ventilate	

FIGURE AA-6 Examples of Verbs Often Used to Write Learning Competencies

Division of Training
Personal Improvement Agreement

Agreement Initiated By: **Today's Date:**

Members Name   **Department ID#:**

I. Concerns/Area Needing Improvement: (Description of behavior, situation or objective causing PIA)

II. Standard Reference: (What is the acceptable level of performance, attitude, conduct or ability?)

III. Shift Commander / Station Officer Action Plan: (What the officer will do help improve the performance)

IV. Personal Action Plan: (Written by the member describing what they will do to improve themselves)

Initial Follow-up date

V. Document Action Plan Progress: (How we are doing and progress check dates)

Completion date

VI. Has the area of concern been corrected? (Has improvement been seen in this area?)

Yes:

No:

VII. On-Going Monitoring: (Describe how monitoring of issue will take place and for what duration)

Firefighter Name (print): _____ Agreement Date:

Station Officer Name (print): _____

Shift Commander Name (print): _____

Training Division Review:

FIGURE AA-7

Outside Training Request Form	Division of Training

Out of Department School Request Form

ATTACH CLASS REGISTRATION FORM TO REQUEST

Member Information		
Member Name	Date of Request	Shift
Station	Company Officer Name	Dates of Course
Course Title	Course Location	Course Sponsor
Class Fee	Hours of Program	Is This A Certification Course?

Member Signature	Signed:	

Training School Request Acknowledgements		
Company Officer Signature	Battalion Chief Signature	Special Teams Leader* (*if applicable)

Training Division Only	Approved	Denied	Hold	Comments
	Training Officer Signature			

- -

Payment Information / Registration Tracking (Training Division Use)		
Department Purchase Order #	Reimbursable (Y / N)	Method of Payment
Other Members Attending (Yes #/No)	Registration Completed	Member Notification Date / Method
FireHouse Class Code Created	Member Instructions Issued	Attendance Confirmation Received
Certificate Received	Certification Received	Monthly Report Entry

Logistics and Out of Town Information on Reverse

FIGURE AA-8A

Outside Training Request Form	Division of Training

Out of Town / Overnight Travel Information (Completed by Training Division)

Equipment Issued		
Helmet		Personal Flotation Device
Eye Protection		Buoyancy Compensator
Hearing Protection		Mask/Snorkel/Fins
Gloves *(Type):*		SCUBA
Bunker Coat		Other Resp. Protection *(Type):*
Hood		HazMat CPC *(Type):*
Bunker Pants		SCBA
Safety Boots		Spare Cylinders
Radio / Spare Batteries		

Transportation		
Department Vehicle Authorized	Personal Vehicle Use	Mileage Reimbursement Eligible *Reimbursement for mileage at approval of Chief
Rental Car Agency	Payment Method	Airfare, Rail, Other
Reimbursements to Employee	Secondary Billing (SUFD, other)	

Out of Town Arrangements		
Hotel Name	Confirmation Number	Direct Billing Arranged (Method)
Hotel Address	Hotel Phone	Check-in / Check-out Dates
Meals in Conference Fee	Per Diem Rate / Days / Total	TOTAL EXPENSES (Estimated)

FIGURE AA-8B

Firefighter Performance Expectations

Work Ethic

- Actively seeks academic and technical knowledge for self-improvement
- Completes tasks assigned without shortcuts and without repeating of tasks
- Work is complete, thorough, and done in a professional manner
- Actively seeks out additional work as it improves the team's ability to thrive
- Accomplishes tasks or goals with a "safety-first" attitude
- Keeps commitments and meets deadlines
- Can be trusted with confidential information
- Can be trusted with the property of others
- Is committed to help at fire department events off duty as necessary

Judgement & Problem Solving

- Makes reasonable and safe decisions when attempting to accomplish a task or solve a problem
- Approaches problems in a safe, logical and well thought-out fashion
- Seeks pro-active solutions to problems
- Applies critical thinking skills to complex and varied situations

Time Management

- Consistently punctual and completes assignments on time
- Manages work so that quality of work is satisactory and not hurried, incomplete or overwhelming to self and team

Teamwork / Interpersonal Skills

- Places the success of the team above self-interest
- Effectively works with others in order to accomplish tasks or solve problems
- Offers to help other company members
- Understands and follows chain-of-command
- Is courteous and respectful of peers and supervisors
- Does not undermine team
- Helps and supports other team members

Adaptability / Stress Management

- Remains calm in stressful situations
- Adapts behavior in order to deal with changing situations in a safe manner
- Adapts behavior in order to accomplish individual and department goals
- Recognizes symptoms of stress in self and others and seeks to deal with stress appropriately
- Communicates with others to resolve problems
- Remains flexible and open to change

FIGURE AA-9

Practical Competence / Physical Ability

- Demonstrates a desire to develop skills that are above minimal performance levels
- Can accomplish multiple tasks in succession
- Can retain and recall previously mastered skills
- Strives to improve practical abilities
- Knows all applicable safety behaviors and actions related to practical skills
- Maintains a high level of physical fitness, dexterity, flexibilty and strength through on-going fitness program participation

Communication

- Uses appropriate tone of voice
- Articulates in a clear, logical and understandable manner
- Displays confidence in message
- Is persuasive and makes a positive impression
- Demonstrates appropriate non-verbal communication techniques
- Avoids letting stress control a communication process or method
- Writes legibly using correct grammar and punctuation
- Listens actively

Initiative / Motivation / Decisiveness

- Accomplishes tasks or goals without being ordered, coerced or motivated by others
- Demonstrates desire for personal and professional development
- Makes decisions definitively and consistently

Empathy

- Shows compassion for others and responds appropriately to heightened emotional responses
- Demonstrates a calm, compassionate and helpful demeanor towards those in need
- Mindful of the impact of his/her demeanor on those in need, family, bystanders, and other members of the public

Community Awareness

- Exercises compassion and willingness to help persons in various situations with varied backgrounds
- Is sensitive to individual and cultural differences
- Knows the role a member of the department represents to the community

Appearance and Personal Hygiene

- Always clean, neat, well-groomed and in good personal hygiene
- Always wears appropriate uniforms in excellent condition

XYZ Fire Department / Academy In-Service Training Program *JPR Skill Sheet*	DESCRIPTION: This JPR Training Guideline references the format identified in NFPA Professional Qualification Standard series. Knowledge, skill, performance and topic description are referenced from the applicable state and local certification objectives to meet the JPR. Local application of each training guideline should be made by instructors utilizing department procedures or guidelines.

Duty Area	Level	Subject		JPR #
Text Reference			Training Module	

Job Performance Requirement:

Objective Number	Skill / Knowledge / Performance / Topic Description	Reference	Standard	Validated
			Pass/Fail	
			Pass/Fail	
			Pass/Fail	
			Pass/Fail	
			Pass/Fail	
			Pass/Fail	
			Pass/Fail	

Critical Safety Points	Required Equipment
• • •	• • •

Instructor Notes:

No.	TASK STEP	FIRST TEST		RETEST	
		Pass	Fail	Pass	Fail
1.					
2.					
3.					
4.					
5.					
6.					
7.					
8.					
9.					
10.					

FIGURE AA-10

Table AA-1	Resources for Training Officers
Driver Training	
USFA Emergency Vehicle Driver Training program	www.usfa.fema.gov—publications section of Web site
Fire apparatus and traffic safety related information	www.respondersafety.com
Driver safety information	www.drivetosurvive.org
Sample apparatus driving SOGs	www.vfis.com
Company Officer Development	
IAFC Officer Development Handbook	http://www.iafc.org/associations/4685/files/OffrsHdbkFINAL3.pdf
Wildland Fire Leadership Development	http://www.fireleadership.gov
Miscellaneous Resources	
National Fire Academy library	www.lrc.fema.gov
Fire Fighter Close Calls	www.firefighterclosecalls.com
National Fire Fighter Near Miss Reporting System	www.firefighternearmiss.com
Everyone Goes Home	www.everyonegoeshome.org
With the Command	www.withthecommand.com
Training Resources and Data Exchange (TRADE)	feti.lsu.edu/municipal/NFA/TRADE
Regulatory Agencies and Reporting Agencies/Investigations	
OSHA	www.osha.gov
U.S. Department of Labor	www.dol.gov
U.S. Environmental Protection Agency	www.epa.gov
NIOSH Fire Fighter Fatality Reports	www.cdc.gov/niosh/fire
NFPA Standards	
National Fire Protection Association (NFPA)	www.nfpa.org
Organizations and Schools	
Fire Department Safety Officers Association	www.fdsoa.org
International Association of Fire Chiefs (IAFC)	www.iafc.org
International Association of Fire Fighters (IAFF)	www.iaff.org
International Society of Fire Service Instructors (ISFSI)	www.isfsi.org
International Fire Service Certification Congress (IFSAC)	www.ifsac.org
National Fire Academy Degree at a Distance Program	www.usfa.dhs.org/hfa/higher_ed/index.shtm
National Volunteer Fire Council	www.nvfc.org
ProBoard Fire Service Professional Qualifications System (NPQS)	theproboard.org
Fire Service Publications	
Fire Engineering Magazine	www.fireengineering.com
Firehouse Magazine	www.firehouse.com
FireRescue Magazine	www.firefighternation.com
Online Magazines and Training Resource Web Sites	
Bshifter	www.bshifter.com/SelectMagazines.aspx
The Fire Training Toolbox	www.firetrainingtoolbox.com
Making a Difference Training Resources	www.thetrainingofficer.com
Green Building Construction	www.greenmaltese.com
Underwriters Laboratory Fire Research	www.ul.com/fireservice
National Institute for Standards and Technology	www.fire.gov www.nist.gov/fire
U.S. Fire Administration	www.usfa.dhs.gov
National Highway Traffic and Safety Administration	www.nhtsa.dot.gov

An Extract from: NFPA 1041, Standard for Fire Service Instructor Professional Qualifications, 2012 Edition

■ Chapter 4 Instructor I

4.1 General.

4.1.1 The Fire Service Instructor I shall meet the JPRs defined in Sections 4.2 through 4.5 of this standard.

4.2 Program Management.

4.2.1 Definition of Duty. The management of basic resources and the records and reports essential to the instructional process.

4.2.2 Assemble course materials, given a specific topic, so that the lesson plan and all materials, resources, and equipment needed to deliver the lesson are obtained.

(A) Requisite Knowledge. Components of a lesson plan, policies and procedures for the procurement of materials and equipment, and resource availability.

(B) Requisite Skills. None required.

4.2.3 Prepare requests for resources, given training goals and current resources, so that the resources required to meet training goals are identified and documented.

(A) Requisite Knowledge. Resource management, sources of instructional resources and equipment.

(B) Requisite Skills. Oral and written communication, forms completion.

4.2.4 Schedule single instructional sessions, given a training assignment, department scheduling procedures, instructional resources, facilities and timeline for delivery, so that the specified sessions are delivered according to department procedure.

(A) Requisite Knowledge. Departmental scheduling procedures and resource management.

(B) Requisite Skills. Training schedule completion.

4.2.5 Complete training records and report forms, given policies and procedures and forms, so that required reports are accurate and submitted in accordance with the procedures.

(A) Requisite Knowledge. Types of records and reports required, and policies and procedures for processing records and reports.

(B) Requisite Skills. Basic report writing and record completion.

4.3 Instructional Development.

4.3.1* Definition of Duty. The review and adaptation of prepared instructional materials.

4.3.2* Review instructional materials, given the materials for a specific topic, target audience, and learning environment, so that elements of the lesson plan, learning environment, and resources that need adaptation are identified.

(A) Requisite Knowledge. Recognition of student limitations and cultural diversity, methods of instruction, types of resource materials, organization of the learning environment, and policies and procedures.

(B) Requisite Skills. Analysis of resources, facilities, and materials.

4.3.3* Adapt a prepared lesson plan, given course materials and an assignment, so that the needs of the student and the objectives of the lesson plan are achieved.

(A)* Requisite Knowledge. Elements of a lesson plan, selection of instructional aids and methods, and organization of the learning environment.

(B) Requisite Skills. Instructor preparation and organizational skills.

4.4 Instructional Delivery.

4.4.1 Definition of Duty. The delivery of instructional sessions utilizing prepared course materials.

4.4.2 Organize the classroom, laboratory, or outdoor learning environment, given a facility and an assignment, so that lighting, distractions, climate control or weather, noise control, seating, audiovisual equipment, teaching aids, and safety are considered.

(A) Requisite Knowledge. Classroom management and safety, advantages and limitations of audiovisual equipment and teaching aids, classroom arrangement, and methods and techniques of instruction.

(B) Requisite Skills. Use of instructional media and teaching aids.

4.4.3 Present prepared lessons, given a prepared lesson plan that specifies the presentation method(s), so that the method(s) indicated in the plan are used and the stated objectives or learning outcomes are achieved, applicable safety standards and practices are followed, and risks are addressed.

(A)* Requisite Knowledge. The laws and principles of learning, methods and techniques of instruction, lesson plan components and elements of the communication process, and lesson plan terminology and definitions; the impact of cultural differences on instructional delivery; safety rules, regulations, and practices; identification of training hazards; elements and limitations of distance learning; distance learning delivery methods; and the instructor's role in distance learning.

(B) Requisite Skills. Oral communication techniques, methods and techniques of instruction, and utilization of lesson plans in an instructional setting.

4.4.4* Adjust presentation, given a lesson plan and changing circumstances in the class environment, so that class continuity and the objectives or learning outcomes are achieved.

(A) Requisite Knowledge. Methods of dealing with changing circumstances.

(B) Requisite Skills. None required.

4.4.5* Adjust to differences in learning styles, abilities, cultures, and behaviors, given the instructional environment, so that lesson objectives are accomplished, disruptive behavior is addressed, and a safe and positive learning environment is maintained.

(A)* Requisite Knowledge. Motivation techniques, learning styles, types of learning disabilities and methods for dealing with them, and methods of dealing with disruptive and unsafe behavior.

(B) Requisite Skills. Basic coaching and motivational techniques, correction of disruptive behaviors, and adaptation of lesson plans or materials to specific instructional situations.

4.4.6 Operate audiovisual equipment and demonstration devices, given a learning environment and equipment, so that the equipment functions properly.

(A) Requisite Knowledge. Components of audiovisual equipment.

(B) Requisite Skills. Use of audiovisual equipment, cleaning, and field level maintenance.

4.4.7 Utilize audiovisual materials, given prepared topical media and equipment, so that the intended objectives are clearly presented, transitions between media and other parts of the presentation are smooth, and media are returned to storage.

(A) Requisite Knowledge. Media types, limitations, and selection criteria.

(B) Requisite Skills. Transition techniques within and between media.

4.5 Evaluation and Testing.

4.5.1* Definition of Duty. The administration and grading of student evaluation instruments.

4.5.2 Administer oral, written, and performance tests, given the lesson plan, evaluation instruments, and evaluation procedures of the agency, so that bias or discrimination is eliminated, the testing is conducted according to procedures, and the security of the materials is maintained.

(A) Requisite Knowledge. Test administration, agency policies, laws and policies pertaining to discrimination during training and testing, methods for eliminating testing bias, laws affecting records and disclosure of training information, purposes of evaluation and testing, and performance skills evaluation.

(B) Requisite Skills. Use of skills checklists and oral questioning techniques.

4.5.3 Grade student oral, written, or performance tests, given class answer sheets or skills checklists and appropriate answer keys, so the examinations are accurately graded and properly secured.

(A) Requisite Knowledge. Grading methods, methods for eliminating bias during grading, and maintaining confidentiality of scores.

(B) Requisite Skills. None required.

4.5.4 Report test results, given a set of test answer sheets or skills checklists, a report form, and policies and procedures for reporting, so that the results are accurately recorded, the forms are forwarded according to procedure, and unusual circumstances are reported.

(A) Requisite Knowledge. Reporting procedures and the interpretation of test results.

(B) Requisite Skills. Communication skills and basic coaching.

4.5.5* Provide evaluation feedback to students, given evaluation data, so that the feedback is timely; specific enough for the student to make efforts to modify behavior; and objective, clear, and relevant; also include suggestions based on the data.

(A) Requisite Knowledge. Reporting procedures and the interpretation of test results.

(B) Requisite Skills. Communication skills and basic coaching.

■ Chapter 5 Instructor II

5.1 General. The Fire Service Instructor II shall meet the requirements for Fire Service Instructor I and the JPRs defined in Sections 5.2 through 5.5 of this standard.

5.2 Program Management.

5.2.1 Definition of Duty. The management of instructional resources, staff, facilities, and records and reports.

5.2.2 Schedule instructional sessions, given department scheduling policy, instructional resources, staff, facilities, and timeline for delivery, so that the specified sessions are delivered according to department policy.

(A) Requisite Knowledge. Departmental policy, scheduling processes, supervision techniques, and resource management.

(B) Requisite Skills. None required.

5.2.3 Formulate budget needs, given training goals, agency budget policy, and current resources, so that the resources required to meet training goals are identified and documented.

(A) Requisite Knowledge. Agency budget policy, resource management, needs analysis, sources of instructional materials, and equipment.

(B) Requisite Skills. Resource analysis and forms completion.

5.2.4 Acquire training resources, given an identified need, so that the resources are obtained within established timelines, budget constraints, and according to agency policy.

(A) Requisite Knowledge. Agency policies, purchasing procedures, and budget management.

(B) Requisite Skills. Forms completion.

5.2.5 Coordinate training record-keeping, given training forms, department policy, and training activity, so that all agency and legal requirements are met.

(A) Requisite Knowledge. Record-keeping processes, departmental policies, laws affecting records and disclosure of training information, professional standards applicable to training records, and databases used for record-keeping.

(B) Requisite Skills. Record auditing procedures.

5.2.6 Evaluate instructors, given an evaluation form, department policy, and JPRs, so that the evaluation identifies areas of strengths and weaknesses, recommends changes in instructional style and communication methods, and provides opportunity for instructor feedback to the evaluator.

(A) Requisite Knowledge. Personnel evaluation methods, supervision techniques, department policy, and effective instructional methods and techniques.

(B) Requisite Skills. Coaching, observation techniques, and completion of evaluation forms.

5.3 Instructional Development.

5.3.1 Definition of Duty. The development of instructional materials for specific topics.

5.3.2 Create a lesson plan, given a topic, audience characteristics, and a standard lesson plan format, so that the JPRs or learning objectives for the topic are addressed, and the plan includes learning objectives, a lesson outline, course materials, instructional aids, and an evaluation plan.

(A) Requisite Knowledge. Elements of a lesson plan, components of learning objectives, methods and techniques of instruction, principles of adult learning, techniques for eliminating bias in instructional materials, types and application of instructional media, evaluation techniques, and sources of references and materials.

(B) Requisite Skills. Basic research, using JPRs to develop behavioral objectives, student needs assessment, development of instructional media, outlining techniques, evaluation techniques, and resource needs analysis.

5.3.3 Modify an existing lesson plan, given a topic, audience characteristics, and a lesson plan, so that the JPRs or learning objectives for the topic are addressed and the plan includes learning objectives, a lesson outline, course materials, instructional aids, and an evaluation plan.

(A) Requisite Knowledge. Elements of a lesson plan, components of learning objectives, methods and techniques of instruction, principles of adult learning, techniques for eliminating bias in instructional materials, types and application of instructional media, evaluation techniques, and sources of references and materials.

(B) Requisite Skills. Basic research, using JPRs to develop behavioral objectives, student needs assessment, development of instructional media, outlining techniques, evaluation techniques, and resource needs analysis.

5.4 Instructional Delivery.

5.4.1 Definition of Duty. Conducting classes using a lesson plan.

5.4.2 Conduct a class using a lesson plan that the instructor has prepared and that involves the utilization of multiple teaching methods and techniques, given a topic and a target audience, so that the lesson objectives are achieved.

(A) Requisite Knowledge. Use and limitations of teaching methods and techniques.

(B)* Requisite Skills. Transition between different teaching methods.

5.4.3* Supervise other instructors and students during training, given a training scenario with increased hazard exposure, so that applicable safety standards and practices are followed, and instructional goals are met.

(A) Requisite Knowledge. Safety rules, regulations, and practices; the incident command system used by the agency; and leadership techniques.

(B) Requisite Skills. Implementation of an incident management system used by the agency.

5.5 Evaluation and Testing.

5.5.1 Definition of Duty. The development of student evaluation instruments to support instruction and the evaluation of test results.

5.5.2 Develop student evaluation instruments, given learning objectives, audience characteristics, and training goals, so that the evaluation instrument determines if the student has achieved the learning objectives; the instrument evaluates relevant performance in an objective, reliable, and verifiable manner; and the evaluation instrument is bias-free to any audience or group.

(A) Requisite Knowledge. Evaluation methods, development of forms, effective instructional methods, and techniques.

(B) Requisite Skills. Evaluation item construction and assembly of evaluation instruments.

5.5.3 Develop a class evaluation instrument, given agency policy and evaluation goals, so that students have the ability to provide feedback to the instructor on instructional methods, communication techniques, learning environment, course content, and student materials.

(A) Requisite Knowledge. Evaluation methods and test validity.

(B) Requisite Skills. Development of evaluation forms.

■ Chapter 6 Instructor III

6.1 General. The Fire Service Instructor III shall meet the requirements for Fire Service Instructor II and the JPRs defined in Sections 6.2 through 6.5 of this standard.

6.2 Program Management.

6.2.1 Definition of Duty. The administration of agency policies and procedures for the management of instructional resources, staff, facilities, records, and reports.

6.2.2* Administer a training record system, given agency policy and type of training activity to be documented, so that the information captured is concise, meets all agency and legal requirements, and can be readily accessed.

(A) Requisite Knowledge. Agency policy, record-keeping systems, professional standards addressing training records, legal requirements affecting record-keeping, and disclosure of information.

(B) Requisite Skills. Development of forms and report generation.

6.2.3 Develop recommendations for policies to support the training program, given agency policies and procedures and the training program goals, so that the training and agency goals are achieved.

(A) Requisite Knowledge. Agency procedures and training program goals, and format for agency policies.

(B) Requisite Skills. Technical writing.

6.2.4 Select instructional staff, given personnel qualifications, instructional requirements, and agency policies and procedures, so that staff selection meets agency policies and achievement of agency and instructional goals.

(A) Requisite Knowledge. Agency policies regarding staff selection, instructional requirements, selection methods, the capabilities of instructional staff, and agency goals.

(B) Requisite Skills. Evaluation techniques.

6.2.5 Construct a performance-based instructor evaluation plan, given agency policies and procedures and job requirements, so that instructors are evaluated at regular intervals, following agency policies.

(A) Requisite Knowledge. Evaluation methods, agency policies, staff schedules, and job requirements.

(B) Requisite Skills. Evaluation techniques.

6.2.6 Write equipment purchasing specifications, given curriculum information, training goals, and agency guidelines, so that the equipment is appropriate and supports the curriculum.

(A) Requisite Knowledge. Equipment purchasing procedures, available department resources, and curriculum needs.

(B) Requisite Skills. Evaluation methods to select the equipment that is most effective and preparation of procurement forms.

6.2.7 Present evaluation findings, conclusions, and recommendations to agency administrator, given data summaries and target audience, so that recommendations are unbiased, supported, and reflect agency goals, policies, and procedures.

(A) Requisite Knowledge. Statistical evaluation procedures and agency goals.

(B) Requisite Skills. Presentation skills and report preparation following agency guidelines.

6.3 Instructional Development.

6.3.1 Definition of Duty. Plans, develops, and implements comprehensive programs and curricula.

6.3.2 Conduct an agency needs analysis, given agency goals, so that instructional needs are identified and solutions are recommended.

(A) Requisite Knowledge. Needs analysis, task analysis, development of JPRs, lesson planning, instructional methods for classroom, training ground, and distance learning, characteristics of adult learners, instructional media, curriculum development, and development of evaluation instruments.

(B) Requisite Skills. Conducting research, committee meetings, and needs and task analysis; organizing information into functional groupings; and interpreting data.

6.3.3 Design programs or curricula, given needs analysis and agency goals, so that the agency goals are supported, the knowledge and skills are job-related, the design is performance-based, adult learning principles are utilized, and the program meets time and budget constraints.

(A) Requisite Knowledge. Instructional design, adult learning principles, principles of performance-based education, research, and fire service terminology.

(B) Requisite Skills. Technical writing and selecting course reference materials.

6.3.4 Modify an existing curriculum, given the curriculum, audience characteristics, learning objectives, instructional resources, and agency training requirements, so that the curriculum meets the requirements of the agency, and the learning objectives are achieved.

(A) Requisite Knowledge. Instructional design, adult learning principles, principles of performance-based education, research, and fire service terminology.

(B) Requisite Skills. Technical writing and selecting course reference materials.

6.3.5 Write program and course goals, given JPRs and needs analysis information, so that the goals are clear, concise, measurable, and correlate to agency goals.

(A) Requisite Knowledge. Components and characteristics of goals, and correlation of JPRs to program and course goals.

(B) Requisite Skills. Writing goal statements.

6.3.6 Write course objectives, given JPRs, so that objectives are clear, concise, measurable, and reflect specific tasks.

(A) Requisite Knowledge. Components of objectives and correlation between JPRs and objectives.

(B) Requisite Skills. Writing course objectives and correlating them to JPRs.

6.3.7 Construct a course content outline, given course objectives, reference sources, functional groupings and the agency structure, so that the content supports the agency structure and reflects current acceptable practices.

(A) Requisite Knowledge. Correlation between course goals, course outline, objectives, JPRs, instructor lesson plans, and instructional methods.

(B) Requisite Skills. None required.

6.4 Instructional Delivery. No JPRs at the Instructor III Level.

6.5 Evaluation and Testing.

6.5.1 Definition of Duty. Develops an evaluation plan; collects, analyzes, and reports data; and utilizes data for program validation and student feedback.

6.5.2 Develop a system for the acquisition, storage, and dissemination of evaluation results, given agency goals and policies, so that the goals are supported and so that those affected by the information receive feedback consistent with agency policies and federal, state, and local laws.

(A) Requisite Knowledge. Record-keeping systems, agency goals, data acquisition techniques, applicable laws, and methods of providing feedback.

(B) Requisite Skills. The evaluation, development, and use of information systems.

6.5.3 Develop course evaluation plan, given course objectives and agency policies, so that objectives are measured and agency policies are followed.

(A) Requisite Knowledge. Evaluation techniques, agency constraints, and resources.

(B) Requisite Skills. Decision making.

6.5.4 Create a program evaluation plan, given agency policies and procedures, so that instructors, course components, and facilities are evaluated and student input is obtained for course improvement.

(A) Requisite Knowledge. Evaluation methods and agency goals.

(B) Requisite Skills. Construction of evaluation instruments.

6.5.5 Analyze student evaluation instruments, given test data, objectives, and agency policies, so that validity is determined and necessary changes are made.

(A) Requisite Knowledge. Test validity, reliability, and item analysis.

(B) Requisite Skills. Item analysis techniques.

Pro Board Assessment Methodology Matrices for NFPA 1041

NFPA 1041 - Fire Service Instructor I - 2012 Edition

INSTRUCTIONS: In the column titled 'Cognitive/Written Test' place the number of questions from the Test Bank that are used to evaluate the applicable JPR, RK, RS, or objective. In the column titled 'Manipulative/Skill Station' identify the skill sheets that are used to evaluate the applicable JPR, RS, or objective. When the Portfolio or Project method is used to evaluate a particular JPR, RK, RS, or objective, identify the applicable section in the appropriate column and provide the procedures to be used as outlined in the NBFSPQ Operational Procedures, COA-5. Evaluation methods that are not cognitive, manipulative, portfolio, or project based should be identified in the 'Other' column.

Objective / JPR, RK, RS		Cognitive	Manipulative			
Section	Abbreviated Text	Written Test	Skills Station	Portfolio	Projects	Other
4.2.2	Assemble course materials					Chapter 1 (p 6), Chapter 6 (pp 132, 134–136, 138–139)
4.2.2(A)	RK: Components of a lesson plan; policies and procedures for the procurement of materials and equipment and resource availability					Chapter 1 (p 6), Chapter 6 (pp 127–132, 134–136)
4.2.3	Prepare request for resources					Chapter 1 (p 6), Chapter 12 (pp 281–282)
4.2.3(A)	RK: Resource management, sources of instructional resources					Chapter 1 (p 6), Chapter 12 (pp 281–282)
4.2.3(B)	RS: Oral and written communications					Chapter 1 (pp 6–10), Chapter 12 (pp 281–282)
4.2.4	Schedule single instructional sessions					Chapter 1 (p 6), Chapter 12 (pp 280–287)
4.2.4(A)	RK: Departmental scheduling procedures					Chapter 12 (pp 280–287)
4.2.4(B)	RS: Training schedule completion					Chapter 12 (pp 281–282)
4.2.5	Complete training records and report forms					Chapter 1 (p 6), Chapter 2 (pp 30–31, 34, 38–39), Chapter 12 (pp 281–282, 287)
4.2.5(A)	RK: Types of records and reports required					Chapter 2 (pp 30, 34, 38–39)
4.2.5(B)	RS: Basic report writing					Chapter 12 (pp 281–282, 287)
4.3.2	Review instructional materials					Chapter 1 (p 6), Chapter 4 (pp 79–81), Chapter 6 (pp 129–132, 134–136), Chapter 7 (pp 161, 163–167, 169–170)
4.3.2(A)	RK: Recognition of student limitations, methods of instruction, types of resource materials; organizing the learning environment; policies and procedures					Chapter 4 (pp 79–81), Chapter 6 (pp 129–132, 134–136), Chapter 7 (pp 161, 163–167, 169–170)
4.3.2(B)	RS: Analysis of resources, facilities, and materials					Chapter 6 (pp 129–132, 134–136), Chapter 7 (pp 163–167)
4.3.3	Adapt a prepared lesson plan					Chapter 1 (p 6), Chapter 6 (pp 138–139)

Section	Abbreviated Text	Written Test	Skills Station	Portfolio	Projects	Other
	Objective / JPR, RK, RS	Cognitive	Manipulative			
4.3.3(A)	RK: Elements of a lesson plan, selection of instructional aids and methods, origination of learning environment					Chapter 6 (pp 138–139), Chapter 8 (pp 178–199)
4.3.3(B)	RS: Instructor preparation and organizational skills					Chapter 6 (pp 138–139), Chapter 8 (pp 178–199)
4.4.2	Organize the classroom, laboratory, or outdoor learning environment					Chapter 1 (pp 6, 10–11), Chapter 7 (pp 163–167), Chapter 8 (pp 188–191, 193–198), Chapter 9 (pp 209–213, 215–216)
4.4.2(A)	RK: Classroom management and safety					Chapter 1 (pp 10–11), Chapter 7 (pp 163–167), Chapter 8 (pp 196, 198), Chapter 9 (pp 209–213, 215–216)
4.4.2(B)	RS: Use of instructional media and materials					Chapter 1 (pp 10–11), Chapter 8 (pp 178–199)
4.4.3	Present prepared lessons					Chapter 1 (p 6), Chapter 3 (pp 61–68), Chapter 4 (pp 79–81), Chapter 5 (pp 105–107, 109–110, 112–114), Chapter 6 (pp 129–132)
4.4.3(A)	RK: Laws and principles of learning					Chapter 3 (pp 61–68), Chapter 4 (pp 79–81),Chapter 5 (pp 104–106, 115–116)
4.4.3(B)	RS: Oral communication techniques, teaching methods and techniques					Chapter 3 (pp 61–68), Chapter 4 (pp 79–81), Chapter 5 (pp 105–107, 109–110, 112–114), Chapter 6 (pp 129–132)
4.4.4	Adjust presentation					Chapter 1 (p 6), Chapter 3 (pp 61, 65), Chapter 6 (pp 138–139)
4.4.4(A)	RK: Methods of dealing with changing circumstances					Chapter 3 (pp 61, 65), Chapter 6 (pp 138–139)
4.4.5	Adjust to differences in learning styles, abilities, and behaviors					Chapter 1 (p 6), Chapter 3 (pp 57–59, 61), Chapter 4 (pp 89–93), Chapter 9 (pp 209–210)
4.4.5(A)	RK: Motivation techniques, learning styles, types of learning disabilities and methods for dealing with them					Chapter 3 (pp 52–54, 57–59, 61, 66–68), Chapter 4 (pp 89–93), Chapter 9 (pp 209–210)
4.4.5(B)	RS: Basic coaching and motivational techniques, adaptation of lesson plans or materials to specific instructional situations					Chapter 3 (pp 52–54, 66–68), Chapter 4 (pp 91–93)
4.4.6	Operate audiovisual equipment and demonstration devices					Chapter 1 (p 6), Chapter 8 (pp 185, 187–189, 197)
4.4.6(A)	RK: Components of audiovisual equipment					Chapter 8 (pp 189–191, 193–198)
4.4.6(B)	RS: Use of audiovisual equipment, cleaning, and field level maintenance					Chapter 8 (pp 189–191, 193–199)
4.4.7	Utilize audiovisual materials					Chapter 1 (p 6), Chapter 8 (pp 197–199)
4.4.7(A)	RK: Media types, limitations, and selection criteria					Chapter 8 (pp 197–199)
4.4.7(B)	RS: Transition techniques within and between media					Chapter 8 (pp 187–188)
4.5.2	Administer oral, written, and performance tests					Chapter 1 (p 6), Chapter 10 (p 232)

Objective / JPR, RK, RS		Cognitive	Manipulative			
Section	Abbreviated Text	Written Test	Skills Station	Portfolio	Projects	Other
4.5.2.(A)	RK: Test administration, agency policies, laws affecting records and disclosure of training information					Chapter 10 (pp 230–234)
4.5.2(B)	RS: Use of skills checklists and oral questioning techniques					Chapter 10 (pp 232–234)
4.5.3	Grade student oral, written, or performance tests					Chapter 1 (p 6), Chapter 10 (p 234)
4.5.3(A)	RK: Grading and maintaining confidentiality of scores					Chapter 10 (p 234)
4.5.4	Report test results					Chapter 1 (p 6), Chapter 10 (p 234)
4.5.4(A)	RK: Reporting procedures, the interpretation of test results					Chapter 10 (p 234)
4.5.4(B)	RS: Communication skills, basic coaching					Chapter 10 (p 234)
4.5.5	Provide evaluation feedback to students					Chapter 1 (p 6), Chapter 10 (pp 234–235)
4.5.5(A)	RK: Reporting procedures, the interpretation of test results					Chapter 10 (pp 234–235)
4.5.5(B)	RS: Communication skills, basic coaching					Chapter 10 (pp 234–235)

NFPA 1041 - Fire Service Instructor II - 2012 Edition

INSTRUCTIONS: In the column titled 'Cognitive/Written Test' place the number of questions from the Test Bank that are used to evaluate the applicable JPR, RK, RS, or objective. In the column titled 'Manipulative/Skill Station' identify the skill sheets that are used to evaluate the applicable JPR, RS, or objective. When the Portfolio or Project method is used to evaluate a particular JPR, RK, RS, or objective, identify the applicable section in the appropriate column and provide the procedures to be used as outlined in the NBFSPQ Operational Procedures, COA-5. Evaluation methods that are not cognitive, manipulative, portfolio, or project based should be identified in the 'Other' column.

Objective / JPR, RK, RS		Cognitive	Manipulative			
Section	Abbreviated Text	Written Test	Skills Station	Portfolio	Projects	Other
5.1	Fire Instructor I					Chapter 1 (pp 6–7)
5.2.2	Schedule instructional sessions					Chapter 1 (p 6), Chapter 12 (pp 280–291)
5.2.2(A)	RK: Departmental policy, scheduling processes, supervision techniques, and resource management					Chapter 12 (pp 280–291)
5.2.3	Formulate budget needs					Chapter 1 (p 7), Chapter 15 (pp 350–353)
5.2.3(A)	RK: Agency budget policy, resource management, needs analysis, sources of instructional materials, and equipment					Chapter 15 (pp 350–353)
5.2.3(B)	RS: Resource analysis and forms completion					Chapter 15 (pp 350–353)
5.2.4	Acquire training resources					Chapter 1 (p 7), Chapter 12 (pp 281–282), Chapter 15 (pp 353–358)
5.2.4(A)	RK: Agency policies, purchasing procedures, budget management					Chapter 15 (pp 350–358)
5.2.4(B)	RS: Forms completion					Chapter 15 (pp 356–358)
5.2.5	Coordinate training record keeping					Chapter 1 (p 7), Chapter 2 (pp 30–31, 34, 38–39)

	Objective / JPR, RK, RS	Cognitive	Manipulative			
Section	Abbreviated Text	Written Test	Skills Station	Portfolio	Projects	Other
5.2.5(A)	RK: Record keeping processes, departmental policies, laws affecting records and disclosure of training information, professional standards applicable to training records, databases used for record keeping					Chapter 2 (pp 30–31, 34, 38–39)
5.2.5(B)	RS: Record auditing procedures					Chapter 2 (pp 30, 34–35, 38)
5.2.6	Evaluate instructors					Chapter 1 (p 7), Chapter 11 (pp 264–267, 269)
5.2.6(A)	RK: Personnel evaluation methods, supervision techniques, department policy, effective instructional methods and techniques					Chapter 11 (pp 264–267, 269)
5.2.6(B)	RS: Coaching, observation techniques, completion of evaluation forms					Chapter 11 (pp 264–267, 269–271)
5.3.2	Create a lesson plan					Chapter 1 (p 7), Chapter 6 (pp 140–149), Chapter 8 (pp 186–187)
5.3.2(A)	RK: Elements of a lesson plan, components of learning objectives					Chapter 6 (pp 140–148), Chapter 8 (pp 186–187)
5.3.2(B)	RS: Basic research, using job performance requirements to develop behavioral objectives					Chapter 6 (pp 144–145)
5.3.3	Modify an existing lesson plan					Chapter 1 (p 7), Chapter 6 (p 149), Chapter 8 (pp 197–198), Chapter 12 (pp 281–282)
5.3.3(A)	RK: Elements of a lesson plan, components of learning objectives					Chapter 6 (pp 140–148)
5.3.3(B)	RS: Basic research, using job performance requirements to develop behavioral objectives					Chapter 6 (pp 144–145)
5.4.2	Conduct a class using a lesson plan					Chapter 1 (p 7), Chapter 3 (pp 61–68), Chapter 7 (pp 161, 169–170), Chapter 8 (pp 185, 187–189)
5.4.2(A)	RK: Use and limitations of teaching methods and techniques					Chapter 3 (pp 61–68), Chapter 7 (pp 161, 163–167, 169–170), Chapter 8 (pp 185, 187–189)
5.4.2(B)	RS: Transition between different teaching methods					Chapter 3 (pp 64–65, 68), Chapter 7 (pp 161, 169–170), Chapter 8 (pp 187–188)
5.4.3	Supervise other instructors and students during high hazard training					Chapter 1 (p 7), Chapter 9 (pp 211–213, 215–216, 218–221), Chapter 12 (pp 291, 293)
5.4.3(A)	RK: Safety rules, regulations and practices, the incident command system used by the agency, and leadership techniques					Chapter 9 (pp 211–213, 215–216, 218–221), Chapter 12 (pp 291, 293)
5.4.3(B)	RS: ICS implementation					Chapter 9 (pp 211, 218, 221)
5.5.2	Develop student evaluation instruments					Chapter 1 (p 7), Chapter 10 (pp 238–240, 242–247)
5.5.2(A)	RK: Evaluation methods, development of forms, effective instructional methods, and techniques					Chapter 10 (pp 238–240, 242–247)

Objective / JPR, RK, RS		Cognitive	Manipulative			
Section	Abbreviated Text	Written Test	Skills Station	Portfolio	Projects	Other
5.5.2(B)	RS: Evaluation item construction and assembly of evaluation instruments					Chapter 10 (pp 238–240, 242–247)
5.5.3	Develop a class evaluation instrument					Chapter 1 (p 7), Chapter 11 (pp 269–271)
5.5.3(A)	RK: Evaluation methods, test validity					Chapter 11 (pp 269–271)
5.5.3(B)	RS: Development of evaluation forms					Chapter 11 (pp 269–271)

NFPA 1041 - Fire Service Instructor III - 2012 Edition

INSTRUCTIONS: In the column titled 'Cognitive/Written Test' place the number of questions from the Test Bank that are used to evaluate the applicable JPR, RK, RS, or objective. In the column titled 'Manipulative/Skill Station' identify the skill sheets that are used to evaluate the applicable JPR, RS, or objective. When the Portfolio or Project method is used to evaluate a particular JPR, RK, RS, or objective, identify the applicable section in the appropriate column and provide the procedures to be used as outlined in the NBFSPQ Operational Procedures, COA-5. Evaluation methods that are not cognitive, manipulative, portfolio, or project based should be identified in the 'Other' column.

Objective / JPR, RK, RS		Cognitive	Manipulative			
Section	Abbreviated Text	Written Test	Skills Station	Portfolio	Projects	Other
6.1	Fire Instructor II					Chapter 1 (p 7)
6.2.2	Administer a training record system					Chapter 1 (p 7), Chapter 2 (pp 30–31, 34, 38–39), Chapter 15 (pp 359–369)
6.2.2(A)	RK: Agency policy, record keeping systems, professional standards addressing training records, legal requirements affecting record keeping, and disclosure of information					Chapter 2 (pp 30–31, 34, 38–39), Chapter 15 (pp 359–369)
6.2.2(B)	RS: Development of forms, report generation					Chapter 15 (pp 361–367)
6.2.3	Development of recommendations for policies					Chapter 1 (p 7), Chapter 15 (pp 365–367, 369)
6.2.3(A)	RK: Agency procedures and training program goals, format for agency policies					Chapter 15 (pp 365–367, 369)
6.2.3(B)	RS: Technical writing					Chapter 15 (pp 365–367, 369)
6.2.4	Select instructional staff					Chapter 1 (p 7), Chapter 15 (pp 369, 371–373)
6.2.4(A)	RK: Agency polices regarding staff selection, instructional requirements, selection methods, the capabilities of instructional staff and agency goals					Chapter 15 (pp 369, 371–373)
6.2.4(B)	RS: Evaluation techniques					Chapter 15 (pp 369, 371–373)
6.2.5	Construct a performance based instructor evaluation plan					Chapter 1 (p 7), Chapter 15 (pp 365–367, 373)
6.2.5(A)	RK: Evaluation methods, agency policies, staff schedules, and job requirements					Chapter 15 (pp 365–367, 373)
6.2.5(B)	RS: Evaluation techniques					Chapter 15 (pp 365–367, 373)
6.2.6	Write equipment purchasing specifications					Chapter 1 (p 7), Chapter 15 (pp 353–358)
6.2.6(A)	RK: Equipment purchasing procedures, available department resources, and curriculum needs					Chapter 15 (pp 353–358)

Objective / JPR, RK, RS		Cognitive	Manipulative			
Section	Abbreviated Text	Written Test	Skills Station	Portfolio	Projects	Other
6.2.6(B)	RS: Evaluation methods to select the equipment that is most effective and preparation of procurement forms					Chapter 15 (pp 353–358)
6.2.7	Present evaluation findings, conclusions, and recommendations to agency administrator					Chapter 1 (p 7), Chapter 15 (p 369)
6.2.7(A)	RK: Statistical evaluation procedures and agency goals					Chapter 15 (p 369)
6.2.7(B)	RS: Presentation skills and report preparation following agency guidelines					Chapter 15 (p 369)
6.3.2	Conduct an agency needs analysis					Chapter 1 (p 7), Chapter 13 (pp 301, 303)
6.3.2(A)	RK: Needs analysis, task analysis, development of job performance requirements					Chapter 13 (pp 301, 303)
6.3.2(B)	RS: Conducting research, committee meetings, and needs and task analysis					Chapter 13 (pp 301, 303)
6.3.3	Design programs or curriculums					Chapter 1 (p 7), Chapter 13 (pp 303–307, 310–312)
6.3.3(A)	RK: Instructional design, adult learning principles, principles of performance based education					Chapter 13 (p 313)
6.3.3(B)	RS: Technical writing, selecting course reference materials					Chapter 13 (pp 314–315)
6.3.4	Modify an existing curriculum					Chapter 1 (p 7), Chapter 13 (pp 307, 309–310)
6.3.4(A)	RK: Instructional design, adult learning principles, principles of performance based education					Chapter 13 (p 313)
6.3.4(B)	RS: Technical writing, selecting course reference material					Chapter 13 (pp 314–315)
6.3.5	Write program and course goals					Chapter 1 (p 7), Chapter 13 (pp 306–307)
6.3.5(A)	RK: Components and characteristics of goals, and correlation of JPRs to program and course goals					Chapter 13 (pp 306–307)
6.3.5(B)	RS: Writing goal statements					Chapter 13 (pp 306–307)
6.3.6	Write course objectives					Chapter 1 (p 7), Chapter 13 (pp 310–312)
6.3.6(A)	RK: Components of objectives and correlation between JPRs and objectives					Chapter 13 (pp 310–312)
6.3.6(B)	RS: Writing course objectives and correlating them to JPRs					Chapter 13 (pp 310–312)
6.3.7	Construct a course content outline					Chapter 1 (p 7), Chapter 13 (pp 303–305, 313)
6.3.7(A)	RK: Correlation between course goals, course outline, objectives, JPRs, instructor lesson plans, and instructional methods					Chapter 13 (pp 300–315)
6.5.2	Develop a system for the acquisition, storage, and dissemination of evaluation results					Chapter 1 (p 7), Chapter 14 (pp 324–325)

Objective / JPR, RK, RS		Cognitive	Manipulative			
Section	Abbreviated Text	Written Test	Skills Station	Portfolio	Projects	Other
6.5.2(A)	RK: Record keeping systems, agency goals, data acquisition techniques, applicable laws, and methods of providing feedback					Chapter 14 (pp 324–325)
6.5.2(B)	RS: Evaluation, development, and use of information systems					Chapter 14 (pp 324–325, 332, 334–339)
6.5.3	Develop course evaluation plan					Chapter 1 (p 7), Chapter 14 (pp 328–332)
6.5.3(A)	RK: Evaluation techniques, agency constraints, and resources					Chapter 14 (pp 325, 327–332)
6.5.3(B)	RS: Decision-making					Chapter 14 (pp 325, 327–332)
6.5.4	Create a program evaluation plan					Chapter 1 (p 7), Chapter 14 (pp 325, 327–328)
6.5.4(A)	RK: Evaluation methods, agency goals					Chapter 14 (pp 325, 327–328)
6.5.4(B)	RS: Construction of evaluation instruments					Chapter 14 (pp 325, 327–328)
6.5.5	Analyze student evaluation instruments					Chapter 1 (p 7), Chapter 14 (pp 332, 334–339)
6.5.5(A)	RK: Test validity, reliability					Chapter 14 (pp 332, 334–339)
6.5.5(B)	RS: Item analysis techniques					Chapter 14 (pp 332, 334–339)

An Extract from: NFPA 1401, *Recommended Practice for Fire Service Training Reports and Records*, 2012 Edition

■ Chapter 4 Elements of Training Documents

4.1 General.

4.1.1 Training records and reports should be utilized by the training officer and line officers for analysis of the effectiveness of the training program in terms of time, staffing, individual performance rating, and financing.

4.1.2* Training records and reports should be utilized to develop specific training objectives and to evaluate compliance with, or deficiencies in, the training program.

4.1.3 Compliance with mandated training requirements should be documented.

4.1.4 The management of training functions should be performed in a closed-feedback loop.

4.1.5 The training functions should not operate as an open-ended cycle.

4.1.6 The closed-feedback loop should consist of the following:

(1) Planning
(2) Organization
(3) Implementation
(4) Operation
(5) Review
(6) Feedback/alteration

4.1.7* In each phase of the cycle, information should be provided for management to perform effectively.

4.1.7.1 The information is provided through various types of records, reports, and studies; therefore, records should be designed to fit into the overall training management cycle.

4.1.8 In order to be most effective, these records should contribute to the overall organization information cycle.

4.2 Elements of Information.

4.2.1 Training documents, regardless of their intent or level of sophistication, should focus on content, accuracy, and clarity.

4.2.2 These documents should relay to the reader at least five specific elements of information as follows (*see Annex B for examples of training record forms*):

(1) Who
  (a) Who was the instructor?
  (b) Who participated?
  (c) Who was in attendance?
  (d) Who is affected by the documents?
  (e) Who was included in the training (individuals, company, multi-company, or organization)?

(2) What
  (a) What was the subject covered?
  (b) What equipment was utilized?
  (c) What operation was evaluated or affected?
  (d) What was the stated objective, and was it met?

(3) When
  (a) When will the training take place? or
  (b) When did the training take place?

(4) Where
  (a) Where will the training take place? or
  (b) Where did the training take place?

(5) Why
  (a) Why is the training necessary? or
  (b) Why did the training occur?

4.3 Additional Information.
Additional information or detail, which should include but not be limited to the following, should be included to explain or clarify the document as necessary:

(1) Source of the information used as a basis for the training
  (a) Textbook title and edition
  (b) Lesson plan title and edition
  (c) Policy name and reference number
  (d) Videotapes, CDs, and DVDs
  (e) Distance learning sources
  (f) Internet address
  (g) Industry best practices
  (h) Post-incident analysis (PIA)
  (i) Other

(2) Method of training used for delivery
  (a) Lecture
  (b) Demonstration
  (c) Skills training
  (d) Self-study
  (e) Video presentation
  (f) Mentoring
  (g) Drill(s)
  (h) Other

(3) Evaluation of training objectives
  (a) Written test
  (b) Skills examination
  (c) Other

■ Chapter 5 Types of Training Documents

5.1 Training Schedules.

5.1.1 Need for Training Schedules.

5.1.1.1 All members of a fire department should receive standardized instruction and training.

5.1.1.2 Standardized training should include considerable planning; however, standardization can be improved through the preparation of training schedules for use by department personnel.

5.1.1.3 Standardized training schedules should be prepared and published for both short-term scheduling (considerable detail), intermediate-term scheduling (less detail), and long-term scheduling (little detail) to facilitate long-term planning by the training staff, instructional staff, company officers, and personnel.

5.1.2* Types of Training Schedules. Training schedules should be prepared for all training ground and classroom sessions.

5.1.2.1 Periodic Training Schedule—Station Training. The station training schedule, which is prepared by the training officer, should designate specific subjects that are to be covered by company or station officers in conducting their station training.

5.1.2.1.1 The company officers should use this schedule to set their own in-station training schedule.

5.1.2.1.2 A balance between manipulative skills training and classroom sessions should be considered in the preparation of training schedules.

5.1.2.1.3 Such training schedules should include all of the topics necessary to satisfy job knowledge requirements and to maintain skills already learned.

5.1.2.2* Periodic Training Schedule—Training Facility Activities. The training facility activities schedule details when companies should report to the training facility for evolutions or classes.

5.1.2.2.1 Days also should be set aside for make-up sessions.

5.1.2.2.2 Training activities conducted outside the training facility or by outside agencies also should be shown on this schedule.

5.1.3* All Other Training. Schedules should be prepared for all training, including, but not limited to, the following:

(1) Recruit or entry-level training

(2) In-service training

(3) Special training

(4) Officer training

(5) Advanced training

(6) Mandated training

(7) Medical training

(8) Safety training

5.2* Training Reports.

5.2.1* Logical Sequence. A training report should be complete and should follow a logical sequence.

5.2.1.1 A report should clearly and concisely present the essentials so those conclusions can be grasped with a minimum of effort and delay.

5.2.1.2 Furthermore, a report should provide sufficient discussion to ensure the correct interpretation of the findings, which should indicate the nature of the analysis and the process of reasoning that leads to those findings.

5.2.2 Purpose. Each item of a report should serve a definite purpose.

5.2.2.1 Each table and chart in a report should be within the scope of the report.

5.2.2.2 The tables and charts should enhance the information stated or shown elsewhere, and they should be accurate and free of the possibility of misunderstanding, within reason.

5.2.3 Organization.

5.2.3.1* The process of writing reports should include five steps that are generally used in identifying, investigating, evaluating, and solving a problem.

5.2.3.2 These five steps, which should be accomplished before the report is written, are as follows:

(1) The purpose and scope of the report should be obtained.

(2) The method or procedure should be outlined.

(3) The essential facts should be collected.

(4) These facts should be analyzed and categorized.

(5) The correct conclusions should be arrived at and the proper recommendations should be made.

5.2.4* While there are differing needs among fire departments, certain reports should be common to most departments.

5.2.4.1 Typical recommended training reports should include the following:

(1) A complete inventory of apparatus and equipment assigned to the training division

(2) Detailed plans for training improvements that include all equipment and facility needs and cost figures

(3) A detailed periodic report on and evaluation of the training of all probationary fire fighters

(4) A monthly summary of all activities of the training division

(5) An annual report of all activities of the training division

(6) A complete inventory of training aids and reference materials available to be used for department training

5.2.4.2 The annual report should describe the accomplishments during the year, restate the goals and objectives of the training division, and describe the projected plans for the upcoming year.

5.2.5 Narrative Report. There are times when a narrative report should be necessary.

5.2.5.1 Before writing a narrative report, the writer should consider the audience for the report.

5.2.5.2 The comprehensiveness of the report should be determined by the recipients' knowledge of the subject.

5.3* Training Records.

5.3.1 Training records should be kept to document department training and should assist in determining the program's effectiveness. Information derived from such records should, for example, provide the supporting data needed to justify additional training personnel and equipment.

5.3.1.1 Training records can include paper or electronic media.

5.3.2 Performance tests, examinations, and personnel evaluations should contribute to the development of the training program if the results are analyzed, filed, and properly applied.

5.3.2.1 Training records should be kept current and should provide the status and progress of all personnel receiving training.

5.3.2.2 Frequent review of training records should provide a clear picture of the success of the training program and document lessons learned.

5.3.3 Properly designed training records should be developed to meet the specific needs of each fire department.

5.3.3.1 Training records should be detailed enough to enable factual reporting while remaining as simple as possible.

5.3.3.2 The number of records should be kept at a minimum to avoid confusion and duplication of effort.

5.3.4 Typical training records should include an evaluation of the competency of the student, as well as hours attended.

5.4* State Certification Records.

5.4.1 Minimum Information.

5.4.1.1 Information and documentation that should serve as a foundation for submission to state certification programs should include, as a minimum, the following:

(1) A single file that includes all training accomplished by the individual fire fighter during his/her career

(2) Dates, hours, locations, and instructors of all special courses or seminars attended

(3) Monthly summaries of all departmental training

5.4.1.2 These records should require the signatures of the instructor and the person instructed as a valid record of an individual's participation in the training.

5.4.2 The format used for state certification should be different from that utilized by an individual department. Otherwise, this is likely to cause considerable problems with accurate record submission and should be addressed on the state level by all parties concerned. Various state certification forms are contained in Annex B.

■ Chapter 6 Evaluating the Effectiveness of Training Records Systems

6.1 Evaluating Records of Individuals.

6.1.1 The evaluation of training records should be done at specified intervals by the local department training officer or training committee.

6.1.2 Each training record should be evaluated to determine the following:

(1) Has the individual taken all the required training?

(2) If not, has the individual been scheduled for missed classes?

(3) Do performance deficiencies show up on the individual's training record?

(4) If performance deficiencies exist, what kind of program is being developed to overcome them?

(5) Have companies met all the required job performance standards established by the department?

(6) If job performance standards have not been met, have the problems been identified and a program developed to overcome them?

(7) Are there areas of training that are being overlooked completely?

(8) Is the cycle of training sufficient to maintain skill levels?

6.2 Evaluating the Record-Keeping System.

6.2.1 All training records and the record-keeping system should be evaluated at least annually.

6.2.2 During the evaluation process, the following questions should be applied to each record:

(1) What is the purpose of the record?

(2) Who uses the information compiled?

(3) Is the record providing the necessary information?

(4) Do other records duplicate the material being compiled?

(5) How long should records be retained?

(6) Can training trends be determined from a compilation of the records?

(7) Is there a simpler and more efficient way of recording the information?

■ Chapter 7 Legal Aspects of Record Keeping

7.1 Privacy of Personal Information.

7.1.1* Employee training and educational records and other examination data included in an individual's training file should be disclosed only with written permission of the employee, unless required by law or statute, or by a court order.

7.1.2 The fire chief and the training officer should verify with legal counsel the federal, state, provincial, and local laws and ordinances regulating the disclosure of confidential information, and ensure adequate control measures are in place for the privacy of personal information.

7.1.2.1 Training records should not use the student's Social Security number for identification purposes.

7.1.2.2 The fire chief or training officer should ensure that training records do not include any confidential medical information.

7.1.2.3 All medical records should be kept in a completely separate file and not mixed with any other records or personnel files.

7.1.2.4 Access to any personally identifiable or proprietary information should be restricted.

7.1.3 Length of Time for Keeping Records or Reports.

7.1.3.1* Legal counsel should be contacted concerning the length of time records or reports, or both, need to be kept available and documented in a records retention schedule. [See Figure B.1(m) for a sample schedule.]

7.1.3.2 Documents should be maintained for a period of time as specified by law or as required by certain agencies and organizations.

7.1.4 Most training records should be maintained in their entirety in a computerized form, thus greatly reducing the amount of paper that needs to be stored.

7.1.4.1* Some training records should be maintained in their original hard-copy form, as required by certain agencies and organizations.

7.1.4.2* Computerized records should be backed up periodically and stored in an off-site location to avoid destruction.

7.2* Record Keeping and Risk Management. Agencies that conduct multijurisdictional training should have a signed release form for those individuals who participate in certain training activities.

■ Chapter 5 Training, Education, and Professional Development

5.1 General Requirements.

5.1.1* The fire department shall establish and maintain a training, education, and professional development program with a goal of preventing occupational deaths, injuries, and illnesses.

5.1.2 The fire department shall provide training, education, and professional development for all department members commensurate with the duties and functions that they are expected to perform.

5.1.3 The fire department shall establish training and education programs that provide new members initial training, proficiency opportunities, and a method of skill and knowledge evaluation for duties assigned to the member prior to engaging in emergency operations.

5.1.4* The fire department shall restrict the activities of new members during emergency operations until the member has demonstrated the skills and abilities to complete the tasks expected.

5.1.5 The fire department shall provide all members with training and education on the department's risk management plan.

5.1.6 The fire department shall provide all members with training and education on the department's written procedures.

5.1.7 The fire department shall provide all members with a training, education, and professional development program commensurate with the emergency medical services that are provided by the department.

5.1.8* The fire department shall provide all members with a documented training and education program that covers the selection, use, maintenance, retirement, and special incidents procedures operation, limitation, maintenance, and retirement criteria for all assigned personal protective equipment (PPE).

5.1.8.1 Training shall comply with applicable governing standards and follow the manufacturer's instructions and guidelines to include the following topics:

(1) The organization's overall program for the selection and use of protective ensembles, ensemble elements, and SCBAs

(2) Technical data package (TDP) where applicable

(3) Proper overlap and fit

(4) Proper donning and doffing (including emergency doffing)

(5) Construction features and function

(6) Performance limitations (including physiological effects on user and effects of heat transfer on the protective ensemble)

(7) Recognizing and responding to indications of protective ensemble and SCBA failure

(8) Routine inspection

(9) Routine cleaning

(10) Proper storage

5.1.8.2 For maintenance of structural and proximity protective ensembles and ensemble elements, refer to NFPA 1851, *Standard on Selection, Care, and Maintenance of Protective Ensembles for Structural Fire Fighting and Proximity Fire Fighting.*

5.1.8.3 For maintenance of SCBA, refer to NFPA 1852, *Standard on Selection, Care, and Maintenance of Open-Circuit Self-Contained Breathing Apparatus (SCBA).*

5.1.8.4 For maintenance of protective ensemble for technical rescue incidents, refer to NFPA 1855, *Standard for Selection, Care, and Maintenance of Protective Ensembles for Technical Rescue Incidents.*

5.1.9 As a duty function, members shall be responsible to maintain proficiency in their skills and knowledge, and to avail themselves of the professional development provided to the members through department training and education programs.

5.1.10 Training programs for all members engaged in emergency operations shall include procedures for the safe exit and accountability of members during rapid evacuation, equipment failure, or other dangerous situations and events.

5.1.11 All members who are likely to be involved in emergency operations shall be trained in the incident management and accountability system used by the fire department.

5.2 Member Qualifications.

5.2.1 All members who engage in structural fire fighting shall meet the requirements of NFPA 1001, *Standard for Fire Fighter Professional Qualifications.*

5.2.2* All driver/operators shall meet the requirements of NFPA 1002, *Standard for Fire Apparatus Driver/Operator Professional Qualifications.*

5.2.3 All aircraft rescue fire fighters (ARFF) shall meet the requirements of NFPA 1003, *Standard for Airport Fire Fighter Professional Qualifications.*

5.2.4 All fire officers shall meet the requirements of NFPA 1021, *Standard for Fire Officer Professional Qualifications.*

5.2.5 All wildland fire fighters shall meet the requirements of NFPA 1051, *Standard for Wildland Fire Fighter Professional Qualifications.*

5.2.6* All members responding to hazardous materials incidents shall meet the operations level as required in NFPA 472, *Standard for Professional Competence of Responders to Hazardous Materials Incidents.*

5.3 Training Requirements.

5.3.1* The fire department shall adopt or develop training and education curriculums that meet the minimum requirements outlined in professional qualification standards covering a member's assigned function.

5.3.2 The fire department shall provide training, education, and professional development programs as required to support the minimum qualifications and certifications expected of its members.

5.3.3 Members shall practice assigned skill sets on a regular basis but not less than annually.

5.3.4 The fire department shall provide specific training to members when written policies, practices, procedures, or guidelines are changed and/or updated.

5.3.5* The respiratory protection training program shall meet the requirements of NFPA 1404, *Standard for Fire Service Respiratory Protection Training*.

5.3.6 Members who perform wildland fire fighting shall be trained at least annually in the proper deployment of an approved fire shelter.

5.3.7* All live fire training and exercises shall be conducted in accordance with NFPA 1403, *Standard on Live Fire Training Evolutions*.

5.3.8* All training and exercises shall be conducted under the direct supervision of a qualified instructor.

5.3.9* All members who are likely to be involved in emergency medical services shall meet the training requirements of the AHJ.

5.3.10* Members shall be fully trained in the use, limitations, care, and maintenance of the protective ensembles and ensemble elements assigned to them or available for their use.

5.3.11 All members shall meet the training requirements as outlined in NFPA 1561, *Standard on Emergency Services Incident Management System*.

5.3.12 All members shall meet the training requirements as outlined in NFPA 1581, *Standard on Fire Department Infection Control Program*.

5.4 Special Operations Training.

5.4.1 The fire department shall provide specific and advanced training to members who engage in special operations as a technician.

5.4.2 The fire department shall provide specific training to members who are likely to respond to special operations incidents in a support role to special operations technicians.

5.4.3 Members expected to perform hazardous materials mitigation activities shall meet the training requirements of a technician as outlined in NFPA 472.

5.4.4 Members expected to perform technical operations at the technician level as defined in NFPA 1670, *Standard on Operations and Training for Technical Search and Rescue Incidents*, shall meet the training requirements specified in NFPA 1006, *Standard for Rescue Technician Professional Qualifications*.

5.5 Member Proficiency.

5.5.1 The fire department shall develop a recurring proficiency cycle with the goal of preventing skill degradation and potential for injury and death of members.

5.5.2 The fire department shall develop and maintain a system to monitor and measure training progress and activities of its members.

5.5.3* The fire department shall provide an annual skills check to verify minimum professional qualifications of its members.

Glossary

ABCD method Process for writing lesson plan objectives that includes four components: audience, behavior, condition, and degree.

Active listening The process of hearing and understanding the communication sent; demonstrating that you are listening and have understood the message.

Adapt To make fit (as for a specific use or situation).

Adult learning The integration of new information into the values, beliefs, and behaviors of adults.

Affective domain The domain of learning that affects attitudes, emotions, or values. It may be associated with a student's perspective or belief being changed as a result of training in this domain.

Agency training needs assessment A needs assessment performed at the direction of the department administration, which helps to identify any regulatory compliance matters that must be included in the training schedule.

Ambient noise The general level of background sound.

Americans with Disabilities Act of 1990 (ADA) A federal civil rights law that prohibits discrimination on the basis of disability.

Andragogy The identification of characteristics associated with adult learning.

Application step The third step of the four-step method of instruction, in which the student applies the information learned during the presentation step.

Assignment The part of the lesson plan that provides the student with opportunities for additional application or exploration of the lesson topic, often in the form of homework that is completed outside of the classroom.

Asynchronous learning An online course format in which the instructor provides material, lectures, tests, and assignments that can be accessed at any time. Learners are not consuming content at the same time; instead, they are given a time frame during which they need to connect to the course and complete assignments, and they work at their own pace, adhering to established submission deadlines set by the instructor.

Attention-deficit/hyperactivity disorder (ADHD) A disorder in which a person has a chronic level of inattention and an impulsive hyperactivity that affects daily functions.

Audience analysis The determination of characteristics common to a group of people; it can be used to choose the best instructional approach.

Audience Who the students are.

Baby boomers The generation born after World War II (1946–1964).

Behavioral objectives Goals that are achieved through the attainment of a skill, knowledge, or both, and that can be measured or observed; also called learning objectives.

Behavior An observable and measurable action for the student to complete.

Behaviorist perspective The theory that learning is a relatively permanent change in behavior that arises from experience.

Blended learning An instruction method that combines online/independent study with face-to-face meetings with the instructor.

Bloom's Taxonomy A classification of the different objectives and skills that educators set for students (learning objectives).

Budget An itemized summary of estimated or intended revenues and expenditures.

Certificate Document given for the completion of a training course or event.

Certification Document awarded for the successful completion of a testing process based on a standard.

Class of instruction An individual lesson or drill that is a smaller part of the entire course. In an academy-style course, you will receive many classes of instruction that prepare you for completion of the course or certification.

Coaching The process of helping individuals develop skills and talents.

Code *See Regulation*

Cognitive domain The domain of learning that effects a change in knowledge. It is most often associated with learning new information.

Cognitive perspective An intellectual process by which experience contributes to relatively permanent changes (learning). It may be associated with learning by experience.

Communication process The process of conveying an intended message from the sender to the receiver and getting feedback to ensure accuracy.

Competency-based learning Learning that is intended to create or improve professional competencies.

Condition The situation in which the student will perform the behavior.

Confidentiality The requirement that, with very limited exceptions, employers must keep medical and other personal information about employees and applicants private.

Continuing education Education or training obtained to maintain skills, proficiency, or certification in a specific position.

Course goal The end result and desired outcome of multiple classes of instruction and lessons.

Courseware Any educational content that is delivered via a computer.

Degree (1) Document awarded by an institution for the completion of required coursework. (2) The last part of a learning objective, which indicates how well the student is expected to perform the behavior in the listed conditions.

Delegation Transfer of authority and responsibility to another person for the purpose of teaching new job skills or as a means of time management. You can delegate authority but never responsibility.

Demographics Characteristics of a given population, possibly including such information as age, race, gender, education, marital status, family structure, and location of agency.

Direct threat A situation in which an individual's disability presents a serious risk to his or her own safety or the safety of his or her co-workers.

Disability A physical or mental condition that interferes with a major life activity.

Discrimination index The value given to an assessment that differentiates between high and low scorers.

Dyscalculia A learning disability in which students have difficulty with math and related subjects.

Dyslexia A learning disability in which students have difficulty reading due to an inability to interpret spatial relationships and integrate visual information.

Dysphasia A learning disability in which students lack the ability to write, spell, or place words together to complete a sentence.

Dyspraxia Lack of physical coordination with motor skills.

Enabling objective An intermediate learning objective that allows the student to meet the terminal objective identified in a course design document. It is usually part of a series that directs the instructor on what he or she needs to instruct and what the learner will learn to accomplish the terminal objective. Several enabling objectives may be required to understand or perform the terminal objective.

Essay test Test that requires students to form a structured argument using materials presented in class or from required reading.

Ethics Principles used to define behavior that is not specifically governed by rules of law but rather in many cases by public perceptions of right and wrong. Ethics is often defined on a regional or local level within the community.

Evaluation step The fourth step of the four-step method of instruction, in which the student is evaluated by the instructor.

Expenditures Moneys spent for goods or services that are considered appropriations.

Face validity A type of validity achieved when a test item has been derived from an area of technical information by an experienced subject-matter expert who can attest to its technical accuracy.

Feedback The fifth and final link of the communication chain. Feedback allows the sender (the instructor) to determine whether the receiver (the student) understood the message.

Formal communication An official fire department communication. The letter or report is presented on stationery with the fire department letterhead and generally is signed by a chief officer or headquarters staff member.

Formative evaluation Process conducted to improve the fire service instructor's performance by identifying his or her strengths and weaknesses.

Four-step method of instruction The most commonly used method of instruction in the fire service. The four steps are preparation, presentation, application, and evaluation.

General orders Short-term documents signed by the fire chief that remain in force for a period of days to one year or more.

Generation X People born after the baby boomers; they are today's adult learners.

Generation Y People born immediately after Generation X.

Generation Z People born immediately after Generation Y, starting in the late 1990s.

Ghosting When viewing text, faint shadows that appear to the right of each letter or number.

Gross negligence An act, or a failure to act, that is so reckless that it shows a conscious, voluntary disregard for the safety of others.

Hold harmless An agreement or contract wherein one party holds the other party free from responsibility for liability or damage that could arise from the transaction between the two parties.

Hostile work environment A general work environment characterized by unwelcome physical or verbal sexual conduct that interferes with an employee's performance.

Hybrid learning Learning environment in which some content is available for completion online and other learning requires face-to-face instruction or demonstration; sometimes referred to as "blended" or "distributed" learning.

Icon A small, pictorial, on-screen representation of an object used to indicate the existence of documents, file folders, and software.

Identifying The process of selecting those persons whom the fire service instructor would like to mentor, coach, and develop.

Ignition officer An individual who is responsible for igniting and controlling the material being burned at a live fire training.

In-service drill A training session scheduled as part of a regular shift schedule.

Indemnification Agreement An agreement or contract wherein one party assumes liability from another party in the event of a claim or loss.

Independent study A method of distance learning in which students order the course materials, complete them, and then return them to the instructor.

Informal communications Internal memos, e-mails, instant messages, and computer-aided dispatch/mobile data terminal messages. Informal reports have a short life and are not archived as permanent records.

Instructor-in-charge An individual who is qualified as an instructor and designated by the authority having jurisdiction to be in charge of fire fighter training.

Instructor-led training (ILT) Learning environment in which a human instructor facilitates both the in-person sections and the online components of coursework. Instructors usually set deadlines for submission, create quizzes and assignments, and track student progress.

Interrogatory A series of formal written questions sent to the opposing side of a legal argument. The opposition must provide written answers under oath.

Item analysis A listing of each student's answer to a particular question used for evaluation.

Job-content/criterion-referenced validity A type of validity obtained through the use of a technical committee of job incumbents who certify the knowledge being measured is required on the job and referenced to known standards.

Job performance requirement (JPR) A statement that describes a specific job task, lists the items necessary to complete that task, and defines measurable or observable outcomes and evaluation areas for the specific task.

Kinesthetic learning Learning that is based on doing or experiencing the information that is being taught.

Learning A relatively permanent change in behavior potential that is traceable to experience and practice.

Learning domains Categories that describe how learning takes place—specifically, the cognitive, psychomotor, and affective domains.

Learning environment A combination of a physical location (classroom or training ground) and the proper emotional elements of both an instructor and a student.

Learning management system (LMS) A Web-based software application that allows online courseware and content to be delivered to learners, wrapped with classroom administration tools such as activity tracking, grades, communications, and calendars.

Learning objective A goal that is achieved through the attainment of a skill, knowledge, or both, and that can be measured or observed.

Learning style The way in which an individual prefers to learn.

Lesson outline The main body of the lesson plan; a chronological listing of the information presented in the lesson plan.

Lesson plan A detailed guide used by an instructor for preparing and delivering instruction.

Lesson summary The part of the lesson plan that briefly reviews the information from the presentation and application sections.

Lesson title or topic The part of the lesson plan that indicates the name or main subject of the lesson plan.

Level of instruction The part of the lesson plan that indicates the difficulty or appropriateness of the lesson for students.

Liability Responsibility; the assignment of blame. It often occurs after a breach of duty.

Major life activity Basic functions of an individual's daily life, including, but not limited to, caring for oneself, performing manual tasks, breathing, walking, learning, seeing, working, and hearing.

Malfeasance Dishonest, intentionally illegal, or immoral actions.

Master training schedule Form used to identify and arrange training topics by the number of times they must be trained on or by the type of regulatory authority that requires the training to be completed.

Mean A value calculated by adding up all the scores from an examination and dividing by the total number of students who took the examination.

Median The score in the middle of the score distribution for an examination.

Medium The third link of the communication chain, which describes how you convey the message.

Mentoring A relationship of trust between an experienced person and a person of less experience for the purpose of growth and career development.

Message The second and most complex link of the communication chain; it describes what you are trying to convey to your students.

Method of instruction The process of teaching material to students.

Misfeasance Mistaken, careless, or inadvertent actions that result in a violation of law.

Mode The most commonly occurring value in a set of values (e.g., scores on an examination).

Modify To make basic or fundamental changes.

Motivational factors States of the person that are relatively temporary and reversible and that tend to energize or activate the behavior of the individual.

Motivation The activator or energizer for an activity or behavior.

Negligence An unintentional breach of duty that is the proximate cause of harm.

Networking An activity or process of like-minded individuals meeting and developing relationships for professional and personal growth through sharing of ideas and beliefs.

Oral tests Tests in which the answers are spoken in essay form in response to direct or open-ended questions. They may accompany a presentation or demonstration.

Organizational chart A graphic display of the fire department's chain of command and operational functions.

P+ value Number of correct responses to the test item. Example: P = 67 means that 67 percent of test takers answered correctly.

Passive listening Listening with your eyes and other senses without reacting to the message.

Pedagogy The art and science of teaching children.

Performance tests Tests that measure a student's ability to do a task under specified conditions and to a specific level of competence. Also known as a skills evaluation.

Pilot course The first offering of a new course, which is designed to allow the developers to test models, applications, evaluations, and course content. Many pilot courses do not fully represent the final course product, as much feedback is exchanged between students and instructors in the pilot program to improve the final product.

Power The ability to influence the actions of others through organizational position, expertise, the ability to reward or punish, or a role modeling of oneself to a subordinate.

Preparation step The first step of the four-step method of instruction, in which the instructor prepares to deliver the class and provides motivation for the students.

Prerequisite A condition that must be met before a student is allowed to receive the instruction contained within a lesson plan—often a certification, rank, or attendance of another class.

Presentation step The second step of the four-step method of instruction, in which the instructor delivers the class to the students.

Psychomotor domain The domain of learning that requires the physical use of knowledge. It represents the ability to physically manipulate an object or move the body to accomplish a task or use a skill. This domain is most often associated with hands-on training or drills.

Qualitative analysis An in-depth research study performed to categorize data into patterns to help determine which test items are acceptable.

Quantitative analysis Use of statistics to determine the acceptability of a test item.

Quid pro quo sexual harassment A situation in which an employee is forced to tolerate sexual harassment so as to keep or obtain a job, benefit, raise, or promotion.

Reasonable accommodation An employer's attempt to make its facilities, programs, policies, and other aspects of the work environment more accessible and usable for a person with a disability.

Receiver The fourth link of the communication chain. In the fire service classroom, the receiver is the student.

Recommendation report A decision document prepared by a fire officer for the senior staff. The goal is support for a decision or an action.

Reference blank Where the current job-relevant source of the test-item content is identified.

Regulation A law that can be established by legislative action, but is most commonly created by an administrative agency or a local entity.

Reliability index Value that refers to the reliability of a test as a whole in terms of consistently measuring the intended material.

Reliability The characteristic that a test measures what it is intended to measure on a consistent basis.

Revenues The income of a government from all sources, which is appropriated for the payment of public expenses and is stated as estimates.

Safety officer An individual who is appointed by the authority having jurisdiction and is qualified to maintain a safe working environment at all training evolutions.

Self-directed study Learning environment in which learners can stop and start as they desire, progress to completion of a module, and submit a final score to determine whether they passed or failed the module, all without an instructor's involvement.

Sender The first link of the communication chain. In the fire service classroom, the sender is the instructor.

Sexual harassment Unwelcome physical or verbal sexual conduct in the workplace that violates federal law.

Sharing The basic concept of giving to others with nothing expected in return.

Single source Materials that are manufactured or distributed by only one vendor.

Skills evaluations *See performance tests.*

Social learning An informal method of learning using technologies that enable collaborative content creation. Examples include blogs, social media (e.g., Twitter, Facebook), and wikis.

Standard deviation The value to which data should be expected to vary from the average.

Standards A set of guidelines outlining behaviors or qualifications of positions or specifications for equipment or processes. Often developed by individuals within the regulated profession, they may be applied voluntarily or referenced within a rule or law.

Statute A law created by legislative action that embodies the law of the land at both the federal and state levels.

Subject-matter expert (SME) A person who is technically competent and who works in the job for which test items are being developed.

Succession planning The act of ensuring the continuity of the organization by preparing its future leaders.

Summative evaluation Process that measures the students' achievements to determine the fire service instructor's strengths and weaknesses.

Supplemental training schedule Form used to identify and arrange training topics available in case of a change in the original training schedule.

Synchronous learning An online learning environment in which learners are all engaged in or receiving educational content simultaneously. Students and instructors are required to be online at the same time. Lectures, discussions, and presentations occur at a specific hour and students must be online at that time to participate and receive credit for attendance. An example might be a live Webcast, Webinar, or online chat session.

Systems approach to training (SAT) process A training process that relies on learning objectives and outcome-based learning.

Tailboard chat An informal gathering where fire fighters discuss various issues.

Technical-content validity A type of validity that occurs when a test item is developed by a subject-matter expert and is documented in current, job-relevant technical resources and training materials.

Technology-based instruction (TBI) Training and education that uses the Internet or a multimedia tool such as a DVD or CD-ROM.

Terminal objective A broad way of framing a JPR or outcome of a set of enabling objectives or learning steps, requiring the learner to have a specific set of skills or knowledge after a learning process. If several enabling objectives are mastered, then the terminal or end point knowledge or skill is achieved.

Title VII The section of the Civil Rights Act of 1964 that prohibits employment discrimination based on personal characteristics such as race, color, religion, sex, and national origin.

Undue hardship A situation in which accommodating an individual's disability would be too expensive or too difficult for the employer, given its size, resources, and the nature of its business.

Validity The documentation and evidence that supports the test item's relationship to a standard of performance in the learning objective and/or performance required on the job.

VARK Preferences A tool that measures a person's learning preferences along visual, aural, read/write, and kinesthetic sensory modalities.

Virtual classroom A digital environment where content can be posted and shared, and where instructors can create quizzes and assignments to be completed and tracked online via an Internet-connected computer.

Vision Having an alertness to the future, recognition of potential, and expectations of improvement.

Visual, auditory, and kinesthetic (VAK) characteristics Learning styles based on the idea that we all have a learning style preference based on sensory intake of information (visual, auditory, and kinesthetic).

Webinar An online meeting, occurring in real time, which usually allows for participants to share their desktop screens, Web browsers, and documents live as the meeting is occurring.

Willful and wanton conduct An act that shows utter indifference or conscious disregard for the safety of others.

Written tests Tests made up of several types of test items, such as multiple-choice, true/false, matching, short-answer essay, long-answer essay, arrangement, completion, and identification test items. Answers are provided on the test or a scannable form used for machine scoring.

Index